The ARRL
Extra Class
License Manual

Edited By:

Larry D. Wolfgang, WR1B

Production Staff:

David Pingree, **N1NAS**, Senior Technical Illustrator:
Technical Illustrations

Paul Lappen, Production Assistant: Layout

Jayne Pratt-Lovelace, Proofreader

Sue Fagan, Graphics Design Supervisor: Cover Design

Michael Daniels, Graphics

Michelle Bloom, **WB1ENT**, Production Supervisor: Layout

Published By:

ARRL—The national association for Amateur Radio
Newington, CT 06111-1494
ARRLWeb: **http://www.arrl.org/**

This book may be used for Amateur Extra class license exams given beginning April 15, 2000. The Extra class (Element 4) question pool in this book is expected to be used for exams given until July 1, 2002. (This ending date assumes no FCC Rules changes to the licensing structure or privileges for this license class, which would force the VEC Question Pool Committee to modify the question pool.) *QST* and *ARRLWeb* (http://www.arrl.org) will have news about any rules changes.

See **At Press Time** on Page *viii*

TABLE OF CONTENTS

FOREWORD

We radio amateurs can look back with pride at our history as experimenters and innovators. We should only pause for a moment, though, before returning our gaze to the future.

As 1999 came to a close, the FCC released a Report and Order that made significant changes to the Amateur Radio license structure. With only three written exams and a single 5-wpm Morse code exam, there are new opportunities to climb the ladder all the way to the Amateur Extra top rung. However, as anyone who picks up this book will discover, the 50-question written exam for an Amateur Extra class license represents a significant challenge.

Earning the license, though, is just the beginning of the adventure. Other challenges lie in fulfilling the other bases and purposes of the Amateur Radio Service: providing the public with voluntary emergency communications services, continuing and extending the amateur's proven ability to contribute to the advancement of the radio art, expansion of the existing reservoir of trained operators, technicians and electronics experts, and the continuation and extension of the amateur's unique ability to enhance international goodwill.

With this seventh edition of *The ARRL Extra Class License Manual*, we continue ARRL's proud tradition of providing the most complete license preparation materials available. When the VEC Question Pool Committee combined the old Advanced and Extra class syllabi and question pools to produce the new Element 4 question pool, we were faced with the task of combining that material into a single *License Manual* that would live up to that tradition. Compared to previous editions, this book may seem huge. Condensed into these pages you will find the detailed explanations, step-by-step math solutions and helpful suggestions that you have come to expect from ARRL *License Manuals*.

You will still find all the material you need to pass the Element 4 written exam! There is study material for every question (more than 650 of them) in the Extra class question pool. The text in Chapter 1 covers all the FCC Rules questions in the pool. Appropriate sections of the current FCC Rules, Part 97, are quoted where the actual rules text will help you understand the material. Chapters 2 through 9 explain the electronics theory and amateur operating practices that you'll need to know to pass your exam.

Chapter 10 is the complete Extra class question pool, as released by the Volunteer-Examiner Coordinators' Question Pool Committee in February 2000 for use beginning April 15, 2000. Use the question pool to drill yourself on the actual examination questions and to check your understanding of the study material. The answer key is included next to each question, for your convenience. If you do well answering the questions from the pool, you can be confident when you go to take your exam.

Of course, you aren't studying just to pass a license exam, and we aren't satisfied just to help you pass. We hold up to you the challenge of continued growth, education and experimentation — along with all the fun of Amateur Radio! ARRL's new

Certification and Continuing Education Program promises many learning opportunities beyond studying for a license. Check *ARRLWeb* (**http://www.arrl.org**) to learn more about this program.

The ARRL provides plenty of technical material and operating aids in the many books and supplies that make up our "Radio Amateur's Library." Check *QST* each month or contact the Publications Sales Office at ARRL Headquarters to request the latest publications catalog or to place an order. (You can reach our sales staff by phone — 888-277-5289; by fax — 860-594-0303; by electronic mail — **pubsales@arrl.org**; and through *ARRLWeb*.)

This seventh edition of *The ARRL Extra Class License Manual* is not just the work of the many ARRL staff members who have helped bring the book to you. Readers of earlier editions sent comments and suggestions: you can, too. First, use the book to prepare for your exam. Then, write your suggestions (or any corrections you think need to be made) on the Feedback Form at the back of this book, and send the form to us. (You can also e-mail your comments to: **pubsfdbk@arrl.org**.) Your comments and suggestions are important to us. Thanks, and good luck!

David Sumner, K1ZZ
Executive Vice President, ARRL
Newington, Connecticut
July 2000

HOW TO USE THIS BOOK

To earn an Amateur Extra class Amateur Radio license, you will have to know some electronics theory and the rules and regulations governing the Amateur Service, as contained in Part 97 of the FCC Rules. This book provides a brief description of the Amateur Service, and the Extra class license in particular. Applicants for the Amateur Extra class license must pass a 50-question written exam drawn from the Element 4 question pool, as released by the Volunteer-Examiner Coordinators' Question Pool Committee (QPC). You'll also have to be able to send and receive the international Morse code at a rate of 5 wpm.

Please note that each class of Amateur Radio license requires that you pass all of the exam elements below that class as well. So if you hold a Technician class license you will also have to pass the Element 3 — General — exam. (If you don't already have an Amateur Radio license, you will also have to pass the Element 2 — Technician — exam. Chapter 10 of this book contains the complete Element 4 question pool and multiple-choice answers released by the VEC QPC in February 2000 for use starting April 15, 2000.

At the beginning of each chapter, you will find a list of **Key Words** that appear in that chapter, along with a simple definition for each word or phrase. As you read the text, you will find these words printed in **boldface type** the first time they appear. You may want to refer to the **Key Words** list at the beginning of the chapter when you come to a **boldface** word, to review the definition. After you have studied the material in that chapter, you may also want to review those definitions. Most of the key words are terms that will appear on exam questions, so a quick review of the definitions just before you go to take the test may also be helpful. Appendix C is a glossary of all the key words used in the book. They are arranged in alphabetical order for your convenience as you review before the exam.

The Question Pool

As you study the material, you will be instructed to turn to sections of the questions in Chapter 10. Be sure to use these questions to review your understanding of the material at the suggested times. This breaks the material into bite-sized pieces, and makes it easier for you to learn. Do not try to memorize all of the questions and answers. With over 650 questions, that will be nearly impossible! Instead, by using the questions for review, you will be familiar with them when you take the test, but you will also understand the electronics theory or Rules point behind the questions.

Most people learn more easily when they are actively involved in the learning process. Study the text, rather than passively reading it. Use the questions to review your progress rather than just reading the question and looking at the correct answer letter. Fold the answer-key column under at the dashed line down the page before you begin answering questions on that page. Look at the answer key only to verify your answer. This will help you check your progress and ensure you know the answer to each question. If you missed one, go back to the supporting text and review that material. Page numbers are included in the answer key for each question. These indicate where to turn in the book to find the explanatory text for that question. You may have to read more than one page for the complete explanation. Paper clips make excellent place markers to help you find your spot in the text and question pool.

Other ARRL Study Materials

If you have not passed the 5-wpm Morse code exam, ARRL offers a set of two audio CDs or cassette tapes to teach Morse code: *Your Introduction to Morse Code*. This package introduces each of the required characters one at a time, and drills you on each character as it is introduced. Then the character is used in words and text before proceeding to the next character. After all characters have been introduced, there is plenty of practice at 5 wpm to help you prepare for the Element 1 (5 wpm) exam. ARRL also offers several audio-CD or cassette-tape sets to help you increase your speed, should you wish to do so: *Increasing Your Code Speed*; *5 to 10 WPM*, *10 to 15 WPM* and *15 to 22 WPM*. Each set includes two 74-minute audio CDs or cassettes with a variety of practice at gradually increasing speeds over the range for that set. The Morse code on these tapes is sent using 18-wpm characters, with extra space between characters to slow the overall code speed, for code up to 18 wpm. This same technique is used on ARRL/VEC Morse code exams.

For those who prefer a computer program to learn and practice Morse code, ARRL offers the excellent GGTE *Morse Tutor Gold* program for IBM PC and compatible computers.

Even with the tapes or computer program, you'll want to tune in the code-practice sessions transmitted by W1AW, the ARRL Headquarters station. As you gain more code proficiency, you may find it helpful to listen to code at faster speeds at the beginning of your practice session, and then decrease the speed. The W1AW fast-code-practice sessions start at 35 wpm and decrease to 30, 25 and 20 wpm before dropping to 15, 13 and 10-wpm practice. You might also want to try copying the W1AW CW bulletins, which are sent at 18 wpm. (There is a W1AW code-practice schedule on page 7 in the Introduction.) For more information about W1AW or how to order any ARRL publication or set of code tapes, write to ARRL Headquarters, 225 Main St, Newington, CT 06111-1494, tel 888-277-5289 or visit ARRLWeb: **http://www.arrl.org**. You can also e-mail us at **pubsales@arrl.org**.

AT PRESS TIME

ARRL Files Partial Reconsideration Petition in Restructuring Proceeding

The ARRL has formally asked the FCC to reconsider and modify two aspects of its December 30, 1999, Report and Order that restructured the Amateur Radio rules. The League wants the FCC to continue to maintain records that indicate whether a Technician licensee has Morse code element credit. It also seeks permanent Morse element credit for any Amateur Radio applicant who has ever passed an FCC-recognized Morse exam of at least 5 wpm.

The ARRL's Petition for Partial Reconsideration in the WT Docket 98-143 proceeding was filed March 13, 2000.

The League suggested that it would be less of an administrative burden for the FCC to maintain the Technician database as it had been doing before April 15, 2000. The database identifies Technician and Tech Plus licensees by encoding the records with a "T" or a "P" respectively. The ARRL also said the inability to identify those Technicians who have HF privileges and those who do not could hamper voluntary enforcement efforts. It further suggested it would be wrong to put the burden of proof of having passed the Morse examination on licensees.

The ARRL cited the demands of fairness in asking the FCC to afford Morse element credit to all applicants who have ever passed an FCC-recognized 5 wpm code exam. The rules already grant Element 1 credit to those holding an expired or unexpired FCC-issued Novice license or an expired or unexpired Technician class operator license document granted before February 14, 1991. It also grants Element 1 credit to applicants possessing an FCC-issued commercial radiotelegraph operator license or permit that's valid or expired less than 5 years.

The ARRL has asked the FCC to "conform the rules" to give similar credit to those who once held General, Advanced or Amateur Extra class licenses.

FCC Launches CORES

The FCC has begun implementing the Commission Registration System, known as CORES. While the action has few immediate implications for Amateur Radio licensees, CORES registration eventually will replace Universal Licensing System (ULS) registration.

Described as an agency-wide registration system for anyone filing applications with or making payments to the FCC, CORES will assign a unique 10-digit FCC Registration Number, or FRN to all registrants. Once the system is fully deployed, all Commission systems that handle financial, authorization of service, and enforcement activities will use the FRN. The FCC says use of the FRN will allow it to more rapidly verify fee payment. Amateurs mailing payments to the FCC—for example, as part of their vanity call sign application—would include their FRN—once assigned—on the revised FCC Form 159.

The on-line filing system and further information on CORES is available by visiting the FCC Web site, **http://www.fcc.gov** and clicking on the CORES registration link.

For now, using an FRN is voluntary, although the Commission says it will consider making it mandatory in the future for anyone doing business with the FCC. The FCC says it will modify its licensing and filing systems—including ULS—over the next several months to accept and use the FRN.

CORES registration will supplant ULS registration, but the ULS itself will remain the licensing database system for Wireless Telecommunications Bureau licensees, including amateurs. For now, the ULS remains available to new registrants. Amateurs who registered in the ULS prior to June 22, 2000 automatically have been registered in CORES and will receive an FCC Registration Number in the mail. ULS registrants also may search for their FRN on-line at the FCC's CORES site. The FCC says ULS passwords will become CORES passwords in most cases. It's possible to register on CORES using a paper Form 160.

As with the ULS, those registering with CORES must supply a Taxpayer Identification Number, or TIN. For individuals, this is usually a Social Security Number. Club stations must obtain an Assigned Taxpayer Identification Number (ATIN) before registering on CORES or ULS.

Anyone can register via CORES and obtain an FRN. CORES/FRN is entity registration. You don't need a license to be registered. The FCC is making every attempt to minimize the impact of CORES/FRN on Amateur Radio licensees, and no action will be required on the part of amateur licensees already registered in ULS.

A copy of the FCC Public Notice on CORES/FRN is available at **http://www.arrl.org/announce/regulatory/da001596.pdf**—*ARRL News Release, July 25, 2000.*

Watch *QST, The ARRL Letter* and other Amateur Radio publications for the latest news.

Changeover to CORES Registration to be Transparent

When the FCC moves its Taxpayer Information Number/Social Security Number registration system for amateurs from the Universal Licensing System to the new FCC Commission Registration System, the changeover will be largely transparent to users, an FCC official said.

Steve Linn of the Commission's Wireless Telecommunications Bureau says once the changeover is in effect, the CORES/FRN system will be linked from the ULS home page. In addition, those already registered in the ULS will—in most cases—still be able to use their ULS password to access CORES.

Amateurs will not have to start signing up in CORES until sometime in 2001, however, and those already registered in ULS won't have to do a thing. For now, Linn says hams should stick simply with ULS "TIN/Call Sign" registration until CORES registration becomes mandatory.

"Don't even worry about CORES," Linn said. "If you have a letter from CORES, hang onto it for your FCC Registration Number." The FCC sent letters to every licensee who was registered in the ULS as of June 22, 2000. The letter contains the individual's new FCC Registration Number—or FRN—and a few words about CORES.

Described as an agency-wide registration system for anyone filing applications with or making payments to the FCC, CORES assigns registrants a unique 10-digit FCC Registration Number. The FCC says it will modify its licensing and filing systems—including ULS—over the next several months to accept and use the FRN.

Once the system is fully deployed, all Commission systems that handle financial, authorization of service, and enforcement activities will use the FRN. The FCC says use of the FRN will allow it to more rapidly verify fee payment.

Until CORES assumes the registration function, Linn encouraged hams to register in ULS, which will continue to house the FCC's Amateur Service database even after CORES registration is implemented. Just when in 2001 that will happen Linn was not able to say. "There are a lot of factors involved," he said.

Amateurs who have moved or who have perhaps acquired a new name should update their ULS licensee information via the ULS home page (go to "Online Filing" and perform an "administrative update"). Amateurs with new addresses, names or call signs also should update their ULS registration information (go to TIN/Call Sign Registration and click on "update registration information" or "update call sign information," as appropriate).

The CORES on-line filing system and further information on CORES is available by visiting the FCC Web site, and clicking on the CORES registration link.—*ARRL News Release, September 1, 2000*

INTRODUCTION

THE AMATEUR EXTRA LICENSE

Every Amateur Radio operator thinks about trying to earn his or her Amateur Extra class license at one time or another. It certainly is a worthy goal to work toward! Maybe you hesitate to actually take the exam, however, because you think the theory is much too hard for you to understand. Once you make the commitment to study and learn what it takes to pass the exam, however, you *will* be able to do it. It may take more than one attempt to pass the exam, but many amateurs do succeed the first time. The key is that you must make the commitment, and be willing to study.

Segments of the 80, 40, 20 and 15-meter bands are reserved for the exclusive use of Amateur Extra class operators. These frequencies are where some of the juiciest DX stations hang out, and you will also find many contest stations in these band segments. So if you want to be sure you aren't missing out on those rare DX QSL cards and extra contest points, you'll want to earn your Amateur Extra class license! Of course, there are many other reasons for wanting the highest class license.

This License Manual was carefully designed to teach you everything you need to know to pass the Element 4 written exam. If you don't already hold a General class license, you will also have to pass the Element 3 written exam. If you don't have a Technician license, you will have to pass the Element 2 written exam, too. To reach any rung on the Amateur Radio license ladder, you must pass all the lower exam elements.

If you have not yet passed the 5 wpm (Element 1) Morse code exam, you will have to pass that also to earn your Extra class license. There are many good Morse code training techniques, including the ARRL code tapes, W1AW code practice and computer programs like GGTE *Morse Tutor Gold*, available from ARRL. Remember, the most enjoyable way to practice your Morse code skills and increase your code speed is with on-the-air operating!

Chapter 1 of this book, **Commission's Rules**, covers those sections of the

FCC Part 97 Rules that you will be tested on for your Amateur Extra class license. We recommend that you also obtain a copy of *The ARRL's FCC Rule Book*. This book includes a complete copy of Part 97, along with detailed explanations for all the rules governing Amateur Radio.

Chapters 2 through 9 cover the remaining eight subelements for the Amateur Extra class license exam. All of the questions in the Extra class question pool are covered in this book. Chapter 10 contains the complete question pool, with answers, for the Extra class license exam.

Whether you now hold a Novice, Technician, Technician Plus (Technician with Morse code credit), General or Advanced license, or even if you don't have any license yet, you will find the exclusive operating privileges available to an Amateur Extra class licensee to be worth the time spent learning about your hobby. After passing the FCC Element 4 written exam, you will be allowed to operate on any frequency assigned to the Amateur Radio Service.

IF YOU'RE A NEWCOMER TO AMATEUR RADIO

Earning an Amateur Radio license, at whatever level, is a special achievement. Nearly ¾ of a million people in the US call themselves Amateur Radio operators, or hams, and we are part of a much larger global fraternity. There are nearly 3 million Amateur Radio operators around the world! Radio amateurs serve the public as a voluntary, noncommercial, communication service. This is especially true during natural disasters or other emergencies. Hams have made many important contributions to the field of electronics and communications, and this tradition continues today. Amateur Radio experimentation is yet another reason many people become part of this self-disciplined group of trained operators, technicians and electronics experts — an asset to any country. Hams pursue their hobby purely for personal enrichment in technical and operating skills, without consideration of any type of

Figure 1 — Many amateurs enjoy operating the digital modes. Here, Bill Price, WA4MCZ operates in the ARRL RTTY Roundup contest.

payment except the personal satisfaction they feel from a job well done!

Radio signals do not know territorial boundaries, so hams have a unique ability to enhance international goodwill. Hams become ambassadors of their country every time they put their stations on the air.

Amateur Radio has been around since before World War I, and hams have always been at the forefront of technology. Today, hams relay signals through their own satellites in the OSCAR (Orbiting Satellite Carrying Amateur Radio) series, bounce signals off the moon, relay messages automatically through computerized radio networks and send pictures by television. Amateurs talk from hand-held transceivers through mountaintop repeater stations that can relay their signals to transceivers in other hams' cars or homes. Hams chat with other hams around the world by voice or tap out messages in Morse code. The "code" represents a distinctive traditional skill, which often proves to be the only way to communicate during an emergency or when signals are weak and interference is present. When emergencies arise, radio amateurs are on the spot to relay information to and from disaster-stricken areas that have lost normal lines of communication.

The US government, through the Federal Communications Commission (FCC), grants all US Amateur Radio licenses. This licensing procedure ensures operating skill and electronics know-how. Without this skill, radio operators might unknowingly cause interference to other services using the radio spectrum because of improperly adjusted equipment or neglected regulations.

Who Can Be a Ham?

The FCC doesn't care how old you are or whether you're a US citizen: If you pass the examination, the Commission will issue you an amateur license. Any person (except the agent of a foreign government) may take the exam, and, if successful, receive an amateur license. It's important to understand that if a citizen of a foreign country received an amateur license in this manner, he or she is a US Amateur Radio operator. (This should not be confused with reciprocal operating authority, which allows visitors from certain countries who hold valid amateur licenses in their homelands to operate their own stations in the US without having to take an FCC exam.)

Licensing Structure

By examining **Table 1**, you'll see that new amateurs can earn three license classes in the US. These are the Technician, General and Amateur Extra license. They vary in the degree of knowledge required and frequency privileges granted. Higher class licenses have more comprehensive examinations. In return for passing a more difficult exam you earn more frequency privileges (frequency space and modes of operation).

There are also several other amateur license classes, but the FCC is no longer issuing new licenses for these classes. The Novice license was long considered the beginner's license. Exams for this license were discontinued as of April 15, 2000. The FCC also stopped issuing new Advanced class licenses on that date. They will continue to renew previously issued licenses, however, so you will meet Novice and Advanced class operators on the air. You will also find certain frequency privileges designated for Novice and Advanced class licensees.

Table 1
Amateur Operator Licenses†

Class	Code Test	Written Examination	Privileges
Technician	(None)	Basic theory and regulations. (Element 2)*	All amateur privileges above 50.0 MHz.
Technician With Morse Code Credit	5 wpm (Element 1)	Basic theory and regulations. (Element 2)*	All "Novice" HF privileges in addition to all Technician privileges.
General	5 wpm (Element 1)	Basic theory and regulations; General theory and regulations. (Elements 2 and 3)	All amateur privileges except those reserved for Advanced and Amateur Extra class; see Table 2.
Amateur Extra	5 wpm (Element 1)	All lower exam elements, plus Extra-class theory (Elements 2, 3 and 4)	All amateur privileges.

†A licensed radio amateur will be required to pass only those elements that are not included in the examination for the amateur license currently held.
*If you have a Technician-class license issued before March 21, 1987, you also have credit for Elements 1 and 3. You must be able to prove your Technician license was issued before March 21, 1987 to claim this credit.

Technician licensees can pass an exam to demonstrate their knowledge in international Morse code at 5 wpm. With proof of passing the Morse code exam, a Technician licensee gains some frequency privileges on four of the amateur HF bands. This license was previously called the Technician Plus license, and many amateurs will refer to it by that name. The HF privileges earned are known as the Novice privileges.

The FCC requires proof of your ability to operate an Amateur Radio station properly. The required knowledge is in line with the privileges of the license you hold. Higher license classes require more knowledge — and offer greater operating privileges. So as you upgrade your license class, you must pass more challenging written examinations. The specific operating privileges for Extra class licensees are shown on the charts in **Table 2**. Amateur Extra licensees have frequency privileges not authorized to Generals. To gain those additional privileges, a General class licensee will have to upgrade to an Amateur Extra class license.

In addition to passing the written exams, you must demonstrate an ability to send and receive international Morse code at 5 wpm for the General and Amateur Extra licenses. It's important to stress that although you may intend to use voice rather than code, this doesn't excuse you from the code test. By international treaty, knowing the international Morse code is a basic requirement for operating on any amateur band below 30 MHz.

The FCC allows Volunteer Examiners to use a range of procedures to accommodate applicants with various disabilities. If you believe you may qualify for special examination procedures for the Morse code examination, contact your local Volunteer Examiners or the Volunteer Examiner Coordinator responsible for

Table 2
Amateur Operating Privileges

US Amateur Bands
April 15, 2000

160 METERS

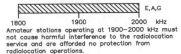

E,A,G

1800   1900   2000   kHz

Amateur stations operating at 1900–2000 kHz must not cause harmful interference to the radiolocation service and are afforded no protection from radiolocation operations.

Novice, Advanced and Technician Plus Allocations

New Novice, Advanced and Technician Plus licenses will not be issued after April 15, 2000. However, the FCC has allowed the frequency allocations for these license classes to remain in effect. They will continue to renew existing licenses for those classes.

80 METERS

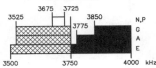

3525   3675 3725
              3850
         3775
                        N,P
                        G
                        A
                        E
3500   3750   4000   kHz

5167.5 kHz (SSB only): Alaska emergency use only.

12 METERS

E,A,G

24,890   24,930   24,990   kHz

40 METERS

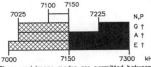

7025   7100 7150
                7225
                        N,P
                        G †
                        A †
                        E †
7000   7150   7300   kHz

† Phone and image modes are permitted between 7075 and 7100 kHz for FCC licensed stations in ITU Regions 1 and 3 and by FCC licensed stations in ITU Region 2 West of 130 degrees West longitude or South of 20 degrees North latitude. See Sections 97.305(c) and 97.307(f)(11). Novice and Technician Plus licensees outside ITU Region 2 may use CW only between 7050 and 7075 kHz. See Section 97.301(e). These exemptions do not apply to stations in the continental US.

10 METERS

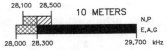

28,100 28,500
                        N,P
                        E,A,G
28,000 28,300   29,700   kHz

Novices and Technician Plus licensees are limited to 200 watts PEP output on 10 meters.

30 METERS

E,A,G

10,100   10,150   kHz

Maximum power on 30 meters is 200 watts PEP output. Amateurs must avoid interference to the fixed service outside the US.

6 METERS

50.1
                        E,A,G,P,T
50.0   54.0   MHz

20 METERS

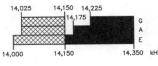

14,025   14,150 14,225
              14,175
                        G
                        A
                        E
14,000   14,150   14,350   kHz

2 METERS

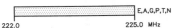

144.1
                        E,A,G,P,T
144.0   148.0   MHz

17 METERS

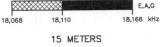

E,A,G

18,068   18,110   18,168   kHz

1.25 METERS

222.0   225.0   MHz

E,A,G,P,T,N

Novices are limited to 25 watts PEP output from 222 to 225 MHz.

15 METERS

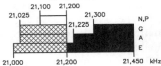

21,025 21,100 21,200
                21,300
                21,225
                        N,P
                        G
                        A
                        E
21,000   21,200   21,450   kHz

70 CENTIMETERS **

E,A,G,P,T

420.0   450.0   MHz

33 CENTIMETERS **

E,A,G,P,T

902.0   928.0   MHz

23 CENTIMETERS **

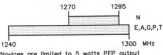

1270   1295
                        N
                        E,A,G,P,T
1240   1300   MHz

Novices are limited to 5 watts PEP output from 1270 to 1295 MHz.

US AMATEUR POWER LIMITS

At all times, transmitter power should be kept down to that necessary to carry out the desired communications. Power is rated in watts PEP output. Unless otherwise stated, the maximum power output is 1500 W. Power for all license classes is limited to 200 W in the 10,100–10,150 kHz band and in all Novice subbands below 28,100 kHz. Novices and Technicians with Morse code credit are restricted to 200 W in the 28,100–28,500 kHz subbands. In addition, Novices are restricted to 25 W in the 222–225 MHz band and 5 W in the 1270–1295 MHz subband.

Operators with Technician class licenses and above may operate on all bands above 50 MHz. For more detailed information see *The FCC Rule Book*.

KEY

⊠ = CW, RTTY and data
▨ = CW, RTTY, data, MCW, test, phone and image
■ = CW, phone and image
▥ = CW and SSB phone
▨ = CW, RTTY, data, phone, and image
☐ = CW only

E = EXTRA CLASS
A = ADVANCED
G = GENERAL
P = TECHNICIAN PLUS
T = TECHNICIAN
N = NOVICE

* Effective April 15, 2000, Technicians passing the Morse code exam will gain HF Novice privileges, although they still hold a Technician license.

** Geographical and power restrictions apply to these bands. See *The FCC Rule Book* for more information about your area.

Above 23 Centimeters:

All licensees except Novices are authorized all modes on the following frequencies:
2300–2310 MHz
2390–2450 MHz
3300–3500 MHz
5650–5925 MHz
10.0–10.5 GHz
24.0–24.25 GHz
47.0–47.2 GHz
75.5–81.0 GHz
119.98–120.02 GHz
142–149 GHz
241–250 GHz
All above 300 GHz

For band plans and sharing arrangements, see *The ARRL Operating Manual* or *The FCC Rule Book*.

the test session you will be attending. You can contact the ARRL/VEC Office, 225 Main Street, Newington, CT 06111-1494 (860-594-0200) for assistance. Ask for more information about the special exam procedures their Volunteer Examiners will use to administer the code exam to you.

Learning the Morse code is a matter of practice. As mentioned earlier, the ARRL publishes Morse code practice audio CDs and cassette tapes to help you pass the code exam. *Your Introduction to Morse Code* teaches all the characters required by the FCC and gives plenty of practice at 5 wpm. This set includes two audio CDs or 74-minute cassette tapes and a handy booklet that includes instructions and a transcript of the text. An entire chapter of *The ARRL General Class License Manual* is dedicated to helping you learn Morse code. The ARRL also publishes *Morse Code: The Essential Language*, which describes much of the history of Morse code and includes plenty of hints on learning and using the code. All of these materials are available from the ARRL, 225 Main Street, Newington, CT 06111 and from many local Amateur Radio equipment dealers. You can learn more about these products and order them directly at *ARRLWeb*: **http:// www.arrl.org**. The ARRL's Maxim Memorial Station, W1AW also transmits Morse code practice on weekdays. See **Table 3**.

Figure 2—Many active Amateur Radio operators enjoy collecting colorful QSL cards, many of them from countries around the world.

Code Practice

Earning an Amateur Extra class license does not require a Morse code exam beyond the General class 5 wpm exam. Many Extra class licensees want to improve their Morse code skills beyond that level, however. Exclusive Extra class CW segments on the 80, 40, 20 and 15 meter bands invite you to enjoy Morse code conversations with other Extra class hams. This is also where you will find some of the juiciest DX stations!

Besides listening to code tapes or a computer program, some on-the-air operating experience will be a great help in building your code speed. When you are in the middle of a contact via Amateur Radio, and must copy the code the other station is sending to continue the conversation, your copying ability will improve quickly! Most operators will be very happy to repeat any information you miss, especially if you tell them you are trying to increase your code speed for a license upgrade.

ARRL's Hiram Percy Maxim Memorial Station, W1AW, transmits code practice and information bulletins of interest to all amateurs. These code-practice sessions and Morse code bulletins provide an excellent opportunity for code practice. Table 3 is a W1AW operating schedule. When we change from Standard Time to

Table 3

W1AW's schedule is at the same local time throughout the year. The schedule according to your local time will change if your local time does not have seasonal adjustments that are made at the same time as North American time changes between standard time and daylight time. From the first Sunday in April to the last Sunday in October, UTC = Eastern Time + 4 hours. For the rest of the year, UTC = Eastern Time + 5 hours.

• **Morse code transmissions:**
Frequencies are 1.818, 3.5815, 7.0475, 14.0475, 18.0975, 21.0675, 28.0675 and 147.555 MHz.
Slow Code = practice sent at 5, 7½, 10, and 15 wpm.

W1AW SCHEDULE								
Pacific	**Mtn**	**Cent**	**East**	**Mon**	**Tue**	**Wed**	**Thu**	**Fri**
6 AM	7 AM	8 AM	9 AM		Fast Code	Slow Code	Fast Code	Slow Code
7 AM-1 PM	8 AM-2 PM	9 AM-3 PM	10 AM-4 PM	Visiting Operator Time (12 PM - 1 PM closed for lunch)				
1 PM	2 PM	3 PM	4 PM	Fast Code	Slow Code	Fast Code	Slow Code	Fast Code
2 PM	3 PM	4 PM	5 PM	Code Bulletin				
3 PM	4 PM	5 PM	6 PM	Teleprinter Bulletin				
4 PM	5 PM	6 PM	7 PM	Slow Code	Fast Code	Slow Code	Fast Code	Slow Code
5 PM	6 PM	7 PM	8 PM	Code Bulletin				
6 PM	7 PM	8 PM	9 PM	Teleprinter Bulletin				
6⁴⁵ PM	7⁴⁵ PM	8⁴⁵ PM	9⁴⁵ PM	Voice Bulletin				
7 PM	8 PM	9 PM	10 PM	Fast Code	Slow Code	Fast Code	Slow Code	Fast Code
8 PM	9 PM	10 PM	11 PM	Code Bulletin				

Fast Code = practice sent at 35, 30, 25, 20, 15 and 10 wpm.
Code practice text is from the pages of *QST*. The source is given at the beginning of each practice session and alternate speeds within each session. For example, "Text is from July 1992 *QST*, pages 9 and 81," indicates that the plain text is from the article on page 9 and mixed number/letter groups are from page 81.
Code bulletins are sent at 18 wpm.
W1AW qualifying runs are sent on the same frequencies as the Morse code transmissions. West Coast qualifying runs are transmitted on approximately 3.590 MHz by K6YR. At the beginning of each code practice session, the schedule for the next qualifying run is presented. Underline one minute of the highest speed you copied, certify that your copy was made without aid, and send it to ARRL for grading. Please include your name, call sign (if any) and complete mailing address. Send a 9×12-inch SASE for a certificate, or a business-size SASE for an endorsement.

• **Teleprinter transmissions:**
Frequencies are 3.625, 7.095, 14.095, 18.1025, 21.095, 28.095 and 147.555 MHz.
Bulletins are sent at 45.45-baud Baudot and 100-baud AMTOR, FEC Mode B. 110-baud ASCII will be sent only as time allows.
On Tuesdays and Fridays at 6:30 PM Eastern Time, Keplerian elements for many amateur satellites are sent on the regular teleprinter frequencies.

• **Voice transmissions:**
Frequencies are 1.855, 3.99, 7.29, 14.29, 18.16, 21.39, 28.59 and 147.555 MHz.

• **Miscellanea:**
On Fridays, UTC, a DX bulletin replaces the regular bulletins.
W1AW is open to visitors from 10 AM until noon and from 1 PM until 3:45 PM on Monday through Friday. FCC licensed amateurs may operate the station during that time. Be sure to bring your current FCC amateur license or a photocopy.
In a communication emergency, monitor W1AW for special bulletins as follows: voice on the hour, teleprinter at 15 minutes past the hour, and CW on the half hour.
Headquarters and W1AW are closed on New Year's Day, President's Day, Good Friday, Memorial Day, Independence Day, Labor Day, Thanksgiving and the following Friday, and Christmas Day.

Daylight Saving Time, the same local times are used.

The information bulletins are sent at 18 wpm, so this is excellent practice as you try to increase your speed. The fast code-practice sessions begin with code sent at 35 wpm, and then decreases to 30, 25, 20, 15 and 10 wpm. Slow code practice is sent at 5, 7½, 10 and 15 wpm. Many people find that as their code speed increases, copying the faster speeds first helps them make more rapid progress toward their goal of conversational CW.

Station Call Signs

Many years ago, by international agreement, the nations of the world decided to allocate certain call-sign prefixes to each country. This means that if you hear a radio station call sign beginning with W, N or K, for example, you know the station is licensed by the United States. A call sign beginning with the letter G is licensed by Great Britain and one that begins with the letters VK is from Australia.

Table 4 is the International Telecommunication Union (ITU) International Call Sign Allocation Table, which will help you determine what country a certain call sign is from. *The ARRL DXCC List* is an operating aid no ham who is active on the HF bands should be without! That booklet, available from ARRL, includes the common call-sign prefixes used by amateurs in virtually every location in the world. It also includes a check-off list to help you keep track of the countries you contact as you work toward collecting QSL cards from 100 or more countries to earn the prestigious DX Century Club award. (DX is ham lingo for distance, generally taken to mean any country outside the one you are operating from on the HF bands.)

The International Telecommunication Union (ITU) radio regulations outline the basic principles used in forming amateur call signs. According to these regulations, an amateur call sign must be made of one or two characters (the first one may be a numeral) as a prefix, followed by a numeral, and then a suffix of not more than three letters. The prefixes W, K, N and A are used in the United States. (When A is used for a US call sign there are two letters for the prefix, AA through AL.) The continental US is divided into 10 Amateur Radio call districts (sometimes called areas), numbered 0 through 9). **Figure 4** is a map showing the US call districts.

All US Amateur Radio call signs are issued systematically by computer at the FCC. In addition, there is a program for "vanity" call signs, in which licensed amateurs may request special call signs. For further information on the FCC's call sign assignment system, and a table listing the blocks of call signs for each license class as well as information about

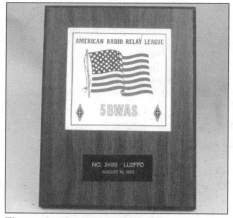

Figure 3—As you collect QSL cards, you may want to work toward the ARRL's prestigious 5-Band Worked All States award, earned by collecting QSL cards from each of the 50 United States on five different amateur bands.

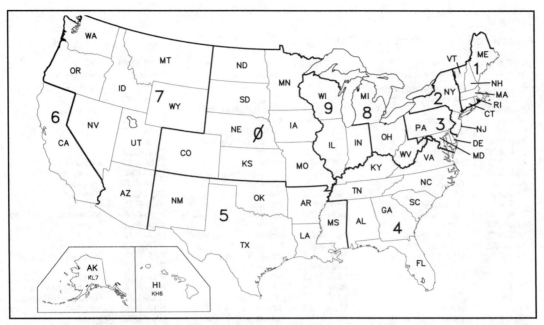

Figure 4 — There are 10 US call areas. Hawaii is part of the sixth call area, and Alaska is part of the seventh.

the vanity call sign system, see *The ARRL's FCC Rule Book*. You can also contact the ARRL/VEC Office for information and forms about the vanity call sign system.

If you already have an amateur call sign, you may keep the same one when you change license class, if you wish. If you wish to receive a new call sign, you must check the "**CHANGE** my **call sign** systematically" box and initial your selection on the Form 605 when you apply for your exam. You can also change your call sign later by filing an FCC Form 605 with the FCC. (Be careful *not* to check this box if you want to keep your present call sign.)

EARNING A LICENSE

Forms and Procedures

To renew or modify a license, you can file a copy of FCC Form 605. In addition, hams who have held a valid license that has expired within the past two years may apply for reinstatement with an FCC Form 605.

Licenses are normally good for ten years. Your application for a license renewal must be submitted to the FCC no more than 90 days before the license expires. (We recommend you submit the application for renewal between 90 and 60 days before your license expires.) If the FCC receives your renewal application before the license expires, you may continue to operate until your new license arrives, even if it is past the expiration date. If you forget to apply before your license expires, you may

Table 4
Allocation of International Call Signs

Call Sign Series	Allocated to	Call Sign Series	Allocated to	Call Sign Series	Allocated to
AAA-ALZ	United States of America	E2A-E2Z	Thailand		Northern Ireland
AMA-AOZ	Spain	E3A-E3Z	Eritrea	NAA-NZZ	United States of America
APA-ASZ	Pakistan	† E4A-E4Z	Palestinian Authority	OAA-OCZ	Peru
ATA-AWZ	India	FAA-FZZ	France	ODA-ODZ	Lebanon
AXA-AXZ	Australia	GAA-GZZ	United Kingdom of Great Britain and Northern Ireland	OEA-OEZ	Austria
AYA-AZZ	Argentina			OFA-OJZ	Finland
A2A-A2Z	Botswana	HAA-HAZ	Hungary	OKA-OLZ	Czech Republic
A3A-A3Z	Tonga	HBA-HBZ	Switzerland	OMA-OMZ	Slovak Republic
A4A-A4Z	Oman	HCA-HDZ	Ecuador	ONA-OTZ	Belgium
A5A-A5Z	Bhutan	HEA-HEZ	Switzerland	OUA-OZZ	Denmark
A6A-A6Z	United Arab Emirates	HFA-HFZ	Poland	PAA-PIZ	Netherlands
A7A-A7Z	Qatar	HGA-HGZ	Hungary	PJA-PJZ	Netherlands Antilles
A8A-A8Z	Liberia	HHA-HHZ	Haiti	PKA-POZ	Indonesia
A9A-A9Z	Bahrain	HIA-HIZ	Dominican Republic	PPA-PYZ	Brazil
BAA-BZZ	China (People's Republic of)	HJA-HKZ	Colombia	PZA-PZZ	Suriname
CAA-CEZ	Chile	HLA-HLZ	South Korea	P2A-P2Z	Papua New Guinea
CFA-CKZ	Canada	HMA-HMZ	North Korea	P3A-P3Z	Cyprus
CLA-CMZ	Cuba	HNA-HNZ	Iraq	P4A-P4Z	Aruba
CNA-CNZ	Morocco	HOA-HPZ	Panama	P5A-P9Z	North Korea
COA-COZ	Cuba	HQA-HRZ	Honduras	RAA-RZZ	Russian Federation
CPA-CPZ	Bolivia	HSA-HSZ	Thailand	SAA-SMZ	Sweden
CQA-CUZ	Portugal	HTA-HTZ	Nicaragua	SNA-SRZ	Poland
CVA-CXZ	Uruguay	HUA-HUZ	El Salvador	• SSA-SSM	Egypt
CYA-CZZ	Canada	HVA-HVZ	Vatican City State	• SSN-STZ	Sudan
C2A-C2Z	Nauru	HWA-HYZ	France	SUA-SUZ	Egypt
C3A-C3Z	Andorra	HZA-HZZ	Saudi Arabia	SVA-SZZ	Greece
C4A-C4Z	Cyprus	H2A-H2Z	Cyprus	S2A-S3Z	Bangladesh
C5A-C5Z	Gambia	H3A-H3Z	Panama	S5A-S5Z	Slovenia
C6A-C6Z	Bahamas	H4A-H4Z	Solomon Islands	S6A-S6Z	Singapore
* C7A-C7Z	World Meteorological Organization	H6A-H7Z	Nicaragua	S7A-S7Z	Seychelles
		H8A-H9Z	Panama	S8A-S8Z	South Africa
C8A-C9Z	Mozambique	IAA-IZZ	Italy	S9A-S9Z	Sao Tome and Principe
DAA-DRZ	Germany	JAA-JSZ	Japan	TAA-TCZ	Turkey
DSA-DTZ	South Korea	JTA-JVZ	Mongolia	TDA-TDZ	Guatemala
DUA-DZZ	Philippines	JWA-JXZ	Norway	TEA-TEZ	Costa Rica
D2A-D3Z	Angola	JYA-JYZ	Jordan	TFA-TFZ	Iceland
D4A-D4Z	Cape Verde	JZA-JZZ	Indonesia	TGA-TGZ	Guatemala
D5A-D5Z	Liberia	J2A-J2Z	Djibouti	THA-THZ	France
D6A-D6Z	Comoros	J3A-J3Z	Grenada	TIA-TIZ	Costa Rica
D7A-D9Z	South Korea	J4A-J4Z	Greece	TJA-TJZ	Cameroon
EAA-EHZ	Spain	J5A-J5Z	Guinea-Bissau	TKA-TKZ	France
EIA-EJZ	Ireland	J6A-J6Z	Saint Lucia	TLA-TLZ	Central Africa
EKA-EKZ	Armenia	J7A-J7Z	Dominica	TMA-TMZ	France
ELA-ELZ	Liberia	J8A-J8Z	St. Vincent and the Grenadines	TNA-TNZ	Congo
EMA-EOZ	Ukraine			TOA-TQZ	France
EPA-EQZ	Iran	KAA-KZZ	United States of America	TRA-TRZ	Gabon
ERA-ERZ	Moldova	LAA-LNZ	Norway	TSA-TSZ	Tunisia
ESA-ESZ	Estonia	LOA-LWZ	Argentina	TTA-TTZ	Chad
ETA-ETZ	Ethiopia	LXA-LXZ	Luxembourg	TUA-TUZ	Ivory Coast
EUA-EWZ	Belarus	LYA-LYZ	Lithuania	TVA-TXZ	France
EXA-EXZ	Krygyzstan	LZA-LZZ	Bulgaria	TYA-TYZ	Benin
EYA-EYZ	Tajikistan	L2A-L9Z	Argentina	TZA-TZZ	Mali
EZA-EZZ	Turkmenistan	MAA-MZZ	United Kingdom of Great Britain and	T2A-T2Z	Tuvalu
				T3A-T3Z	Kiribati

Call Sign Series	Allocated to	Call Sign Series	Allocated to	Call Sign Series	Allocated to
T4A-T4Z	Cuba	Y2A-Y9Z	Germany	5PA-5QZ	Denmark
T5A-T5Z	Somalia	ZAA-ZAZ	Albania	5RA-5SZ	Madagascar
T6A-T6Z	Afghanistan	ZBA-ZJZ	United Kingdom of	5TA-5TZ	Mauritania
T7A-T7Z	San Marino		Great Britain and	5UA-5UZ	Niger
T8A-T8Z	Palau		Northern Ireland	5VA-5VZ	Togo
T9A-T9Z	Bosnia and	ZKA-ZMZ	New Zealand	5WA-5WZ	Western Samoa
	Herzegovina	ZNA-ZOZ	United Kingdom of	5XA-5XZ	Uganda
UAA-UIZ	Russian Federation		Great Britain and	5YA-5ZZ	Kenya
UJA-UMZ	Uzbekistan		Northern Ireland	6AA-6BZ	Egypt
UNA-UQZ	Kazakhstan	ZPA-ZPZ	Paraguay	6CA-6CZ	Syria
URA-UTZ	Ukraine	ZQA-ZQZ	United Kingdom of	6DA-6JZ	Mexico
UUA-UZZ	Ukraine		Great Britain and	6KA-6NZ	South Korea
VAA-VGZ	Canada		Northern Ireland	6OA-6OZ	Somalia
VHA-VNZ	Australia	ZRA-ZUZ	South Africa	6PA-6SZ	Pakistan
VOA-VOZ	Canada	ZVA-ZZZ	Brazil	6TA-6UZ	Sudan
VPA-VQZ	United Kingdom of	Z2A-Z2Z	Zimbabwe	6VA-6WZ	Senegal
	Great Britain and	Z3A-Z3Z	Macedonia (Former	6XA-6XZ	Madagascar
	Northern Ireland		Yugoslav Republic)	6YA-6YZ	Jamaica
† VRA-VRZ	China (People's	2AA-2ZZ	United Kingdom of	6ZA-6ZZ	Liberia
	Republic of)-Hong		Great Britain and	7AA-7IZ	Indonesia
	Kong		Northern Ireland	7JA-7NZ	Japan
VSA-VSZ	United Kingdom of	3AA-3AZ	Monaco	7OA-7OZ	Yemen
	Great Britain and	3BA-3BZ	Mauritius	7PA-7PZ	Lesotho
	Northern Ireland	3CA-3CZ	Equatorial Guinea	7QA-7QZ	Malawi
VTA-VWZ	India	• 3DA-3DM	Swaziland	7RA-7RZ	Algeria
VXA-VYZ	Canada	• 3DN-3DZ	Fiji	7SA-7SZ	Sweden
VZA-VZZ	Australia	3EA-3FZ	Panama	7TA-7YZ	Algeria
V2A-V2Z	Antigua and	3GA-3GZ	Chile	7ZA-7ZZ	Saudi Arabia
	Barbuda	3HA-3UZ	China	8AA-8IZ	Indonesia
V3A-V3Z	Belize	3VA-3VZ	Tunisia	8JA-8NZ	Japan
V4A-V4Z	Saint Kitts and Nevis	3WA-3WZ	Viet Nam	8OA-8OZ	Botswana
V5A-V5Z	Namibia	3XA-3XZ	Guinea	8PA-8PZ	Barbados
V6A-V6Z	Micronesia	3YA-3YZ	Norway	8QA-8QZ	Maldives
V7A-V7Z	Marshall Islands	3ZA-3ZZ	Poland	8RA-8RZ	Guyana
V8A-V8Z	Brunei	4AA-4CZ	Mexico	8SA-8SZ	Sweden
WAA-WZZ	United States of	4DA-4IZ	Philippines	8TA-8YZ	India
	America	4JA-4KZ	Azerbaijan	8ZA-8ZZ	Saudi Arabia
XAA-XIZ	Mexico	4LA-4LZ	Georgia	9AA-9AZ	Croatia
XJA-XOZ	Canada	4MA-4MZ	Venezuela	9BA-9DZ	Iran
XPA-XPZ	Denmark	4NA-4OZ	Yugoslavia	9EA-9FZ	Ethiopia
XQA-XRZ	Chile	4PA-4SZ	Sri Lanka	9GA-9GZ	Ghana
XSA-XSZ	China	4TA-4TZ	Peru	9HA-9HZ	Malta
XTA-XTZ	Burkina Faso	* 4UA-4UZ	United Nations	9IA-9JZ	Zambia
XUA-XUZ	Cambodia	4VA-4VZ	Haiti	9KA-9KZ	Kuwait
XVA-XVZ	Viet Nam	4XA-4XZ	Israel	9LA-9LZ	Sierra Leone
XWA-XWZ	Laos	4WA-4WZ	UNTAET (E. Timor)	9MA-9MZ	Malaysia
XXA-XXZ	Portugal	* 4YA-4YZ	International Civil	9NA-9NZ	Nepal
XYA-XZZ	Myanmar		Aviation	9OA-9TZ	Democratic
YAA-YAZ	Afghanistan		Organization		Republic of the
YBA-YHZ	Indonesia	4ZA-4ZZ	Israel		Congo
YIA-YIZ	Iraq	5AA-5AZ	Libya	9UA-9UZ	Burundi
YJA-YJZ	Vanuatu	5BA-5BZ	Cyprus	9VA-9VZ	Singapore
YKA-YKZ	Syria	5CA-5GZ	Morocco	9WA-9WZ	Malaysia
YLA-YLZ	Latvia	5HA-5IZ	Tanzania	9XA-9XZ	Rwanda
YMA-YMZ	Turkey	5JA-5KZ	Colombia	9YZ-9ZZ	Trinidad and
YNA-YNZ	Nicaragua	5LA-5MZ	Liberia		Tobago
YOA-YRZ	Romania	5NA-5OZ	Nigeria		
YSA-YSZ	El Salvador				
YTA-YUZ	Yugoslavia	• Half-series			
YVA-YYZ	Venezuela	* Series allocated to an international organization			
YZA-YZZ	Yugoslavia	† Provisional allocation in accordance with S-19.33			

FCC 605
Approved by OMB
Main Form

Quick-Form Application for Authorization in the Ship, Aircraft,

Amateur, Restricted and Commercial Operator, and the
General Mobile Radio Services

3060 - 0850

See instructions for
public burden estimate

1) Radio Service Code: **H A**

Application Purpose (Select only one) **(MD)**

2)	**NE** - New	**RO** - Renewal Only	**WD** - Withdrawal of Application
	MD - Modification	**RM** - Renewal/Modification	**DU** - Duplicate License
	AM - Amendment	**CA** - Cancellation of License	**AU** - Administrative Update

3)	If this request is for a <u>D</u>evelopmental License or <u>S</u>TA (Special Temporary Authorization) enter the appropriate code and attach the required exhibit as described in the instructions. Otherwise enter <u>N</u> (Not Applicable).	(**N**) D S N/A
4)	If this request is for an Amendment or Withdrawal of Application, enter the file number of the pending application currently on file with the FCC.	File Number
5)	If this request is for a Modification, Renewal Only, Renewal/Modification, Cancellation of License, Duplicate License, or Administrative Update, enter the call sign of the existing FCC license.	Call Sign **W R 1 B**
6)	If this request is for a New, Amendment, Renewal Only, or Renewal/Modification, enter the requested authorization expiration date (this item is optional).	MM DD
7)	Does this filing request a Waiver of the Commission's rules? If 'Y', attach the required showing as described in the instructions.	(**N**)<u>Y</u>es <u>N</u>o
8)	Are attachments (other than associated schedules) being filed with this application?	(**N**)<u>Y</u>es <u>N</u>o

Applicant Information

9a) Taxpayer Identification Number: **1 2 3 4 5 6 7 8 9**	9b) SGIN:

10) Applicant/Licensee is a(n): (**I**)	Individual Corporation	Unincorporated Association Limited Liability Corporation	Trust Partnership	Government Entity Joint Venture Consortium

11) First Name (if individual): **Larry**	MI: **D**	Last Name: **Wolfgang**	Suffix:

12) Entity Name (if other than individual):

13) Attention To:

14) P.O. Box:	And/Or	15) Street Address: **225 Main St**

16) City: **Newington**	17) State: **CT**	18) Zip: **06111**	19) Country: **USA**

20) Telephone Number:	21) FAX:

22) E-Mail Address:

FCC 605- Main Form
July 1999 - Page 1

Figure 5 — This sample FCC Form 605 shows the sections you should complete to notify the FCC of a change in your address.

Fee Status

23) Is the applicant exempt from FCC application fees?	(**N**)Yes No
24) Is the applicant exempt from FCC regulatory fees?	(**N**)Yes No

General Certification Statements

1) The Applicant waives any claim to the use of any particular frequency or of the electromagnetic spectrum as against the regulatory power of the United States because of the previous use of the same, whether by license or otherwise, and requests an authorization in accordance with this application.

2) The applicant certifies that all statements made in this application and in the exhibits, attachments, or documents incorporated by reference are material, are part of this application, and are true, complete, correct, and made in good faith.

3) Neither the Applicant nor any member thereof is a foreign government or a representative thereof.

4) The applicant certifies that neither the applicant nor any other party to the application is subject to a denial of Federal benefits pursuant to Section 5301 of the Anti-Drug Abuse Act of 1988, 21 U.S.C. § 862, because of a conviction for possession or distribution of a controlled substance. **This certification does not apply to applications filed in services exempted under Section 1.2002(c) of the rules, 47 CFR § 1.2002(c).** See Section 1.2002(b) of the rules, 47 CFR § 1.2002(b), for the definition of "party to the application" as used in this certification.

5) Amateur or GMRS Applicant certifies that the construction of the station would NOT be an action which is likely to have a significant environmental effect (see the Commission's Rules 47 CFR Sections 1.1301-1.1319 and Section 97.13(a).

6) Amateur Applicant certifies that they have READ and WILL COMPLY WITH Section 97.13(c) of the Commission's Rules regarding RADIOFREQUENCY (RF) RADIATION SAFETY and the amateur service section of OST/OET Bulletin Number 65.

Certification Statements For GMRS Applicants

1) Applicant certifies that he or she is claiming eligibility under Rule Section 95.5 of the Commission's Rules.

2) Applicant certifies that he or she is at least 18 years of age.

3) Applicant certifies that he or she will comply with the requirement that use of frequencies 462.650, 467.650, 462.700 and 467.700 MHz is not permitted near the Canadian border North of Line A and East of Line C. These frequencies are used throughout Canada and harmful interference is anticipated.

Signature

25) Typed or Printed Name of Party Authorized to Sign

First Name: *Larry*	MI: *D*	Last Name: *Wolfgang*	Suffix:

26) Title:

Signature: *Larry D. Wolfgang*	27) Date: 09/15/1999

Failure To Sign This Application May Result In Dismissal Of The Application And Forfeiture Of Any Fees Paid

---

WILLFUL FALSE STATEMENTS MADE ON THIS FORM OR ANY ATTACHMENTS ARE PUNISHABLE BY FINE AND/OR IMPRISONMENT (U.S. Code, Title 18, Section 1001) AND/OR REVOCATION OF ANY STATION LICENSE OR CONSTRUCTION PERMIT (U.S. Code, Title 47, Section 312(a)(1)), AND/OR FORFEITURE (U.S. Code, Title 47, Section 503).

still be able to renew your license without taking another exam. There is a two-year grace period, during which you may apply for renewal of your expired license. Use an FCC Form 605 to apply for reinstatement (and your old call sign). If you apply for reinstatement of your expired license under this two-year grace period, you may not operate your station until your new license is issued. If you move or change addresses you should use an FCC Form 605 to notify the FCC of the change. If your license is lost or destroyed, however, just write a letter to the FCC explaining why you are requesting a new copy of your license.

You can ask one of the Volunteer Examiner Coordinators' offices to file your renewal application electronically if you don't want to mail the form to the FCC. You must still mail the form to the VEC, however. The ARRL/VEC Office will electronically file applications forms for any ARRL member free of charge.

You can also file your license renewal or address modification using the Universal Licensing System (ULS) on the World Wide Web. Go to **http:// www.fcc.gov/wtb/uls** and click on the "TIN/Call Sign Registration" button. Follow the directions to register with the Universal Licensing System. Next click on the "Connecting to ULS" button and follow the directions given there to connect to the FCC's ULS database.

The FCC has a set of detailed instructions for the Form 605, which are included with the form. To obtain a new Form 605, call the FCC Forms Distribution Center at 800-418-3676. You can also write to: Federal Communications Commission, Forms Distribution Center, 2803 52nd Avenue, Hyattsville, MD 20781 (specify "Form 605" on the envelope). The Form 605 also is available from the FCC's fax on demand service. Call 202-418-0177 and ask for form number 000605. Form 605 also is available via the Internet. The World Wide Web location is: **http://www.fcc.gov/formpage.html** or you can receive the form via ftp to: **ftp.fcc.gov/pub/Forms/Form605**.

The ARRL/VEC has created a package that includes the portions of Form 605 that are needed for amateur applications, as well as a condensed set of instructions for completing the form. Write to: ARRL/VEC, Form 605, 225 Main Street, Newington, CT 06111-1494. (Please include a large business sized stamped, self-addressed envelope with your request.) **Figure 5** is a sample of those portions of an FCC Form 605 that you would complete to submit a change of address to the FCC.

Most of the form is simple to fill out. You will need to know that the Radio Service Code for box 1 is HA for Amateur Radio. (Just remember HAM radio.) You will have to include a "Taxpayer Identification Number" on the Form. This is normally your Social Security Number. If you don't want to write your Social Security Number on this form, then you can register with the ULS as described above. Then you will receive a ULS Registration Number from the FCC, and you can use that number instead of your Social Security Number on the Form. Of course, you will have to supply your Social Security Number to register with the ULS.

The telephone number, fax number and e-mail address information is optional. The FCC will use that information to contact you in case there is a problem with your application.

Page two includes six General Certification Statements. Statement five may seem confusing. Basically, this statement means that you do not plan to install an antenna over 200 feet high, and that your permanent station location will not be in a designated wilderness area, wildlife preserve or nationally recognized scenic and recreational area.

The sixth statement indicates that you are familiar with the FCC RF Safety Rules, and that you will obey them. The Technician and General license exams (Elements 2 and 3) include questions about RF safety.

Sometime during the second half of 2000, the FCC is expected to announce a new Commission Registration System (CORES). This system is expected to replace ULS registration, but not to replace the entire ULS. Each licensee will receive a CORES FCC Registration Number (FRN) that will allow access to the ULS database. The FCC web site will provide instructions when the new system goes into use.

Volunteer Examiner Program

Before you can take an FCC exam, you'll have to fill out a copy of the National Conference of Volunteer Examiner Coordinators' (NCVEC) Quick Form 605. This form is used as an application for a new license or an upgraded license. The NCVEC Quick Form 605 is only used at license exam sessions. This form includes some information that the Volunteer Examiner Coordinator's office will need to process your application with the FCC. See **Figure 6**. You should not use an NCVEC Quick Form 605 to apply for a license renewal or modification with the FCC. *Never* mail these forms to the FCC, because that will result in a rejection of the application. Likewise, an FCC Form 605 can't be used for an exam application.

All US amateur exams are administered by Volunteer Examiners who are certified by a Volunteer-Examiner Coordinator (VEC). Program. *The ARRL's FCC Rule Book* contains more details about the Volunteer-Examiner program.

To qualify for an Amateur Extra license you must pass Elements 1, 2, 3 and 4. If you have a Novice license, then you have credit for passing Element 1. If you hold a Technician license you will receive credit for passing Element 2. If you also have passed a Morse code exam along with your Technician license, then you have credit for Elements 1 and 2, and will not have to retake those exam elements. If you have a Technician license issued before March 21, 1987, and you wish to upgrade to General, you also have credit for passing Element 3. In that case, all you need to do is go to an exam session with your old original license. If you no longer have the original license, you may be able to obtain suitable proof of having held such a license. Such proof can come from an old *Radio Amateurs Callbook*, the data from an old edition of the *QRZ* CD ROM (available on the Internet at **http:// www.qrz.com**) and other sources. Contact the ARRL/VEC Office if you have questions about this. If you have a Technician license issued after March 21, 1987, you must study the questions in the Element 3 question pool as well as the Element 4 question pool, and pass both exams. See Table 1 for details.

The Element 4 exam consists of 50 questions taken from a pool of more than 650. The question pools for all amateur exams are maintained by a Question Pool Committee selected by the Volunteer Examiner Coordinators. The FCC allows Volunteer Examiners to select the questions for an amateur exam, but they must use the questions exactly as they are released by the VEC that coordinates the test session. If you attend a test session coordinated by the ARRL/VEC, your test will be designed by the ARRL/VEC or by a computer program designed by the VEC. The questions and answers will be exactly as they are printed in Chapter 10 of this book.

NCVEC QUICK-FORM 605 APPLICATION FOR
AMATEUR OPERATOR/PRIMARY STATION LICENSE

SECTION 1 - TO BE COMPLETED BY APPLICANT

PRINT LAST NAME	SUFFIX	FIRST NAME	INITIAL	STATION CALL SIGN (IF ANY)
Tracy		Michael	D	KC1SX

MAILING ADDRESS (Number and Street or P.O. Box)	SOCIAL SECURITY NUMBER/TIN (OR FCC LICENSEE ID #)
83 Main St. Apt 9B	123 456 789

CITY	STATE CODE	ZIP CODE (5 or 9 Numbers)	E-MAIL ADDRESS (OPTIONAL)
Newington	CT	06111	mtracy@arrl.org

DAYTIME TELEPHONE NUMBER (Include Area Code) OPTIONAL	FAX NUMBER (Include Area Code) OPTIONAL	ENTITY NAME (IF CLUB, MILITARY RECREATION, RACES)

Type of Applicant: ☒ Individual ☐ Amateur Club ☐ Military Recreation ☐ RACES (Modify Only)

TRUSTEE OR CUSTODIAN CALL SIGN

I HEREBY APPLY FOR (Make an X in the appropriate box(es))

SIGNATURE OF RESPONSIBLE CLUB OFFICIAL

☐ EXAMINATION for a **new** license grant

☒ EXAMINATION for **upgrade** of my license class

☐ CHANGE my **name** on my license to my new name

Former Name: _____
(Last name) (Suffix) (First name) (MI)

☐ CHANGE my mailing address to **above** address

☐ CHANGE my station **call sign** systematically

Applicant's Initials: _____

☐ RENEWAL of my license grant.

Do you have another license application on file with the FCC which has not been acted upon?	PURPOSE OF OTHER APPLICATION	PENDING FILE NUMBER (FOR VEC USE ONLY)

I certify that:
* I waive any claim to the use of any particular frequency regardless of prior use by license or otherwise;
* All statements and attachments are true, complete and correct to the best of my knowledge and belief and are made in good faith;
* I am not a representative of a foreign government;
* I am not subject to a denial of Federal benefits pursuant to Section 5301of the Anti-Drug Abuse Act of 1988, 21 U.S.C. § 862;
* The construction of my station will NOT be an action which is likely to have a significant environmental effect (See 47 CFR Sections 1.301-1.319 and Section 97.13(a));
* I have read and WILL COMPLY with Section 97.13(c) of the Commission's Rules regarding RADIOFREQUENCY (RF) RADIATION SAFETY and the amateur service section of OST/OET Bulletin Number 65.

Signature of applicant (Do not print, type, or stamp. Must match applicant's name above.)

X Michael Tracy   Date Signed: April 15, 2000

SECTION 2 - TO BE COMPLETED BY ALL ADMINISTERING VEs

Applicant is qualified for operator license class:

☐ NO NEW LICENSE OR UPGRADE WAS EARNED

☐ TECHNICIAN   Element 2

☐ GENERAL   Elements 1, 2 and 3

☒ AMATEUR EXTRA   Elements 1, 2, 3 and 4

DATE OF EXAMINATION SESSION
4 / 15 / 2000

EXAMINATION SESSION LOCATION
Newington, CT

VEC ORGANIZATION
D O l

VEC RECEIPT DATE
APR 20 2000

I CERTIFY THAT I HAVE COMPLIED WITH THE ADMINISTERING VE REQUIRMENTS IN PART 97 OF THE COMMISSION'S RULES AND WITH THE INSTRUCTIONS PROVIDED BY THE COORDINATING VEC AND THE FCC.

1st VEs NAME (Print First, MI, Last, Suffix)	VEs STATION CALL SIGN	VEs SIGNATURE (Must match name)	DATE SIGNED
DAVID C. PATTON	N1IN	David C. Patton	4/15/00
2nd VEs NAME (Print First, MI, Last, Suffix)	VEs STATION CALL SIGN	VEs SIGNATURE (Must match name)	DATE SIGNED
Larry D. Wolfgang	WR1B	Larry D. Wolfgang	4/15/00
3rd VEs NAME (Print First, MI, Last, Suffix)	VEs STATION CALL SIGN	VEs SIGNATURE (Must match name)	DATE SIGNED
WAYNE K. IRWIN	W1KI	Wayne K. Irwin	04/15/2000

NCVEC FORM 605 - APRIL 2000
FOR VE/VEC USE ONLY - Page 1

Figure 6 — This sample NCVEC Quick Form 605 shows how your form will look after you have completed your upgrade to Amateur Extra.

INSTRUCTIONS FOR COMPLETING APPLICATION FORM NCVEC FORM 605

ARE WRITTEN TESTS AN FCC-LICENSE REQUIREMENT? ARE THERE EXEMPTIONS?

Beginning April 15, 2000, you may be examined on only three classes of operator licenses, each authorizing varying levels of privileges. The class for which each examinee is qualified is determined by the degree of skill and knowledge in operating a station that the examinee demonstrates to volunteer examiners (VEs) in his or her community. The demonstration of this know-ledge is required in order to obtain an Amateur Operator/Primary Station License. There is no exemption from the written exam requirements for persons with difficulty in reading, writing, or because of a handicap or disability. There are exam accommo-dations that can be afforded examinees (see ACCOMMODATING A HANDICAPPED PERSON below). Most new amateur operators start at the Technician class and then advance one class at a time. The VEs give examination credit for the license class currently (and in some cases, previously) held so that examinations required for that license need not be repeated. The written exami-nations are constructed from question pools that have been made public (see: <http://www.arrl.org/arrlvec/pools.html>.) Helpful study guides and training courses are also widely avail-able. To locate examination opportunities in your area, contact your local club, VE group, one of the 14 VECs or see the online listings at: <http://www.w5yi.org/vol-exam.htm> or <http//www.arrl.org/arrlvec/examsearch.phtml>.

IS KNOWLEDGE OF MORSE CODE AN FCC-LICENSE REQUIREMENT? ARE THERE EXEMPTIONS?

Some persons have difficulty in taking Morse code tests because of a handicap or disability. There is available to all otherwise qualified persons, handicapped or not, the Technician Class operator license that does not require passing a Morse code examination. Because of international regulations, how-ever, any US FCC licensee seeking access to the HF bands (frequencies below 30 MHz) must have demonstrated proficiency in Morse code. If a US FCC licensee wishes to gain access to the HF bands, there is no exemption available from this Morse code proficiency requirement. If licensed as a Tech-nician class, upon passing a Morse code examination operation on certain HF bands is permitted.

THE REASON FOR THE MORSE CODE EXAMINATION

Telegraphy is a method of electrical communication that the Amateur Radio Service community strongly desires to preserve. The FCC supports this objective by authorizing additional operating privileges to amateur operators who pass a Morse Code examination. Normally, to attain this skill, intense practice is required. Annually, thousands of amateur operators prove, by passing examinations, that they have acquired the skill. These examinations are prepared and administered by amateur ope-rators in the local community who volunteer their time and effort.

THE EXAMINATION PROCEDURE

The volunteer examiners (VEs) send a short message in the Morse code. The examinee must decipher a series of audible dots and dashes into 43 different alphabetic, numeric, and punctuation characters used in the message. Usually a 10-question quiz is then administered asking questions about items contained in the mes-sage.

ACCOMMODATING A HANDICAPPED PERSON

Many handicapped persons accept and benefit from the personal challenge of passing the examination in spite of their hardships. For handicapped persons who have difficulty in proving that they can decipher messages sent in the Morse code, the VEs make exceptionally accommodative arrangements. To assist such persons, the VEs will:

- adjust the tone in frequency and volume to suit the examinee.
- administer the examination at a place convenient and com-fortable to the examinee, even at bedside.
- for a deaf person, they will send the dots and dashes to a vibrating surface or flashing light.
- write the examinee's dictation.
- where warranted, they will pause in sending the message after each sentence, each phrase, each word, or in extreme cases they will pause the exam message character-by-character to allow the examinee additional time to absorb, to interpret or even to speak out what was sent.
- or they will even allow the examinee to send the message, rather than receive it.

Should you have any questions, please contact your local volunteer examiner team, or contact one of the 14 volunteer examiner coordina-tor (VEC) organizations. For contact information for VECs, or to contact the FCC, call 888-225-5322 (weekdays), or write to FCC, 1270 Fairfield Road, Gettysburg PA 17325-7245. Fax 717-338-2696. Also see the FCC web at: <http//www.fcc.gov/wtb/amateur/>.

RENEWING, MODIFYING OR REINSTATING YOUR AMATEUR RADIO OPERATOR/PRIMARY STATION LICENSE

RENEWING YOUR AMATEUR LICENSE

The NCVEC Form 605 may also be used to renew or modify your Amateur Radio Operator/Primary Station license. License renewal may only be completed during the final 90 days prior to license expiration, or up to two years after expiration. Changes to your mailing address, name and requests for a sequential change of your station call sign appropriate for your current license class may be requested at any time. This form may not be used to apply for a specific ("Vanity") station call sign.

REINSTATING YOUR AMATEUR LICENSE

This form may also be used to reinstate your Amateur Radio Operator/Station license if it has been expired less than the two year grace period for renewal. After the two year grace period you must retake the amateur license examinations to become relicensed. You will be issued a new systematic call sign.

RENEWING OR MODIFYING YOUR LICENSE

On-line renewal: You can submit your renewal or license modifica-tions to FCC on-line via the internet/WWW at: <http://www.fcc.gov/wtb/uls>. To do so, you must first register in ULS by following the "TIN/Call Sign Registration" tab procedures, then choose the "Connecting to ULS" tab procedures and use their special dial-in to an FCC 800# modem-only access system.

Renewal by mail: If you choose to renew by mail, you can mail the "FCC Form 605" to FCC. You can obtain FCC Form 605 via the internet at <http://www.fcc.gov/formpage.html> or <ftp://ftp.fcc.gov/pub/Forms/Form605/>. It's available by fax at 202-418-0177 (request Form 000605). The FCC Forms Distribution Center will accept form orders by calling 800-418-3676. FCC Form 605 has a main form, plus a Schedule D. The main form is all that is needed for renewals. Mail FCC Form 605 to: FCC, 1270 Fairfield Rd, Gettysburg PA 17325-7245. This is a free FCC service.

The NCVEC Form 605 application can be used for a license renewal, modification or reinstatement. NCVEC Form 605 can be processed by VECs, but not all VECs provide this as a routine service. ARRL Members can submit NCVEC Form 605 to the ARRL/VEC for process-ing. ARRL Members or others can choose to submit their NCVEC Form 605 to a local VEC (check with the VEC office before forwarding), or it can be returned with a $6.00 application fee to: The W5YI Group, Inc., P.O. Box 565101, Dallas, Texas 75356 (a portion of this fee goes to the National Conference of VECs to help defray their expenses). The NCVEC Form 605 may not be returned to the FCC since it is an internal VEC form. Once again, the service provided by FCC is free.

THE FCC APPLICATION FORM 605

The FCC version of the Form 605 may not be used for applications submitted to a VE team or a VEC since it does not request information needed by the administering VEs. The FCC Form 605 may, however, be used to routinely renew or modify your license without charge. It should be sent to the FCC, 1270 Fairfield Rd., Gettysburg PA 17325-7245.

CLUB AND MILITARY RECREATION CALL SIGN ADMINISTRATORS

The NCVEC Form 605 may also be used for the processing of applications for Amateur Service club and military recreation station call signs and for the modification of RACES stations. No fee may be charged by an administrator for this service. As of March 9, 2000, FCC had not yet implemented the Call Sign Administrator System.

NCVEC FORM 605 - APRIL 2000
FOR VE/VEC USE ONLY - Page 2

Finding an Exam Opportunity

To determine where and when an exam will be given, contact the ARRL/VEC Office, or watch for announcements in the Hamfest Calendar and Coming Conventions columns in *QST*. Many local clubs sponsor exams, so they are another good source of information on exam opportunities. ARRL officials such as Directors, Vice Directors and Section Managers receive notices about test sessions in their area. See the latest issue of *QST* for the names and addresses of your local ARRL officials. You can also use the ARRL Exam Session Search feature of *ARRLWeb*, at: **http://www.arrl.org/arrlvec/examsearch.phtml**.

To register for an exam, send a completed NCVEC Quick Form 605 to the Volunteer Examining team responsible for the exam session if preregistration is required. Otherwise, bring the form to the session. Registration deadlines, and the time and location of the exams, are mentioned prominently in publicity releases about upcoming sessions.

Taking The Exam

By the time examination day rolls around, you should have already prepared yourself. This means getting your schedule, supplies and mental attitude ready. Plan your schedule so you'll get to the examination site with plenty of time to spare. There's no harm in being early. In fact, you might have time to discuss hamming with another applicant, which is a great way to calm pre-test nerves. Try not to discuss the material that will be on the examination, as this may make you even more nervous. By this time, it's too late to study anyway!

What supplies will you need? First, be sure you bring your current *original* Amateur Radio license, if you have one. Bring a photocopy of your license, too, as well as the original and a photocopy of any Certificates of Successful Completion of Examination (CSCE) that you plan to use for exam credit. Bring along several sharpened number 2 pencils and two pens (blue or black ink). Be sure to have a good eraser. A pocket calculator will also come in handy. You may use a programmable calculator if that is the kind you have, but take it into your exam "empty." Don't program a lot of equations ahead of time, because you may be asked to demonstrate that there is nothing in the calculator's memory. If you use a slide rule, that should also be allowed, although you will probably *not* be allowed to take math tables, such as trigonometry tables or logarithm tables into the exam with you.

The Volunteer Examining Team is required to check two forms of identification before you enter the test room. This includes your *original* Amateur Radio license, if you have one. A photo ID of some type is best for the second form of ID, but is not required by FCC. Other acceptable forms of identification include a driver's license, a piece of mail addressed to you, a birth certificate, or some other such document.

The following description of the testing procedure applies to exams coordinated by the ARRL/VEC, although many other VECs use a similar procedure.

Code Tests

The 5 wpm code test is usually given before the written exams. Before you take the code test, you'll be handed a piece of paper to copy the code as it is sent. The test will begin with about a minute of practice copy. Then comes the actual test: at least five minutes of Morse code. You are responsible for knowing the 26 letters of the alphabet, the numerals 0 through 9, the period, comma, question mark, and procedural signals $\overline{AR}$ (+), $\overline{SK}$, $\overline{BT}$ (= also called double dash) and $\overline{DN}$ (/ or fraction bar, sometimes called the "slant bar"). You may copy the entire text word for word, or just take notes on the content. At the end of the transmission, the examiner will hand you 10 questions about the text. Depending on the test format, answer the multiple-choice questions or simply fill in the blanks with your answers. (You must spell each answer exactly as it was sent.) If you get at least seven correct, you pass! Alternatively, the exam team has the option to look at your copy sheet if you fail the 10-question exam. If you have one minute of solid copy (25 characters without an error), the examiners can certify that you passed the test on that basis. The format of the test transmission is similar to one side of a normal on-the-air amateur conversation.

The National Conference of VECs has voted to eliminate the multiple-choice type of code exam. In addition, all code exams would use the Farnsworth method, in which characters are sent faster than the 5-wpm overall speed. Farnsworth "character speed" would be in the range of 13 to 15 wpm. These changes will take effect by July 1, 2001, although VECs may implement them sooner.

A sending test probably won't be required. The FCC has decided that if applicants can demonstrate receiving ability, they most likely can also send at that speed. But be prepared, just in case! Subpart 97.503(a) of the FCC Rules says, "A telegraphy examination must be sufficient to prove that the examinee has the ability to send correctly by hand and to receive correctly by ear texts in the international Morse code at not less than the prescribed speed...."

Written Tests

After the code test has been administered, you'll then take the written examination. The examiner will give all applicants a test booklet, an answer sheet and scratch paper. After that, you're on your own. The first thing to do is read the instructions. Be sure to sign your name every place it's called for. Do all of this at the beginning to get it out of the way.

Next, check the examination to see that all pages and questions are there. If not, report this to the examiner immediately. When filling in your answer sheet, make sure your answers are marked next to the numbers that correspond to each question.

Go through the entire exam, and answer the easy questions first. Next, go back to the beginning and try the harder questions. Leave the really tough questions for last. Guessing can only help, as there is no additional penalty for answering incorrectly.

If you have to guess, do it intelligently: At first glance, you may find that you can eliminate one or more distractors. Of the remaining responses, more than one may seem correct; only one is the *best* answer, however. To the applicant who is fully prepared, incorrect distractors to each question are obvious. Nothing beats preparation!

After you've finished, check the examination thoroughly. You may have read a question wrong or goofed in your arithmetic. Don't be overconfident. There's no rush, so take your time. Think, and check your answer sheet. When you feel you've done your best and can do no more, return the test booklet, answer sheet and scratch pad to the examiner.

The Volunteer Examiner team will grade the exam right away. The passing mark is 74%. (That means no more than 13 incorrect answers on the 50-question exam.) You will receive a Certificate of Successful Completion of Examination (CSCE) showing all the exam elements that you pass at that exam session. If you are already licensed, and you pass the exam elements required to earn a higher license class, the CSCE authorizes you to operate with your new privileges. When you use these new privileges, you must sign your call sign, followed by the slant mark ("/"; on voice, say "slant" or "stroke") and the letters "AE" for an upgrade to Extra. You only have to use this special identification procedure until the FCC issues your new license, however.

If you pass only some of the exam elements required for a higher class license, you will still receive a CSCE. That certificate shows what exam elements you passed, and is valid for 365 days. Use it as proof that you passed those exam elements so you won't have to take them over again the next time you try for the upgrade.

AND NOW, LET'S BEGIN

The complete Amateur Extra class question pool (Element 4) is printed in Chapter 10. Chapters 1 through 9 explain the material covered in subelements E1 through E9 of the question pool.

This book provides the background in FCC Rules, operating procedures, radio-wave propagation, Amateur Radio practice, electrical principles, circuit components, practical circuits, signals and emissions, and antennas and feed lines that you will need to pass the Element 4 Amateur Extra class written exam.

Table 5 shows the study guide or syllabus for the Element 4, Amateur Extra exam. This study guide was released by the Volunteer Examiner Coordinators' Question Pool Committee in February 2000. The syllabus lists the topics to be covered by the Amateur Extra exam, and so forms a basic outline for the remainder of this book. Use the syllabus to guide your study, and to ensure that you have studied the material for all of the topics listed.

The question numbers used in the question pool refer to this syllabus. Each question number begins with a syllabus-point number (E1D, E2A, E3B and so on). The question numbers end with a two-digit number. For example, question E1D03 is the third question in the series about syllabus point E1D. Question E9C10 is the tenth question about syllabus point E9C.

The Question Pool Committee designed the syllabus and question pool so there are the same number of points in each subelement as there are exam questions from that subelement. For example, seven questions on the Amateur Extra exam must come from the "Commission's Rules" subelement. There are seven groups of questions for this subelement, E1A through E1G. Three exam questions must be from the "Radio-Wave Propagation" subelement, so there are three groups for that subelement. These are numbered E3A, E3B and E3C. While not a requirement of the FCC Rules, the Question Pool Committee recommends that one question be taken from each group to make the best possible Amateur Extra class (Element 4) exam.

Good luck with your studies!

Table 5

Amateur Extra Class (Element 4) Syllabus

(Required for Amateur Extra Licenses)

SUBELEMENT E1 — COMMISSION'S RULES
[7 Exam Questions — 7 Groups]

E1A Operating standards: frequency privileges for Extra class amateurs; emission standards; message forwarding; frequency sharing between ITU Regions; FCC modification of station license; 30-meter band sharing; stations aboard ships or aircraft; telemetry; telecommand of an amateur station; authorized telecommand transmissions; definitions of image, pulse and test

E1B Station restrictions: restrictions on station locations; restricted operation; teacher as control operator; station antenna structures; definition and operation of remote control and automatic control; control link

E1C Reciprocal operating: definition of reciprocal operating permit; purpose of reciprocal agreement rules; alien control operator privileges; identification; application for reciprocal permit; reciprocal permit license term (Note: This includes CEPT and IARP.)

E1D Radio Amateur Civil Emergency Service (RACES): definition; purpose; station registration; station license required; control operator requirements; control operator privileges; frequencies available; limitations on use of RACES frequencies; points of communication for RACES operation; permissible communications

E1E Amateur Satellite Service: definition; purpose; station license required for space station; frequencies available; telecommand operation: definition; eligibility; telecommand station (definition); space telecommand station; special provisions; telemetry: definition; special provisions; space station: definition; eligibility; special provisions; authorized frequencies (space station); notification requirements; earth operation: definition; eligibility {97.209(a)}; authorized frequencies (Earth station)

E1F Volunteer Examiner Coordinators (VECs): definition; VEC qualifications; VEC agreement; scheduling examinations; coordinating VEs; reimbursement for expenses {97.527}; accrediting VEs; question pools; Volunteer Examiners (VEs): definition; requirements; accreditation; reimbursement for expenses; VE conduct; preparing an examination; examination elements; definition of code and written elements; preparation responsibility; examination requirements; examination credit; examination procedure; examination administration; temporary operating authority

E1G Certification of external RF power amplifiers and external RF power amplifier kits; Line A; National Radio Quiet Zone; business communications; definition and operation of spread spectrum; auxiliary station operation

SUBELEMENT E2 — OPERATING PROCEDURES
[4 Exam Questions — 4 Groups]

E2A Amateur Satellites: Orbital mechanics; Frequencies available for satellite operation; Satellite hardware; Operating through amateur satellites

E2B Television: fast scan television (FSTV) standards; slow scan television (SSTV) standards; facsimile (fax) communications

E2C Contest and DX operating; spread-spectrum transmissions; automatic HF forwarding

E2D Digital Operating: HF digital communications (ie, PacTOR, CLOVER, AMTOR, PSK31, HF packet); packet clusters; HF digital bulletin boards

SUBELEMENT E3 — RADIO WAVE PROPAGATION
[3 Exam Questions — 3 Groups]

E3A Earth-Moon-Earth (EME or moonbounce) communications; meteor scatter

E3B Transequatorial; long path; gray line

E3C Auroral propagation; selective fading; radio-path horizon; take-off angle over flat or sloping terrain; earth effects on propagation

SUBELEMENT E4 — AMATEUR RADIO PRACTICES
[5 Exam Questions — 5 Groups]

E4A Test equipment: spectrum analyzers (interpreting spectrum analyzer displays; transmitter output spectrum); logic probes (indications of high and low states in digital circuits; indications of pulse conditions in digital circuits)

E4B Frequency measurement devices (i.e., frequency counter, oscilloscope Lissajous figures, dip meter); meter performance limitations; oscilloscope performance limitations; frequency counter performance limitations

E4C Receiver performance characteristics (i.e., phase noise, desensitization, capture effect, intercept point, noise floor, dynamic range {blocking and IMD}, image rejection, MDS, signal-to-noise-ratio); intermodulation and cross-modulation interference

E4D Noise suppression: ignition noise; alternator noise (whine); electronic motor noise; static; line noise

E4E Component mounting techniques (i.e., surface, dead bug {raised}, circuit board); direction finding: techniques and equipment; fox hunting

SUBELEMENT E5 — ELECTRICAL PRINCIPLES
[9 Exam Questions — 9 Groups]

E5A Characteristics of resonant circuits: Series resonance (capacitor and inductor to resonate at a specific frequency); Parallel resonance (capacitor and inductor to resonate at a specific frequency); half-power bandwidth

E5B Exponential charge/discharge curves (time constants): definition; time constants in RL and RC circuits

E5C Impedance diagrams: Basic principles of Smith charts; impedance of RLC networks at specified frequencies

E5D Phase angle between voltage and current; impedances and phase angles of series and parallel circuits; algebraic operations using complex numbers: rectangular coordinates (real and imaginary parts); polar coordinates (magnitude and angle)

E5E Skin effect; electrostatic and electromagnetic fields

E5F Circuit Q; reactive power; power factor

E5G Effective radiated power; system gains and losses

E5H Replacement of voltage source and resistive voltage divider with equivalent voltage source and one resistor (Thevenin's Theorem)

E5I Photoconductive principles and effects

SUBELEMENT E6 — CIRCUIT COMPONENTS
[5 Exam Questions — 5 Groups]

E6A Semiconductor material: Germanium, Silicon, P-type, N-type; Transistor types: NPN, PNP, junction, unijunction, power; field-effect transistors (FETs): enhancement mode; depletion mode; MOS; CMOS; N-channel; P-channel

E6B Diodes: Zener, tunnel, varactor, hot-carrier, junction, point contact, PIN and light emitting; operational amplifiers (inverting amplifiers, noninverting amplifiers, voltage gain, frequency response, FET amplifier circuits, single-stage amplifier applications); phase-locked loops

E6C TTL digital integrated circuits; CMOS digital integrated circuits; gates

E6D Vidicon and cathode-ray tube devices; charge-coupled devices (CCDs); liquid crystal displays (LCDs); toroids: permeability, core material, selecting, winding

E6E Quartz crystal (frequency determining properties as used in oscillators and filters); monolithic amplifiers (MMICs)

SUBELEMENT E7 — PRACTICAL CIRCUITS
[7 Exam Questions — 7 Groups]

E7A Digital logic circuits: Flip flops; Astable and monostable multivibrators; Gates (AND, NAND, OR, NOR); Positive and negative logic

E7B Amplifier circuits: Class A, Class AB, Class B, Class C, amplifier operating efficiency (ie, DC input versus PEP), transmitter final amplifiers; amplifier circuits: tube, bipolar transistor, FET

E7C Impedance-matching networks: Pi, L, Pi-L; filter circuits: constant K, M-derived, band-stop, notch, crystal lattice, pi-section, T-section, L-section, Butterworth, Chebyshev, elliptical; filter applications (audio, IF, digital signal processing {DSP})

E7D Oscillators: types, applications, stability; voltage-regulator circuits: discrete, integrated and switched mode

E7E Modulators: reactance, phase, balanced; detectors; mixer stages; frequency synthesizers

E7F Digital frequency divider circuits; frequency marker generators; frequency counters

E7G Active audio filters: characteristics; basic circuit design; preselector applications

SUBELEMENT E8 — SIGNALS AND EMISSIONS
[5 Exam Questions — 5 Groups]

E8A AC waveforms: sine wave, square wave, sawtooth wave; AC measurements: peak, peak-to-peak and root-mean-square (RMS) value, peak-envelope-power (PEP) relative to average

E8B FCC emission designators versus emission types; modulation symbols and transmission characteristics; modulation methods; modulation index; deviation ratio; pulse modulation: width; position

E8C Digital signals: CW; baudot; ASCII; packet; AMTOR; Clover; information rate vs bandwidth

E8D Amplitude compandored single-sideband (ACSSB); spread-spectrum communications

E8E Peak amplitude (positive and negative); peak-to-peak values: measurements; Electromagnetic radiation; wave polarization; signal-to-noise (S/N) ratio

SUBELEMENT E9 — ANTENNAS AND FEED LINES
[5 Exam Questions — 5 Groups]

E9A Isotropic radiators: definition; used as a standard for comparison; radiation pattern; basic antenna parameters: radiation resistance and reactance (including wire dipole, folded dipole), gain, beamwidth, efficiency

E9B Free-space antenna patterns: E and H plane patterns (ie, azimuth and elevation in free-space); gain as a function of pattern; antenna design (computer modeling of antennas)

E9C Phased vertical antennas; radiation patterns; beverage antennas; rhombic antennas: resonant; nonresonant; radiation pattern; antenna patterns: elevation above real ground, ground effects as related to polarization, take-off angles as a function of height above ground

E9D Space and satellite communications antennas: gain; beamwidth; tracking; losses in real antennas and matching: resistivity losses, losses in resonating elements (loading coils, matching networks, etc. {ie, mobile, trap}); SWR bandwidth; efficiency

E9E Matching antennas to feed lines; characteristics of open and shorted feed lines: 1/8 wave-length; 1/4 wavelength; 3/8 wavelength; 1/2 wavelength; 1/4 wavelength matching transform-ers; feed lines: coax versus open-wire; velocity factor; electrical length; transformation characteristics of line terminated in impedance not equal to characteristic impedance

CHAPTER 1
KEYWORDS
KEYWORDS
KEYWORDS

Accreditation — The process by which a **Volunteer Examiner Coordinator (VEC)** certifies that their **Volunteer Examiners (VEs)** are qualified to administer Amateur Radio license exams.

Amateur Satellite Service — A radiocommunication service using stations on Earth satellites for the same purpose as those of the amateur service.

Authorization for alien reciprocal operation — FCC authorization to someone holding an amateur license issued by certain foreign governments to operate an amateur station in the US.

Automatic control — The operation of an amateur station without a control operator present at the control point. In §97.3 (a) (6), the FCC defines automatic control as "The use of devices and procedures for control of a station when it is transmitting so that compliance with the FCC Rules is achieved without the control operator being present at a control point."

CEPT (European Conference of Postal and Telecommunications Administrations) arrangement — A multilateral operating arrangement that allows US amateurs to operate in many European countries, and amateurs from many European countries to operate in the US.

Certificate of Successful Completion of Examination (CSCE) — A document issued by a Volunteer Examiner Team to certify that a candidate has passed specific exam elements at their test session. If the candidate qualified for a license upgrade at the exam session, this CSCE provides the authority to operate using the newly earned license privileges, with special identification procedures.

Certification — Equipment authorization granted by the FCC used to ensure that the equipment will function properly in the service for which it has been accepted.

Control link — A device used by a control operator to manipulate the station adjustment controls from a location other than the station location. A control link provides the means of control between a control point and a remotely controlled station.

Earth station — An amateur station located on, or within 50 km of the Earth's surface intended for communications with space stations or with other Earth stations by means of one or more other objects in space.

Emissions — Any signals produced by a transmitter that reach the antenna connector to be radiated.

Examination elements — Any of the telegraphy or written exam sections required for an Amateur Radio operator's license.

IARP (International Amateur Radio Permit) — A multilateral operating arrangement that allows US amateurs to operate in many Central and South American countries, and amateurs from many Central and South American countries to operate in the US.

International Telecommunication Union (ITU) — The international organization with responsibility for dividing the range of communications frequencies between the various services for the entire world.

Radio Amateur Civil Emergency Service (RACES) — A radio service using amateur stations for civil defense communications during periods of local, regional or national civil emergencies.

Remote control — The operation of an Amateur Radio station using a **control link** to manipulate the station operating adjustments from somewhere other than the station location.

Space station — An amateur station located more than 50 km above the Earth's surface.

Spurious emissions — Any **emission** that is not part of the desired signal. The FCC defines this term as "an emission, on frequencies outside the necessary bandwidth of a transmission, the level of which may be reduced without affecting the information being transmitted."

Telecommand operation — A one-way transmission to initiate, modify, or terminate functions of a device at a distance.

Telecommand station — An amateur station that transmits communications to initiate, modify, or terminate functions of a space station.

Telemetry — A one-way transmission of measurements at a distance from the measuring instrument.

Volunteer Examiner (VE) — A licensed amateur who is accredited by a **Volunteer Examiner Coordinator (VEC)** to administer amateur license exams.

Volunteer Examiner Coordinator (VEC) — An organization that has entered into an agreement with the FCC to coordinate amateur license examinations.

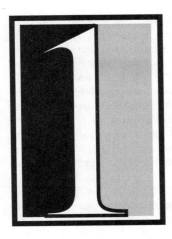

COMMISSION'S RULES

Part 97 of the Federal Communications Commission's Rules governs the Amateur Radio Service in the United States. Each Amateur Radio license exam includes questions covering sections of these rules. Your Amateur Extra license exam will include seven questions about FCC Rules. These questions will come from the seven groups of questions in Commission's Rules, Subelement 1 of the Extra Class Question Pool. This chapter of *The ARRL Extra Class License Manual* covers the specific rules covered on the Extra class (Element 4) license exam.

This book does not contain a complete listing of the Part 97 Rules. The FCC modifies Part 97 on an irregular basis, and it would be impossible to keep an up-to-date set of rules in a book designed to be revised with the release of each new Extra Class Question Pool. We do recommend, however, that every Amateur Radio operator have an up-to-date copy of the rules in their station for reference. *The ARRL's FCC Rule Book* contains the complete text of Part 97, along with detailed explanations of all the regulations. *The ARRL's FCC Rule Book* is updated as necessary to keep it current with the latest rule changes.

The seven rules questions on your Amateur Extra class license exam will come from seven exam-question groups. Each group forms a section of the Commission's Rules subelement of the Element 4 syllabus or study guide, and the question pool. The seven groups for this subelement are:

E1A Operating standards: frequency privileges for Extra class amateurs; emission standards; message forwarding; frequency sharing between ITU Regions; FCC modification of station license; 30-meter band sharing; stations aboard ships or aircraft; telemetry; telecommand of an amateur station; authorized telecommand transmissions; definitions of image, pulse and test

E1B Station restrictions: restrictions on station locations; restricted operation; teacher as control operator; station antenna structures; definition and operation of remote control and automatic control; control link

E1C Reciprocal operating: definition of reciprocal operating permit; purpose of reciprocal agreement rules; alien control operator privileges; identification; application for reciprocal permit; reciprocal permit license term (Note: This includes CEPT and IARP.)

E1D Radio Amateur Civil Emergency Service (RACES): definition; purpose; station registration; station license required; control operator requirements; control operator privileges; frequencies available; limitations on use of RACES frequencies; points of communication for RACES operation; permissible communications

E1E Amateur Satellite Service: definition; purpose; station license required for space station; frequencies available; telecommand operation: definition; eligibility; telecommand station (definition); space telecommand station; special provisions; telemetry: definition; special provisions; space station: definition; eligibility; special provisions; authorized frequencies (space station); notification requirements; earth operation: definition; eligibility {97.209(a)}; authorized frequencies (Earth station)

E1F Volunteer Examiner Coordinators (VECs): definition; VEC qualifications; VEC agreement; scheduling examinations; coordinating VEs; reimbursement for expenses {97.527}; accrediting VEs; question pools; Volunteer Examiners (VEs): definition; requirements; accreditation; reimbursement for expenses; VE conduct; preparing an examination; examination elements; definition of code and written elements; preparation responsibility; examination requirements; examination credit; examination procedure; examination administration; temporary operating authority

E1G Certification of external RF power amplifiers and external RF power amplifier kits; Line A; National Radio Quiet Zone; business communications; definition and operation of spread spectrum; auxiliary station operation

AMATEUR EXTRA CONTROL OPERATOR FREQUENCY PRIVILEGES

As an Extra-class Amateur Radio licensee, you will have all Amateur Radio frequency privileges above 50 MHz, just as a Technician, Technician Plus, General and Advanced class licensee has. These frequency privileges are listed in Section 97.301(a) of the FCC Rules. (Section is often represented by the § character when referencing a specific portion of the FCC Rules.) You will also have extensive privileges on the high-frequency (HF) bands, as listed in §97.301(b). **Table 1-1** shows this section of the Rules, along with the appropriate notes from §97.303.

Table 1-1
Amateur Extra Class HF Bands and Sharing Requirements

§97.301 Authorized frequency bands.

The following transmitting frequency bands are available to an amateur station located within 50 km of the Earth's surface, within the specified ITU Region, and outside any area where the amateur service is regulated by any authority other than the FCC.

(b) For a station having a control operator who has been granted an Amateur Extra Class operator license or who holds a CEPT radio-amateur Class 1 license or Class 1 IARP:

Wavelength Band	ITU Region 1	ITU Region 2	ITU Region 3	Sharing requirements See §97.303, Paragraph:
MF	kHz	kHz	kHz	
160 m	1810 - 1850	1800 - 2000	1800 - 2000	(a), (b), (c)
HF	MHz	MHz	MHz	
80 m	3.50 - 3.75	3.50 - 3.75	3.50 - 3.75	(a)
75 m	3.75 - 3.80	3.75 - 4.00	3.75 - 3.90	(a)
40 m	7.0 - 7.1	7.0 - 7.3	7.0 - 7.1	(a)
30 m	10.10 - 10.15	10.10 - 10.15	10.10 - 10.15	(d)
20 m	14.00 - 14.35	14.00 - 14.35	14.00 - 14.35	
17 m	18.068 - 18.168	18.068 - 18.168	18.068 - 18.168	
15 m	21.00 - 21.45	21.00 - 21.45	21.00 - 21.45	
12 m	24.89 - 24.99	24.89 - 24.99	24.89 - 24.99	
10 m	28.0 - 29.7	28.0 - 29.7	28.0 - 29.7	

§97.303 Frequency sharing requirements.

The following is a summary of the frequency sharing requirements that apply to amateur station transmissions on the frequency bands specified in §97.301 of this Part. (For each ITU Region, each frequency band allocated to the amateur service is designated as either a secondary service or a primary service. A station in a secondary service must not cause harmful interference to, and must accept interference from, stations in a primary service. See §§2.105 and 2.106 of the FCC Rules, *United States Table of Frequency Allocations* for complete requirements.)

(a) Where, in adjacent ITU Regions or Subregions, a band of frequencies is allocated to different services of the same category, the basic principle is the equality of right to operate. The stations of each service in one region must operate so as not to cause harmful interference to services in the other Regions or Subregions. (See *ITU Radio Regulations*, No. 346 (Geneva, 1979).)

(b) No amateur station transmitting in the 1900-2000 kHz segment, the 70 cm band, the 33 cm band, the 13 cm band, the 9 cm band, the 5 cm band, the 3 cm band, the 24.05-24.25 GHz segment, the 76-81 GHz segment, the 144-149 GHz segment and the 241-248 GHz segment shall cause harmful interference to, nor is protected from interference due to the operation of, the Government radiolocation service.

(c) No amateur station transmitting in the 1900-2000 kHz segment, the 3 cm band, the 76-81 GHz segment, the 144-149 GHz segment and the 241-248 GHz segment shall cause harmful interference to, nor is protected from interference due to the operation of, stations in the non-Government radiolocation service.

(d) No amateur station transmitting in the 30 meter band shall cause harmful interference to stations authorized by other nations in the fixed service. The licensee of the amateur station must make all necessary adjustments, including termination of transmissions, if harmful interference is caused.

Table 1-2
Exclusive HF Privileges for Amateur Extra Class Operators

Band	Operating Privileges	Frequency Privileges
80 m	CW, RTTY, Data	3500 - 3525 kHz
75 m	CW, Phone, Image	3750 - 3775 kHz
40 m	CW, RTTY, Data	7000 - 7025 kHz
20 m	CW, RTTY, Data	14.000 - 14.025 MHz
	CW, Phone, Image	14.150 - 14.175 MHz
15 m	CW, RTTY, Data	21.000 - 21.025 MHz
	CW, Phone, Image	21.200 - 21.225 MHz

Table 1-2 summarizes the *exclusive* privileges that Amateur Extra class operators have on the HF bands for ITU Region 2. That table will help you memorize the frequencies for your exam. The table also shows the emission types authorized for use on certain portions of these bands. An emission is any type of signal transmitted from an Amateur Radio station.

Study the frequency limits for each band shown in Table 1-2, and learn the segments designated for CW, RTTY and data as well as the segments designated for CW, phone and image.

Although you can operate CW on any frequency you are authorized to operate on, it is common practice to refer to the lower-frequency section of each band as the *CW Band* and the higher-frequency segment as the *Phone Band*. §97.3(c)(1, 2, 3, 5 and 7) list the definitions of these emission types.

[After you have learned the HF frequency privileges described in Table 1-2, turn to Chapter 10 and study questions E1A01 through E1A05. Review this section of the chapter if you have difficulty with any of these questions. As you study the questions, turn the edge of the question-pool page under to hide the answer key.]

STATION OPERATING STANDARDS

The FCC Rules establish certain standards to govern the operation of your Amateur Radio station. Within those standards you have plenty of freedom to experiment with new modes of operation and communications techniques. As technology and the needs of amateurs change, so do the FCC Rules.

Packet radio is a good example of such changing rules. As amateurs pioneered packet bulletin boards and message-forwarding systems, the rules regarding responsibility for messages in these systems have evolved. Amateurs have always been held accountable for any message transmitted from their stations. When all third-party communications were relayed by individual amateurs, such rules made sense, and were easy for amateurs to obey. With the advent of automatic forwarding of messages through the packet radio system, however, requiring every individual to personally screen every message caused problems. So the rules were changed, and now the FCC recognizes that the original station is primarily responsible. The first station to forward an illegal message into the packet system also bears some responsibility, but other stations that automatically forward the message are not responsible. If you become aware that your packet bulletin board station has inadvertently forwarded a communication that violates FCC Rules, you should, of course, immediately discontinue forwarding that message to other stations.

§97.219 Message forwarding system.

(a) Any amateur station may participate in a message forwarding system, subject to the privileges of the class of operator license held.

(b) For stations participating in a message forwarding system, the control operator of the station originating a message is primarily accountable for any violation of the rules in this Part contained in the message.

(c) Except as noted in paragraph (d) of this section, for stations participating in a message forwarding system, the control operators of forwarding stations that retransmit inadvertently communications that violate the rules in this Part are not accountable for the violative communications. They are, however, responsible for discontinuing such communications once they become aware of their presence.

(d) For stations participating in a message forwarding system, the control operator of the first forwarding station must:

(1) Authenticate the identity of the station from which it accepts communication on behalf of the system; or

(2) Accept accountability for any violation of the rules in this Part contained in messages it retransmits to the system.

EMISSION STANDARDS

The signals produced by your transmitter are called **emissions**. These emissions must meet certain technical standards established by the FCC in the Part 97 Rules. These emission standards are given in §97.307 of the Rules.

The FCC Rules specify limits to the strength of spurious signals, or unwanted emissions from an amateur transmitter. In general you should reduce out-of-band signals to the minimum possible level. The primary reason for such regulations is to ensure that your amateur station doesn't interfere with other radio stations. If a spurious emission from your station does cause such interference, you must take steps to eliminate the interference, in accordance with good engineering practice. §97.307 (c) covers this situation.

Paragraphs a, b and c of §97.307 establish some general guidelines:

§97.307 Emission standards.

(a) No amateur station transmission shall occupy more bandwidth than necessary for the information rate and emission type being

transmitted, in accordance with good amateur practice.
(b) Emissions resulting from modulation must be confined to the band or segment available to the control operator. Emissions outside the necessary bandwidth must not cause splatter or keyclick interference to operations on adjacent frequencies.
(c) All spurious emissions from a station transmitter must be reduced to the greatest extent practicable. If any spurious emission, including chassis or power line radiation, causes harmful interference to the reception of another radio station, the licensee of the interfering amateur station is required to take steps to eliminate the interference, in accordance with good engineering practice.

In addition, paragraphs d and e list some more specific requirements:

(d) The mean power of any spurious emission from a station transmitter or external RF power amplifier transmitting on a frequency below 30 MHz must not exceed 50 mW and must be at least 40 dB below the mean power of the fundamental emission. For a transmitter of mean power less than 5 W, the attenuation must be at least 30 dB. A transmitter built before April 15, 1977, or first marketed before January 1, 1978, is exempt from this requirement.
(e) The mean power of any spurious emission from a station transmitter or external RF power amplifier transmitting on a frequency between 30-225 MHz must be at least 60 dB below the mean power of the fundamental. For a transmitter having a mean power of 25 W or less, the mean power of any spurious emission supplied to the antenna transmission line must not exceed 25 µW and must be at least 40 dB below the mean power of the fundamental emission, but need not be reduced below the power of 10 µW. A transmitter built before April 15, 1977, or first marketed before January 1, 1978, is exempt from this requirement.

Notice that there are two frequency ranges specified in this Rule point, below 30 MHz and between 30 and 225 MHz. Also notice that both frequency ranges have **spurious emissions** limits that vary with transmitter power. Let's take a look at what those numbers really mean.

Suppose you have an HF transmitter that produces a mean power of 500 watts. What is the maximum mean power permitted for any spurious emission from this transmitter? There are two limits listed in §97.307 (d) for transmitters with power above 5 W. Which will apply here? Let's run through a calculation to find out. First we'll assume that the spurious emissions are 50 mW, and calculate how many decibels below the mean power of the fundamental emission this is.

$$dB = 10 \log \left(\frac{P_1}{P_2} \right)$$

(Equation 1-1)

where:

P_1 = mean power of the fundamental.
P_2 = mean power of spurious emission.

$$dB = 10 \log \left(\frac{P_1}{P_2} \right) = 10 \log \left(\frac{500 \text{ W}}{50 \times 10^{-3} \text{ W}} \right) = 10 \log \left(1 \times 10^4 \right) = 10 \times 4 = 40 \text{ dB}$$

In this case the spurious emissions meet both the 50 mW maximum and they are at least 40 dB below the mean power of the fundamental signal.

Suppose your transmitter (or amplifier) produces 1500 W. A spurious emission that is 40 dB down from the fundamental will have a mean power of 150 mW! That does not meet the maximum 50 mW spec, so we'll need more filtering to suppress the spurious signals some more. In fact, we'll need almost 45 dB of suppression in this case, to keep the unwanted signals below the 50 mW level. (You can calculate all these numbers using Equation 1-1. To find the mean power of the spurious signals for 40 dB of suppression, you'll have to solve that equation for P_2.)

For a 100 W transmitter, with 40-dB suppression of the spurious signals, they will have no more than 10 mW of mean power. If the transmitter power is below 5 W, we only need 30 dB of suppression. In that case the maximum strength of the spurious signals will be 5 mW.

For the frequency range between 30 and 225 MHz, spurious emissions from high-power signals must be at least 60 dB below the mean power of the fundamental. If the transmitter power is 25 W or less, spurious signals must be no stronger than 25 μW, and at least 40 dB below the mean power of the fundamental signal.

[Turn to Chapter 10 and study questions E1A06 through E1A12 to check your understanding of this material. Review this section if you have trouble with any of the questions.]

STATION RESTRICTIONS

In general, you can operate your Amateur Radio station anytime you want to. You can operate from nearly any location in the US. (Of course the owners of private property, such as a business, may not permit you to operate from their property.) Under certain conditions, the FCC may modify the terms of your amateur station license.

If the Commission determines it will promote the public interest, convenience and necessity, they may modify your license. For example, suppose a spurious emission from your station is causing interference to the reception of a domestic broadcast radio station. If a receiver of good engineering design is being used, and you are unable to resolve the problem, the FCC might restrict your operating hours. Such a restriction would apply to operation on any amateur frequencies that cause the interference. According to §97.121 (a) of the Rules, you would not be permitted to operate your station on the affected frequencies between 8 PM to 10:30 PM local time daily. In addition, you would not be permitted to operate between 10:30 AM and 1:00 PM on Sunday.

If you want to operate your Amateur Radio station from on board a ship or aircraft you must have the radio installation approved by the master of the ship or the pilot in command of the aircraft. §97.11 lists some specific requirements for such a radio installation. You don't need any other special permit or permission from the FCC for such an operation, though. Your FCC Amateur Radio license, or a reciprocal permit for alien amateur licensee is all you need.

§97.11 Stations aboard ships or aircraft.

(a) The installation and operation of an amateur station on a ship or aircraft must be approved by the master of the ship or pilot in command of the aircraft.

(b) The station must be separate from and independent of all other radio apparatus installed on the ship or aircraft, except a common antenna may be shared with a voluntary ship radio installation. The station's transmissions must not cause interference to any other apparatus installed on the ship or aircraft.

(c) The station must not constitute a hazard to the safety of life or property. For a station aboard an aircraft, the apparatus shall not be operated while the aircraft is operating under Instrument Flight Rules, as defined by the FAA, unless the station has been found to comply with all applicable FAA Rules.

You should also be aware that if you are operating from international waters (or international air space) the vessel or plane must be registered in the US. If the boat or plane travels into an area controlled by another country then you must have a license or permission from that country to operate. Also, you must obey the frequency restrictions that apply to the ITU Region from which you are operating.

There are certain conditions that may restrict the physical location of your amateur station. If the land has environmental importance, or is significant in American history, architecture or culture you "may be required to take certain actions prescribed by §1.1305 - 1.319 of [the FCC Rules.] [§97.13 (a)] For example, if your station will be located within the boundaries of an officially designated wilderness area, wildlife preserve or an area listed in the National Register of Historical Places you must submit an Environmental Assessment to the FCC.

If your station will be located within 1 mile of an FCC monitoring facility, you must protect that facility from harmful interference. If you do cause interference to such a facility the FCC Engineer in Charge may impose operating restrictions on your station. *The ARRL's FCC Rule Book* contains a list of these monitoring stations, with addresses and contact information.

The FCC Rules also place some limitations and restrictions on the construction of antennas for Amateur Radio stations. Without prior FCC approval, you may not build an antenna structure, including a tower or other support structure, higher than 200 feet. This applies even if your station is located in a valley or canyon. Unless it is approved, the antenna may not be more than 200 feet above the ground level at the station location.

If your antenna is located near an airport, then further height limitations may apply. You must obtain approval from both the FCC and the Federal Aviation Administration (FAA) if your antenna is within certain distances of an airport. If your antenna is no more than 20 feet above any natural or existing man-made structure then you do not need approval for the antenna.

On the other hand, the FCC does *not* impose any special restrictions on specific types of amateur station antennas. For this reason, there are no special restrictions on amateur antennas mounted on motor vehicles. (You should ensure that no one can be injured by your mobile antenna, however. For example, be sure that an antenna element doesn't extend beyond the vehicle, so that it might hit someone standing near the vehicle as you drive by.)

The FCC has even issued a public notice (called PRB-1), which specifies that any legislation passed by state and local authorities must reasonably accommodate the installation of amateur antenna structures. This notice was intended to counteract zoning ordinances specifically designed to limit antenna heights based only on aesthetic reasons.

§97.15 Station antenna structures.

(a) Owners of certain antenna structures more than 60.96 meters (200 feet) above ground level at the site or located near or at a public use airport must notify the Federal Aviation Administration and register with the Commission as required by Part 17 of this chapter.

(b) Except as otherwise provided herein, a station antenna structure may be erected at heights and dimensions sufficient to accommodate amateur service communications. [State and local regulation of a station antenna structure must not preclude amateur service communications. Rather, it must reasonably accommodate such communications and must constitute the minimum practicable regulation to accomplish the state or local authority's legitimate purpose. See PRB-1, 101 FCC 2d 952 (1985) for details.]

There is a general rule that you may not be paid to operate your Amateur Radio station. In most cases this also means that you cannot operate an Amateur Radio station during the time that you are being paid by your employer. There are a few significant exceptions written into the FCC Rules. For example, an operator may be paid to operate a club station during the time that club station is transmitting information bulletins that are of general interest to other amateurs, or while the station is transmitting Morse code practice. There are a number of requirements to be met before this exception applies. The station must operate on a schedule that is published in advance, the station transmits such bulletins and code practice at least 40 hours per week, and the station transmits on at least six amateur bands in the MF and HF range.

Another significant exception is that a paid professional teacher may be the control operator of a station in that teacher's classroom. The teacher must be using the Amateur Radio station as part of their classroom instruction at an educational institution.

§97.113 Prohibited transmissions.

(c) A control operator may accept compensation as an incident of a teaching position during periods of time when an amateur station is used by that teacher as a part of classroom instruction at an educational institution.

[Before you go on to the next section you should turn to Chapter 10 and study questions E1A13 through E1A15 and questions E1B01 through E1B12. Review this section if any of these questions give you difficulty.]

REMOTELY AND AUTOMATICALLY CONTROLLED STATIONS

When you studied for your Technician class written exam, you learned that every Amateur Radio station must have a control operator whenever it is transmitting. There are times, however, when that control operator need not be located physically at the radio itself.

A station operating under **remote control** uses a **control link** to manipulate the station operating adjustments. A remotely controlled station is operated indirectly through a control link. The control link may be through a wire line, such as a telephone-line link, or through a radio link. Remote control through a radio link is called *telecommand*.

Automatic control refers to amateur station operation in which the control operator is not present at the control point. The station *does* have a control operator, as do all amateur stations when they are transmitting. A station under remote control *may not* also be under automatic control at the same time. You can use remote control for the operation and control of a model craft, but such stations may not operate under automatic control.

Probably the most common type of amateur operation involving remote and automatic control is repeater operation. There are several other types of amateur stations that may be automatically controlled. These include beacon stations and auxiliary stations.

[Before you go on to the next section, turn to Chapter 10 and study questions E1B13 through E1B18. Review this section as needed.]

RECIPROCAL OPERATING

The FCC Rules provide a way for a person licensed as an Amateur Radio operator by a foreign government to enjoy Amateur Radio while they are in the United States. If the other amateur's government has entered into a prior agreement with the US, that person may qualify for an FCC **authorization for alien reciprocal operation**. This authorization allows the foreign amateur to operate an amateur station in the US.

Any non-US citizen who holds a valid amateur license from a country that has entered into a reciprocal operating arrangement with the US is eligible. The only stipulation is that the non-US citizen must be a citizen of the country that issued the amateur license.

Canadian citizens who are licensed amateurs and who wish to operate in the US do not need an FCC authorization or any other document. The US and Canadian governments have an agreement that allows amateurs in each country to operate in the other country using only their own amateur license.

Unless the FCC specifies otherwise, someone holding an authorization for alien reciprocal operation can operate on the frequencies authorized to US amateurs that the person would have in their own country. For example, if a Costa Rican amateur with FCC authorization for alien reciprocal operation can operate on 14.225 MHz at home, she may also do so in the US. Privileges cannot exceed those of an Amateur Extra class licensee, however.

FCC Rules state that foreign amateurs, other than Canadians operating in the US under reciprocal operating authority, must identify by using an indicator consisting of the appropriate letter-numeral designating the station location. It is to be included *before* the call sign issued to the station by the licensing country and separated by a slant bar ("/" if on a non-phone mode or "stroke" or similar words if on phone, such as W1/G5RV).

There are no restrictions on the locations where the holder of an FCC authorization for alien operation can serve as a control operator. The alien operator can transmit from anyplace under the jurisdiction of the FCC.

US citizens, even those holding a foreign amateur license, are *not* eligible to obtain an FCC authorization for alien operation. US citizens must qualify for a US license issued by the FCC before they can operate a station in the US.

In the past, US amateurs who traveled to foreign countries had to obtain a reciprocal operating permit from the government of the country visited. This changed when the US became a participant in the **CEPT arrangement**. The European Conference of Postal and Telecommunications Administrations (CEPT) Recommendation T/R 61-01 allows US amateurs to operate in certain European countries. It also allows amateurs from many European countries to operate in the US.

The FCC and US State Department authorize citizens of certain countries in the Americas to operate their amateur stations while on short visits to the US. Likewise, citizens of the US may operate amateur stations in certain countries in the Americas while on short visits. US amateurs must first apply for and obtain an International Amateur Radio Permit (**IARP**) from the ARRL.

§97.5 Station license grant required.

(c) The person named in the station license grant or who is authorized for alien reciprocal operation by §97.107 of this Part may use, in accordance with the applicable rules of this Part, the transmitting apparatus under the physical control of the person at places where the amateur service is regulated by the FCC.

(d) A CEPT radio-amateur license is issued to the person by the country of which the person is a citizen. The person must not:

(1) Be a resident alien or citizen of the United States, regardless of any other citizenship also held;

(2) Hold an FCC-issued amateur operator license nor reciprocal permit for alien amateur licensee;

(3) Be a prior amateur service licensee whose FCC-issued license was revoked, suspended for less than the balance of the license term and the suspension is still in effect, suspended for the balance of the license term and relicensing has not taken place, or surrendered for cancellation following notice of revocation, suspension or monetary forfeiture proceedings; or

(4) Be the subject of a cease and desist order that relates to amateur service operation and which is still in effect.

(e) An IARP is issued to the person by the country of which the person is a citizen. The person must not:

(1) Be a resident alien or citizen of the United States, regardless of any other citizenship also held;

(2) Hold an FCC-issued amateur operator license nor reciprocal permit for alien amateur licensee;

(3) Be a prior amateur service licensee whose FCC-issued license was revoked, suspended for less than the balance of the license term and the suspension is still in effect, suspended for the balance of the license term and relicensing has not taken place, or surrendered for cancellation following notice of revocation, suspension or monetary forfeiture proceedings; or

(4) Be the subject of a cease and desist order that relates to amateur service operation and which is still in effect.

§97.107 Reciprocal operating authority.

A non-citizen of the United States ("alien") holding an amateur service authorization granted by the alien's government is authorized to be the control operator of an amateur station located at places where the amateur service is regulated by the FCC, provided there is in effect a multilateral or bilateral reciprocal operating arrangement, to which the United States and the alien's government are parties, for amateur service operation on a reciprocal basis. The FCC will issue public announcements listing the countries with which the United States has such an arrangement. No citizen of the United States or person holding an FCC amateur operator/primary station license grant is eligible for the reciprocal operating authority granted by this section. The privileges granted to a control operator under this authorization are:

(a) For an amateur service license issued by the Government of Canada:

(1) The terms of the *Convention Between the United States and Canada (TIAS No. 2508) Relating to the Operation by Citizens of Either Country of Certain Radio Equipment or Stations in the Other Country;*

(2) The operating terms and conditions of the amateur service license issued by the Government of Canada; and

(3) The applicable rules of this part, but not to exceed the control operator privileges of an FCC-issued Amateur Extra Class operator license.

(b) For an amateur service license granted by any country, other than Canada, with which the United States has a multilateral or bilateral agreement:

    (1) The terms of the agreement between the alien's government and the United States;

    (2) The operating terms and conditions of the amateur service license granted by the alien's government;

    (3) The applicable rules of this part, but not to exceed the control operator privileges of an FCC-issued Amateur Extra Class operator license; and

(c) At any time the FCC may, in its discretion, modify, suspend, or cancel the reciprocal operating authority granted to any person by this section.

§97.119 Station identification.

(g) When the station is transmitting under the authority of §97.107 of this part, an indicator consisting of the appropriate letter-numeral designating the station location must be included before the call sign that was issued to the station by the country granting the license. For an amateur service license granted by the Government of Canada, however, the indicator must be included after the call sign. At least once during each intercommunication, the identification announcement must include the geographical location as nearly as possible by city and state, commonwealth or possession.

[Turn to Chapter 10 now and study exam questions E1C01 through E1C11. Review this section if any of those questions give you difficulty.]

RADIO AMATEUR CIVIL EMERGENCY SERVICE (RACES)

The **Radio Amateur Civil Emergency Service (RACES)** is a radio service intended for civil defense purposes. Before you can register your Amateur Radio station with RACES you must register with your local civil defense organization.

At one time the FCC issued RACES licenses to civil defense organizations. In an action that became effective April 15, 2000, the FCC will no longer renew RACES licenses. (New RACES licenses have not been granted since 1978.) Remaining RACES licenses are held only by the person who is designated the *license custodian*.

§97.3 Definitions.

(a) The definitions of terms used in Part 97 are:
  (37) *RACES* (radio amateur civil emergency service). A radio
  service using amateur stations for civil defense communications
  during periods of local, regional or national civil emergencies.

§97.21 Application for a modified or renewed license.

(a) A person holding a valid amateur station license grant:
  (1) Must apply to the FCC for a modification of the license grant
  as necessary to show the correct mailing address, licensee
  name, club name, license trustee name or license custodian
  name in accordance with §1.913 of this chapter. For a club,
  military recreation or RACES station license grant, it must be
  presented in document form to a Club Station Call Sign Administrator who must submit the information thereon to the FCC in an
  electronic batch file. The Club Station Call Sign Administrator
  must retain the collected information for at least 15 months and
  make it available to the FCC upon request.
  (3)(iii). . .RACES station license grants will not be renewed.

Any licensed amateur station may be operated in RACES, as long as the station is certified by the local civil defense organization. Likewise, any licensed
amateur may be the control operator of a RACES station if they are also certified
by the local civil defense organization.

RACES station operators do not receive any additional operator privileges
because of their RACES registration. So a Technician class operator may only use
Technician frequencies when serving as the control operator. Extra class operators
must also follow the operator privileges granted by their license. In general, all
amateur frequencies are available for RACES operation.

RACES stations may communicate with any RACES station as well as certain
other stations authorized by the responsible civil defense official. This might occur
if the civil defense organization had some stations operating on frequencies outside the amateur bands, and there was a need for communications between the two
groups. RACES communications normally involve national defense or immediate
safety of people and property, or other communications authorized by the civil
defense organization.

If there is an emergency that causes the President of the United States to in-
voke the War Emergency Powers under the Communications Act of 1934, there
may be limits placed on the frequencies available for RACES operation. Of course
this would also result in limits on the frequencies available for normal amateur
communication. There may even be further restrictions on amateur communications under such conditions.

§97.407 Radio Amateur Civil Emergency Service (RACES).

(a) No station may transmit in RACES unless it is an FCC-licensed primary, club, or military recreation station and it is certified by a civil defense organization as registered with that organization, or it is an FCC-licensed RACES station. No person may be the control operator of a RACES station, or may be the control operator of an amateur station transmitting in RACES unless that person holds a FCC-issued amateur operator license and is certified by a civil defense organization as enrolled in that organization.

(b) The frequency bands and segments and emissions authorized to the control operator are available to stations transmitting communications in RACES on a shared basis with the amateur service. In the event of an emergency which necessitates the invoking of the President's War Emergency Powers under the provisions of §706 of the Communications Act of 1934, as amended, 47 U.S.C. §606, RACES stations and amateur stations participating in RACES may only transmit on the following frequency segments:

(1) The 1800-1825 kHz, 1975-2000 kHz, 3.50-3.55 MHz, 3.93-3.98 MHz, 3.984-4.000 MHz, 7.079-7.125 MHz, 7.245-7.255 MHz, 10.10-10.15 MHz, 14.047-14.053 MHz, 14.22-14.23 MHz, 14.331-14.350 MHz, 21.047-21.053 MHz, 21.228-21.267 MHz, 28.55-28.75 MHz, 29.237-29.273 MHz, 29.45-29.65 MHz, 50.35-50.75 MHz, 52-54 MHz, 144.50-145.71 MHz, 146-148 MHz, 2390-2450 MHz segments;

(2) The 1.25 m, 70 cm and 23 cm bands; and

(3) The channels at 3.997 MHz and 53.30 MHz may be used in emergency areas when required to make initial contact with a military unit and for communications with military stations on matters requiring coordination.

(c) A RACES station may only communicate with:

(1) Another RACES station;

(2) An amateur station registered with a civil defense organization;

(3) A United States Government station authorized by the responsible agency to communicate with RACES stations;

(4) A station in a service regulated by the FCC whenever such communication is authorized by the FCC.

(d) An amateur station registered with a civil defense organization may only communicate with:

(1) A RACES station licensed to the civil defense organization with which the amateur station is registered;

(2) The following stations upon authorization of the responsible civil defense official for the organization with which the amateur station is registered:

(i) A RACES station licensed to another civil defense organization;

(ii) An amateur station registered with the same or another civil defense organization;

   (iii) A United States Government station authorized by the
    responsible agency to communicate with RACES stations; and
   (iv) A station in a service regulated by the FCC whenever such
    communication is authorized by the FCC.
  (e) All communications transmitted in RACES must be specifically
  authorized by the civil defense organization for the area served.
  Only civil defense communications of the following types may be
  transmitted:
   (1) Messages concerning impending or actual conditions jeopardiz-
    ing the public safety, or affecting the national defense or security
    during periods of local, regional, or national civil emergencies;
   (2) Messages directly concerning the immediate safety of life of
    individuals, the immediate protection of property, maintenance
    of law and order, alleviation of human suffering and need, and
    the combating of armed attack or sabotage;
   (3) Messages directly concerning the accumulation and dissemi-
    nation of public information or instructions to the civilian
    population essential to the activities of the civil defense
    organization or other authorized governmental or relief
    agencies; and
   (4) Communications for RACES training drills and tests neces-
    sary to ensure the establishment and maintenance of orderly
    and efficient operation of the RACES as ordered by the
    responsible civil defense organizations served. Such drills and
    tests may not exceed a total time of 1 hour per week. With the
    approval of the chief officer for emergency planning in the
    applicable State, Commonwealth, District or territory, however,
    such tests and drills may be conducted for a period not to
    exceed 72 hour no more than twice in any calendar year.

[Before you go on to the next section, turn to Chapter 10 and study exam questions E1D01 through E1D11. If you have difficulty answering any of those questions you should review the material presented here.]

THE AMATEUR SATELLITE SERVICE (ASAT)

  The **Amateur Satellite Service** is a radio service within the Amateur Radio Service. It uses amateur stations onboard satellites orbiting the Earth to provide Amateur Radio communications. Stations on the Earth (or within 50 km of the Earth) are called **Earth stations** and stations more than 50 km above the Earth are called **space stations**.

  Since most space stations are unmanned, amateurs must have some way to control the various satellite functions. The process of transmitting communications to a satellite to initiate, modify or terminate the various functions of a space station is called **telecommand operation**. Stations that transmit these command communications are called **telecommand stations**. Any amateur station that is designated by the space station licensee may serve as a telecommand station.

  Obviously, not every licensed Amateur Radio operator should be able to send telecommand communications to a satellite. Unauthorized operators might change satellite functions accidentally, or to suit their own operating needs. For this reason, the FCC allows telecommand stations to use special codes that are intended to obscure the meaning of telecommand messages. This is one of the few times an amateur may intentionally obscure the meaning of a message. Otherwise, anyone who copied the transmission could learn the proper control codes for the satellite.

Space stations provide amateurs with a unique opportunity to learn about conditions in space. For example, a satellite might record the temperature, amount of solar radiation or other measurements and then transmit that information back to Earth. The satellite might also measure its own operating parameters and transmit that information. The state of its battery charge, transmitter temperature or other spacecraft conditions can be sent back to Earth stations. These measurements are called **telemetry**.

§97.3 Definitions.

(a) The definitions of terms used in Part 97 are:
   (3) *Amateur-satellite service*. A radiocommunication service using stations on Earth satellites for the same purpose as those of the amateur service.
   (16) *Earth station*. An amateur station located on, or within 50 km of the Earth's surface intended for communications with space stations or with other Earth stations by means of one or more other objects in space.
   (40) *Space station*. An amateur station located more than 50 km above the Earth's surface.
   (43) *Telecommand*. A one-way transmission to initiate, modify, or terminate functions of a device at a distance.
   (44) *Telecommand station*. An amateur station that transmits communications to initiate, modify, or terminate functions of a space station.
   (45) *Telemetry*. A one-way transmission of measurements at a distance from the measuring instrument.

§97.211 Space telecommand station.

(a) Any amateur station designated by the licensee of a space station is eligible to transmit as a telecommand station for that space station, subject to the privileges of the class of operator license held by the control operator.
(b) A telecommand station may transmit special codes intended to obscure the meaning of telecommand messages to the station in space operation.

Any licensed Amateur Radio station can be a space station. That means you don't need a special license, and any class of operator license allows you to be the station licensee. Of course there is a little more involved in operating a space station than there is in operating a repeater or packet radio bulletin board! You have to consider how you will get the satellite into space — or how you will get yourself into space, among other things.

Even if you can find a way to have NASA or some space agency launch your satellite, or if you manage to earn a ride on the space shuttle, there are some notification requirements that you must take care of. For example, you must notify the FCC of your intention to operate a space station 27 months before the operation begins. You must also give them another notice 5 months before the operation.

When your space operations begin, you must notify the FCC in writing within 7 days. After the space station transmissions are terminated you must provide a post-space notification to the FCC. That notification must be given no more than 3 months after the operation ends.

There are certain operating-frequency restrictions that you must follow for your space-station operation. On the HF bands, your space station may only operate on portions of the 40 and 20-meter bands, as well as the 17, 15, 12 and 10-meter bands. Segments of several VHF, UHF and microwave bands are also available for space operation.

§97.207 Space station.

(a) Any amateur station may be a space station. A holder of any class operator license may be the control operator of a space station, subject to the privileges of the class of operator license held by the control operator.

(b) A space station must be capable of effecting a cessation of transmissions by telecommand whenever such cessation is ordered by the FCC.

(c) The following frequency bands and segments are authorized to space stations:
  (1) The 17 m, 15 m, 12 m and 10 m bands, 6 mm, 4 mm, 2 mm and 1 mm bands; and
  (2) The 7.0-7.1 MHz, 14.00-14.25 MHz, 144-146 MHz, 435-438 MHz, 1260-1270 MHz and 2400-2450 MHz, 3.40-3.41 GHz, 5.83-5.85 GHz, 10.45-10.50 GHz and 24.00-24.05 GHz segments.

(d) A space station may automatically retransmit the radio signals of Earth stations and other space stations.

(e) A space station may transmit one-way communications.

(f) Space telemetry transmissions may consist of specially coded messages intended to facilitate communications or related to the function of the spacecraft.

(g) The license grantee of each space station must make two written, pre-space station notifications to the International Bureau, FCC, Washington, DC 20554. Each notification must be in accord with the provisions of Articles 11 and 13 of the Radio Regulations.
  (1) The first notification is required no less than 27 months prior to initiating space station transmissions and must specify the information required by Appendix 4, and Resolution No. 642 of

the Radio Regulations.

(2) The second notification is required no less than 5 months prior to initiating space station transmissions and must specify the information required by Appendix 3 and Resolution No. 642 of the Radio Regulations.

(h) The license grantee of each space station must make a written, in-space station notification to the International Bureau no later than 7 days following initiation of space station transmissions. The notification must update the information contained in the pre-space notification.

(i) The license grantee of each space station must make a written, post-space station notification to the International Bureau no later than 3 months after termination of the space station transmissions. When the termination is ordered by the FCC, notification is required no later than 24 hours after termination.

[You should turn to Chapter 10 now and study question E1A16. Also study questions E1E01 through E1E12. Review this section as needed.]

THE VOLUNTEER EXAMINER COORDINATOR (VEC)

A **Volunteer Examiner Coordinator (VEC)** is an organization that has entered into an agreement with the FCC to coordinate amateur license examinations. A VEC does not administer or grade the actual examinations, however. The VEC accredits licensed Amateur Radio operators — known as **Volunteer Examiners (VEs)** — to administer exams.

An organization that wants to coordinate exams must meet certain criteria before they can become a VEC. The organization must exist for the purpose of furthering the amateur service. It should be more than just a local radio club or group of hams, though. A VEC is expected to coordinate exams at least throughout an entire call district. The organization must also agree to coordinate exams for all classes of amateur operator license, and to ensure that anyone desiring an amateur license can register and take the exams without regard to race, sex, religion or national origin.

§97.521 VEC qualifications.

No organization may serve as a VEC unless it has entered into a written agreement with the FCC. The VEC must abide by the terms of the agreement. In order to be eligible to be a VEC, the entity must:

(a) Be an organization that exists for the purpose of furthering the amateur service;

(b) Be capable of serving as a VEC in at least the VEC region (see Appendix 2) proposed;

(c) Agree to coordinate examinations for any class of amateur operator license;

(d) Agree to assure that, for any examination, every examinee qualified under these rules is registered without regard to race, sex, religion, national origin or membership (or lack thereof) in any amateur service organization.

Coordinating amateur exams involves a bit more responsibility than simply recruiting amateurs to administer the exams. A VEC coordinates the preparation and administration of exams. Some VECs actually prepare the exams and provide their examiners with the necessary test forms while others require their VEs to prepare their own exams or purchase exams prepared by someone else. After the test is completed, the VEC must collect the application documents (NCVEC Form 605) and test results. After reviewing the materials to ensure accuracy, the VEC must forward the documentation to the FCC for qualified applicants.

§97.519 Coordinating examination sessions.

(a) A VEC must coordinate the efforts of VEs in preparing and administering examinations.
(b) At the completion of each examination session, the coordinating VEC must collect applicant information and tests results from the administering VEs. Within 10 days of collection, the coordinating VEC must:
   (1) Screen collected information;
   (2) Resolve all discrepancies and verify that the VE's certifications are properly completed; and
   (3) For qualified examinees, forward electronically all required data to the FCC. All data forwarded must be retained for at least 15 months and must be made available to the FCC upon request.
(c) Each VEC must make any examination records available to the FCC, upon request.
(d) The FCC may:
   (1) Administer any examination element itself;
   (2) Readminister any examination element previously administered by VEs, either itself or under the supervision of a VEC or VEs designated by the FCC; or
   (3) Cancel the operator/primary station license of any licensee who fails to appear for readministration of an examination when directed by the FCC, or who does not successfully complete any required element that is readministered. In an instance of such cancellation, the person will be granted an operator/primary station license consistent with completed examination elements that have not been invalidated by not appearing for, or by failing, the examination upon readministration.

To prepare any written Amateur Radio license exam, the VEC or its VEs select questions from the appropriate question pool. All VECs must cooperate to maintain one question pool for each written exam element. Section 97.523 of the FCC Rules specifies additional question-pool requirements.

§97.523 Question pools.

All VECs must cooperate in maintaining one question pool for each written examination element. Each question pool must contain at least 10 times the number of questions required for a single examination. Each question pool must be published and made available to the public prior to its use for making a question set. Each question on each VEC question pool must be prepared by a VE holding the required FCC-issued operator license. See §97.507(a) of this Part.

[Now turn to Chapter 10 and study examination questions E1F01 through E1F04. Also study question E1F06. Review this section if you have difficulty with any of those questions.]

THE VOLUNTEER EXAMINER (VE)

Each VEC is responsible for recruiting and training Volunteer Examiners to administer amateur examinations under their program. The Volunteer Examiners determine where and when examinations for amateur operator licenses will be administered. When a VEC accredits a Volunteer Examiner, it is certifying that the amateur is qualified to perform all the duties of a VE. A VEC has the responsibility to refuse to accredit a person as a VE if the VEC determines that the person's integrity or honesty could compromise amateur license exams.

The **accreditation** process is simply the steps that each VEC takes to ensure their VEs meet all the FCC requirements to serve in the Volunteer Examiner program. As an Extra class licensee, you will be qualified to administer all classes of license exams. We hope you will be interested in becoming a Volunteer Examiner with one of the VECs that serve your local region. The ARRL VEC coordinates exams in all of the regions, and would be pleased to have you apply for accreditation. You do not have to be an ARRL member to serve as an ARRL VE. In fact, one of the requirements of VECs is that they not demand membership in any organization as a prerequisite to serving as a VE! **Figure 1-1** is a sample ARRL VE Application Form.

If you are at least 18 years of age and hold at least a General class license you meet the basic FCC requirements to be a VE. In addition, you must never have had your amateur license suspended or revoked.

§97.509 Administering VE requirements.

(a) Each examination for an amateur operator license must be administered by a team of at least 3 VEs at an examination session coordinated by a VEC. Before the session, the administering VEs or the VE sesion manager must ensure that a public announcement is made giving the location and time of the session. The number of examinees at the session may be limited.
(b) Each administering VE must:
(1) Be accredited by the coordinating VEC;

ARRL / VEC
VOLUNTEER EXAMINER APPLICATION FORM

Please type or print clearly in ink

Control Number
(ARRL / VEC will assign)

☐ General
☐ Advanced
Call: _____ ☐ Extra

License expiration date: _____

Name: _____
        (First)   MI   Last)

Mailing Address: _____

City: _____ State: _____ ZIP: _____

Day Phone: (_____) _____ Night Phone: (_____) _____

Was your license ever suspended or revoked?   ☐ Yes   ☐ No
Have you ever been disaccredited by another VEC?   ☐ Yes   ☐ No
If yes, which VEC(s) and when?
Do you have a call sign change pending with FCC?   ☐ Yes   ☐ No
Do you have any kind of Form 610 pending action with the FCC?   ☐ Yes   ☐ No

Person to contact if you cannot be reached? _____   _____
                                                (name)   (phone)
Mailing address where UPS or daytime delivery is *reliably* possible:

_____   _____
           (name)   (street address)

_____   _____
           (city)   (state)   (zip)

Please list any foreign countries that you will be serving: _____
For instant accreditation, have you participated as a VE in another VEC program,
and is your accreditation in that program current?   ☐ Yes   ☐ No
If yes, which VEC? _____

_____

CERTIFICATION

By signing this Application Form, I certify that to the best of my knowledge the above information AND the following statements are true:

1) I am at least 18 years of age.
2) I agree to comply with the FCC Rules-(see especially Subpart F—Section 97.515 [b]).
3) I agree to comply with examination procedures established by the ARRL as Volunteer Examiner Coordinator.
4) I understand that violation of the FCC Rules or willful noncompliance with the VEC will result in the loss of my VE accreditation, and could result in loss of my Amateur Radio operator and/or station licenses, or both.
5) I understand that even though I may be accredited as a VE, if I am not able or competent to perform certain VE functions required for any particular examination, I should not administer that examination (Section 97.525[a][3]).

_____   _____   _____
        (signature)   (call sign)   (date)

(Please attach a photocopy of your Amateur Radio license, and if applicable a photocopy of any other VEC accreditation held, to this application.)

Figure 1-1 — You can copy this Volunteer Examiner Application Form, fill it out and send it to the ARRL/VEC, 225 Main Street, Newington, CT 06111-1494. Include a note that this form came from *The ARRL Extra Class License Manual*, and request a copy of the VE Manual and additional information about becoming a Certified ARRL Volunteer Examiner.

(2) Be at least 18 years of age;

(3) Be a person who holds an FCC amateur operator license of the class specified below:

  (i) Amateur Extra, Advanced, or General Class in order to administer a Technician Class operator license examination;

  (ii) Amateur Extra or Advanced Class in order to administer a General Class operator license examination.

  (iii) Amateur Extra Class in order to administer an Amateur Extra Class operator license examination.

(4) Not be a person whose grant of an amateur station license or amateur operator license has ever been revoked or suspended.

§97.525 Accrediting VEs.

(a) No VEC may accredit a person as a VE if:

(1) The person does not meet minimum VE statutory qualifications or minimum qualifications as prescribed by this Part;

(2) The FCC does not accept the voluntary and uncompensated services of the person;

(3) The VEC determines that the person is not competent to perform the VE functions; or

(4) The VEC determines that questions of the person's integrity or honesty could compromise the examinations.

(b) Each VEC must seek a broad representation of amateur operators to be VEs. No VEC may discriminate in accrediting VEs on the basis of race, sex, religion or national origin; nor on the basis of membership (or lack thereof) in an amateur service organization; nor on the basis of the person accepting or declining to accept reimbursement.

The Amateur Radio license-exam system depends on the services of volunteers. As licensed amateurs and accredited VEs, these volunteers may not charge a fee to administer exams. They may not receive any type of payment for the services they provide. VECs may not charge a fee for coordinating exam sessions, either. Printing exams and forms, mailing paperwork and securing a suitable location to administer exams may all cost money. The VEC nor the VEs should have to bear these costs out of their own pockets, however. The FCC Rules provide a means for those being examined to reimburse the VEs and VEC for certain costs involved with the program. These costs include expenses involved with preparing, processing and administering license exams.

§97.527 Reimbursement for expenses.

(a) VEs and VECs may be reimbursed by examinees for out-of-pocket expenses incurred in preparing, processing, administering, or coordinating an examination for an amateur operator license.

(b) The maximum amount of reimbursement from any one

examinee for any one examination at a particular session regardless of the number of examination elements taken must not exceed that announced by the FCC in a Public Notice. (The basis for the maximum fee is $4.00 for 1984, adjusted annually each January 1 thereafter for changes in the Department of Labor Consumer Price Index.)

[Before you go on to the last section in this chapter you should turn to Chapter 10 and study exam questions E1F05, and E1F07 through E1F11. If you have difficulty answering any of those questions you should review this section.]

EXAMINATIONS

Volunteer Examiners may prepare written exams for all classes of Amateur Radio operator license. Section 97.507 of the FCC Rules gives detailed instructions about who may prepare the various **examination elements**. You must hold a Technician (including Tech Plus), General, Advanced or Amateur Extra class of license to prepare an Element 2 written exam, for the Technician class license. Only Advanced and Amateur Extra class licensees may prepare the Element 3 exam, and you must hold an Amateur Extra class license to prepare an Element 4 exam.

Some VECs, such as the ARRL VEC, prepare exams for use by their VEs, while other VECs recommend that VEs obtain exams from a qualified supplier if they don't want to prepare their own exams. If the VEC or a qualified supplier prepares the exams, they must still use amateurs with the proper license class to prepare the exams. In every case, the exams are prepared by selecting questions from the appropriate question pool.

The general rule for determining who may administer the various exam elements is that you must hold a higher class of license than the elements you are administering. Section 97.509 of the FCC Rules provides detailed instructions about the qualifications of a Volunteer Examiner. To administer a Technician license exam, for example, you must hold at least a General, Advanced or Amateur Extra class of license. To administer a General exam, you must hold an Advanced or Amateur Extra class of license. Only Amateur Extra class licensees may administer exams for the Extra class license. Of course the Volunteer Examiners must also be accredited by a VEC before they can participate in the administration of a license exam.

§97.507 Preparing an examination.

(a) Each telegraphy message and each written question set administered to an examinee must be prepared by a VE who has been granted an Amateur Extra Class operator license. A telegraphy message or written question set, however, may also be prepared for the following elements by a VE holding an operator license of the class indicated:
(1) Element 3: Advanced Class operator.
(2) Elements 1 and 2: Advanced, General, or Technician (including Technician Plus) Class operators.

(b) Each question set administered to an examinee must utilize questions taken from the applicable question pool.

(c) Each telegraphy message and each written question set administered to an examinee for an amateur operator license must be prepared, or obtained from a supplier, by the administering VEs according to instructions from the coordinating VEC.

Every Amateur Radio license exam session must be coordinated by a VEC, and must be administered by at least three accredited VEs. All three examiners must be present during the entire exam session, observing the candidates to ensure that the session is conducted properly. All three VEs are responsible for supervising the exam session.

Before actually beginning to administer an examination, the VEs should determine what exam credit, if any, the candidates should be given. For example, any candidates who already hold an amateur operator license must receive credit for having passed all of the exam elements necessary for the class of license they already hold. In addition, any candidate who presents a valid **Certificate of Successful Completion of Examination (CSCE)** must be given credit for each exam element that the CSCE indicates the examinee has passed. The combination of element credits and exam elements passed at the current exam session will determine if a candidate qualifies for a higher class of license.

During the exam session, the candidates must follow all instructions given by the Volunteer Examiners. If any candidate fails to comply with a VE's instructions during an exam, the VE Team should immediately terminate that candidate's exam.

When the candidates have completed their exams, the VEs must collect the test papers and grade them immediately. They then notify the candidates of whether they passed or failed the exam. If any candidates did not pass all the exam elements needed to complete their license upgrade, then the examiners must return their applications to the candidates and inform them of their grades.

Of course everyone hopes all the candidates pass the exam elements and upgrade their licenses. After grading the exams of those candidates who do upgrade, the VE Team will certify their qualifications for new licenses on their application forms and issue each a CSCE for their upgrade.

After a successful exam session, the VE Team must submit the application forms and test papers for all the candidates who passed to the coordinating VEC. They must do this within 10 days of the test session. This is to ensure that the VEC can review the paperwork and forward the information to the FCC in a timely fashion.

---

§97.505 Element credit.

(a) The administering VEs must give credit as specified below to an examinee holding any of the following license grants or license documents:

(6) A CSCE: Each element the CSCE indicates the examinee passed within the previous 365 days.

---

§97.509 Administering VE requirements.

(c) Each administering VE must be present and observing the examinee throughout the entire examination. The administering VEs are responsible for the proper conduct and necessary supervision of each examination. The administering VEs must immediately terminate the examination upon failure of the examinee to comply with their instructions.

(d) No VE may administer an examination to his or her spouse, children, grandchildren, stepchildren, parents, grandparents, stepparents, brothers, sisters, stepbrothers, stepsisters, aunts, uncles, nieces, nephews, and in-laws.

(e) No VE may administer or certify any examination by fraudulent means or for monetary or other consideration including reimbursement in any amount in excess of that permitted. Violation of this provision may result in the revocation of the grant of the VE's amateur station license and the suspension of the grant of the VE's amateur operator license.

(f) No examination that has been compromised shall be administered to any examinee. Neither the same telegraphy message nor the same question set may be re-administered to the same examinee.

(g) Passing a telegraphy receiving examination is adequate proof of an examinee's ability to both send and receive telegraphy. The administering VEs, however, may also include a sending segment in a telegraphy examination.

(h) Upon completion of each examination element, the administering VEs must immediately grade the examinee's answers. The administering VEs are responsible for determining the correctness of the examinee's answers.

(i) When the examinee is credited for all examination elements required for the operator license sought, the administering VEs must certify on the examinee's application document that the applicant is qualified for the license.

(j) When the examinee does not score a passing grade on an examination element, the administering VEs must return the application document to the examinee and inform the examinee of the grade.

(k) The administering VEs must accommodate an examinee whose physical disabilities require a special examination procedure. The administering VEs may require a physician's certification indicating the nature of the disability before determining which, if any, special procedures must be used.

(l) The administering VEs must issue a CSCE to an examinee who scores a passing grade on an examination element.

(m) Within 10 days of the administration of a successful examination for an amateur operator license, the administering VEs must submit the application document to the coordinating VEC.

VEC: American Radio Relay League/VEC
CERTIFICATE of SUCCESSFUL COMPLETION of EXAMINATION

Test Site (city/state): *CONCORD, NH*   Test Date: *JUNE 20, 2000*

The applicant named herein has presented the following valid exam element credit(s) in order to qualify for the license earned category indicated below: Circle the **bold** text from one or more of these examples:
-for pre 3/21/87 Technicians circle **3/21/87 Tech-EL 1+3**;
-for pre 2/14/91 Technicans circle **2/14/91 Tech-EI 1**;
-for lifetime Novice code credit circle **Novice-EI 1**;
-for a valid or expired-less-than-5-years FCC Radiotelegraph license/permit circle FCC **Telegraph-EL 1**;
-for CSCEs issued within 365 days of the presentation date circle
CSCE-EL 3B, or CSCE-EL **4B**; or **CSCE-EL 4A+4B**, or for any valid Morse code credit CSCEs circle **CSCE-EL 1**

CREDIT for ELEMENTS PASSED
You have passed the telegraphy and/or written element(s) indicated at right. You will be given credit for the appropriate examination element(s), for up to 365 days from the date shown at the top of this certificate, if you wish to upgrade your license class again while a newly-upgraded license application is pending with the FCC.

LICENSE UPGRADE NOTICE
If you also hold a valid FCC-issued Amateur radio license grant, this Certificate validates temporary operation with the operating privileges of your new operator class (see Sction 97.9[b] of the FCC's Rules) until you are granted the license for your new operator class, or for a period of 365 days from the test date stated above on this certificate, whichever comes first. **Note:** If you hold a current FCC-granted (codeless) Technician class operator license, and if this certificate indicates Element 1 credit, this certificate indefinitely permits you HF operating privileges as specified in Section 97.301(e) of the FCC rules. This document must be kept indefinitely with your Technician class operator license in order to use these privileges.

NOTE TO VE TEAM: COMPLETELY CROSS OUT ALL BOXES BELOW THAT DO NOT APPLY TO THIS CANDIDATE.

EXAM ELEMENTS EARNED
~~passed 5 wpm code element 1~~
~~passed written element 2~~
~~passed written element 3~~
(passed written element 4)

LICENSE STATUS INQUIRIES
You can find out if a new license or upgrade has been "granted" by the FCC. For on-line inquiries see the FCC Web at http://www.fcc.gov/wtb/uls ("License Search" tab), or see the ARRL Web at http://www.arrl.org/fcc/fcclook.php3; or by calling FCC toll free at 888-225-5322; or by calling the ARRL at 1-860-594-0300 during business hours. Allow 15 days from the test date before calling.

NEW LICENSE CLASS EARNED
~~TECHNICIAN~~
~~TECHNICIAN w/HF~~
~~GENERAL~~
(EXTRA)

THIS CERTIFICATE IS NOT A LICENSE, PERMIT, OR ANY OTHER KIND OF OPERATING AUTHORITY IN AND OF ITSELF. THE ELEMENT CREDITS AND/OR OPERATING PRIVILEGES THAT MAY BE INDICATED IN THE LICENSE UPGRADE NOTICE ARE VALID FOR 365 DAYS FROM THE TEST DATE. THE HOLDER NAMED HEREON MUST ALSO HAVE BEEN GRANTED AN AMATEUR RADIO LICENSE ISSUED BY THE FCC TO OPERATE ON THE AIR.

Candidate's signature *James McCall*
Candidate's name *JAMES M. McCALL* Call sign *WB1TC* (if none, write none)
Address *123 LAWRENCE ST*
City *ANDOVER* State *NH* ZIP *03216*

VE #1 *John P. Hamm* *WN1ZE* call sign
VE #2 *Paul Jones* *W1QQX* call sign
VE #3 *Richard Smith* *WA3UIL* call sign

Candidate's copy=white•ARRL/VEC's copy=pink•VE Team's copy=yellow

Figure 1-2 — This is an example of the CSCE you will get after you pass the Element 4 exam to upgrade from an Advanced class license if your test session is coordinated by the ARRL/VEC. The certificates issued by other VECs will differ from this, but will serve the same purpose.

The Certificate of Successful Completion of Examination is the document that allows amateurs who have passed an examination for a higher class of license to operate using their new privileges while waiting for the FCC to issue their new licenses. **Figure 1-2** shows a sample of the form used by the ARRL/VEC.

When you use your new privileges after upgrading your license, you must use special identification procedures until the FCC issues your new license. You will add a special identifier after your call sign when you give your station identification. If you hold an Advanced class license and upgrade to Amateur Extra, you will add the identifier code "AE" as a suffix to your call sign. For example, KB1GW would give his station ID as: "DE KB1GW/AE" on CW or as "This is KB1GW slant AE" on phone. (Any word used to denote the slant bar may be used, such as "stroke" or "temporary.")

---

§97.119 Station identification.

(f) When the control operator who is exercising the rights and privileges authorized by §97.9(b) of this Part, an indicator must be included after the call sign as follows:

(4) For a control operator who has requested a license

modification from Novice, Technician, General, or Advanced
Class operator to Amateur Extra Class: AE.

[You should now turn to Chapter 10 and study exam questions E1F12 through
E1F20. Review the material in this section if you have difficulty with any of these
questions.]

RF POWER AMPLIFIERS

RF power amplifiers capable of operating on frequencies below 144 MHz may
require FCC **certification**. Sections 97.315 and 97.317 describe the conditions
under which certification is required, and set out the standards to be met for certification. See **Table 1-3**.

Many of these rules apply to manufacturers of amplifiers or kits, but several
points are important for individual amateurs. Amateurs may build their own amplifiers or modify amplifiers for use in their own station without concern for the certification rules. If you build or modify an amplifier for use by another amateur
operator, then §97.315(a) applies. This rule says you cannot build or modify more
than one amplifier of any particular model during any calendar year without obtaining a grant of certification from the FCC. An unlicensed person may not build
or modify any amplifier capable of operating below 144 MHz without a grant of
FCC certification.

To receive a grant of certification, an amplifier must satisfy the spurious emission
standards specified in §97.307(d) or (e) when operated at full power output. The amplifier must also meet the spurious emission standards when it is being driven with at least
50 W mean RF input power (unless a higher drive level is specified.) In addition, the
amplifier must meet the spurious emission standards when it is placed in the "standby"
or "off" position but is still connected to the transmitter. The amplifier must not be
capable of reaching its designed output power when driven with less than 40 watts. The
amplifier must also not be capable of operation on any frequency between 24 and
35 MHz. (This is to prevent the amplifier from being used illegally on the Citizen's
Band frequencies.) The amplifier may be capable of operation on all amateur bands
with frequencies below 24 MHz, however.

A manufacturer must obtain a separate grant of certification for each amplifier
model. The manufacturer must obtain another grant of certification for future
amplifier models as they are developed.

An amplifier may not be built with accessible wiring that can be modified to
permit the amplifier to operate in a manner contrary to FCC Rules, and must not
include instructions for operation or modification of the amplifier in a way that
would violate FCC Rules. The amplifier may not have any features designed to be
used in a telecommunication service other than the Amateur Service. The amplifier
must not produce 3 dB or more gain for input signals between 26 MHz and
28 MHz. Failure to meet any of these conditions would disqualify the amplifier
from being granted FCC certification.

There is one condition by which an equipment dealer may sell an amplifier
that has not been granted FCC certification. A dealer may purchase a used ampli-

fier from an amateur operator and sell it to another licensed amateur for use at that operator's station.

> [Congratulations! You have completed your study of the FCC Rules and Regulations for your Amateur Extra class exam. Before you go on to Chapter 2, however, turn to Chapter 10 and review questions E1G01 through E1G11. If any of these questions give you difficulty, review this section.]

Table 1-3
RF Amplifier Certification Standards

§97.315 Certification of external RF power amplifiers.

(a) No more than 1 unit of 1 model of an external RF power amplifier capable of operation below 144 MHz may be constructed or modified during any calendar year by an amateur operator for use at a station without a grant of certification. No amplifier capable of operation below 144 MHz may be constructed or modified by a non-amateur operator without a grant of certification from the FCC.

(b) Any external RF power amplifier or external RF power amplifier kit (see §2.815 of the FCC Rules), manufactured, imported or modified for use in a station or attached at any station must be certificated for use in the amateur service in accordance with Subpart J of Part 2 of the FCC Rules. This requirement does not apply if one or more of the following conditions are met:

(1) The amplifier is not capable of operation on frequencies below 144 MHz. For the purpose of this part, an amplifier will be deemed to be incapable of operation below 144 MHz if it is not capable of being easily modified to increase its amplification characteristics below 120 MHz and either:

(i) The mean output power of the amplifier decreases, as frequency decreases from 144 MHz, to a point where 0 dB or less gain is exhibited at 120 MHz; or

(ii) The amplifier is not capable of amplifying signals below 120 MHz even for brief periods without sustaining permanent damage to its amplification circuitry.

(2) The amplifier was manufactured before April 28, 1978, and has been issued a marketing waiver by the FCC, or the amplifier was purchased before April 28, 1978, by an amateur operator for use at that amateur operator's station.

(3) The amplifier was:

(i) Constructed by the licensee, not from an external RF power amplifier kit, for use at the licensee's station; or

(ii) Modified by the licensee for use at the licensee's station.

(4) The amplifier is sold by an amateur operator to another amateur operator or to a dealer.

(5) The amplifier is purchased in used condition by an equipment dealer from an amateur operator and the amplifier is further sold to another amateur operator for use at that operator's station.

(c) Any external RF power amplifier appearing in the Commission's database as certificated for use in the amateur service may be marketed for use in the amateur service.

§97.317 Standards for certification of external RF power amplifiers.

(a) To receive a grant of certification, the amplifier must satisfy the spurious emission standards of §97.307(d) or (e) of this Part, as applicable, when the amplifier is:

(1) Operated at its full output power;

(2) Placed in the "standby" or "off" positions, but still connected to the transmitter; and

(3) Driven with at least 50 W mean RF input power (unless higher drive level is specified).

(b) To receive a grant of certification, the amplifier must not be capable of operation on any frequency or frequencies between 24 MHz and 35 MHz. The amplifier will be deemed incapable of such operation if it:

(1) Exhibits no more than 6 dB gain between 24 MHz and 26 MHz and between 28 MHz and 35 MHz. (This gain will be determined by the ratio of the input RF driving signal (mean power measurement) to the mean RF output power of the amplifier); and

(2) Exhibits no amplification (0 dB gain) between 26 MHz and 28 MHz.

(c) Certification may be denied when denial would prevent the use of these amplifiers in services other than the amateur service. The following features will result in dismissal or denial of an application for certification:

(1) Any accessible wiring which, when altered, would permit operation of the amplifier in a manner contrary to the FCC Rules;

(2) Circuit boards or similar circuitry to facilitate the addition of components to change the amplifier's operating characteristics in a manner contrary to the FCC Rules;

(3) Instructions for operation or modification of the amplifier in a manner contrary to the FCC Rules;

(4) Any internal or external controls or adjustments to facilitate operation of the amplifier in a manner contrary to the FCC Rules;

(5) Any internal RF sensing circuitry or any external switch, the purpose of which is to place the amplifier in the transmit mode;

(6) The incorporation of more gain in the amplifier than is necessary to operate in the amateur service; for purposes of this paragraph, the amplifier must:

(i) Not be capable of achieving designed output power when driven with less than 40 W mean RF input power;

(ii) Not be capable of amplifying the input RF driving signal by more than 15 dB, unless the amplifier has a designed transmitter power of less than 1.5 kW (in such a case, gain must be reduced by the same number of dB as the transmitter power relationship to 1.5 kW; This gain limitation is determined by the ratio of the input RF driving signal to the RF output power of the amplifier where both signals are expressed in peak envelope power or mean power);

(iii) Not exhibit more gain than permitted by paragraph (c)(6)(ii) of this Section when driven by an RF input signal of less than 50 W mean power; and

(iv) Be capable of sustained operation at its designed power level.

(7) Any attenuation in the input of the amplifier which, when removed or modified, would permit the amplifier to function at its designed transmitter power when driven by an RF frequency input signal of less than 50 W mean power; or

(8) Any other features designed to facilitate operation in a telecommunication service other than the Amateur Radio Services, such as the Citizens Band (CB) Radio Service.

CHAPTER 2 KEYWORDS KEYWORDS KEYWORDS

Ascending Pass — With respect to a particular ground station, a satellite pass during which the spacecraft is headed in a northerly direction while it is in range.

ATV (amateur television) — A wideband TV system that can use commercial transmission standards. ATV is only permitted on the 70-cm band and higher frequencies.

Bandwidth — The frequency range (measured in hertz — Hz) over which a signal is stronger than some specified amount below the peak signal level. For example, a certain signal is at least half as strong as the peak power level over a range of ±3 kHz, so it has a 3-dB bandwidth of 6 kHz.

Baud — A unit of signaling speed equal to the number of discrete conditions or events per second. (For example, if the duration of a pulse is 3.33 milliseconds, the signaling rate is 300 bauds or the reciprocal of 0.00333 seconds.)

Baudot code — A digital code used for radioteletype operation, and also known as the International Telegraph Alphabet Number 2 (ITA2). Each character is represented by five data bits, plus additional start and stop bits.

Blanking — Portion of a video signal that is "blacker than black," used to be certain that the return trace is invisible.

Contest — An Amateur Radio operating activity in which operators try to contact as many other stations as possible. While each contest has its own particular rules and information to be exchanged, all contests have the common purpose of enhancing the communication and operating skills of amateurs, and helping ensure their readiness for emergency communications.

Descending Pass — With respect to a particular ground station, a satellite pass during which the spacecraft is headed in a southerly direction while it is in range.

Doppler shift — A change in the observed frequency of a signal, as compared with the transmitted frequency, caused by satellite movement toward or away from you.

DX — Distance. On HF, often used to describe stations in countries outside your own.

Faraday rotation — A rotation of the polarization plane of radio waves when the waves travel through the ionized magnetic field of the ionosphere.

Fast-scan TV (FSTV) — Another name for ATV, used because a new frame is transmitted every $1/30$ of a second, as compared to every 8 seconds for slow-scan TV.

Facsimile (fax) — The process of scanning pictures or images and converting the information into signals that can be used to form a likeness of the copy in another location. The pictures are often printed on paper for permanent display.

Forward error correction (FEC) — One form of AMTOR operation, in which each character is sent twice. The receiving station compares the mark/space ratio of the characters to determine if any errors occurred in the reception.

Gray scale — A photographic term that defines a series of neutral densities (based on the percentage of incident light that is reflected from a surface), ranging from white to black.

Grid square locator — A 2° longitude by 1° latitude square, as part of a world wide numbering system. Grid square locators are exchanged in some contests, and are used as the basis for some VHF/UHF awards.

Node — A point where a satellite crosses the plane passing through the Earth's equator. It is an ascending node if the satellite is moving from south to north, and a descending node if the satellite is moving from north to south.

Packet cluster — A packet radio system that is devoted to serving a special interest group, such as DXers or contest operators.

Packet radio — A form of digital communication that includes error checking and correction, to ensure virtually error-free information exchange.

Period — The time it takes for a complete orbit, usually measured from one equator crossing to the next. The higher the altitude, the longer the period.

Photocell — A solid-state device in which the voltage and current-conducting characteristics change as the amount of light striking the cell changes.

Photodetector — A device that produces an amplified signal that changes with the amount of light striking a light-sensitive surface.

Phototransistor — A bipolar transistor constructed so the base-emitter junction is exposed to incident light. When light strikes this surface, current is generated at the junction, and this current is then amplified by transistor action

Polarization — A property of an electromagnetic wave that describes the orientation of the electric field of the wave with respect to earth.

Scanning — The process of analyzing or synthesizing, in a predetermined manner, the light values or equivalent characteristics of elements constituting a picture area. Also the process of recreating those values to produce a picture on a CRT screen.

Slow-scan television (SSTV) — A television system used by amateurs to transmit pictures within a signal bandwidth allowed on the HF bands by the FCC. It takes approximately 8 seconds to send a single black and white SSTV frame, and between 12 seconds and 4½ minutes for the various color systems currently in use.

Spin modulation — Periodic amplitude fade-and-peak variations resulting from a Phase 3 satellite's 60 r/min spin.

Spread-spectrum (SS) communication — A communications method in which the RF bandwidth of the transmitted signal is much larger than that needed for traditional modulation schemes, and in which the RF bandwidth is independent of the modulation content. The frequency or phase of the RF carrier changes very rapidly according to a particular pseudorandom sequence. SS systems are resistant to interference because signals not using the same spreading sequence code are suppressed in the receiver.

Transponder — A repeater aboard a satellite that retransmits, on another frequency band, any type of signals it receives. Signals within a certain receiver bandwidth are translated to a new frequency band, so many signals can share a transponder simultaneously.

Vestigial sideband (VSB) — A signal-transmission method in which one sideband, the carrier and part of the second sideband are transmitted. The bandwidth is not as wide as for a double-sideband AM signal, but not as narrow as a single-sideband signal.

OPERATING PROCEDURES

There will be four questions on your Amateur Extra class exam from the Operating Procedures subelement, and the question pool includes four general topics. The following syllabus points describe these four topics.

E2A   Amateur Satellites: Orbital mechanics; Frequencies available for satellite operation; Satellite hardware; Operating through amateur satellites

E2B   Television: Fast scan television (FSTV) standards; Slow scan television (SSTV) standards; facsimile (fax) communications

E2C   Contest and DX Operating; spread-spectrum transmissions; automatic HF forwarding

E2D   Digital Operating: HF digital communications (ie, PacTOR, CLOVER, AMTOR, PSK31, HF packet); packet clusters; HF digital bulletin boards

When you have studied the information in each section of the text, you will be directed to the examination questions in Chapter 10 to check your understanding of the material. If you are unable to answer a question correctly, go back and review the appropriate part of this chapter.

AMATEUR RADIO SATELLITES

Terrestrial communication is limited by the spherical shape of the Earth and other factors. There are numerous propagation mechanisms that can be used to transmit a signal around the Earth at HF. Long-haul communication at VHF and UHF, however, may require the use of higher effective radiated power or may not be possible at all. The communication range of amateur stations is increased greatly by using relay equipment mounted in satellites orbiting the Earth.

If you haven't already explored amateur satellite communications, this is a good time to learn more about our Orbiting Satellites Carrying Amateur Radio (OSCARs), or other types of space stations. Sections 97.207 through 97.211 of the FCC rules cover the Amateur Satellite Service. These rules outline *space operation*, *Earth operation* and *telecommand operation*. They also list frequencies available for space operation and describe the requirements for notifying the FCC about your space operation. You should be especially familiar with this section of the rules.

Astronauts aboard the space shuttles have found Amateur Radio to be an interesting pastime. If one or more crew members on a particular flight is a ham, SAREX — the Shuttle Amateur Radio Experiment — is likely to be part of the mission. The international space station includes a permanent Amateur Radio station. The Amateur Satellite Service rules govern this "space mobile" operation as well!

Understanding Satellite Orbits

Two factors affect a body in orbit around the Earth. These factors are forward motion and the force of gravitational attraction. Forward motion tends to move the body away from the Earth in a straight line in the direction it is moving at that instant. Gravity tends to pull the body toward the Earth. Motion and gravity must balance to maintain an orbit.

Johannes Kepler described the planetary orbits of our solar system. His three laws of planetary motion also describe the lunar orbit and the orbits of artificial Earth satellites. A brief summary of Kepler's Laws will help you understand the motion of artificial Earth satellites.

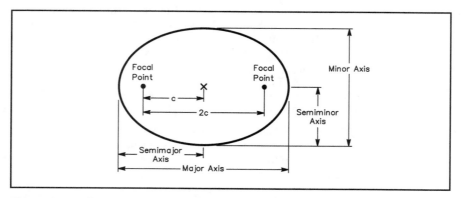

Figure 2-1 — Geometry of an ellipse. Labeled "c," the distance from the center to either focal point is called linear eccentricity. This distance should not be confused with numerical eccentricity, which is the distance c divided by the semimajor axis. Being a ratio, numerical eccentricity is a unitless number — unless otherwise stated this is simply referred to as eccentricity.

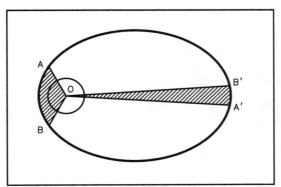

Figure 2-2 — A graphical representation of Kepler's second and third laws. The two shaded areas are equal. The time for the satellite to move along its orbit from A to B or from A' to B' is the same.

Kepler's Laws

Kepler's first law tells us that any satellite orbit is shaped like an ellipse, and the Earth will be at one of the focal points (**Figure 2-1**). The distance through the thickest part of an ellipse is called the major axis — through the thinnest it is called the minor axis. The semimajor axis is half the length of the major axis. The semiminor axis is also one half the minor-axis length. Eccentricity is the distance from the center to one of the focal points divided by the semimajor axis. Eccentricity ranges from 0 (a circle) to 1 (a straight line) — the larger the number the "thinner" the ellipse.

Kepler's second law is illustrated in **Figure 2-2**. The time required for a satellite to move in its orbit from point A to point B is the same as the time required to move from A' to B'. Another way of saying this is that a satellite moves faster in its elliptical orbit when it is closer to the Earth, and slower when it is farther away. The area of section AOB is the same as the area of section A'OB'.

The third law tells us that the greater the average distance from the Earth, the longer it takes for a satellite to complete each orbit. The time required for a satellite to make a complete orbit is called the orbital **period**. Low-flying amateur satellites typically have periods of approximately 90 minutes. The space shuttle usually flies in a fairly low, nearly circular orbit, and it usually has a period of about 90 minutes. OSCAR 10 has a high, elliptical orbit, and has a period of more than $11\frac{1}{2}$ hours.

Kepler's Laws can be expressed mathematically, and if you know the values of a set of measurements about the satellite orbit (called Keplerian Elements) you can calculate the position of the satellite at any other time. Most amateurs use one of several computer programs to perform such orbital calculations. The results can even be displayed over a map on the computer monitor, with position updates calculated and shown every few seconds. You should get a new set of Keplerian Elements about once a month for satellites you want to track. These Keplerian Elements for the current OSCAR satellites are available on the Internet. Go to the satellite page at *ARRLWeb*: **http://www.arrl.org/w1aw/kep/**.

Orbital Terminology

Inclination is the angle of a satellite orbit with respect to Earth. Inclination is measured between the plane of the orbit and the plane of the equator (**Figure 2-3**). If a satellite is always over the equator as it travels through its orbit, the orbit has an inclination of 0°. Should the orbit take the satellite over the poles (if it goes over one it will go over the other), the inclination is 90°. The inclination angle is always measured from the equator counterclockwise to the satellite path. See **Figure 2-4**.

A **node** is the point where a satellite crosses the equator. The ascending node is where it crosses the equator when it is traveling from south to north. Inclination

Figure 2-3 — A graphical representation of some basic satellite-orbit terminology.

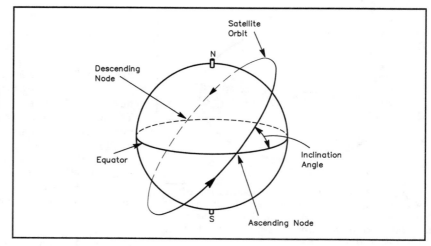

Figure 2-4 — Satellite orbit inclination angles are measured at an ascending node. The angle is measured from the equator line to the orbital plane, on the eastern side of the ascending node.

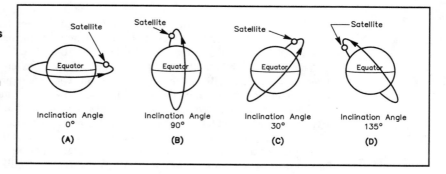

Figure 2-5 — A graphical representation of satellite-orbit terminology. The very elliptical orbit of a Phase 3 satellite is shown.

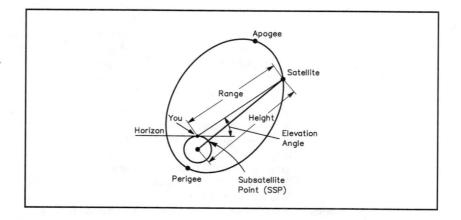

is specified at the ascending node. Equator crossing (EQX) is usually given in time (UTC) of crossing, and in degrees west longitude. The descending node is the point where the satellite crosses the equator traveling from north to south. When the satellite is within range of your location, it is common to describe the pass as either an **ascending pass** or a **descending pass**. This depends on whether the satellite is traveling from north to south or south to north over your area.

The point of greatest height in a satellite orbit is called the *apogee* (**Figure 2-5**). *Perigee* is the point of least height. Half the distance between the apogee and perigee is equal to the semimajor axis of the satellite orbit.

More information on satellite orbits and methods of tracking satellites can be found in *The Radio Amateur's Satellite Handbook*, published by ARRL.

[Now turn to Chapter 10 and study examination questions E2A01 through E2A03. Review this section as needed.]

Satellite Systems and Hardware

Present communications satellites are functionally integrated systems. Rechargeable batteries, solar cell arrays, voltage regulators, command decoders, antenna-deployment mechanisms, stabilization systems, sensors, telemetry encoders and even on-board computers and kick motors each serve a unique and indispensable purpose. But to the radio amateur interested in communicating through a satellite, the **transponder** is of primary importance.

The earliest OSCARs simply orbited the Earth and transmitted information about the satellite battery voltage, temperature and other physical conditions. These had relatively low, nearly circular orbits, and are sometimes referred to as the Phase 1 satellites. Next came active two-way communications satellites, also with relatively low, nearly circular orbits. There are several satellites in current use that fit this category, sometimes called Phase 2 satellites. Their orbits are approximately 500 to 700 km above the Earth. The orbital period of Phase 1 and Phase 2 satellites is around 90 minutes.

The latest generation of OSCARs have very elliptical orbits that take them more than 35,000 km from Earth at apogee and about 4000 km at perigee. Known as Phase 3 satellites, this group has orbital periods of 8 to 12 hours, and provide greater communications range.

Transponders

By convention, **transponder** is the name given to any linear translator that is installed in a satellite. The translator or transponder is similar to a repeater in many ways. Each is a combination of a receiver and a transmitter that is used to extend the range of mobile, portable and fixed stations. A typical FM voice repeater receives on a single frequency or channel and retransmits what it receives on another channel. By contrast, a transponder receive passband includes enough spectrum for many channels. An amateur satellite transponder does not use channels in the way that voice repeaters do. Received signals from a band segment are amplified, shifted to a new frequency range and retransmitted by the transponder. **Figure 2-6** shows block diagrams for a simple voice repeater and a simple transponder.

By comparing block diagrams, you can see that the major hardware difference between a repeater and a transponder is signal detection. In a repeater, the signal is

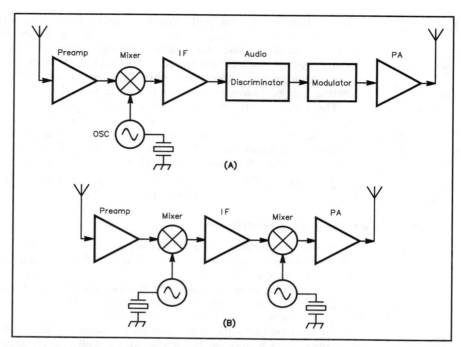

Figure 2-6 — Part A shows the block diagram of an FM voice repeater; Part B shows the block diagram of a linear transponder.

reduced to baseband (audio) before it is retransmitted. In a transponder, signals in the passband are moved to an IF for amplification and retransmission.

Operationally, the contrast is much greater. An FM voice repeater is a one-signal, one-mode-input and one-signal, one-mode-output device. A transponder can receive several signals at once and convert them to a new range. Further, a transponder can be thought of as a multimode repeater. Whatever goes in is what comes out. The same transponder can simultaneously handle SSB, ACSSB and CW signals. The use of a transponder rather than a channelized repeater allows more stations to use the satellite at one time. In fact, the number of different stations using a transponder at any one time is limited only by mutual interference, and the fact that the output power of the satellite (a couple of watts on the low-orbit satellites and about 50 watts on Phase 3) is divided between the users. Because all users must share the power output, continuous-carrier modes such as FM and RTTY are not used through the amateur satellites.

Satellite Operating Frequencies

Repeaters are often referred to by their operating frequencies. For example, a 34/94 two-meter repeater receives on 146.34 MHz and transmits on 146.94 MHz. A different convention is used with transponders: The input band is given, followed by the corresponding output band. For example, a 2-m/10-m transponder would have an input passband centered near 146 MHz and an output passband centered near 29 MHz. Transponders usually are identified by mode — but not mode of transmission such as SSB or CW. Mode has an entirely different meaning in this case.

Satellites used for two-way communication generally use one amateur band to receive signals from Earth (uplink) and another to transmit back to Earth (downlink). The amateur satellites use a variety of uplink and downlink band combinations. These band combinations are called operating *modes*. Don't be confused by this term, however. It is not used in the way we normally refer to modes when we talk about CW, SSB or RTTY, for example. The operating modes of a satellite simply refer to the uplink and downlink frequency bands that the satellite uses. **Table 2-1** lists these modes by their common letter designations. An OSCAR satellite normally operates on several of these modes.

Table 2-1

Bands Used for Satellite Communications

Mode	Uplink	Downlink
A	2 m	10 m
B	70 cm	2 m
J	2 m	70 cm
K	15 m	10 m
L	23 cm	70 cm
S	70 cm	13 cm
T	15 m	2 m

Sections 97.207 and 97.209 of the FCC Rules list the frequencies available to the Amateur Satellite Service. We covered those sections of the Rules in Chapter 1. You can find the complete text of Part 97, the rules for the Amateur Radio Service, in *The ARRL's FCC Rule Book*. The text of Part 97 is also available on *ARRLWeb*.

[Questions E2A04 through E2A08 cover satellite operating modes and transponders. Turn to Chapter 10 and study those examination questions now. Review Table 2-1 as needed.]

Doppler Shift

Doppler shift is caused by the relative motion between you and the satellite. If there were no relative motion, you could predict precise downlink frequencies coming from the satellite. In operation, as the satellite is moving toward you, the frequency of a downlink signal appears to be increased by a small amount. When the satellite passes overhead and begins to move away from you, there will be a sudden frequency drop of a few kilohertz, in much the same way as the tone of a car horn or train whistle drops as the vehicle moves past you. The result is that signals passing through the satellite move around the expected or calculated downlink frequency, depending on whether the satellite is moving toward you or away from you. Those signal frequencies appear higher when the satellite is moving toward you, lower when it is moving away. Locating your own signal is, therefore, a little more difficult than simply computing the relation between input and output frequency; the effects of Doppler must be taken into account.

Since the speed of the satellite, relative to Earth, is greater for a satellite that is close to Earth, Doppler shift is generally more noticeable with low-orbit satellites than with high-orbit ones like OSCAR 10. Doppler shift is also frequency dependent. The higher the operating frequency, the greater the shift.

Faraday Rotation

The **polarization** of a radio signal passing through the ionosphere does not remain constant. A "horizontally polarized" signal leaving a satellite will not be horizontally polarized when it reaches Earth. That signal will in fact seem to be changing polarization at a receiving station. This effect is called **Faraday rotation**. The best way to deal with Faraday rotation is to use circularly polarized antennas for transmitting and receiving.

Spin Modulation

Spin modulation is a phenomenon that has emerged with the introduction of the AMSAT Phase 3 satellites. As the satellite orbits overhead, the on-board computer pulses an electromagnet that works against the magnetic field of the Earth. This spins the spacecraft at approximately 1 revolution per second, thereby stabilizing it. A side effect, however, is the relatively rapid (3 Hz) periodic fade of the transmitted signal amplitude, called spin modulation. The 3-Hz spin modulation is caused by having a transmitting antenna on each satellite "arm," so three antennas spin past you each second. It is important to note that the passband is not electronically modulated in the sense to which amateurs are accustomed; rather, the apparent modulation is a residual effect of physical rotation.

Use of linear antennas (horizontal or vertical polarization) will deepen the spin-modulation fades to a point where they may become annoying. Circularly polarized antennas of the proper sense will minimize the effect.

[Now turn to Chapter 10 and study examination questions E2A09 through E2A11. Review this section as needed.]

FAST-SCAN AMATEUR TELEVISION

Fast-scan TV (**FSTV**) can be used by any amateur holding a Technician or higher-class license. FSTV closely resembles broadcast-quality television, because it normally uses the same technical standards. Amateurs may use commercial transmission standards for TV. They are not *limited* to commercial standards, however.

Popularly known as **ATV** (**amateur television**), this mode is permitted in the 420 to 450-MHz band and higher frequencies. Because the power density is comparatively low, typically 10 to 100 watts spread across 6 MHz, reliable amateur coverage is only on the order of 20 miles. Nevertheless, you might find yourself exchanging pictures with stations up to 200 miles away when tropospheric conditions are good.

Most wide-band TV activity occurs in the 420 to 450-MHz band. The exact frequency used depends on local custom. Some population centers have ATV repeaters. ATVers try to avoid interfering with the weak-signal work (moonbounce, for example) being done around 432 MHz and repeater operation above 442 MHz. The FCC uses emission designators A3F or C3F to describe FSTV. Under an old system of emission designators this was known as A5. Many long-time ATVers still refer to Amateur TV as A5.

The Audio Channel

There are at least three ways to transmit voice information with a TV signal. The most popular method is by talking on another band, often 2-meter FM. This has the advantage of letting other local hams listen in on what you are doing — a good way to pick up some ATV converts! Rather than tie up a busy repeater for this, it is best to use a simplex frequency.

Commercial TV has an FM voice subcarrier 4.5 MHz above the TV picture carrier. If you provide FM audio at a frequency 4.5 MHz above the video frequency, the audio can easily be received in the usual way on a regular TV set. Many of the surplus FM rigs (tube type) that are available do not have enough bandwidth to pass both the picture carrier and the voice subcarrier at the same time, however, so other methods must be used.

Another way to go is to frequency modulate the video carrier. Since the video is amplitude modulated, it should not interfere with the FM audio, or vice versa. The usual way to receive this is with the FM receiver section of a UHF rig. It is also possible to use a lowband police/fire monitor coupled into the 44-MHz IF of a TV set.

The Scanning Process

A picture is divided sequentially into pieces for transmission or presentation (viewing); this process is called **scanning**. A total of 525 scan lines comprise a frame (complete picture) in the US television system. Thirty frames are generated each second. Each frame consists of two fields, each field containing $262\frac{1}{2}$ lines. Sixty fields are generated each second, therefore. Scan lines from one field fall between (interlace) lines from the other field. The scanned area is called the *television raster*. If all 525 scan lines are numbered from top to bottom, then one field scans all the even-numbered lines and the second field scans all the odd-numbered lines. **Table 2-2** lists some common ATV standards.

Figure 2-7 illustrates the principle involved in scanning an electron beam (in the pickup or receiving device) across and down to produce the television raster. Field-one scanning begins in the upper left corner of the picture area. The electron beam sweeps across the picture to the right side. At the end of the line, the beam is turned off, or blanked and returned to the left side where the process repeats. In the meantime, the beam has also been moved slightly downward. At the end of $262\frac{1}{2}$ horizontal scans (lines), the beam is blanked and rapidly returned to the top of the picture area. At that point, scanning of field two begins. Notice that this time the beam starts scanning from the middle of the picture. For that reason, the scanning lines of field two will fall halfway between the lines of field one. At the end of $262\frac{1}{2}$ lines the beam is rapidly returned to the top of the picture again. Scanning continues — this time with field one of the next frame.

The picture area is scanned from top to bottom 60 times each second. That is frequent enough to avoid visible flicker. Because the entire picture area is scanned only 30 times each second, bandwidth is reduced.

It should be obvious that the horizontal and vertical oscillators that control the electron-beam movement must be "locked together" for the two fields to interlace properly. If the frequencies of these oscillators are not locked, proper interlace will be lost and vertical resolution or detail will be degraded.

Table 2-2

ATV Standards

Line rate	15,750 Hz
Field rate	60 Hz
Frame rate	30 Hz
Horizontal lines	$262\frac{1}{2}$ / field
	525 / frame
Sound subcarrier	4.5 MHz
Channel bandwidth (VSB — C3F)	6 MHz

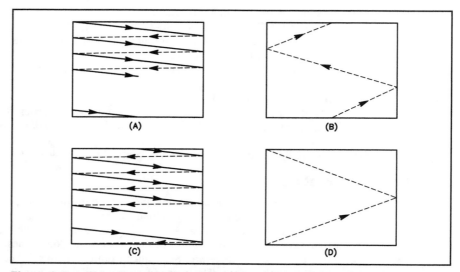

Figure 2-7 — This diagram shows the interlaced scanning used in TV. In field one, 262½ lines are scanned (A). At the end of field one, the electron scanning beam is returned to the top of the picture area (B). Scanning lines in field two (C) fall between the lines of field one. At the end of field two, the scanning beam is again returned to the top, where scanning continues with field one (D).

Deflection

In most TV applications, the electron beam is scanned (deflected) by means of two coil pairs. Because the deflection of the electron beam is accomplished magnetically, coils for horizontal deflection are located above and below the beam. Vertical deflection coils are located on either side of the beam. See **Figure 2-8**.

The electron beam is deflected as a result of a "sawtooth" current passing through the deflection coils (**Figure 2-9**). The frequency of the horizontal sawtooth current is 15,750 Hz. A similar waveform with a frequency of 60 Hz causes vertical deflection. The electron beam is turned off during beam retrace by a process called **blanking**. Blanking is only associated with the electron beam and does not affect the deflection coil current or the resulting magnetic fields. The video voltage level is made smaller than the signal for black, so the retrace line is invisible on the screen as it moves from right to left and bottom to top on the screen.

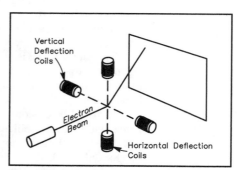

Figure 2-8 — Electron beam deflection in TV cameras and receivers is usually accomplished by using two sets of coils. This is called electromagnetic deflection.

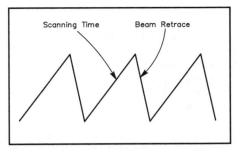

Figure 2-9 — Electromagnetic deflection coils use a sawtooth current waveform for deflection of the electron scanning beam.

Table 2-3
Standard Video Levels

	IEEE Units	% PEV
Zero carrier	120	0
White	100	12.5
Black	7.5	70
Blanking	0	75
Sync tip	− 40	100

PEV is peak envelope voltage and corresponds to levels as seen on an oscilloscope.

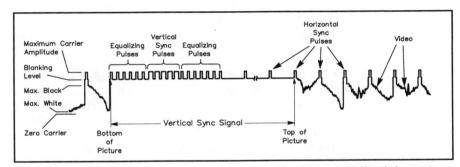

Figure 2-10 — This waveform shows a composite video signal as it is sent from a TV transmitter. The horizontal sync pulses, superimposed on the video (picture information) signal "tell" the TV receiver when to start a new horizontal line. The vertical sync signal is sent at the bottom of each vertical sweep to tell the TV set to go to the top of the screen to start a new field.

Video Transmission

Video from the TV camera normally has white positive and sync negative. The standard video voltage level between white and the sync tip is 1-V peak to peak. Monitors are made for sync-negative video. For transmission, however, the polarity or sense of the video is inverted — the sync is positive. That puts the sync tip at peak power output from the transmitter. The video sense is inverted again in the receiver. **Figure 2-10** shows the waveform of a transmitted composite video signal.

Television engineers measure video levels in *IEEE Units* or as a percentage of the *peak-envelope voltage (% PEV)*. **Table 2-3** compares these two measurement systems.

Because the sync tip corresponds to maximum transmitter output, a receiver is better able to hold a stable picture. This is especially true under adverse reception conditions such as noisy or weak signals. It is far better to have a bit more noise in the picture than to have it be unstable.

Channel Bandwidth

The video portion of color TV pictures requires a bandwidth of about 4 MHz. Satisfactory black and white pictures can be realized with less bandwidth. If a color TV picture is to be transmitted with double sideband AM (DSB), more than 8 MHz of spectrum space is required. It is not necessary to use that much spectrum — although amateurs are permitted to, and some do, use DSB. The emission designator for double-sideband emissions on a single channel containing analog video information is A3F.

Commercial TV stations and many amateurs use **vestigial sideband (VSB)** for transmission. VSB is like SSB with full carrier except a portion (vestige) of the unwanted sideband is retained. In the case of VSB TV, approximately 1 MHz of the lower sideband and all of the upper sideband plus full carrier comprise the transmitted picture signal. The emission designator for vestigial-sideband emissions on a single channel containing analog video information is C3F. Both A3F and C3F emissions were designated A5 under the old system of emissions designators, and you may still hear some ATVers refer to A5.

The sound carrier is 4.5 MHz above the picture carrier, as shown in **Figure 2-11**. This figure also shows that the channel bandwidth for VSB TV with sound carrier is 6 MHz.

Most ATV operators use amplitude-modulation systems for video transmission. It is also possible to use frequency modulation systems to transmit the video signal. FM systems have a bandwidth of 17 to 21 MHz, which is significantly greater than the bandwidth for AM systems. Most of the available FMTV equipment operates in the 1.2, 2.4 and 10.25-GHz bands. FMTV receiving systems are more complex than AMTV systems. AM systems give better weak-signal performance,

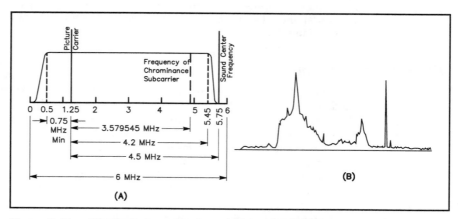

Figure 2-11 — The frequency spectrum of a color TV signal is shown in diagram A. B represents a spectrum analyzer display. Spectrum power density will vary with picture content, but typically 90% of the sideband power is within the first megahertz.

while FM systems give better picture quality with stronger signals. The FM systems do *not* provide immunity from signal fading, however.

[Before proceeding to Chapter 3, turn to Chapter 10 and study examination questions E2B01 through E2B08. Review this section as needed.]

FACSIMILE SYSTEMS

The FCC permits image transmissions in the voice segments of all amateur bands. The bandwidth of an image transmission must be no greater than the bandwidth of a voice signal of the same modulation type. Note that there is no voice — or image — operation allowed on 30 meters, though.

Facsimile (fax) is the earliest operating system used for image transmission over radio. Originally, fax systems used a printed picture or image, which was scanned to produce an electrical signal for transmission. The received signals were converted to a printed picture again, for permanent display. Today, many amateurs use computer systems to operate fax, and the pictures can be displayed on a computer monitor and stored digitally. The pictures have high resolution, which is achieved by using from 500 to several thousand lines per frame.

Amateurs also use **slow-scan television (SSTV)** systems to transmit images on the HF bands. Many SSTV systems use a computer and connect to a transceiver through some type of interface. Computer systems can take advantage of all the graphics editing features of the computer, providing a nearly unlimited supply of image possibilities. These range from cartoons to scanned photos, digital cameras and video cameras with frame capture capabilities. Many fax systems also use computers. Both fax and SSTV systems transmit the images as a series of audio tones, usually ranging between 1500 and 2300 Hz. These frequencies carry the information about the image shading between black and white or color and brightness.

To keep the transmitted-signal bandwidth within the narrow limits allowed by the FCC on the HF bands, relatively long transmission times are used for fax pictures. Depending on the system in use, it may take from a bit more than three minutes up to 15 minutes to receive a single frame. Of course if it takes longer than 10 minutes for an amateur station to transmit the frame, there has to be a pause for station identification. That presents additional problems, and hams seldom use such systems.

Many amateurs use fax to receive weather-satellite photos that are transmitted direct from space. News services also use this mode to distribute "wirephotos" throughout the world, and fax systems are becoming increasingly common in business and law-enforcement offices to transfer documents, signatures, fingerprints and photographs.

Transmission

A typical modern fax system uses electronics to scan the image and produce an electrical signal that can be transmitted over a telephone line or used to modulate an RF signal for radio transmission. A small spot of light is focused on the printed material, and reflected light is picked up by a **photocell, photodetector** or **phototransistor**. The photodetector converts variations in picture brightness and darkness into voltage variations. The light source and sensor scan the entire picture area.

Table 2-4

Standards for Various Fax Services

Service	Transmission Speed (lines per minute)	Size (inches)	Scan Density (lines per inch)
WEFAX Satellite	240	11	75
APT Satellite	240	11	166
Weather Charts	120	19	96
Wire photos	90	11	96
Wire photos	180	11	166

Voltage variations from the light pickup are amplified and used to modulate an audio subcarrier signal. Either amplitude modulation (AM) (A3C) or frequency modulation (FM) (F3C or G3C) methods may be used, with FM being the standard for amateur HF operation and some of the press services using AM.

The important characteristics for any fax system include the number of scan lines per frame, the scanning speed or transmission speed (which determines the scan density of the picture) and the modulation characteristics. These characteristics must be the same for the receiving and transmitting stations. **Table 2-4** lists standards used by some of the common fax services. The 240 lines-per-minute speed is a good rate for amateur HF use because a detailed picture (800 lines) can be transmitted in about 3.3 minutes. If a speed of 120 lines-per-minute is used, then it takes approximately 6 minutes to transmit a complete picture.

A computerized fax system using 480 scan lines transmitted at a speed of 138.3 seconds per frame is popular with amateurs. The 480-line system provides good detail on a computer VGA monitor set to 640 pixels per line, 480 scan lines and 16 levels of gray. A simple interface and free software is all you need to experience this mode. The FAX480 system was developed by Ralph Taggart, WB8DQT, and first described in the February 1993 issue of *QST*.

Scan density is usually expressed as a number of lines per inch. This parameter determines the aspect ratio of the picture (image width/height) and the transmission time for one frame.

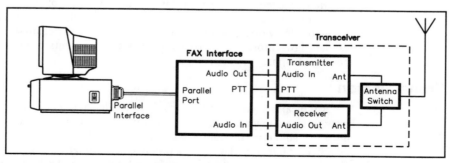

Figure 2-12 — This block diagram shows a computerized system for transmitting and receiving pictures by facsimile (fax).

Figure 2-13 — This is an example of a fax picture received by Ralph Taggart, WB8DQT, using his computerized FAX480 system.

Reception

To receive fax pictures, the characteristics of your system must match those of the transmitting station. Amateurs commonly use the 240 line-per-minute standard or the "FAX480" system.

There are several methods available for displaying a received fax picture. Most amateurs who operate fax use some type of computerized system. Some have dedicated microprocessor-based units while others have an interface and software for a personal computer of some type. The principal advantages over old-fashioned mechanical recorders are the ability to process the signals digitally during reception, and the ability to store the images digitally for later display or printing. **Figure 2-12** shows a simplified block diagram of a computerized fax system.

Computer systems use high-resolution TV-monitor displays to provide a 16-level (or more) **gray scale** and good resolution. Permanent hard copy is possible with a graphics printer. The computer makes it possible to display inverted images, perform various types of image enhancement, and even produce "color" images. **Figure 2-13** shows a fax picture received by Ralph Taggart, WB8DQT, using his FAX480 format.

[Before proceeding, study questions E2B09 through E2B12 in Chapter 10. Also study questions E2D07 and E2D08. Review this section as needed.]

CONTEST AND DX OPERATING

Two very popular activities on Amateur Radio are "chasing DX" and **contest** operating. **DX** is Amateur Radio jargon for distant stations. Your definition of DX will probably depend upon the bands on which you are operating. For HF operators, DX usually means any stations outside of your own country. On the VHF and UHF bands, however, DX may mean any station more than 50 or 100 miles away.

Contests are Amateur Radio on-the-air operating events, usually held on weekends, in which operators try to make as many contacts as possible within the rules of the particular contest. Contests provide an outlet for the competitive side of your personality, allow you to quickly add to your state or country totals, and offer a level of excitement that's hard to imagine until you've tried it!

While each contest has its own particular purpose and operating rules, the main purpose for all Amateur Radio contests is to enhance the communications and operating skills of amateurs in readiness for an emergency. When you optimize your station for best operating efficiency "in the heat of battle" and learn to pull out those weak stations to make the last available contacts, you are honing skills that may someday save a life. The best way to maintain emergency communications readiness is to practice those skills on a regular basis. Contests provide a fun way to keep a keen edge on your equipment capabilities and your operating skills.

Of course every contest operator must also follow the best operating practices. When you are looking for a frequency to call "CQ CONTEST," you should always listen before you transmit and be courteous to other operators. Do not cause harmful interference to other communications.

Good contest operators are aware of the various "Band Plans" and "Gentlemen's Agreements" that designate certain portions of the bands for specific operations. Although the FCC Rules may not place restrictions on a certain operating mode and frequency, the band plans try to accommodate all communications. If you are in the US and operating on the 160-meter band, for example, you should avoid using frequencies between 1830 kHz and 1850 kHz to call other US stations. These frequencies have been designated a "DX window" by gentlemen's agreement. The 6-meter band also has a "DX window" designated by gentlemen's agreement. Frequencies between 50.100 MHz and 50.125 MHz are reserved for DX contacts on that band.

You should also be familiar with the band plans so you will know where to expect various types of activity. For example, during a VHF or UHF contest you should expect to find most of the contest activity at the bottom, or low-frequency end of each band, in the section designated for weak signal work. Much of the contest activity will probably be on CW or SSB modes, rather than FM (although you will find quite a bit of FM activity in many VHF/UHF contests).

Every contest has its own particular pattern of information to be exchanged during a valid contact. Nearly every contest will require you to exchange call signs and a signal report. The rules and a listing of the information to be exchanged in a particular contest are printed in *QST* and other Amateur Radio publications a month or so prior to the contest each year.

Chasing DX is a popular activity both during DX contests and at other times. For this reason a frequency can become extremely crowded with amateurs calling the DX station. Many DX operators use a technique called operating split to combat this interference and make it possible for them to hear the calling stations and exchange information. They transmit on one frequency, and listen over a range of other frequencies. If you hear the DX station announcing "Up 5," or something similar, that means they are listening on a frequency 5 kHz higher than their transmit frequency.

You should learn how to adjust your transceiver for split-frequency operation so you will be sure to transmit on the frequency to which the DX station is listening, rather than on that station's transmit frequency. During the excitement of the chase, it is easy to set one of your radio controls improperly.

Contest operating and chasing DX are good activities to help you learn more about radio-wave propagation. You must be familiar with various propagation conditions as you plan your operating strategy. For example, if you want to participate

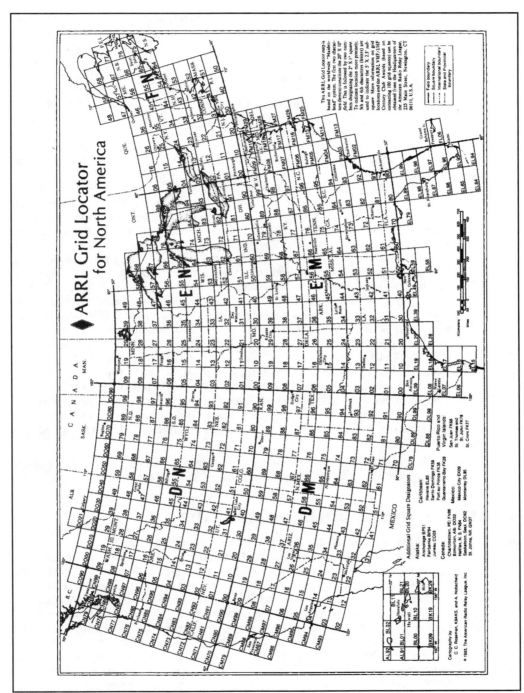

Figure 2-14 — This reduced-size ARRL Grid Locator Map shows how the Maidenhead Grid Square system applies to the US.

in a worldwide DX contest, you might immediately think about operating on the higher-frequency bands, such as 20, 15 and even 10 meters. During a sunspot (solar) minimum, however, a more ideal operating strategy would be to emphasize operation on the lower-frequency bands, such as 160 to 40 meters during the evening and nighttime hours, and then changing to 20 meters during daylight hours.

You can maximize your contest scores and your operating enjoyment by paying attention to the band conditions as you operate, also. This will help you decide when to change bands or modify your strategy. For example, if you notice that DX signals from Europe have become weak and fluttery across an entire band a few hours after sunset, it may be time to change to a lower-frequency band because the MUF has decreased. It may be especially important to notice these conditions and change bands during a period of low sunspot activity because the higher-frequency bands may close entirely after dark, with no DX stations to be heard.

Some DX contests and many VHF/UHF contests use the "Maidenhead grid locator" as part of the contest exchange. Even during casual contacts you may be asked for your "**grid square locator**," because many operators enjoy collecting these designators. (They are called Maidenhead grid squares because the system was developed at a conference in Maidenhead, England.) The Earth is divided into "squares" that are 2° of longitude by 1° of latitude. Each square is designated by two letters and two numbers. For example, W1AW is located in grid square FN31. Each grid square can be further divided into smaller segments, described by two more letters. These subsquares are only a few kilometers wide. (The actual dimensions vary with latitude.) Maidenhead grid squares are always based on latitude and longitude. **Figure 2-14** shows a small version of the ARRL Grid Locator Map.

The ARRL *UHF/Microwave Experimenter's Software* diskette (sold separately from the book) includes a program called GRIDLOC.BAS, which will determine your grid square given your latitude and longitude. GRIDLOC.BAS is also available on *ARRLWeb*: **http://www.arrl.org/locate/gridinfo.html#programs**. For additional information about the Maidenhead Grid Locator system see *The ARRL Operating Manual*, *The ARRL US Grid Locator Map* and *The ARRL World Grid Locator Atlas*.

[Now turn to Chapter 10 and study questions E2C01 through E2C09. Review this section if you have difficulty with any of these questions.]

SPREAD-SPECTRUM COMMUNICATIONS

When Amateur Radio communication is disturbed by interference from other stations on nearby frequencies, the most common "solutions" are to increase transmitter power or use narrower bandwidth receive filters. The increased transmitter power is intended to "punch through" the interfering signals, while the narrower bandwidth receive filters are intended to reject more of the unwanted signals.

Amateurs can take a different approach to solving this problem, however, by using **spread-spectrum communication**. SS, as it is often called, actually reduces the average power transmitted on any one frequency, and distributes the signals over a much wider bandwidth. **Figure 2-15** illustrates the effects of spreading the same transmitter power over a wider and wider bandwidth. The average power transmitted on any one frequency decreases, as long as the total transmitter power remains the same.

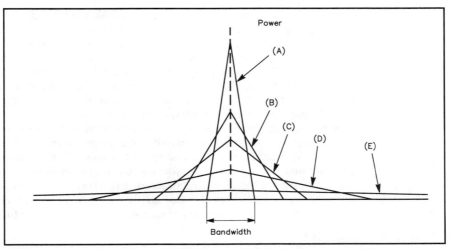

Figure 2-15 — This graph represents the average power distribution over a frequency range as the signal bandwidth increases. Signal A contains most of its energy in a narrow range around the center frequency. As the bandwidth increases and the transmitter power remains the same, the power at the center frequency decreases, as at B. Signals C and D show how the energy is distributed across the spread signal bandwidth. At E, the signal energy is spread over a very wide bandwidth, and there is little power at any one frequency.

There are many ways to cause a carrier to spread over a range of frequencies, but all SS systems can be considered as a combination of two modulation processes. First, the information to be transmitted is applied to the carrier. A conventional form of analog or digital modulation is used for this step. Second, the carrier is modulated by a "spreading code." The spreading code distributes the carrier over a large bandwidth. Four spreading techniques are commonly used in military and space communications, but amateurs are only authorized to use two of them: frequency hopping and direct sequence.

Frequency Hopping

Frequency hopping (FH) spread spectrum is a form of spreading in which the center frequency of a conventional carrier is altered many times per second in accordance with a list of frequency channels. This list is called a *pseudorandom* list because the channel-selection order appears to be random. The order is not truly random, however, because the receiver has to be able to follow it. In other words, there doesn't appear to be a set pattern, such as going through the frequencies in direct increasing order. The frequency channels are selected to avoid interference to fixed-frequency users such as repeaters. FH is generally used to transmit single-sideband (SSB) signals.

As an example, suppose a transmitter can transmit on any of 100 discrete frequencies, which we can call F1 through F100. Now suppose the transmitter operates for 1 second on F1, then jumps to F62 and transmits for another second before going to F33 to transmit for a second and then on to F47, and so on. The frequency pattern seems random, but as long as the receiver knows the sequence, it can also jump from F1 to F62 to F33 to F47 in sync with the transmitter. There may be a signal on F33 that will interfere with the SS station, but that will only last for 1 second. Likewise, the SS signal may cause some interference to the station on F33, but only for a second. In either case the interference is barely perceptible.

Frequency hopping spread spectrum works because the frequency of the RF carrier is changed very rapidly according to a particular pseudorandom sequence of frequencies. There are many possible sequences for the same set of frequencies, so many SS contacts can take place simultaneously without interference. Signals from stations not using SS and from stations using a different spreading sequence are suppressed in the receiver, because the receiver is changing frequency in step with the transmitter. Spread spectrum signals present little or no interference to other stations because they only remain on any one frequency for a brief instant.

Direct Sequence

In direct sequence (DS) spread spectrum, a binary bit stream is used to shift the phase of an RF carrier very rapidly in a particular pattern. The binary sequence is designed to appear to be random (that is, a mix of approximately equal numbers of zeros and ones), but the sequence is generated by a digital circuit. The binary sequence can be duplicated and synchronized at the transmitter and receiver. This sequence is called pseudonoise, or PN, because the signal only *appears* to be random. The transmitting and receiving systems must both use the same spreading codes, so the receiver will know where to look next for the transmitted signal. DS spread spectrum is typically used to transmit digital information.

[Now turn to Chapter 10 and study questions E2C10 and E2C11. Review this section if either of these questions give you trouble.]

DIGITAL OPERATING MODES AND PRACTICES

Radio amateurs communicate using a number of "digital" operating modes. These communications modes involve signals transmitted in digital form, such as a series of 1s and 0s or ON and OFF conditions. Strictly speaking, Morse code is a digital communications mode, but for this discussion we will limit our study to those modes that normally require a digital computer or other type of processor as part of the system.

Many hams exchange printed messages via radioteletype (RTTY). The message is typed on a computer keyboard (or teleprinter machine) and displayed on the monitor screen or printed on paper. Traditional RTTY communication uses the **Baudot code**. This digital code, also called the International Telegraph Alphabet Number 2 (ITA2), uses five data bits to represent each character, with additional start and stop bits to indicate the beginning and end of each character. Each character is represented by a unique pattern of 1s and 0s.

The 1s and 0s of the digital codes are transmitted as two tones, or frequencies, usually called *mark* and *space*. For HF RTTY and other digital emissions, the frequency of the RF carrier is usually shifted from one tone to the other, in the sequence of mark and space data bits to represent each character. This technique is called frequency-shift keying (FSK). On the VHF and UHF bands, most hams use FM radios for digital communications, and in that case two audio frequency tones are fed into the microphone input to form the mark and space signals. This is called audio-frequency-shift keying (AFSK).

AMTOR is a digital communications mode that uses a variation of the Baudot code. When two AMTOR stations are in direct contact they employ an error detection method called *A*utomatic *R*epeat re*Q*uest (ARQ) — also sometimes called AMTOR Mode A. Three characters are sent at a time, and the receiving station determines if the mark/space ratio is correct. (Each character contains four mark elements and three space elements.) The receiving station then sends an acknowledgment to the sending station. If the receiving station doesn't acknowledge a block of characters, the transmitting station repeats the same characters until they are confirmed.

AMTOR can also be used to send messages to a group of stations, such as for transmitting information bulletins. In this case, Mode B, or **Forward Error Correction (FEC)** is used. In this case the transmitting station sends each character twice and the receiving station checks each one for the correct mark/space ratio. Both methods provide a reasonable assurance that the received data is correct, although neither is foolproof. In general, AMTOR provides more accurate copy than Baudot RTTY, however. Some newer digital communications modes use other, more effective error correction methods.

Some AMTOR bulletin-transmission systems include a special message header that allows the receiving system to determine if the bulletin has been previously received. This system is commonly known as NAVTEX because it was first developed for use with transmissions to ships at sea. The amateur adaptation of this system is sometimes called AMTEX. You can set up your computer system to copy such bulletin transmissions automatically. When your station receives the message header it checks to see if this message has been previously received. If the message has not been previously received, then your system will copy the message and save it for you to read later. If the message has been received, however, your system will ignore it.

Some operators like to connect the mark and space tones from their TNC or RTTY terminal unit to an oscilloscope. The signals form two crossed ellipses on the scope when they are tuned properly. If you are looking at such a display as you operate RTTY and notice that one of the ellipses suddenly disappears, you can probably conclude that the received signal has gone through selective fading. This happens when one frequency (mark or space) of the received signal fades away momentarily, while the other frequency does not fade.

Packet radio is a form of digital communication that is used most often on the VHF and UHF bands. Packet is also used on HF, but there are a few differences in the way this mode is used on HF. HF packet typically uses an FSK signal with a data rate of 300 **bauds**, while 2-meter packet uses an AFSK signal with a data rate of 1200 bauds. On the HF bands packet operation is limited to only those band segments reserved for CW and data modes by FCC Rules while 2-meter packet is allowed wherever FM operations are allowed. (In addition to these FCC Rules, however, you should also follow the accepted band plans when selecting a packet operating frequency.) Packet radio uses the American National Standard Code for Information Interchange (ASCII) to transfer information between stations, no matter what bands the stations operate on.

When you first turn on your computer system or terminal and terminal node controller (TNC), you should see "CMD:" on your video monitor. The TNC is letting you know it is in "command mode" and is ready to receive instructions from the keyboard.

Many amateurs interested in chasing DX and operating in DX contests find that they enjoy using their local **packet cluster** system and the packet cluster bulletin board. A packet cluster is normally devoted to serving a special interest group, such as DXers. These systems allow many stations to connect to the cluster and communicate with each other and all stations on the cluster. For example, a DX packet cluster might relay "DX spots" to all current stations. These spots include a station call sign, operating frequency, the time the station was spotted and other information. This helps identify stations you may want to try contacting.

[This completes your study of amateur operating procedures for your Amateur Extra class license. Before turning to Chapter 3, you should go to Chapter 10 and review question E2C12. Also study questions E2D01 through E2D06 and E2D09 through E2D11. Review this section if any of those questions give you difficulty.]

Absorption — The loss of energy from an electromagnetic wave as it travels through any material. The energy may be converted to heat or other forms. Absorption usually refers to energy lost as the wave travels through the ionosphere.

Apogee — That point in a satellite's orbit (such as the Moon) when it is farthest from the Earth.

Aurora — A disturbance of the atmosphere at high latitudes, which results from an interaction between electrically charged particles from the sun and the magnetic field of the Earth. Often a display of colored lights is produced, which is visible to those who are close enough to the magnetic-polar regions. Auroras can disrupt HF radio communication and enhance VHF communication. They are classified as visible auroras and radio auroras.

Earth-Moon-Earth (EME) — A method of communicating with other stations by reflecting radio signals off the Moon's surface.

Equinoxes — One of two spots on the orbital path of the Earth around the sun, at which the Earth crosses a horizontal plane extending through the equator of the sun. The *vernal equinox* marks the beginning of spring and the *autumnal equinox* marks the beginning of autumn.

Gray line — A band around the Earth that is the transition region between daylight and darkness.

Gray-line propagation — A generally north-south enhancement of propagation that occurs along the gray line, when D layer absorption is rapidly decreasing at sunset, or has not yet built up around sunrise.

Great circle — An imaginary circle around the surface of the Earth formed by the intersection of the surface with a plane passing through the center of the Earth.

Great-circle path — The shortest distance between two points on the surface of the Earth, which follows the arc of a great circle passing through both points.

K index — A geomagnetic-field measurement that is updated every three hours at Boulder, Colorado. Changes in the K index can be used to indicate HF propagation conditions. Rising values generally indicate disturbed conditions while falling values indicate improving conditions.

Libration fading — A fluttery, rapid fading of EME signals, caused by short-term motion of the Moon's surface relative to an observer on Earth.

Long-path propagation — Propagation between two points on the Earth's surface that follows a path along the great circle between them, but is in a direction opposite from the shortest distance between them.

Meteor — A particle of mineral or metallic material that is in a highly elliptical orbit around the Sun. As the Earth's orbit crosses the orbital path of a meteor, it is attracted by the Earth's gravitational field, and enters the atmosphere. A typical meteor is about the size of a grain of sand.

Meteor-scatter communication — A method of radio communication that uses the ionized trail of a meteor, which has burned up in the Earth's atmosphere, to reflect radio signals back to Earth.

Moonbounce — A common name for EME communication.

Multipath — A fading effect caused by the transmitted signal traveling to the receiving station over more than one path.

Path loss — The total signal loss between transmitting and receiving stations relative to the total radiated signal energy.

Pedersen ray — A high-angle radio wave that penetrates deeper into the F region of the ionosphere, so the wave is bent less than a lower-angle wave, and thus travels for some distance through the F region, returning to Earth at a distance farther than normally expected for single-hop propagation.

Perigee — That point in the orbit of a satellite (such as the Moon) when it is closest to the Earth.

Polarization — A property of an electromagnetic wave that tells whether the electric field of the wave is oriented vertically or horizontally. The polarization sense can change from vertical to horizontal under some conditions, and can even be gradually rotating either in a clockwise (right-hand-circular polarization) or a counterclockwise (left-hand-circular polarization) direction.

Radio horizon — The position at which a direct wave radiated from an antenna becomes tangent to the surface of the Earth. Note that as the wave continues past the horizon, the wave gets higher and higher above the surface.

Selective fading — A variation of radio-wave field intensity that is different over small frequency changes. It may be caused by changes in the material that the wave is traveling through or changes in transmission path, among other things.

Solar wind — Electrically charged particles emitted by the sun, and traveling through space. The wind strength depends on how severe the disturbance on the sun was. These charged particles may have a sudden impact on radio communications when they arrive at the atmosphere of the Earth.

Terminator — A band around the Earth that separates night from day.

Transequatorial propagation — A form of F-layer ionospheric propagation, in which signals of higher frequency than the expected MUF are propagated across the Earth's magnetic equator.

Tropospheric ducting — A type of radio-wave propagation whereby the VHF communications range is greatly extended. Certain weather conditions cause portions of the troposphere to act like a duct or waveguide for the radio signals.

RADIO-WAVE PROPAGATION

By the time you are ready to study for an Amateur Extra class license, you should have a pretty good understanding of the basic modes of propagation for radio waves. Now it is time to learn about ways to take advantage of some exotic propagation methods. There are two groups of questions in the Element 4 question pool about the Radio-Wave Propagation subelement. Your exam will include two questions, selected from those two groups. The Syllabus topics for this subelement are:

E3A   Earth-Moon-Earth (EME or moonbounce) communications; meteor scatter
E3B   Transequatorial; long path; gray line
E3C   Auroral propagation; selective fading; radio-path horizon; take off angle over flat or sloping terrain; Earth effects on propagation

When you have studied the information in each section of this chapter, use the examination questions in Chapter 10 to check your understanding of the material. If you are unable to answer a question correctly, go back and review the appropriate part of this chapter.

RADIO PATH HORIZON

In the early days of VHF amateur communications, it was generally believed that space-wave communications depended on direct line-of-sight paths between the communicating station antennas. After some experiments with good equipment and antennas, however, it became clear that radio waves are bent or scattered in several ways, and that reliable VHF and UHF communications are possible with stations beyond the visual horizon. The farthest point to which space waves will travel directly is called the **radio horizon**.

Under normal conditions, the structure of the atmosphere near the Earth causes radio waves to bend into a curved path that keeps them nearer to the Earth than true straight-line travel would. This bending of the radio waves is why the distance to the radio horizon exceeds the distance to the visual, or geometric, horizon. See **Figure 3-1**. An antenna that is on a high hill or tall building well above any surrounding obstructions has a much farther radio horizon than an antenna located in a valley or shadowed by other obstructions.

The point at the horizon is assumed to be on the ground. If the receiving antenna is also elevated, the maximum space-wave distance between the two anten-

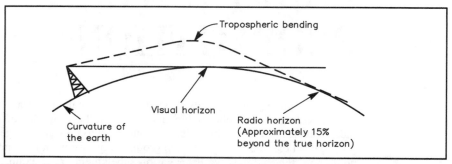

Figure 3-1 — Under normal conditions, tropospheric bending causes VHF and UHF radio waves to be returned to Earth beyond the visible horizon.

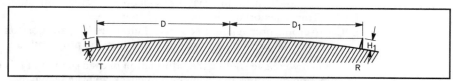

Figure 3-2 — The distance, D, to the radio horizon from an antenna of height H is given by the formula in the text. The maximum distance over which two stations may communicate by space wave is equal to the sum of their distances to the horizon.

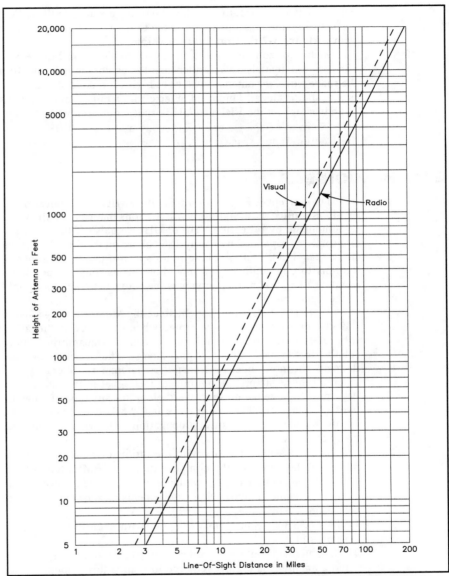

Figure 3-3 — Distance to the radio horizon from an antenna of given height above average terrain is indicated by the solid line. The broken line indicates the distance to the visual, or geometric, horizon. The radio horizon is approximately 15% farther than the visual horizon.

nas is equal to D + D1; that is, the sum of the distance to the radio horizon from the transmitting antenna plus the distance to the radio horizon from the receiving antenna. **Figure 3-2** illustrates this principle.

Radio-horizon distances are shown graphically in **Figure 3-3**. The radio horizon is approximately 15% farther than the geometric horizon. To make best use of the space wave, the antenna must be as high as possible above the surroundings. This is why stations located high and in the clear on hills or mountaintops have a substantial advantage on the VHF and UHF bands, compared with stations in lower areas.

[Turn to Chapter 10 now and study question E3C08. Review the material in this section if necessary.]

EARTH-MOON-EARTH

The concept of **Earth-Moon-Earth (EME)** communications, popularly known as **moonbounce**, is straightforward: Stations that can simultaneously see the Moon communicate by reflecting VHF or UHF signals off the lunar surface. Those stations may be separated by nearly 180° of arc on the Earth's surface — a distance of more than 11,000 miles. There is no specific maximum separation between two stations to communicate via moonbounce, as long as they have a mutual lunar window. (This means the moon must be above the "radio horizon" for both stations at the same time.)

There is a drawback, though; since the Moon's average distance from Earth is 239,000 miles, path losses are huge when compared to "local" VHF paths. **Path loss** refers to the total signal loss between the transmitting and receiving stations relative to the total radiated signal energy. Thus, each station on an EME circuit demands the most out of the transmitter, antenna, receiver and operator skills. Even when all those factors are optimized, the signal in the headphones may be barely perceptible above the noise. Nevertheless, for any type of amateur communication over a distance of 500 miles or more at 432 MHz, for example, moonbounce comes out the winner over terrestrial propagation paths when various factors are weighed on a balance sheet.

EME presents amateurs with the ultimate challenge in radio system performance. Today, most of the components for an EME station on 144, 220, 432 or 1296 MHz are commercially available. Whether one chooses to buy or build station equipment, some system design requirements must be met, because this is extremely weak-signal work.

1) CW or SSB transmissions work best, with as close to the maximum power level as possible.

2) The antenna should have at least 20 dB gain over a dipole. An array of four or more antennas is usually required to realize this amount of gain. Chapter 9 has more information about antenna-system gain.

3) Antenna rotators are needed for both azimuth and elevation. Since the half-power beamwidth of a high-gain antenna is quite sharp, the rotators must have an appropriate accuracy.

4) Transmission-line losses should be held to a minimum.

5) The receiving system should have a very low noise figure.

A low-noise receiving setup is essential for successful EME work. Since many of the signals to be copied on EME are barely, but not always, out of the noise, a low-noise-figure receiver is a must. The mark to shoot for at 144 MHz is something under 0.5 dB, as the cosmic noise will then be the limiting factor in the system. Noise figures of this level are relatively easy to achieve with inexpensive modern devices. As low a noise figure as can be attained will be usable at 432 and 1296 MHz. Noise figures on the order of 0.5 dB are possible with amplifiers that use GaAsFET devices.

EME Scheduling

The best days to schedule an EME contact are usually when the Moon is at **perigee** (closest to the Earth) since the path loss is typically 2 dB less than when the Moon is at **apogee** (farthest from the Earth). The Moon's perigee and apogee dates may be determined from publications such as *The Nautical Almanac* by inspecting the section of the tables headed "S.D." (*semidiameter* of the Moon in minutes of arc). An S.D. of 16.53 equates to an approximate Earth-to-Moon distance of

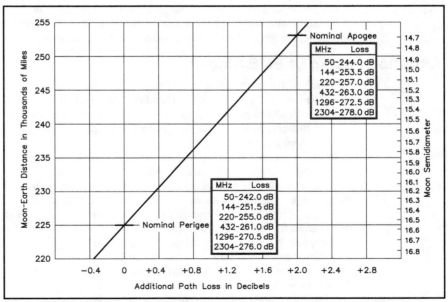

Figure 3-4—Variations in EME path loss can be determined from this graph. S.D. refers to the semidiameter of the Moon, which is indicated for each day of the year in *The Nautical Almanac*.

225,000 miles, typical perigee, and an S.D. of 14.7 to an approximate distance of 252,500 miles, typical apogee. Once you know the semidiameter of the Moon for a given date, locate the value on the graph of **Figure 3-4**. Determine the EME path loss in decibels by reading the loss at nominal perigee for the frequency of interest from the chart near the bottom of the graph. Then add the additional path loss value as read from the graph.

The Moon's orbit is slightly elliptical. Hence, the day-to-day path-loss changes at apogee and perigee are minor. The greatest changes take place at the time when the Moon is traversing between apogee and perigee. Several other factors in addition to the path losses must be considered for optimum scheduling.

One or two days will be unusable near the time of a new Moon, since the Sun is behind the Moon and will cause increased Sun-noise pickup. This noise will mask weak signals. Therefore, avoid schedules when the Moon is within 10° of the Sun (and farther if your antenna has a wide beamwidth or strong side lobes). The Moon's orbit follows a cycle of 18 to 19 years, so the relationships between perigee and new Moon will not be the same from one year to the next.

High Moon declinations and high antenna elevation angles should yield best results. By contrast, low Moon declinations and low aiming elevations generally produce poor results and should be avoided if possible. Generally, low elevation angles increase antenna-noise pickup and increase tropospheric absorption, especially above 420 MHz, where the galactic noise is very low. This situation cannot be avoided when one station is unable to elevate the antenna above the horizon or when there is great terrestrial distance between stations.

Libration Fading of EME Signals

One of the most troublesome aspects of receiving a moonbounce signal, besides the enormous path loss and Faraday-rotation fading, is **libration** (pronounced lie-bray-shun) **fading**. **Libration fading** of an EME signal is characterized in general as fluttery, rapid, irregular fading not unlike that observed in tropospheric-scatter propagation. Fading can be very deep, 20 dB or more, and the maximum fading will depend on the operating frequency. At 1296 MHz the maximum fading rate is about 10 Hz, and is directly proportional to frequency.

On a weak CW EME signal, libration fading gives the impression of a randomly keyed signal. In fact on very slow CW telegraphy the effect is as though the keying is being done at a much faster speed. On very weak signals only the peaks of libration fading are heard in the form of occasional short bursts or "pings."

Figure 3-5 shows samples of a typical EME echo signal at 1296 MHz. These recordings show the wild fading characteristics with sufficient signal-to-noise ratio to record the deep fades. Circular polarization was used to eliminate Faraday fading; thus, these recordings are of libration fading only. The recording bandwidth was limited to about 40 Hz to minimize the higher sideband frequency components of libration fading that exist but are much smaller in amplitude. In the recordings shown by Figure 3-5, the average signal-return level computed from path loss and

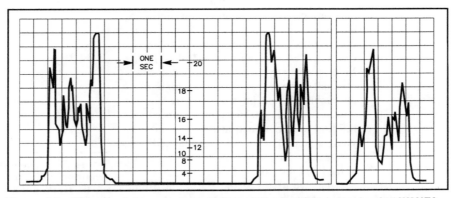

Figure 3-5—This graph is a chart recording of Moon echoes received at W2NFA.

mean reflection coefficient of the Moon is at about the + 15 dB S/N level.

It is clear that enhancement of echoes far in excess of this average level is observed. This point should be kept clearly in mind when attempting to obtain echoes or receive EME signals with marginal equipment. The probability of hearing an occasional peak is quite good since random enhancement by as much as 10 dB is possible. Under these conditions, however, the amount of useful information that can be copied will be near zero. Enthusiastic newcomers to EME communications will be frustrated by this effect. They hear the signal strong enough to copy on peaks but they can't copy enough to make sense.

What causes libration fading? Very simply, multipath scattering of the radio waves from the very large (2000-mile diameter) and rough Moon surface combined with the relative short-term motion between Earth and Moon. These short-term oscillations in the apparent aspect of the Moon relative to Earth are called librations.

To understand these effects, assume first that the Earth and Moon are stationary (no libration) and that a plane wave front arrives at the Moon from your Earth-bound station as shown in **Figure 3-6A**.

The reflected wave shown in Figure 3-6B consists of many scattered contributions from the rough Moon surface. It is perhaps easier to visualize the process as if the scattering were from many small individual flat mirrors on the Moon

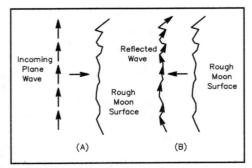

Figure 3-6—This diagram shows how the rough surface of the Moon reflects a radio wave.

that reflect small portions (amplitudes) of the incident wave energy in different directions (paths) and with different path lengths (phase). Those signals reflected from the Moon arrive at your antenna as a collection of small wave fronts (field vectors) of various amplitudes and phases. All these returned waves (we can consider their number to be infinite) are combined at the feed point of your antenna. The level of the final addition, as measured by a receiver, can have any value from zero to some maximum. Remember that we assumed the Earth and Moon were not moving with respect to one another.

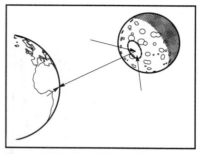

Figure 3-7—The Moon appears to "wander" in its orbit about the Earth. Thus, a fixed marker on the Moon's surface will appear to move about in a circular area.

Consider now that the Earth and Moon are moving relative to each other (as they are in nature), so the incident radio wave "sees" a slightly different surface of the Moon from moment to moment. Since the lunar surface is very irregular, the reflected wave will be equally irregular, changing in amplitude and phase from moment to moment. The result is a number of continuously varying multipath signals at your receiving antenna feed point, which produces the effect called libration fading of the Moon-reflected signal.

The term libration is used to describe small irregularities in the apparent movement of celestial bodies relative to an observer on Earth. Lunar librations result from a variety of factors, which aren't important for our discussion here. If you are interested in learning more about librations of the Moon, many astronomy texts have detailed discussions.

These librations combine to produce a very slight rocking motion of the Moon's surface with respect to an observer on Earth. This rocking motion can be visualized as follows: Imagine a marker on the surface of the Moon at the center of the Moon disc, which is the point closest to the observer, as shown in **Figure 3-7**. Over time, we will observe that this marker wanders around within a small area. This means the surface of the Moon as seen from the Earth is not quite fixed but changes slightly as different areas of the periphery are exposed because of the rocking motion. Moon libration is very slow. Although the libration motions are very small and slow, the larger surface area of the Moon has an infinite number of scattering points, each with a very small area. This means that even slight movements can alter the return multipath echo by a significant amount.

Frequencies

EME contacts are generally made by prearranged schedule, although some contacts are made at random. Many stations, especially those with marginal capability, prefer to set up a specific time and frequency in advance so that they will have a better chance of finding each other. The larger stations, especially on 144 and 432 MHz where there is a good amount of activity, often call CQ during evenings and weekends when the Moon is at perigee, and listen for random replies. Most of the work on 220, 1296 and 2304 MHz, where activity is light, is done by schedule.

An EME net meets on weekends on 14.345 MHz for the purpose of arranging schedules and exchanging pertinent information. Those operating EME at 432 MHz and above meet at 1600 UTC, followed by the 144-MHz operators at 1700 UTC. Both nets carry information on 220-MHz EME operation.

Most amateur EME work on 144 and 220 MHz takes place near the low edge of the band. You may find 2-meter EME activity anywhere between 144.000 and 144.100 MHz, although most of the activity tends to concentrate in the lower 50 kHz except during peak hours. Generally, random activity and CQ calling take place in the lower 10 kHz or so, and schedules are run higher in the band.

On the 70-cm band you may find EME activity anywhere from 432.000 to 432.100 MHz. Formal schedules (that is, schedules arranged well in advance) are run on 432.000, 432.025 and 432.030 MHz. Other schedules are normally run on 432.035 and up to 432.070. For this band, the EME random-calling frequency is 432.010, with random activity spread out between 432.005 and 432.020. Random SSB CQ calling is at 432.015. Terrestrial activity is centered on 432.100 and is, by agreement, limited to 432.075 and above in North America.

Moving up in frequency, formal schedules are run on 1296.000 and 1296.025 MHz. The EME random calling frequency is 1296.010 with random activity spread out between 1296.005 and 1296.020. Terrestrial activity is centered at 1296.100. There is some EME activity on 2304 MHz. Specific frequencies are dictated by equipment availability and are arranged by the stations involved. For EME SSB contacts on 144 and 432 MHz, contact is usually established on CW, and then the stations move up 100 kHz from the CW frequency. (This method was adopted because of the US requirement for CW only below 144.1 MHz.)

Of course, it is obvious that as the number of stations on EME increases, the frequency spread must become greater. Since the Moon is in convenient locations only a few days out of the month, and only a certain number of stations can be scheduled for EME during a given evening, the answer will be in the use of simultaneous schedules, spaced a few kilohertz apart. The time may not be too far away — QRM has already been experienced on each of our three most active EME frequencies.

EME Operating Techniques

EME operators have agreed to a standard calling procedure. For 144 MHz contacts, a two-minute calling sequence is used. The eastern-most station transmits first, for two full minutes, and then that station receives for two full minutes while the western-most station transmits. For 432 MHz EME contacts, operators use two and a half minute calling sequences. Again, the eastern-most station transmits first, for two and a half minutes. Then the eastern station listens for the next two and a half minutes, while the western-most station transmits.

[Now turn to Chapter 10 and study examination questions E3A01 through E3A08. Review this section as needed.]

METEOR-SCATTER PROPAGATION

Meteors are particles of mineral or metallic matter that travel in highly elliptical orbits about the Sun. Most of these are microscopic in size. Every day hundreds of millions of these meteors enter the Earth's atmosphere. Drawn by the Earth's gravitational field, they attain speeds from 6 to 60 mi/s (22,000 to 220,000 mi/h).

As a meteor speeds through the upper atmosphere, it begins to burn or vaporize as it collides with air molecules. This action creates heat and light and leaves a trail of free electrons and positively charged ions behind as the meteor races along its parabolic path. Trail size is directly dependent on meteor size and speed. A typical meteor is the size of a grain of sand. A particle of this size creates a trail about 3 feet in diameter and 12 miles or longer, depending on speed.

The duration of this meteor-produced ionization is directly related to electron density. Ionized air molecules contact and recombine with free electrons over time, gradually lowering the electron density until it returns to its previous state.

Radio waves can be reflected as they encounter the ionized trail of a meteor. The ability of a meteor trail to reflect radio signals depends on electron density — greater density causing greater reflecting ability and reflection at higher frequencies. The electron density in a typical meteor trail will strongly affect radio waves between 28 and 148 MHz. Signal frequencies as low as 20 MHz and as high as 432 MHz will be usable for meteor-scatter communication at times.

Meteor trails are formed at approximately the altitude of the ionospheric E layer, 50 to 75 miles above the Earth. That means the range for meteor-scatter propa-

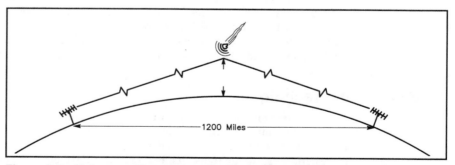

Figure 3-8—Meteor-scatter communication makes extended-range VHF communications possible.

gation is about the same as for single-hop E (or sporadic-E) skip, a maximum of approximately 1200 miles, as **Figure 3-8** shows.

The propagation potential of a "meteor trail" is highly frequency dependent. For example, consider the result of a relatively large (about the size of a peanut) meteor entering the Earth's atmosphere. Using the ionized trail of that meteor, no 432-MHz propagation is possible. At 220 MHz, moderately strong signals are heard over a 1200-mile path for perhaps 12 seconds. The signals gradually fade into the noise over the next four or five seconds. At 144 MHz, the same conditions might result in very strong signals for 20 seconds that slowly fade into the noise during the next 20 seconds. At 28 MHz, propagation might well last for a couple of minutes.

Meteors the size of a peanut are relatively rare; most are much smaller. As a result, the typical meteor burst is much shorter than that cited in the example.

Meteor Showers

At certain times of the year, the Earth encounters greatly increased numbers of meteors. Great swarms of meteors, probably the remnants of a comet, orbit the Sun. Each year the Earth passes through these swarms causing the so-called meteor showers. These showers greatly enhance meteor-scatter communications at VHF. The degree of enhancement depends on the time of day, shower intensity and the frequency in use. The largest meteor showers of the year are the Perseids and the Geminids. The Perseids appear to come from the constellation Perseus; this shower occurs in August. December 11 and 12 is when the Geminids peak; these seem to come from Gemini. **Table 3-1** is a partial list of meteor showers throughout the year.

Meteor-scatter communication (sometimes called meteor-burst communication) is best between midnight and dawn. The part of the Earth that is between those hours, local time, is always on the leading edge as the Earth travels along its orbit. It is at that leading edge where most meteors enter the Earth's atmosphere.

Operating Hints

The secret to successful meteor-burst communication is short transmissions. To call CQ, for example, say (or send) CQ followed by your call sign, repeated two or three times. To answer a

Table 3-1
Major Meteor Showers

Date	Name
January 3-5	Quadrantids
April 19-23	Lyrids
*May 19-21	Cetids
*June 4-6	Perseids
*June 8	Arietids
*June 30-July 2	Taurids
July 26-31	Aquarids
July 27-August 14	Perseids
October 18-23	Orionids
October 26-November 16	Taurids
November 14-18	Leonids
December 10-14	Geminids
December 22	Ursids

All showers occur in the evening except those marked (*), which are daytime showers. Evening showers begin at approximately 2300 local standard time, daylight showers at approximately 0500 local standard time.

CQ, give the other station's call once followed by your call phonetically. Example: "CQ from Whiskey Bravo Five Lima Uniform Alfa." "WB5LUA this is Whiskey One Juliett Romeo." WB5LUA responds: "W1JR 59 Texas." W1JR immediately responds: "Roger 59 Massachusetts." WB5LUA says: "Roger QRZ from WB5LUA." The entire QSO with information exchanged and confirmed in both directions lasted only 12 seconds!

A single meteor may produce a strong enough path to sustain communication long enough to complete a short QSO. At other times multiple bursts are needed to complete the QSO, especially at higher frequencies. Remember, the key is short, concise transmissions. Do not repeat information unnecessarily; that is a waste of time and propagation.

The accepted convention for transmission timing breaks each minute into four 15-second periods. The station at the western end of the path transmits during the first and third period of each minute. During the second and fourth 15-second periods, the eastern station transmits.

Meteor-burst propagation and packet radio seem to be made for each other. High-speed data transfers can take place even during short bursts. Commercial and military meteor-scatter communication using packet radio takes place on the 40 to 50-MHz band. Some experimentation has been done by amateurs using this combination, but more can be learned by further work in this promising field.

[Now turn to Chapter 10 and study examination questions E3A09 through E3A11. Review this section as needed.]

TRANSEQUATORIAL PROPAGATION

Transequatorial propagation (TE) is a form of F-layer ionospheric propagation that was discovered by amateurs in the late 1940s. Amateurs on all continents reported the phenomenon almost simultaneously on various north-south paths. Those amateurs were communicating successfully on 50 MHz during the evening hours. At that time, the predicted MUF was around 40 MHz for daylight hours, when the MUF should have been maximum. Since that time, research carried out by amateurs has shown that the TE mode works on 144 MHz and even to some degree at 432 MHz. The enhanced propagation occurs between stations approximately the same distance north and south of the Earth's magnetic equator.

The ionosphere is directly influenced by solar radiation. You might expect the ionization density to be a maximum over the equator around the vernal (spring) and autumnal (fall) equinoxes. In fact, at the equinoxes there is not one area of maximum ionization but two. These maxima form in the morning, are well established by noon and last until after midnight. The high-density-ionization regions form approximately between 10° and 15° on either side of the Earth's magnetic equator — not the geographic equator. The system does not move north and south with the seasons. As the relative position of the Sun moves away from the equator,

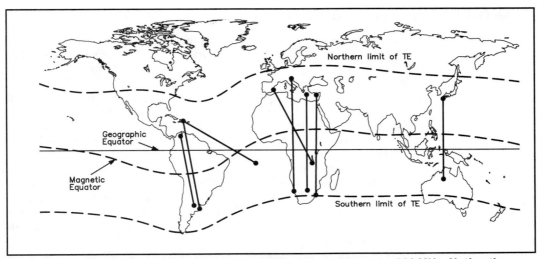

Figure 3-9—This world map shows TE paths worked by amateurs on 144 MHz. Notice the symmetrical distribution of stations with respect to the magnetic equator.

the ionization levels in the Northern and Southern Hemispheres become unbalanced. So the best time of year to look for transequatorial propagation is around March 21 and September 21.

TE may provide very strong signals on the HF bands during the afternoon and early evening, so these are the best times to look for this propagation mode. Later at night, and sometimes in the early morning as well, you will only hear weak and watery signals arriving by TE.

As the signal frequency increases, the communication zones become more restricted to those equidistant from, and perpendicular to, the magnetic equator (**Figure 3-9**). Further, the duration of the opening tends to be shorter and closer to 8 PM local time. The rate of flutter fading and the degree of frequency spreading increase with signal frequency. TE range extends to approximately 5000 miles — 2500 miles on each side of the magnetic equator.

The MUF for TE will be higher during solar activity peaks. The best conditions exist when the Earth's magnetic field is quiet. Seasonal variations favor the periods around the equinoxes.

[Now study examination questions E3B01 through E3B03. Review this section as needed.]

LONG-PATH PROPAGATION

Propagation between any two points on the Earth's surface is usually by the shortest direct route, which is a **great-circle path** between the two points. A **great circle** is an imaginary line drawn around the Earth, formed by a plane passing through the center of the Earth. The diameter of a great circle is equal to the diameter of the Earth. You can find a great-circle path between two points by stretching a string tightly between those two points on a globe. If a rubber band is used to mark the entire great circle, by stretching it around the globe, then you can see that there are really two great-circle paths. See **Figure 3-10**.

One of those paths will usually be longer than the other, and the longer path may be useful for communications when conditions are favorable. Of course you must have a beam antenna that you can point in the desired direction to make effective use of **long-path propagation**. The station at the other end of the path must also point his or her antenna in the long-path direction to your station to make the best use of this propagation.

The long and short-path directions always differ by 180°. Since the circumference of the Earth is 24,000 miles, short-path propagation is always over a path length of less than 12,000 miles. The long-path distance is 24,000 miles minus the short-path distance for a specific communications circuit. For example, the distance from Gordon, Pennsylvania, to the Canary Islands is 3510 miles at a bearing of 85°. The long-path circuit would be a distance of 20,490 miles at a bearing of 265°. (If you want to know how to perform distance and bearing calculations, see Chapter 4 of *The ARRL Operating Manual*.)

Suppose you are in North America, receiving signals from a European station with your beam antenna pointed at that station. If you notice an echo on the signals, you should turn your beam antenna to a direction that is 180° from the direct bearing to that station. If the signals from that station are significantly stronger when you turn your antenna 180°, you are receiving the station via the long path. The reason for the echo is that you are receiving two signals. One signal is coming in over the direct path and the second one is coming in over the long path, which introduces a slight delay because the signal must travel a significantly longer distance.

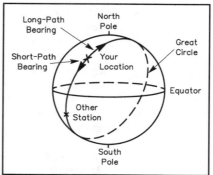

Figure 3-10—This sketch of the Earth shows a great circle drawn between two stations. The short-path and long-path bearings are shown from the Northern Hemisphere station.

Long-path propagation can occur on any band that provides ionospheric propagation, so you may hear this signal enhancement on the 160 to 6-meter bands. You can consistently make use of long-path enhancement on the 20-meter band. All it takes is a modest beam antenna with a relatively high gain compared to a dipole, such as a three-element Yagi or two-element Quad.

Normally, radio signals propagate most effectively along the great-circle path that provides the least absorption. Since the D layer of the ionosphere is most responsible for signal absorption, the amount of sunlight that the signal must travel through helps determine how strong the signals will be. So at times, your signals will be stronger at the receiving stations when the signals travel over the nighttime side of the Earth.

For paths less than about 6000 miles, the short-path signal will almost always be stronger, because of the increased losses caused by multiple-hop ground-reflection losses and ionospheric absorption over the long path. When the short path is more than 6000 miles, however, long-path propagation will usually be observed either along the "**gray line**" between darkness and light, or over the nighttime side of the Earth.

While signals will generally travel best along the great-circle path, there is often some deviation from the exact predicted beam heading. Ionospheric conditions can cause radio signals to reflect or refract in unexpected ways. It's always a good idea to rock your rotator control back and forth around the expected beam heading and listen for the peak signals.

[Turn to Chapter 10 and study examination questions E3B04 through E3B07. Review this section if you have any difficulty.]

GRAY-LINE PROPAGATION

The **gray line** (sometimes called the "twilight zone") is a band around the Earth that separates daylight from darkness. Astronomers call this the **terminator**. It is a somewhat fuzzy region because the Earth's atmosphere tends to diffuse the light into the darkness. **Figure 3-11** illustrates the gray line around the Earth. Notice that on one side of the Earth, the gray line is coming into daylight (sunrise), and on the other side it is coming into darkness (sunset).

Propagation along the gray line is very efficient. One major reason for this is that the D-layer, which absorbs HF signals, disappears rapidly on the sunset side of the gray line, and it has not yet built up on the sunrise side.

Look for **gray-line propagation** at twilight, around sunrise and sunset. Contacts up to 8,000 to 10,000 miles are possible using this propagation enhancement. The three or four lowest-frequency amateur bands (160, 80, 40 and 30 meters) are the most likely to experience gray-line enhancement because they are the most affected by D-layer absorption.

The gray line runs generally north and south, but varies as much as 23° either

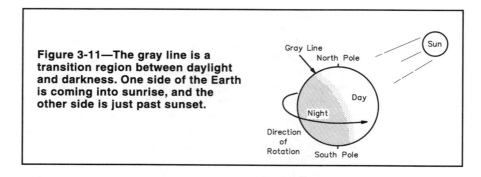

Figure 3-11—The gray line is a transition region between daylight and darkness. One side of the Earth is coming into sunrise, and the other side is just past sunset.

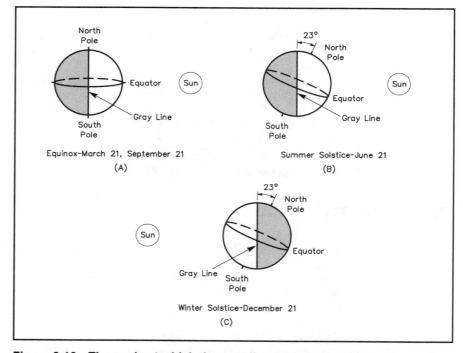

Figure 3-12—The angle at which the gray line crosses the equator depends on the time of year, and on whether it is at sunset or sunrise. Part A shows that the gray line is perpendicular to the equator twice a year, at the vernal equinox and the autumnal equinox. B indicates that the north pole is tilted toward the Sun at a 23° angle at the summer solstice, and C shows that the north pole is tilted away from the Sun at a 23° angle at the winter solstice. These changes in the Sun's position relative to the poles cause the gray line to shift position.

side of the north-south lines. This variation is caused by the tilt of the Earth's axis relative to its orbital plane around the sun. The gray line will be exactly north and south at the **equinoxes** (March 21 and September 21). On the first day of summer in the Northern Hemisphere, June 21, it is tilted a maximum of 23° one way, and on the first day of winter, December 21, it is tilted a maximum of 23° the other way. **Figure 3-12** illustrates the changing gray-line tilt. The tilt angle will be between these extremes during the rest of the year. Knowledge of the tilt angle will be helpful in determining what directions are likely to provide gray-line propagation on a particular day.

[Now turn to Chapter 10 and study examination questions E3B08 through E3B11. Review this section as needed.]

AURORAL PROPAGATION

Auroral propagation occurs when VHF radio waves are reflected from ionization associated with an auroral curtain. It is a VHF and UHF propagation mode that allows contacts up to about 1400 miles. Auroral propagation occurs for stations near the northern and southern polar regions, but the discussion here is limited to auroral propagation in the Northern Hemisphere.

Aurora results from a large-scale interaction between the magnetic field of the Earth and electrically charged particles. During times of enhanced solar activity, electrically charged particles are ejected from the surface of the sun. These particles form a **solar wind**, which travels through space. If this solar wind travels toward Earth, then the charged particles interact with the magnetic field around the Earth. A visible aurora, often called the northern lights, or aurora borealis, is caused by the collision of these solar-wind particles with oxygen and nitrogen molecules in the upper atmosphere.

When the oxygen and nitrogen molecules are struck by the electrically charged particles in the solar wind, they are ionized. When the electrons that were knocked loose recombine with the molecules, light is produced. The extent of the ionization determines how bright the aurora will appear. At times, the ionization is so strong that it is able to reflect radio signals with frequencies above about 20 MHz. This ionization occurs at an altitude of about 70 miles, very near the E layer of the ionosphere. Not all auroral activity is intense enough to reflect radio signals, so a distinction is made between a visible aurora and a radio aurora.

The number of auroras (both visible and radio) varies with geomagnetic latitude. Generally, auroral propagation is available only to stations in the northern states, but, on occasion, extremely intense auroras reflect signals from stations as far south as the Carolinas. Auroral propagation is most common for stations in the northeastern states and adjacent areas of Canada, which are closest to the north magnetic pole. This mode is rare below about 32° north latitude in the southeast and about 38° to 40° N in the southwest. See **Figure 3-13**.

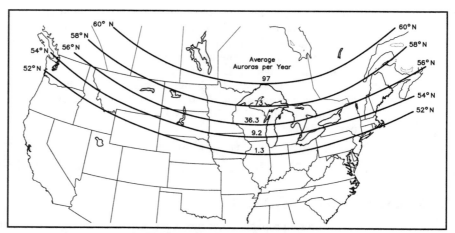

Figure 3-13 — The possibility of auroral propagation decreases as distance from the geomagnetic North Pole increases.

The number and distribution of auroras are related to the solar cycle. Auroras occur most often during sunspot peaks, but the peak of the auroral cycle appears to lag the solar-cycle peak by about two years. Intense auroras can, however, occur at any point in the solar cycle.

Auroras also follow seasonal patterns. Although they may occur at any time, they are most common around the **equinoxes** in March and September. Auroral propagation is most often observed in the late afternoon and early evening hours, and it usually lasts from a few minutes to many hours. Often, it will disappear for a few hours and reappear around midnight. Major auroras often start in the early afternoon and last until early morning the next day.

Using Aurora

Most common on 10, 6 and 2 meters, some auroral work has been done on 222 and 432 MHz. The number and duration of openings decreases rapidly as the operating frequency rises.

The reflecting properties of an aurora vary rapidly, so signals received via this mode are badly distorted by multipath effects. CW is the most effective mode for auroral work. CW signals have a fluttery tone. The tone is distorted, and is most often a buzzing sound rather than a pure tone. For this reason, auroral propagation is often called the "buzz mode." SSB is usable for 6-meter auroral work if signals are strong; voices are often intelligible if the operator speaks slowly and distinctly. SSB is rarely usable at 2 meters or higher frequencies.

In addition to scattering radio signals, auroras have other effects on worldwide radio propagation. Communication below 20 MHz is disrupted in high latitudes, primarily by absorption, and is especially noticeable over polar and near-polar paths. Signals on the AM broadcast band through the 40-meter band late in the afternoon may become weak and "watery" sounding. The 20-meter band may close down altogether. At the same time, the MUF in equatorial regions may temporarily rise dramatically, providing transequatorial paths at frequencies as high as 50 MHz.

All stations point their antennas north during the aurora, and, in effect, "bounce" their signals off the auroral zone. The optimum antenna heading varies with the position of the aurora and may change rapidly, just as the visible aurora does. Constant probing with the antenna is recommended to peak signals, especially if the beamwidth is narrow. Usually, an eastern station will work the greatest distance to the west by aiming as far west of north as possible. The opposite applies for western stations working east. This does not always follow, however, so you should keep your antenna moving to find the best heading. See **Figure 3-14.**

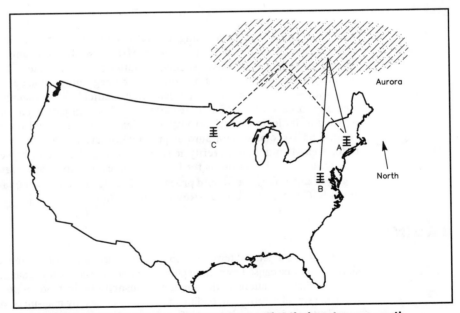

Figure 3-14 — To work the aurora, stations point their antennas north. Station A may have to beam west of north to work station C.

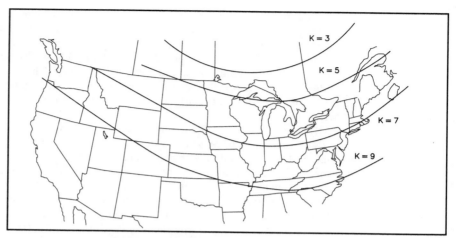

Figure 3-15 — As the intensity of auroral activity increases, stations farther south are able to take advantage of it. The numbers refer to the K index.

You can observe developing auroral conditions by monitoring signals in the region between the broadcast band and 5 MHz or so. If, for example, signals in the 75-meter band begin to waver suddenly (flutter or sound "watery") in the afternoon or early evening hours, a radio aurora may be beginning. Since auroras are associated with solar disturbances, you can often predict one by listening to the WWV Geo alert broadcasts at 18 minutes after each hour. In particular, the **K index** may be used to indicate auroral activity. K-index values of 3, and rising, indicate that conditions associated with auroral propagation are present in the Boulder, Colorado, area. Timing and severity may be different elsewhere, however. Maximum occurrence of radio aurora is for K-index values of 7 to 9. See **Figure 3-15**.

[At this point, you should proceed to Chapter 10 and study examination questions E3C01 through E3C04. Review this section as needed.]

FADING

Fading is a general term used to describe variations in the strength of a received signal. It may be caused by natural phenomena such as constantly changing ionospheric-layer heights, variations in the amount of **absorption**, or random **polarization** shifts when the signal is refracted. Fading may also be caused by man-made phenomena such as reflections from passing aircraft and ionospheric disturbances caused by exhaust from large rocket engines.

Multipath

A common cause of fading is an effect known as **multipath**. Several components of the same transmitted signal may arrive at the receiving antenna from different directions, and the phase relationships between the signals may cancel or reinforce each other. Multipath fading is responsible for the effect known as "picket fencing" in VHF communications, when signals from a mobile station have a rapid fluttering quality. This fluttering is caused by the change in the paths taken by the transmitted signal to reach the receiving station as the mobile station moves. This effect is illustrated in **Figure 3-16**.

Multipath effects can occur whenever the transmitted signal follows more than one path to the receiving station. Some examples of this with HF propagation would be if part of the signal goes through the ionosphere and part follows a ground-wave path, or if the signal is split in the ionosphere and travels through different layers before reaching the receiving station. It is even possible to experience multipath fading if part of the signal follows the long path around the Earth to reach the receiver, while part of the signal follows the direct short-path route. When the transmitted signal reaches the receiver over several paths, the end result is a variable-strength signal.

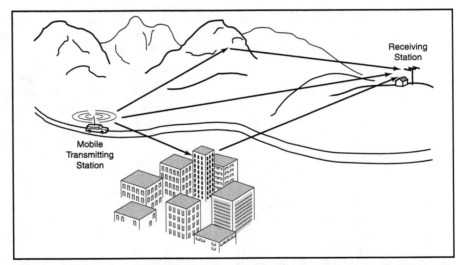

Figure 3-16 — If a signal travels from a transmitter to a receiver over several different paths, the signals may arrive at the receiver slightly out of phase. The out-of-phase signals alternately cancel and reinforce each other, and the result is a fading signal. This effect is known as multipath fading.

Selective Fading

Selective fading is a type of fading that occurs when the wave path from a transmitting station to a receiving station varies with very small changes in frequency. It is possible for components of the same signal that are only a few kilohertz apart (such as the carrier and the sidebands in an AM signal) to be acted upon differently by the ionosphere, causing modulation sidebands to arrive at the receiver out of phase. Selective fading occurs because of phase differences between radio-wave components of the same transmission, as experienced at the receiving station. The result is distortion that may range from mild to severe.

Wideband signals, such as FM and double-sideband AM, suffer the most from selective fading. The sidebands may have different fading rates from each other or from the carrier. Distortion from selective fading is especially bad when the carrier of an FM or AM signal fades while the sidebands do not. In general, the distortion from selective fading is more pronounced for signals with wider bandwidths. It is worse with FM than it is with AM. SSB and CW signals, which have a narrower bandwidth, are affected less by selective fading.

[Now study examination questions E3C05 through E3C07 in Chapter 10. Review this section as needed.]

EARTH EFFECTS ON PROPAGATION

The space wave goes essentially in a straight line between the transmitter and the receiver. Antennas that are low-angle radiators (that is, antennas that concentrate the energy toward the horizon) are best. Energy radiated at angles above the horizon may pass over the receiving antenna.

In general, the radiation takeoff angle from a Yagi antenna with horizontally mounted elements decreases as the antenna height increases above flat ground. If you can raise the height of your antenna, the takeoff angle will decrease.

If the ground under your antenna is not flat, you may be able to use the sloping ground to your advantage. For example, a Yagi antenna mounted on the side of a hill, and pointed *away* from the hill, will have a lower radiation takeoff angle. The steeper the slope, the lower the takeoff angle will be. See **Figure 3-17**. Of course if the antenna is pointed *toward* the hill, the takeoff angle will be increased.

The polarization of both the receiving and transmitting antennas should be the same for VHF and UHF operation because the polarization of a space wave remains constant as it travels. There may be as much as 20 dB of signal loss between two stations that are using antennas with opposite polarizations.

As a radio wave travels in space, it collides with air molecules and other particles. When it collides with these particles, the radio wave gives up some energy. This is why there is a limit to distances that may be covered by space-wave communications.

VHF propagation is usually limited to distances of approximately 500 miles. This is the normal limit for stations using high-gain antennas, high power and

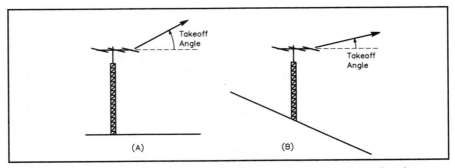

Figure 3-17 — Part A illustrates the takeoff angle for radio waves leaving a Yagi antenna with horizontal elements over flat ground. Higher antenna elevations result in smaller takeoff angles. Part B shows the takeoff angle for a similar antenna over sloping ground. For steeper slopes away from the front of the antenna, the takeoff angle gets smaller.

sensitive receivers. At times, however, VHF communications are possible with stations up to 2000 or more miles away. Certain weather conditions cause ducts in the troposphere, simulating propagation within a waveguide. Such ducts cause VHF radio waves to follow the curvature of the Earth for hundreds, or thousands, of miles. This form of propagation is called **tropospheric ducting**.

The possibility of propagating radio waves by tropospheric ducting increases with frequency. Ducting is rare on 50 MHz, fairly common on 144 MHz and more common on higher frequencies. Gulf Coast states see it often, and the Atlantic Seaboard, Great Lakes and Mississippi Valley areas see it occasionally, usually in September and October.

[Now turn to Chapter 10 and study examination questions E3C09, E3C11 and E3C12. Review this section if any of those questions give you difficulty.]

F-REGION PROPAGATION

The maximum one-hop skip distance for high-frequency radio signals is usually considered to be about 2500 miles. Most HF communication beyond that distance takes place by means of several ionospheric hops. Radio signals return to Earth from the ionosphere, and the surface of the Earth reflects them back into the ionosphere. It is also possible that signals may reflect between the E and F regions, or even be reflected several times within the F region. **Figure 3-18** shows several possible paths for the signals to take in reaching some distant location.

There is a propagation theory that suggests radio waves may at times propagate for some distance through the F region of the ionosphere. This theory is supported by the results of propagation studies showing that a signal radiated at a

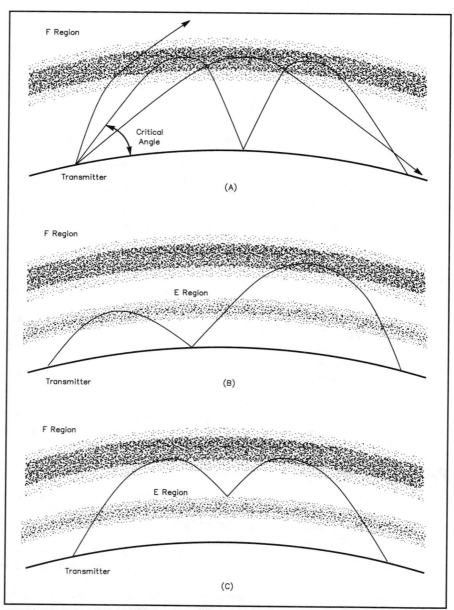

Figure 3-18 — Parts A, B and C show various multihop paths for radio signals that reach farther than 2500 miles.

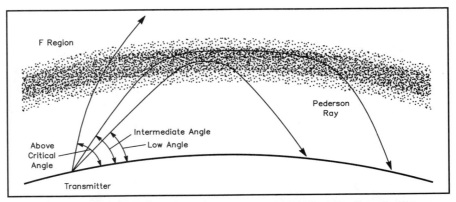

Figure 3-19 — This diagram shows a radio wave entering the F region at an intermediate angle, which penetrates higher than normal into the F region and then follows that region for some distance before being bent enough to return to Earth. A signal that travels for some distance through the F region is called a Pedersen ray.

medium elevation angle sometimes reaches the Earth at a greater distance than a lower-angle wave. This higher-angle wave, called the **Pedersen ray**, is believed to penetrate the F region farther than lower-angle rays. In the less densely ionized upper edge of the region, the amount of refraction is less, nearly equaling the curvature of the region itself as it encircles the Earth. **Figure 3-19** shows how the Pedersen ray could provide propagation beyond the normal single-hop distance.

This Pedersen-ray theory is further supported by studies of propagation times and signal strengths for signals that travel completely around the Earth. The time required is significantly less than would be necessary to hop between the Earth and the ionosphere 10 or more times while circling the Earth. Return signal strengths are also significantly higher than should be expected otherwise. There is less attenuation for a signal that stays in the F region than for one that makes several additional trips through the E and D regions, in addition to the loss produced by reflections off the surface of the Earth.

[Before going on to Chapter 4, turn to Chapter 10 and study examination question E3C10. Review this section if you do not understand the Pedersen ray.]

CHAPTER 4
KEYWORDS
KEYWORDS
KEYWORDS

Adcock array — A radio direction finding antenna array consisting of two vertical elements fed 180° apart and capable of being rotated.

Alternator whine — A common form of conducted interference typified by an audio tone being induced onto the received or transmitted signal. The pitch of the noise varies with alternator speed.

Antenna effect — One of two operational modes of a simple loop antenna wherein the antenna exhibits the characteristics of a small, nondirectional vertical antenna.

Bandwidth — The frequency range (measured in hertz — Hz) over which a signal is stronger than some specified amount below the peak signal level. For example, if a certain signal is at least half as strong as the peak power level over a range of ± 3 kHz, the signal has a 3-dB bandwidth of 6 kHz.

Blocking — A receiver condition in which reception of a desired weak signal is prevented because of a nearby, unwanted strong signal.

Capacitive coupling (of a dip meter) — A method of transferring energy from a dip-meter oscillator to a tuned circuit by means of an electric field.

Capture effect — An effect especially noticed with FM and PM systems whereby the strongest signal to reach the demodulator is the one to be received. You cannot tell whether weaker signals are present.

Cardioid radiation pattern — A heart-shaped antenna pattern characterized by a single, large lobe in one direction, and a deep, narrow null in the opposite direction.

Circulator — A passive device with three or more ports or input/output terminals. It can be used to combine the output from several transmitters to one antenna. A circulator acts as a one-way valve to allow radio waves to travel in one direction (to the antenna) but not in another (to the receiver).

Coaxial capacitor — A cylindrical capacitor used for noise-suppression purposes. The line to be filtered connects to the two ends of the capacitor, and a third connection is made to electrical ground. One side of the capacitor provides a dc path between the ends, while the other side of the capacitor connects to ground.

Conducted noise — Electrical noise that is imparted to a radio receiver or transmitter through the power connections to the radio.

Desensitization — A reduction in receiver sensitivity caused by the receiver front end being overloaded by noise or RF from a local transmitter.

Dip meter — A tunable RF oscillator that supplies energy to another circuit resonant at the frequency that the oscillator is tuned to. A meter indicates when the most energy is being coupled out of the circuit by showing a dip in indicated current.

Directive antenna — An antenna that concentrates the radiated energy to form one or more major lobes in specific directions. The receiving pattern is the same as the transmitting pattern.

Dynamic range — The ability of a receiver to tolerate strong signals outside the band-pass range. Blocking dynamic range and intermodulation distortion (IMD) dynamic range are the two most common dynamic range measurements used to predict receiver performance.

Frequency counter — A digital-electronic device that counts the cycles of an electromagnetic wave for a certain amount of time and gives a digital readout of the frequency.

Frequency domain — A time-independent way to view a complex signal. The various component sine waves that make up a complex waveform are shown by frequency and amplitude on a graph or the CRT display of a spectrum analyzer.

Frequency standard — A circuit or device used to produce a highly accurate reference frequency. The frequency standard may be a crystal oscillator in a marker generator or a radio broadcast, such as from WWV, with a carefully controlled transmit frequency.

Ground-wave signals — Radio signals that are propagated along the ground rather than through the ionosphere or by some other means.

Image signal — An unwanted signal that mixes with a receiver local oscillator to produce a signal at the desired intermediate frequency.

Inductive coupling (of a dip meter) — A method of transferring energy from a dip-meter oscillator to a tuned circuit by means of a magnetic field between two coils.

Intermodulation distortion (IMD) — A type of interference that results from the unwanted mixing of two strong signals, producing a signal on an unintended frequency. The resulting mixing products can interfere with desired signals on those frequencies. "Intermod" usually occurs in a nonlinear stage or device.

Isolator — A passive attenuator in which the loss in one direction is much greater than the loss in the other.

Lissajous figure — An oscilloscope pattern obtained by connecting one sine wave to the vertical amplifier and another sine wave to the horizontal amplifier. The two signals must be harmonically related to produce a stable pattern.

Logic probe — A simple piece of test equipment used to indicate high or low logic states (voltage levels) in digital-electronic circuits.

Loop antenna — An antenna configured in the shape of a loop. If the current in the loop, or in multiple parallel turns, is essentially uniform, and if the loop circumference is small compared with a wavelength, the radiation pattern is symmetrical, with maximum response in either direction of the loop plane.

Minimum discernible signal (MDS) — The smallest input signal level that can just be detected above the receiver internal noise. Also called **noise floor**.

Noise figure — A ratio of the noise output power to the noise input power when the input termination is at a standard temperature of 290 K. It is a measure of the noise generated in the receiver circuitry.

Noise floor — The smallest input signal level that can just be detected above the receiver internal noise. Also called **minimum discernible signal (MDS)**.

Oscilloscope — A device using a cathode-ray tube to display the waveform of an electric signal with respect to time or as compared with another signal.

Phase noise — Undesired variations in the phase of an oscillator signal. Phase noise is usually associated with phase-locked loop (PLL) oscillators.

Radiated noise — Usually referring to a mobile installation, noise that is being radiated from the ignition system or electrical system of a vehicle and causing interference to the reception of radio signals.

Selectivity — A measure of the ability of a receiver to distinguish between a desired signal and an undesired one at some different frequency. Selectivity can be applied to the RF, IF and AF stages.

Sensing antenna — An omnidirectional antenna used in conjunction with an antenna that exhibits a bidirectional pattern to produce a radio direction-finding system with a cardioid pattern.

Sensitivity — A measure of the minimum input-signal level that will produce a certain audio output from a receiver.

Sky-wave signals — Radio signals that travel through the ionosphere to reach the receiving station. Sky-wave signals will cause a variation in the measured received-signal direction, resulting in an error with a radio direction-finding system.

Spectrum analyzer — A test instrument generally used to display the power (or amplitude) distribution of a signal with respect to frequency.

Surface-mount package — An electronic component without wire leads, designed to be soldered directly to copper-foil pads on a circuit board.

Time domain — A method of viewing a complex signal. The amplitude of the complex wave is displayed over changing time. The display shows only the complex waveform, and does not necessarily indicate the sine-wave signals that make up the wave.

Triangulation — A radio direction-finding technique in which compass bearings from two or more locations are taken, and lines are drawn on a map to predict the location of a radio signal source.

AMATEUR RADIO PRACTICE

This chapter contains material on Amateur Radio practices that you must be familiar with to pass your Amateur Extra class license examination. There will be five questions on your exam about these Amateur Radio practices, and those questions will be taken from the five groups of questions under syllabus point E4.

E4A Test equipment: spectrum analyzers (interpreting spectrum analyzer displays; transmitter output spectrum); logic probes (indications of high and low states in digital circuits; indications of pulse conditions in digital circuits)

E4B Frequency measurement devices (i.e., frequency counter, oscilloscope Lissajous figures, dip meter); meter performance limitations; oscilloscope performance limitations; frequency counter performance limitations

E4C Receiver performance characteristics (i.e., phase noise, desensitization, capture effect, intercept point, noise floor, dynamic range {blocking and IMD}, image rejection, MDS, signal-to-noise-ratio); intermodulation and cross-modulation interference

E4D Noise suppression: ignition noise; alternator noise (whine); electronic motor noise; static; line noise

E4E Component mounting techniques (i.e., surface, dead bug {raised}, circuit board); direction finding: techniques and equipment; fox hunting

USE OF TEST EQUIPMENT

Frequency Counter

A **frequency counter**, once considered almost a luxury item, is now an integral part of most commercially made amateur transceivers. The name is completely descriptive: A frequency counter is used to make frequency measurements. It counts the number of cycles per second (hertz) of a signal, and displays that number on a digital readout. Some frequency counters even incorporate voice synthesizers that announce the frequency on command.

Modern transceivers often have a built-in frequency counter to measure and display the operating frequency, although not all digital readouts employ a frequency counter. If a counter is used, then the frequency readout will be as accurate as the counter itself. Some rigs sample some information from the control circuitry and then calculate the operating frequency for display. In such a case, if some part of the radio circuitry is not working properly, you may get an erroneous display.

You can also use a frequency counter to measure signal frequencies throughout a piece of equipment, and to make fine adjustments to tuned circuits. In this case, a frequency counter becomes a valuable piece of test equipment.

Although usually quite accurate, a frequency counter should be checked regularly against WWV, WWVH or some other **frequency standard**. Frequency counters that operate well into the gigahertz range are available. Counters that operate at VHF or UHF, and sometimes those that operate near the top of the HF range, usually employ a prescaler. The prescaler uses logic circuitry to divide the frequency prior to counting, greatly extending the useful range of the frequency counter. A typical counter is illustrated in block-diagram form in **Figure 4-1**.

A frequency counter circuit can only measure time as accurately as the crystal reference oscillator, or time base, built into the circuit. The more accurate this crystal is, the more accurate the readings will be. Close-tolerance crystals are used, and there is usually a trimmer capacitor across the crystal so the frequency can be set exactly once it is in the circuit. One way to increase the accuracy of a frequency counter is to increase the accuracy of the time base oscillator. (Accuracy refers to the closeness of the measured value to the actual value.)

The accuracy of frequency counters is often expressed in parts per million (ppm). Even after checking the counter against WWV, you must take this possible

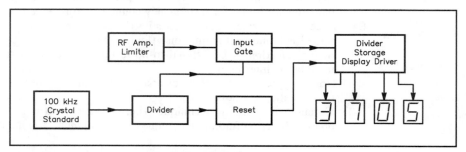

Figure 4-1 — Frequency-counter block diagram.

error into account. The readout error can be as much as:

$$\text{Error} = f(\text{Hz}) \times \frac{\text{counter error}}{1{,}000{,}000}$$ (Equation 4-1)

Suppose you are using a frequency counter with a time base accuracy of 10 parts per million to measure the operating frequency of a 146.52-MHz transceiver. You can use Equation 4-1 to calculate the maximum frequency-readout error:

$$\text{Error} = 146{,}520{,}000 \text{ Hz} \times \frac{10}{1{,}000{,}000} = 1465.20 \text{ Hz}$$

This means the actual operating frequency of this radio could be as high as 146.5214652 MHz or as low as 146.5185348 MHz. You should always take the maximum frequency-readout error into consideration when deciding how close to a band edge you should operate. In addition, you must be sure that the modulation sidebands also stay inside the band edge.

The stability of the time-base oscillator is also very critical to the accuracy and precision of the counter. (Precision refers to the repeatability of a measurement.) Any variation of the time base oscillator frequency affects the counter accuracy and precision. The counter keeps track of the number of RF cycles that occur in a given time interval. Even a slight increase in the time interval could result in a significant increase in the measured frequency. As a simple example, suppose you are using a frequency counter that counts RF cycles for 1 second, and displays the frequency as a result of that count. If the time-base oscillator slows down, it will count RF cycles for a longer period, so the displayed frequency will increase. If the oscillator speeds up, it will count RF cycles for a shorter period, so the displayed frequency decreases. The displayed frequency on this counter will change even if the signal it is counting does not vary!

The speed of the digital logic in the frequency counter limits the upper frequency response of the counter. In an extreme example, if the signal you are trying to measure has two RF cycles in the time it takes the logic in your counter to respond to a single pulse, then the displayed count will only be half the actual frequency. A basic frequency counter may have an upper frequency limit of 10 MHz. By using faster digital logic in the circuit it may be possible to build a similar counter with an upper frequency limit of 100 MHz. The most common way to increase the useful measurement range of a counter is to include a *prescaler*. A prescaler divides the input signal by some value such as 10 or 100, so the counter logic doesn't have to be as fast.

Several factors limit the accuracy, frequency response and stability of a frequency counter. These factors are the time-base oscillator accuracy, the speed of the digital logic and the stability of the time-base oscillator.

[Before you go on to the next section, study the examination questions in Chapter 10 with numbers E4B01 through E4B10. Review this section as needed.]

Dip Meter

Once called a grid-dip meter because it employed a vacuum tube with a meter to indicate grid current, this handy device is actually an RF oscillator. Modern solid-state **dip meters** use FETs.

The principle of operation of a dip meter is that when the meter is brought near a circuit resonant at the meter oscillation frequency, that circuit will take some power from the dip meter. Power taken from the oscillator results in a slight drop in the meter reading of the feedback circuit current. A dip-meter, then, gives an indication of the resonant frequency of a circuit. This will not be a highly accurate frequency measurement, but often all you need is a general indication.

Figure 4-2 is the schematic diagram of a simple dip meter. The plug-in coil assembly includes an inductor (L_1) and capacitors (C_1 and C_2) to form an LC oscillator. The main tuning capacitor (C_3) varies the oscillator frequency over a range set by the plug-in coil assembly. Most dip meters have several plug-in coil assemblies to cover a wide frequency range. A meter in the feedback circuit measures the current to give an indication that power is being coupled out of the oscillator.

The most frequent amateur use of a dip meter is to determine the resonant frequency of an antenna or antenna traps. Dip meters can also be used to determine the resonant frequency of other circuits. The circuit under test should have no signal or power applied to it while you couple the dip meter to the circuit. Sometimes you simply need a small signal at some specific frequency to inject into a circuit. The tunable RF oscillator of a dip meter is ideal for this task.

Although dip meters are relatively easy to use, the touchiest part of their use is coupling the oscillator coil to the circuit being tested. This coupling should be as loose as possible and still provide a definite, but small, dip in the current when

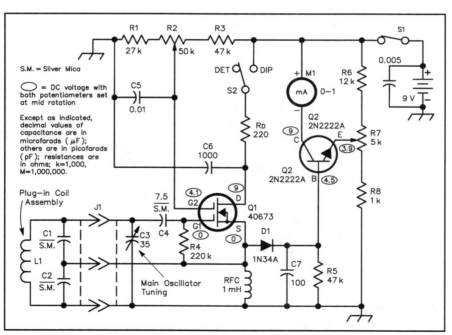

Figure 4-2 — Schematic diagram of a dual-gate MOSFET dip meter. L_1, C_1 and C_2 make up plug-in tuned circuits to change the operating frequency.

coupled to a circuit resonant at the dip-meter-oscillator frequency. Coupling that is too loose will not give a dip sufficient to be a positive indication of resonance.

Whenever two circuits are coupled, however loosely, each circuit affects the other to some extent. If the coupling is loose, the effect will be small and will not create a significant change in the resonant frequency of either circuit. Too tight a coupling, however, almost certainly will yield a false reading on the dip meter.

Dip meters are usually coupled to a circuit by allowing the oscillator-coil field to cut through a coil in the circuit under test. This is called **inductive coupling**. The energy is transferred through the magnetic fields of the coils. Sometimes it is not possible to couple the meter to an inductor, so **capacitive coupling** is used. In this case, the dip-meter coil is simply brought close to an element in the circuit, and the capacitance between the components couples a signal to the circuit. This means that it is the electric fields between the components that transfers the energy. **Figure 4-3** indicates the methods of coupling a dip meter to a circuit.

The procedure for using a dip meter is to bring the dip-meter coil within a few inches of the circuit to be tested and then sweep the oscillator through the frequency band until the meter needle indicates a dip. This dip should be symmetrical — that is, the needle should move downward and upward at about the same speed when the oscillator is tuned through resonance. A jumpy needle may indicate that the coupling is too tight, or that there is a problem with the dip meter itself or the circuit being tested. A jumpy meter needle may also indicate there is a strong influence from yet another circuit active in the vicinity of the test area. The test should be repeated several times, to be sure the dip occurs at the same point each time, and thus accurately indicates the resonance point.

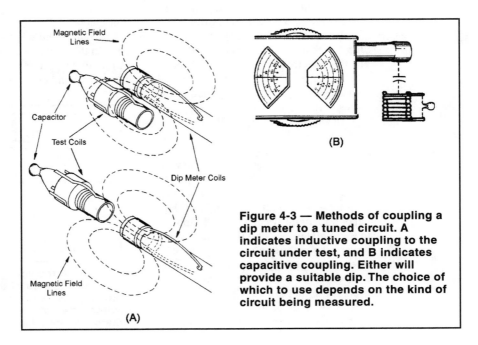

Figure 4-3 — Methods of coupling a dip meter to a tuned circuit. A indicates inductive coupling to the circuit under test, and B indicates capacitive coupling. Either will provide a suitable dip. The choice of which to use depends on the kind of circuit being measured.

If no dip appears during this check, try another coil on the dip meter. The actual resonant frequency of the circuit being tested might be far from that expected — too far even to be within the resonant-frequency range of the coil originally used on the dip meter. Some experience with dip meters usually is required before you can be sure you are not getting a false reading, caused by coupling that is too tight or by another resonant circuit near or connected to the one you are testing.

If you want to check the resonant frequency of an antenna, a noise bridge and receiver can also be used to indicate resonance. This setup will also give an indication of antenna impedance. If a noise bridge is available, it may be better to learn how to use it for checking your antenna resonant frequency. Antenna analyzers are also fairly easy to use and give excellent results for this type of measurement.

Another problem with dip meters is the possibility of reading a harmonic rather than the fundamental frequency. The dip for a harmonic, of course, will not be as deep as that for the fundamental. Nevertheless, it is a good idea to take a reading at double, triple, one-half and one-third the apparent resonant frequency. By doing so, you can be sure the original dip actually occurs at the true resonant frequency.

[Go to the examination questions in Chapter 10 and study questions E4B12 through E4B17 before proceeding. Review this section as needed.]

Oscilloscope

An **oscilloscope** is built around a **cathode-ray tube**. The cathode-ray tube in an oscilloscope differs greatly from that in a television receiver. In fact, about the only thing the tubes have in common is that both use an electron beam focused on a fluorescent screen. The oscilloscope electron beam is controlled by electrostatic charges on the vertical and horizontal deflection plates.

When a voltage is applied to the plates, the beam is pulled toward the plate with the positive charge and repelled by the one with the negative charge. (Remember that electrons are negatively charged ions.) With two pairs of plates at right angles, the beam can be moved from side to side and up and down. Frequencies far into the RF region can be applied to the plates, with the use of very little power.

In the absence of deflection voltages, the oscilloscope controls are adjusted so a small bright spot appears in the center of the screen. Then, an ac voltage applied to the horizontal plates causes the spot to move from side to side (**Figure 4-4**). Usually, the time base signal or horizontal sweep voltage is applied to these plates, causing the spot to move toward the right at a steady speed. The sweep-voltage frequency (a voltage with a sawtooth waveform) is selected by the user. The signal to be analyzed is applied to the vertical deflection plates. The speed at which the spot moves in any

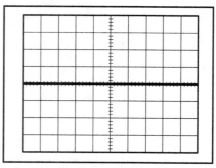

Figure 4-4 — If the voltage applied to the vertical plates of the oscilloscope is zero, only the horizontal line created by the sweep oscillator will be visible.

direction is exactly proportional to the rate at which the voltage is changing.

In the course of one ac cycle, the spot will move upward while the voltage applied to the vertical plates is increasing from zero. When the positive peak of the cycle is reached, and the voltage begins to decrease, the spot will reverse its direction and move downward. It will continue in this direction until the negative peak is reached. Then, as the voltage increases toward zero, the spot moves upward again. If the horizontal sweep voltage is moving the spot horizontally at a uniform speed at the same time, it is easy to analyze the signal waveform applied to the vertical deflection plates. See **Figure 4-5**.

There are many uses for an oscilloscope in an amateur station. This instrument is often used to display the output

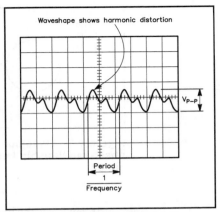

Figure 4-5 — A typical pattern resulting when a complex waveform is applied to the vertical plates, and the sweep-oscillator output is applied to the horizontal plates.

waveform of a transmitter during a two-tone test. Such a test can help you determine if the amplifier stages in your rig are operating in a linear manner. An oscilloscope can also be used to display signal waveforms during troubleshooting procedures.

Another frequent use of the oscilloscope in amateur practice is the comparison of two signals, one of which is applied to the vertical deflection plates, and the other to the horizontal plates. This procedure produces a **Lissajous figure** on the scope. When sinusoidal ac voltages are applied to both sets of deflection plates, the resulting pattern depends on the relative amplitudes, frequencies and phases of the two voltages. If the ratio between the two frequencies is constant and can be expressed as two integers, the Lissajous pattern will be stationary. If the frequency ratio is not an exact integer value, the pattern may seem to shift or rotate around a vertical or horizontal axis.

Examples of some simple Lissajous patterns are shown in **Figure 4-6**. The frequency ratio is found by counting the number of loops along two adjacent edges. Thus, in the pattern shown at C, there are three loops along a horizontal edge and only one along the vertical, so the ratio of the vertical frequency to the horizontal frequency is 3:1. In part E, there are four loops along the horizontal edge and three along the vertical edge, producing a ratio of 4:3. Assuming that a known frequency is applied to the horizontal plates and an unknown frequency is applied to the vertical plates, the relationship is given by:

$$\frac{f_H}{f_V} = \frac{n_V}{n_H}$$ (Equation 4-2)

where:
  f_H is the known frequency, on the horizontal axis
  f_V is the unknown frequency, on the vertical axis
  n_V is the number of loops along a vertical edge
  n_H is the number of loops along a horizontal edge.

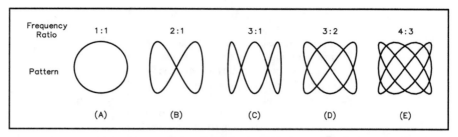

**Figure 4-6—Lissajous figures and corresponding frequency ratios for a 90°
phase relationship between the voltages applied to the two sets of deflection
plates.**

If you know the two signal frequencies being supplied to the oscilloscope, you
can predict the Lissajous pattern that will be produced. Reduce the fraction f_H / f_V
to the lowest denominator. To do this, divide the top (numerator) and bottom (denominator) of the fraction by the same number, until further division does not give
integer (whole number) values. The resulting fraction is the ratio of vertical to
horizontal loops in the Lissajous pattern (n_V / n_H). An example will make this procedure easier to understand: Suppose you feed a 100-Hz signal to the horizontal
input of an oscilloscope and a 150-Hz signal to the vertical input. What type of
Lissajous pattern will this produce? Use Equation 4-2.

$$\frac{f_H}{f_V} = \frac{100 \text{ Hz}}{150 \text{ Hz}} = \frac{n_V}{n_H}$$

We can reduce this fraction by dividing both the numerator and denominator by 50.
Notice that we could divide both by 2, 5 or 10 and still get integer values for the
numerator and denominator, but by choosing 50 we get the smallest possible integer values.

$$\frac{f_H}{f_V} = \frac{\dfrac{100}{50}}{\dfrac{150}{50}} = \frac{2}{3} = \frac{n_V}{n_H}$$

From this calculation, you can see that the Lissajous pattern will have two loops on
the vertical axis and three loops on the horizontal axis. The pattern will look like
the one shown in Figure 4-6D.

Lissajous figures are used more often to measure a signal with an unknown
frequency by comparing it with a signal of known frequency. In that case, you
would look at the Lissajous pattern produced on the oscilloscope and then solve
Equation 4-2 for the unknown frequency. Equation 4-3 is the result of cross multiplying the terms of Equation 4-2, solving for the vertical frequency, which is usually the unknown.

$$f_V = \frac{n_H}{n_V} f_H$$

(Equation 4-3)

For example, suppose the signal applied to the horizontal input has a frequency of 200 Hz. When an unknown signal is applied to the vertical input, a Lissajous pattern like the one shown in Figure 4-6C results. What is the frequency of the signal applied to the vertical input? Equation 4-3 gives the result.

$$f_V = \frac{n_H}{n_V} f_H = \frac{3}{1} \times 200 \text{ Hz} = 600 \text{ Hz}$$

An important application of Lissajous figures is the calibration of audio-frequency signal generators. For very low frequencies, the 60-Hz power-line frequency is used as a reasonably good standard for comparison. This frequency is accurately controlled by the power companies, so although the signal frequency may vary somewhat, the long-term average is very accurate. The medium AF range can be covered by comparison with the 440- and 600-Hz audio modulation on WWV transmissions. It is possible to calibrate over a 10:1 range both upward and downward from these frequencies and thus cover the audio ranges useful for voice communication.

An oscilloscope has both a horizontal and a vertical amplifier (in addition to the horizontal sweep-oscillator circuitry). This is desirable because it is convenient to have a means for adjusting the voltages applied to the deflection plates to secure a suitable pattern size. Several hundred volts is usually required for full-scale deflection in either the vertical or horizontal direction, but the current required is usually somewhere in the microampere range. Thus, the actual power needed for full-scale deflection is extremely low.

One important limitation to the accuracy, frequency response and stability of an oscilloscope is the bandwidth (frequency response) of the scope deflection amplifiers. Another important limitation is the accuracy and linearity of the time base, or horizontal sweep oscillator. If the horizontal sweep oscillator only works in the audio-frequency range, the scope won't be very useful for making RF measurements. Unless the sweep oscillator is stable and the frequency can be set accurately, frequency measurements made with the scope will not be very accurate.

Scopes with a horizontal sweep rate that is limited to audio frequencies are relatively inexpensive and serve many amateur needs. Increasing the bandwidth of the horizontal and vertical amplifier, and increasing the horizontal sweep-oscillator frequency, will increase the useful frequency response of the scope. Of course, it is a distinct advantage for amateurs to have a scope that will handle RF through their most-used range; that is, to 30 MHz, to 150 MHz or even higher. Scopes with such RF capabilities are usually expensive, however.

One way to extend the useful frequency range of a narrow-bandwidth oscilloscope is with an adapter circuit. The idea is to use a mixer and an oscillator set to a frequency that will provide a sum or difference frequency within the useful range of your scope. With a 25-MHz oscillator, for example, you can display signals in the 20- to 30-MHz range on a 5-MHz scope. Other oscillator frequencies will enable you to display a signal with almost any desired frequency. This will not improve the transient-signal response of the narrowband scope, however.

Other limitations amateurs should be aware of are the stability of the scope sweep-frequency oscillator, absence of a horizontal deflection-plate amplifier (which would make it impossible to produce a Lissajous pattern), the ease of varying the horizontal sweep rate, the upper and lower limits of the available sweep rates, and the amount of voltage required for full-scale deflection of the electron beam. All of these factors will affect the frequency response and stability of a scope.

[Now turn to the examination questions in Chapter 10 and study questions E4B11 and E4B18 before proceeding. Review this section if you have difficulty with any of these questions.]

Spectrum Analyzer

A **spectrum analyzer** produces a graphic representation of a dynamic, or changing, electrical signal. In this sense, it is similar to an oscilloscope. The oscilloscope presents complex signals in the **time domain**. That is, it shows amplitude as a function of time. There are, however, signals that cannot be represented properly in the time domain. Amplifiers, oscillators, detectors, modulators, mixers and filters are best characterized in terms of their frequency response. We must observe these signals in the **frequency domain** (amplitude as a function of frequency). The spectrum analyzer is one instrument that displays signals in the frequency domain.

Understanding the Time and Frequency Domains

To better understand the concepts of time and frequency domains, refer to **Figure 4-7**. In Figure 4-7A, the three-dimensional coordinates show time (as the line sloping toward the bottom right), frequency (as the line sloping toward the top right), and amplitude (as the vertical axis). The two frequencies shown are harmonically related (f_1 and $2f_1$). The time domain is represented in Figure 4-7B, where all frequency components are added together. If the two frequencies were applied to the input of an oscilloscope, we would see the solid line (which represents $f_1 + 2f_1$) on the display.

The display shown in Figure 4-7C is typical of a spectrum analyzer presentation of a complex signal (a signal composed of more than one frequency). Here the signal is separated into the individual frequency components, and a measurement may be made of the power level at each frequency.

The frequency domain contains information not found in the time domain, and vice versa. Hence, the spectrum ana-

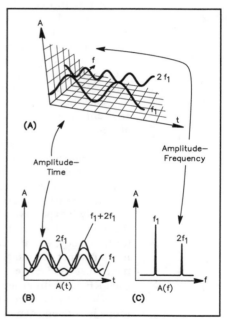

Figure 4-7 — This diagram shows different ways a complex signal may be characterized. Part A is a three-dimensional display of amplitude, time and frequency. At B, this information is shown only in the time domain as would be seen on an oscilloscope. At C, the same information is shown in the frequency domain as it would be viewed on a spectrum analyzer.

lyzer offers advantages over the oscilloscope for certain measurements, but for measurements in the time domain, the oscilloscope is an invaluable instrument.

Spectrum Analyzer Basics

There are several types of spectrum analyzers. The most popular type is the swept superheterodyne. A simplified block diagram of such an analyzer is shown in **Figure 4-8**. The analyzer is basically a narrow-band receiver that is electronically tuned in frequency. Tuning is accomplished by applying a linear ramp voltage to the frequency-controlling element of a voltage-controlled oscillator. The same ramp voltage is simultaneously applied to the horizontal deflection plates of the cathode ray tube (CRT). This means the horizontal axis of the spectrum analyzer displays the frequency. The receiver output is synchronously applied to the vertical deflection plates of the CRT, which means the vertical axis of the spectrum analyzer display shows the signal amplitude. The resulting spectrum analyzer display shows amplitude versus frequency.

Spectrum analyzers are calibrated in both frequency and amplitude for relative and absolute measurements. The frequency range, controlled by the scan-width control, is calibrated in hertz, kilohertz or megahertz per division (graticule marking). Within the limits of the voltage-controlled oscillator, you can test a transmitter over any frequency range. Whether you are testing an HF or a VHF transmitter, the spectrum analyzer displays all frequency components of the transmitted signal.

The vertical axis of the display is calibrated for amplitude. Common calibrations for amplitude are 1 dB, 2 dB or 10 dB per division. For transmitter testing, the 10-dB-per-division range is commonly used, because it allows you to view a wide range of signal strengths, such as those of the fundamental signal, harmonics and spurious signals.

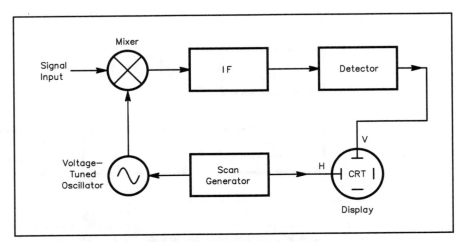

Figure 4-8 — This simplified block diagram illustrates the operation of a spectrum analyzer.

Transmitter Testing with the Spectrum Analyzer

The spectrum analyzer is ideally suited for checking the output from a transmitter or amplifier for spectral purity. You will easily see any spurious signals from the transmitter on a spectrum analyzer display. There are also many other practical uses. **Figure 4-9** shows two test setups commonly used for transmitter testing. The setup at B is the better and more accurate approach for broadband measurements because most line-sampling devices do not exhibit a constant-amplitude output across a broad frequency spectrum. The ARRL Headquarters Laboratory staff uses the setup shown in B. The power attenuator may be a single, high-power type, or it may consist of a string of lower-power attenuators connected in series to reduce the rated output of the transmitter to a level suitable for the analyzer (typically 10 mW or less).

Figure 4-10 is a sketch showing a typical spectrum analyzer display for a transmitter operating key down on the 40-meter band. The full-scale pip at the far left of the display is generated within the spectrum analyzer and represents "zero" frequency. The horizontal scale is 5 MHz per division, and the vertical scale is 10 dB per division. Moving to the right, the next tall pip is seen at roughly 7 MHz. This signal is the fundamental frequency. When the spectrum analyzer is adjusted so that the top of this signal touches the top (reference) line of the display, all other signal levels can be referenced to the power of the fundamental. Moving farther right, the next signal, at 14 MHz, is the second harmonic of the fundamental. Its level is 60 dB down from the fundamental. Even farther to the right is the third harmonic, at 21 MHz, which is 50 dB down. To the left of the fundamental are a couple of small pips. These signals, probably spurious mixing products or oscillator leakage, are more than 60 dB below the fundamental, an acceptable level. This spectrograph is typical of a well-designed multiband rig.

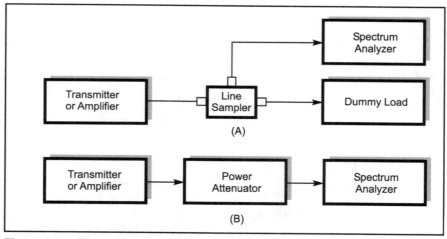

Figure 4-9 — These diagrams show two commonly used test setups to observe the output of a transmitter or amplifier on a spectrum analyzer. The system at A uses a line sampler to pick off a small amount of the transmitter or amplifier power. At B, the majority of the transmitter power is dissipated in the power attenuator.

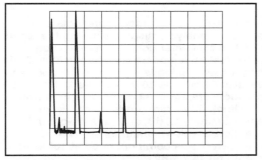

Figure 4-10 — This sketch represents the spectrum analyzer display for a well-designed transmitter operating key down on the 40-meter band. Each horizontal division represents 5 MHz and each vertical division is 10 dB.

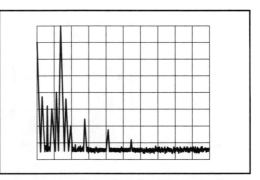

Figure 4-11 — This sketch represents the spectrum analyzer display for a not-so-well-designed transmitter operating key down on the 40-meter band. Compare this display with the one shown in Figure 4-10.

The graph shown in **Figure 4-11** represents the output spectrum from a not-so-well-designed rig. The horizontal and vertical calibration is the same as shown in Figure 4-10. In addition to the higher order harmonics, at about 28 MHz and 35 MHz, a number of mixing products are visible above and below the fundamental. The chances of causing interference are greater with this transmitter. Both transmitters, however, are legal according to the FCC Rules.

Another area of concern in the realm of transmitter spectral purity has to do with the **intermodulation distortion (IMD)** levels associated with SSB transmitters and amplifiers. The test setup shown in **Figure 4-12** is used for transmitter IMD testing. Two equal-amplitude, but not harmonically related, audio tones are fed into the transmitter. (In the ARRL Lab we use 700 and 1900-Hz tones.) The transmitter is first adjusted for rated PEP output using just a single tone, then the

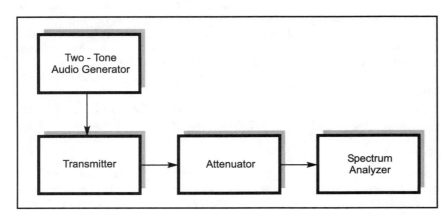

Figure 4-12 — This diagram illustrates the test setup used in the ARRL Laboratory for measuring the IMD performance of transmitters and amplifiers.

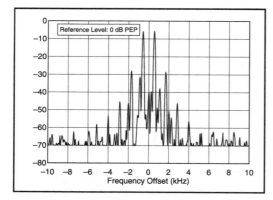

Figure 4-13 — This graph represents a spectrum analyzer display showing the result of a two-tone intermodulation-distortion (IMD) test on a modern SSB transmitter. Each horizontal division is equal to 2 kHz and each vertical division is 10 dB. Third-order products are about 32 dB below PEP output (top line), fifth-order products are down approximately 47 dB, and seventh-order products are down about 52 dB. The transceiver was being operated at 100 W PEP output on 14.2 MHz.

single tone is replaced by the two equal-amplitude tones and the output from the two-tone generator and microphone-level control on the transmitter are adjusted for best IMD performance while maintaining each tone at a level 6 dB below the top line (PEP output). **Figure 4-13** shows a typical display. Responses other than the two individual tones, which appear near the center of the display, are distortion products. In this example, third-order products are down approximately 32 dB, fifth-order products are down about 47 dB, and seventh-order products are down about 52 dB from the PEP output. The two individual tones are 6 dB below PEP output because they are displayed as two discrete frequencies. At the instant when the voltages of the individual tones are in phase, they add to produce a peak in the envelope-waveform pattern that is twice the voltage amplitude of a single tone alone. The power at the peaks of the envelope (PEP) is therefore four times that of a single tone — a 4:1 power ratio being equal to 6 dB. The power ratio is four because doubling the applied voltage will double the current, assuming a constant load resistance.

As mentioned earlier, you can make many types of measurements with a spectrum analyzer. In addition to checking a transmitter spectral output and measuring the two-tone IMD levels, you can determine whether a crystal is operating on its fundamental or overtone frequency by building a simple oscillator circuit and measuring the output frequency. You can even measure the degree of isolation between the input and output ports of a 2-meter duplexer. One thing you could not do with a spectrum analyzer, however, is measure the speed at which a transceiver switches from transmit to receive when it is being used for packet radio.

You can find a more detailed discussion of spectrum analyzer measurement techniques in *The ARRL Handbook*.

[Turn to Chapter 10 now and study exam questions E4A01 through E4A07. Review this section if you have any difficulty with these questions.]

The Logic Probe

Most amateurs are familiar with the use of the standard multimeter for trouble-shooting, but how do you go about testing digital-logic circuits? In digital circuitry there are only two states to worry about — the logical "one" and the logical "zero."

A voltmeter or oscilloscope could be used to monitor these logic states on a particular gate, but most of the time it is necessary to know only if there is either a one or a zero at the input or output of a gate. The **logic probe** provides a means of doing just that, without the problem of carrying a scope or VOM around with you. Logic probes are smaller than voltmeters, and have a simplified readout or indicator to show high and low logic states, rather than a meter scale. **Figure 4-14** shows a circuit for a small, portable logic probe.

Operation of this logic probe is simple — connect the $+ V_{CC}$ lead to the positive terminal of the IC or power supply, and the negative lead to ground. The probe tip is then touched to the part of the circuit you wish to check for a logic state. The readout then indicates 1 for a logical one, 0 for a logical zero, and H for a high-impedance point. Other logic probes may use LEDs to indicate the high and low states, or other conditions, such as a pulsed signal.

Pulsed signals can present some confusing indications on a simple logic probe

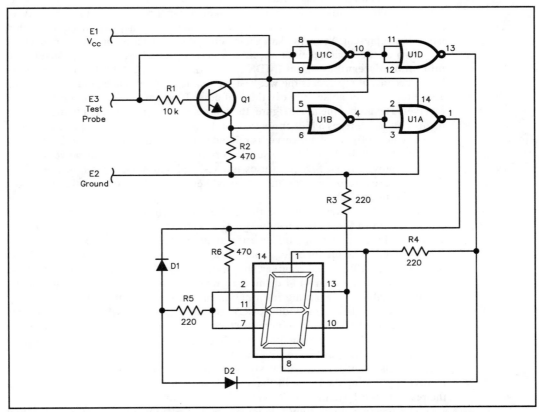

Figure 4-14 — Schematic diagram of a simple TTL logic probe. The readout will indicate "0" for a zero state, "1" for a one state, and "H" for a high-impedance state.

because the signal changes between states zero and one many times each second in some circuits. Many logic probes are designed to indicate these regular variations. Such a probe will indicate a pulse condition in a circuit, although you won't be able to determine the frequency or duty cycle of the pulses as you would with an oscilloscope.

[Now turn to Chapter 10 and study exam questions E4A08 through E4A11. Review this section if you have any difficulty with these questions.]

RECEIVER PERFORMANCE CONCEPTS

Receivers are an important part of any Amateur Radio station. No matter how much transmitter power you have available to you, "You can't work 'em if you can't hear 'em!" Therefore, some discussion of receiver parameters is in order.

Blocking and IMD Dynamic Range

When you begin to evaluate a receiver or transceiver for your Amateur Radio station, you can read about a wide array of measurements and tests in *QST* Product Reviews. The results of those measurements can present a dizzying array of numbers unless you understand how they were made, and what performance characteristics they represent. You will learn about some of these measurements in this section.

Two of the basic receiver specifications are **noise figure** and **sensitivity**, or **minimum discernible signal (MDS)**. The MDS is also called the receiver **noise floor**, because it represents the weakest, or smallest signal that can just be detected above the receiver noise. You can think of noise figure as a "figure of merit" for the receiver. The higher the noise figure, the more noise that is generated in the receiver itself. This also means a higher noise floor. Lower noise figures are more desirable.

The noise figure of a receiver is related to the signal-to-noise ratio of the input and output signals. Signal-to-noise (S/N) ratio is defined as signal power divided by noise power. Input S/N ratio uses input signal and noise powers while output S/N ratio uses output signal and noise powers. Signal-to-noise ratios are often expressed in decibels.

It is useful to know that the theoretical noise power at the input of an ideal receiver, with an input-filter bandwidth of 1 hertz, is −174 dBm. (*The ARRL Handbook* contains more detailed information about how to calculate this number.) This is considered to be the theoretical best (lowest) noise floor a receiver can have. In other words, for this ideal receiver, the strength of any received signal would have to be slightly more than −174 dBm. A receiver bandwidth of 1 Hz is impractical, and undesirable for receiving information on the signal. A more typical receiver might have a 500-Hz bandwidth for CW operation, or even wider for SSB or FM voice. The 500-Hz bandwidth increases the receiver noise by a factor of 500, which corresponds to 27 dB (10 log 500). The theoretical noise floor (minimum discernible signal, or MDS) of this receiver will be −174 dBm + 27 dB = − 147 dBm. If the receiver has a wider bandwidth, you can calculate the theoretical MDS for that receiver by finding the log of the bandwidth and multiplying that value by ten. Add the result to the 1-Hz bandwidth value of −174 dBm.

The receiver noise figure degrades the noise floor, or raises the power that actual signals must have to be heard. You can calculate the actual noise floor of a receiver by adding the noise figure (expressed in dB) to the theoretical best MDS value.

Noise Floor = Theoretical MDS + noise figure   (Equation 4-4)

For example, suppose our 500-Hz-bandwidth receiver has a noise figure of 8 dB. We can use Equation 4-4 to calculate the actual noise floor of this receiver.

Noise Floor = −147 dBm + 8 dB = −139 dBm

Dynamic range is an important receiver parameter. This refers to the ability of the receiver to tolerate strong signals outside the normal passband. We can state a general definition of dynamic range as the ratio between the minimum discernible signal and the largest tolerable signal without causing audible distortion products. There are several types of dynamic range measurements used to describe receiver performance. These are usually calculated from input-signal power levels given in decibels compared to a milliwatt (dBm), with the dynamic-range measurements stated in decibels (dB).

Intermodulation distortion (IMD) dynamic range measures the impact on the receiver of the production of spurious signals that result when two or more signals mix in the receiver. When the IMD dynamic range is exceeded, false signals begin to appear along with the desired signal. (Undesired signals are strong enough to mix with other signals and produce spurious signals that show up in the receiver passband along with the desired signal.)

You can calculate the third-order intercept point by multiplying the IMD dynamic range by 1.5, and then adding the receiver noise floor value to this result. Suppose the example receiver just described has an IMD dynamic range of 94 dB. The third-order intercept point is:

Third-order intercept = 1.5(94 dB) + (−139 dBm) = 141 dB − 139 dBm = 2 dBm

A larger value for the third-order intercept point would represent a better receiver.

Refer to **Figure 4-15**. This plot shows the output power of the desired signal and the output power of the distortion products as a function of the input-tone power levels. Eventually, the input signal levels get large enough so the desired output levels no longer increase linearly. This effect is called gain compression or blocking. Blocking may be defined as desensitization to a desired signal caused by a nearby strong interfering signal.

The linear portion of the curves of Figure 4-15 may be extended or extrapolated to higher power levels even though the receiver is not capable of operating properly at these levels. If this is done (as shown by the dashed line), the two curves will eventually cross each other. At some usually unattainable output power level, the distortion-product level equals that of the desired output levels. This point is commonly referred to as the amplifier *intercept point*. Specifically, the output power level at which the curves intersect is called the amplifier output intercept. Similarly, the input power level corresponding to the point of intersection is called the input intercept.

Blocking dynamic range refers to the difference in signal powers between the noise floor and a signal that causes 1 dB of gain compression in the receiver. When the blocking dynamic range is exceeded, the receiver begins to lose the ability to amplify weak signals.

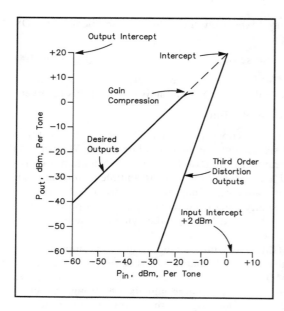

Figure 4-15 — This graph shows how the output power of the desired signal and the output power of the distortion products vary with changes of input power. The input consists of two equal-power sine-wave signals.

Figure 4-16 shows the basic test setup to measure blocking dynamic range. One signal generator is used to feed a weak signal into the receiver input at about 20 dB above the noise floor. The receiver is tuned to this first signal. A second signal generator is set to a frequency 20 kHz away, and the strength of this signal is increased until the receiver audio output of the desired signal drops by 1 dB. At that point the second signal is causing the receiver AGC to limit the audio output of the desired signal. The second signal is desensitizing the receiver, or **blocking** it.

As Figure 4-16 shows, you can calculate the receiver blocking level by adding the undesired signal power (expressed in dBm) with the hybrid combiner loss and the step attenuator loss. The blocking dynamic range is then calculated by subtracting the blocking level from the receiver noise floor, and taking the absolute value of the result. (The absolute value means you take the final result as a positive value, even if the calculation results in a negative value. Keep in mind that when you subtract a negative value, the sign changes, so the operation is the same as adding

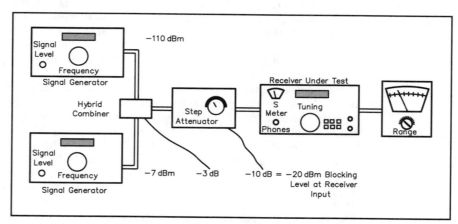

Figure 4-16 — This diagram shows the equipment and its arrangement for measuring receiver blocking dynamic range. Measurements shown are for the example discussed in the text.

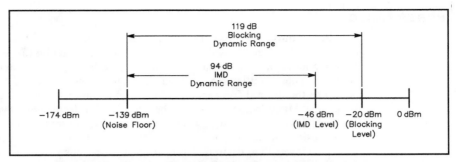

Figure 4-17 — This graph displays the dynamic-range performance of a hypothetical (though typical) receiver. The noise floor is –139 dBm, blocking level is –20 dBm and the IMD level is –46 dBm. This corresponds to a receiver blocking dynamic range of 119 dB and an IMD dynamic range of 94 dB.

a positive value.) Equation 4-5 gives this expression in mathematical terms.

Blocking Dynamic Range = | Noise Floor – Blocking Level | (Equation 4-5)

The vertical bars around the calculation indicate that the result is taken as an absolute value.

Suppose we have a receiver that has a 500-Hz bandwidth, and we measure the noise figure as 8-dB. We find the blocking level is –20 dBm. What will the blocking dynamic range of this receiver be?

The first step is to calculate the receiver noise floor, using Equation 4-4. We know that a theoretically ideal receiver with a 500-Hz bandwidth has an MDS of –147 dBm.

Noise Floor = Theoretical MDS + noise figure   (Equation 4-4)

Noise Floor = –147 dBm + 8 dB = –139 dBm

Next we will use Equation 4-5 to calculate the blocking dynamic range.

Blocking Dynamic Range = | Noise Floor – Blocking Level | (Equation 4-5)

Blocking Dynamic Range = | –139 dBm – (–20 dBm) | = | –139 dBm + 20 dBm | = 119 dB

This blocking dynamic range measurement tells us that signals more than 119 dB above the noise floor will block reception of weak signals, even though the undesired signal is well outside the normal pass band of the receiver.

Figure 4-17 illustrates the relationship between the input signal levels, noise floor, blocking dynamic range and IMD dynamic range. With modern receiver designs, blocking dynamic ranges of a little over 100 dB are possible, and this is an acceptable dynamic range figure. So the receiver we have used in this example could be any typical modern receiver. If a receiver has poor dynamic range there will be cross-modulation (IMD) of the desired signal and desensitization (blocking) from strong adjacent signals.

Phase Noise

Phase noise is a problem that has become more apparent as other improvements have reduced the noise floor and improved dynamic range measurements. Most modern commercial transceivers use phase-locked loop (PLL) frequency synthesizers. These oscillators use a feedback loop to detect the output frequency, compare it to the desired frequency, and make automatic corrections to the oscillator frequency. Many factors can cause the output signal frequency to vary a bit, so the circuit is constantly adjusting itself. On average, the output frequency is very close to the desired value, but at any instant the actual frequency is likely to be a little higher or lower than desired. These slight fluctuations result in a signal in which the phase of any cycle is likely to be slightly different from the phase of adjacent cycles. **Figure 4-18** exaggerates these phase variations of a PLL oscillator.

One result of receiver phase noise is that as you tune towards a strong signal, the receiver noise floor appears to increase. In other words, you hear an increasing amount of noise in an otherwise quiet receiver as you tune towards the strong signal.

Excessive phase noise in a receiver local oscillator allows strong signals on nearby frequencies to interfere with the reception of a weak desired signal. You can understand this if you think about tuning in a weak signal that is barely above the receiver noise floor. While listening to this signal, another station begins transmitting a strong signal on a nearby frequency. The new signal is outside the receiver passband, so you don't hear the interfering signal directly, but there is a sudden increase in receiver noise. This increased receiver noise can cover a weak desired signal, or at least make copying it more difficult. **Figure 4-19** illustrates how this phase noise can cover a weak signal.

Excessive phase noise can cause interference in two ways. The first is the result of receiver phase noise just discussed. The second is caused by a transmitter with excessive phase noise on its transmitted signal. The effect is virtually the same.

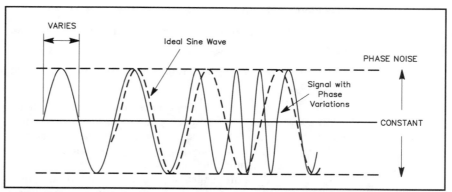

Figure 4-18 — This diagram exaggerates the effects of phase noise from a phase-locked loop oscillator. The dashed line represents the output of an ideal oscillator as it might be viewed on an oscilloscope. The solid line represents the output of an oscillator with a large amount of phase noise.

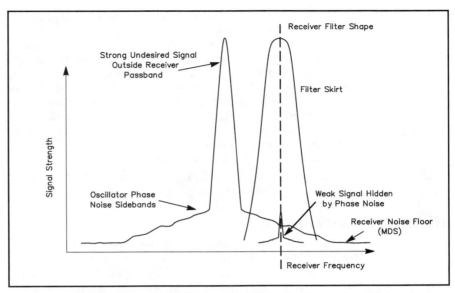

Figure 4-19 — In a receiver with excessive phase noise, a strong signal near the receiver passband can raise the apparent receiver noise floor in the passband. This increased noise can cover a weak signal that you are trying to receive.

The phase-noisy signal "splatters" up and down the band for some range around the desired transmit frequency, and this additional noise can fall within the passband of a receiver tuned to some weak signal. So even if you have a receiver with very low phase noise, you can be bothered by this type of interference!

One final receiver-performance limitation that you need to understand for your Extra class license exam is the concept of images and image-rejection ratio. The process of mixing two signals produces a new signal at the sum of their frequencies and another new signal at the difference between their frequencies. One of these mixing products is the desired one, and the other is undesired. Suppose we want to use the signal at the difference between the two frequencies for further processing in our receiver. For example, a receiver uses a 15-MHz local oscillator frequency to tune a signal at 25 MHz. The difference signal is 10 MHz, which represents an intermediate frequency for this receiver. Now suppose a signal at 5 MHz also gets into the mixer. The sum of the LO frequency and the unwanted signal is also 10 MHz! Both signals are amplified in the IF stage. The unwanted signal is known as an **image signal**.

The input filter to the RF amplifier must reject the unwanted signals and pass the desired signals. Most receivers have separate input filters for each band, which are selected by the bandswitch. Proper filter design is the key to solving an image problem with any receiver. It is not difficult to design a filter for the receiver just described, that will pass the 25-MHz signal and reject the 5-MHz signal. If the 5-MHz signal doesn't reach the mixer input, then it can't produce an image response in the receiver.

Other signals will also produce a response in this receiver. For example, a signal at 27 MHz will mix with the LO to produce a signal at 12 MHz. The RF amplifier input filter may not be able to adequately reject the 27-MHz signal, so the 12-MHz image may be passed to the first IF stage. If that signal were to make it through the first IF filter, and be passed on to the product detector or a second mixer, an image response could be produced. For example, the second IF may be at 1 MHz, and use an 11-MHz local oscillator. Both the desired 10-MHz signal and the undesired 12-MHz signal would produce signals at the 1-MHz IF, in this case. From this example, you can see that the RF amplifier stage determines the receiver image rejection ratio. This stage must include adequate filtering to pass the desired signal while blocking the undesired one.

[At this time, you should turn to Chapter 10 and study examination questions E4C01 and E4C08 through E4C13. Review this section as needed.]

Receiver Desensitization

A problem often encountered with repeater systems is **desensitization** of the receiver, almost invariably by a strong signal from a nearby transmitter (regardless of that transmitter's frequency). This signal so overloads the receiver that it becomes relatively insensitive to the signals it is supposed to receive. Desensitization can also result when a strong signal on a frequency near the received frequency causes a reduction in receiver sensitivity. In that case strong adjacent-channel signals cause the receiver desensitization.

In the case of a desensitized repeater receiver, the offending transmitter is often that of the repeater itself, so the transmitter is physically close to the receiver *and* operating on a frequency near the receive frequency. The key to eliminating desensitization is isolation — that is, isolating the receiver from any transmitters that might be causing or contributing to the problem.

Obviously, the repeater receiver and transmitter must be carefully shielded from each other. Such shielding usually involves physically separating the receiver from the transmitter and enclosing each in a metal box. All connections between the two boxes must be carefully shielded. Double-shielded coaxial cable is recommended. Separation of the transmitting and receiving antennas also is helpful, but often is impractical.

When both receiver and transmitter must use the same antenna, a series of cavities called a duplexer is employed to provide the isolation. The duplexer acts as a notch filter at the receive frequency to effectively attenuate the transmitter signal at the receiver. Total attenuation can easily be as much as 35 dB.

If the desensitization is the result of powerful nearby transmitters not operating on the same band, however, the use of low-pass, band-pass or high-pass filters might solve the problem. Which kind of filter is used depends on the frequency of the offending transmitter. If it is a commercial AM transmitter, a high-pass filter on the repeater receiver input might solve the problem. If not, rearrangement of the receiver antenna might also be necessary. If the problem should be a nearby commercial UHF television transmitter, a low-pass filter on the receiver input probably would help, assuming the receiver input frequency is lower than the TV frequency. If the problem is caused by a nearby amateur transmitter on the same band as the receiver, the best solution is to use a high-dynamic-range front end in your receiver.

Capture Effect

One of the most notable differences between an amplitude-modulated (AM) receiver and a frequency-modulated (FM) receiver is how noise and interference affect an incoming signal.

From the time of the first spark transmitter, "rotten QRM" has been a major problem for amateurs. The limiter and discriminator stages in an FM receiver can eliminate most of the impulse-type noise, except any noise that has frequency-modulation characteristics. For good noise suppression, the receiver IF system and detector phase tuning must be accurately aligned.

FM receivers perform quite differently from AM, SSB and CW receivers when QRM is present, exhibiting a characteristic known as the **capture effect**. The loudest signal received, even if it is only two or three times stronger than other signals on the same frequency, will be the only signal demodulated. On the other hand, an S9 AM, SSB or CW signal suffers noticeable interference from an S2 signal. Capture effect can be an advantage if you are trying to receive a strong station and there are weaker stations on the same frequency. At the same time, this phenomenon will prevent you from receiving one of the weaker signals if that is your desire.

Capture effect is most pronounced with FM emissions. The stronger signal blocks weaker signals, preventing them from being detected.

[Now turn to Chapter 10 and study examination questions E4C02 through E4C07. Review this section as needed.]

Intermodulation Distortion

Intermodulation distortion (IMD) or intermodulation interference occurs when signals from several transmitters, each operating on a different frequency, are mixed in a nonlinear manner. This produces signals at mixing products and may cause severe interference in a nearby receiver. Harmonics can also be generated in the nonlinear stage, and those frequencies will add to the possible mixing combinations. The intermod, as it is called, is transmitted along with the desired signal. Intermodulation distortion can also be produced in a receiver. Nonlinear circuits or devices cause intermodulation distortion in electronic circuits.

For example, suppose an amateur repeater receives on 144.85 MHz. Nearby, are relatively powerful nonamateur transmitters operating on 181.25 MHz and on 36.4 MHz (**Figure 4-20**). Neither of these frequencies is harmonically related to 144.85. The difference between the frequencies of the two nonamateur transmitters, however, is 144.85 MHz. If the signals from these transmitters are somehow mixed, the difference frequency (intermod) could be picked up by the amateur repeater and retransmitted over its own antenna. In this example, the signals may be mixing in one of the transmitters, and the intermod signal may actually be transmitted with the desired signal from that transmitter. Intermodulation interference can be produced when two transmitted signals mix in the final amplifiers of one or both transmitters, and unwanted signals at the sum and difference frequencies of the original signals are generated.

All transmitter operators are morally and legally required to ensure that their stations emit only clean signals. The problem often is that the operators of the other

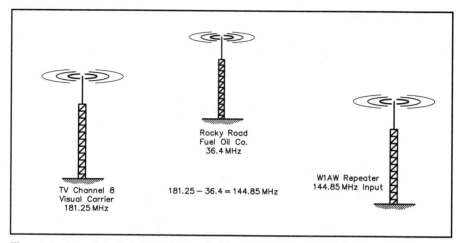

Figure 4-20 — This diagram shows a potential intermod situation. It is possible for some of the 36.4-MHZ Rocky Road transmitter carrier to mix with the channel 8 carrier signal at 181.25 MHz in the output stage of the TV transmitter. This mixing produces intermodulation distortion products at the sum and difference frequencies (181.25 + 36.4 = 217.65 MHz, 181.25 − 36.4 = 144.85 MHz). One of these intermod products is the input frequency of the W1AW repeater, and may key up the repeater. An isolator or circulator (see text) should be installed at the Channel 8 transmitter to solve this problem.

transmitters do not know their equipment is causing intermod with your club's repeater. When they become aware of it, they can take several steps to eliminate the problem.

Push-pull amplifiers are quite effective in eliminating even-numbered harmonics, and it is often the second harmonic of the frequency that is causing intermod. Another solution would be to use a linear (class-AB1) power amplifier instead of a class-C amplifier. (Linear amplifiers present fewer intermod problems than do non-linear amplifiers.) Should this remedy be impractical, at the very least, the class-C amplifier should be operated with the minimum grid current possible for efficient operation. (Chapter 7 contains detailed explanations of all these types of amplifiers and how they work.) All stages in the offending transmitter should be neutralized, to eliminate parasitic oscillations. (Low-pass and band-pass filters usually are ineffective in reducing intermod problems, because at VHF and UHF they are seldom sharp enough to suppress the offending signal without also weakening the wanted one.)

Two other devices that usually are highly effective in eliminating intermod are **isolators** and **circulators**. A terminated circulator is a precisely engineered ferrite component that functions like a one-way valve. It is a three-port device that can combine two or more transmitters for operation on one antenna. Very little transmitter energy is lost as RF travels to the antenna, but a considerable loss is imposed on any energy coming down the feed line to the transmitter. Thus, the circulator effectively reduces intermod problems. Another advantage to its use is that it provides a matched load at the transmitter output regardless of what the antenna-system SWR might be.

A two-port device called an isolator also helps reduce intermod; this device incorporates a built-in termination or load. **Figure 4-21** illustrates how circulators or isolators may be included in your repeater system.

Circulators and isolators are available for 144- and 450-MHz use. They come in power levels up to a few hundred watts. Typical bandwidth for a 150-MHz unit is 3 MHz. Insertion loss is roughly 0.5 dB, and rejection of unwanted energy coming down the feeder is usually 20 to 28 dB. The 450-MHz types have greater bandwidth — about 20 MHz — but otherwise perform about the same as the 150-MHz versions. Isolators and circulators are rather expensive, but, when needed, they are worth considerably more than their price.

Intermod, of course, is not limited to repeaters. Anywhere two relatively powerful and close-by transmitter fundamental-frequency outputs or their harmonics combine to create a sum or a difference signal at the frequency on which any other transmitter or receiver is operating, an intermod problem can develop.

A two-tone IMD test on a receiver is a measure of the range of signals that can be tolerated at the receiver input before it begins to generate spurious mixing products. This specification gives an idea of how the receiver will perform in the presence of strong signals.

A receiver can experience intermodulation distortion interference when it receives two strong signals. In that case, the interference is actually generated in the receiver, so there is nothing the transmitter operators can do to solve the problem. (Well, nothing short of changing frequency or turning it off, at least.) As an example, suppose you have a receiver tuned to 146.70 MHz, to monitor a local repeater. You begin to hear a distorted signal, which you eventually identify as another local amateur operating on

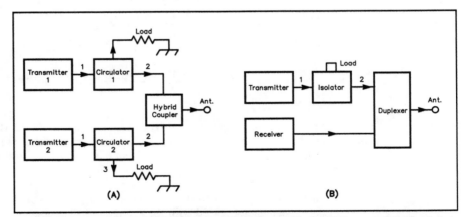

Figure 4-21 — A block diagram showing the use of circulators and an isolator. Two circulators may be used to connect one antenna to two transmitters, as in A. Often, duplexers also are used, along with the circulators. In B, an isolator is placed between the transmitter and the duplexer to reduce intermodulation products.

146.52 MHz. The interference is intermittent, and by listening on another receiver tuned to 146.52 MHz, you determine that the interference does not occur every time the other ham transmits. What are the two most likely frequencies for the second signal that may be causing this intermod in your receiver?

This interference is an example of *third-order IMD*. The interfering signals occur at frequencies corresponding to the sum and difference of two times one signal frequency and a second signal frequency. This statement may be easier to understand if we express it as four equations.

$$f_{IMD\ 1} = 2f_1 + f_2 \qquad\qquad \text{(Equation 4-6)}$$

$$f_{IMD\ 2} = 2f_1 - f_2 \qquad\qquad \text{(Equation 4-7)}$$

$$f_{IMD\ 3} = 2f_2 + f_1 \qquad\qquad \text{(Equation 4-8)}$$

$$f_{IMD\ 4} = 2f_2 - f_1 \qquad\qquad \text{(Equation 4-9)}$$

It turns out that only the subtractive products (given by Equations 4-7 and 4-9) are close enough to the desired signal to cause significant interference. In our example, we know the IMD frequency of interest is 146.70 MHz, and we have found one of the signals producing the interference at 146.52 MHz, and want to find the two most likely frequencies for the second signal. Use Equation 4-7 to calculate one possible unknown frequency (f_2). The first step is to rearrange the terms to solve the equation for f_2.

$$f_{IMD\ 2} = 2f_1 - f_2$$

$$f_2 = 2f_1 - f_{IMD\ 2} = 2 \times 146.52 \text{ MHz} - 146.70 \text{ MHz}$$

$$f_2 = 293.04 \text{ MHz} - 146.70 \text{ MHz} = 146.34 \text{ MHz}$$

Next we will use Equation 4-9 to calculate another possible frequency for the second interfering signal. Again, the first step is to rearrange terms to solve for f_2.

$$f_{IMD\ 4} = 2f_2 - f_1$$

$$2f_2 = f_{IMD\ 4} + f_1$$

$$f_2 = \frac{f_{IMD\ 4} + f_1}{2} = \frac{146.70 \text{ MHz} + 146.52 \text{ MHz}}{2} = \frac{293.22 \text{ MHz}}{2}$$

$$f_2 = 146.61 \text{ MHz}$$

The two most likely frequencies to be mixing with the 146.52 MHz signal and producing this receiver intermodulation distortion interference are 146.34 MHz and 146.61 MHz. Higher-order IMD products can also occur, and they could be causing the interference, but the higher-order products are usually weaker, so they are less likely to be the cause.

Another IMD topic has to do with transmitter spectral-output purity. When several audio signals are mixed with the carrier signal to generate the modulated signal, spurious signals will also be produced. These are normally reduced by filtering after the mixer, but their strength will depend on the level of the signals being mixed, among other things, and they will be present in the transmitter output to some extent. You can perform a transmitter two-tone test by putting two equal-amplitude audio tones into the microphone circuit. Then you can view the transmitter output on an oscilloscope (or spectrum analyzer) to get an indication of how linear the amplifier stages are. This is important if you are to be sure that your transmitted signal is clean.

Excessive intermodulation distortion of an SSB transmitter output signal results in splatter being transmitted over a wide bandwidth. The transmitted signal will be distorted, with spurious (unwanted) signals on adjacent frequencies.

[Before proceeding, turn to Chapter 10 and study questions E4C14, E4C15 and E4C16. Review this section as needed.]

NOISE SUPPRESSION

One of the most significant deterrents to effective signal reception during mobile or portable operation is electrical impulse noise from the automotive ignition system. The problem also arises during use of gasoline-powered portable generators. This form of interference can completely mask a weak signal. Other sources of noise include conducted interference from the vehicle battery-charging system, instrument-caused interference, static and corona discharge from the mobile antenna. Atmospheric static and electrical line noise present some difficult problems for home-station operation. This section offers some suggestions about how to solve these noise problems.

Ignition Noise

Most electrical noise can be eliminated by taking logical steps to suppress it. The first step is to clean up the noise source itself, then to use the receiver's built-in noise-reducing circuit as a last measure to minimize any noise impulses from passing cars or other man-made sources. In general, you can suppress electrical noise in a mobile transceiver by applying shielding and filtering. The exact method of applying these fixes will depend on the type of ignition system your car uses.

Most vehicles manufactured prior to 1975 were equipped with inductive-discharge ignition systems. A variety of noise-suppression methods were devised for these systems, such as: resistor spark plugs, clip-on suppressors, resistive high-voltage spark-plug cable, and even complete shielding. Resistor spark plugs and resistive high-voltage cable provide the most effective noise reduction for inductive-discharge systems at the least cost and effort. Almost all vehicles produced after 1960 had resistance cable as standard equipment. Such cable develops cracks in the insulation after a few years of service, and should be replaced on a regular schedule. Two years is a reasonable replacement schedule.

Most newer automobiles employ sophisticated, high-energy, electronic-ignition systems (sometimes called capacitive-discharge ignition) to reduce exhaust pollution and increase fuel mileage. Solutions to noise problems that are effective for inductive-discharge systems cannot uniformly be applied to the modern electronic systems. Such fixes may be ineffective at best, and at worst may impair engine performance. One significant feature of capacitive-discharge systems is extremely rapid voltage rise, which combats misfiring caused by fouled plugs. Rapid voltage rise depends on a low RC time constant being presented to the output transformer. High-voltage cable designed for capacitive-discharge systems exhibits a distributed resistance of about 600 ohms per foot, compared with 10 kilohms per foot for cable used with inductive-discharge systems. Increasing the RC product by shielding or installing improper spark-plug cable could seriously affect the capacitive-discharge-circuit operation.

Ferrite beads are a possible means for RFI reduction in newer vehicles. Both primary and secondary ignition leads are candidates for beads. Install them liberally, then test the engine under load to ensure adequate spark-plug performance.

Electrical bonding can reduce the level of ignition noise, both inside the vehicle and out. The sheet metal surfaces, frame and body parts may exhibit resonance in one of the amateur bands, and such resonances encourage the reradiation of spark impulse energy. Bonding the metal surfaces together will encourage lower-frequency resonances. Other types of noise, described later, can also be helped by bonding. Bond the following structural members with heavy metal braid: (1) engine to frame; (2) air cleaner to engine block; (3) exhaust lines to frame; (4) battery ground terminal to frame; (5) hood to fire wall; (6) steering column to frame; (7) bumpers to frame; and (8) trunk lid to frame.

Charging-System Noise

Noise from the vehicle battery-charging system can interfere with both reception and transmission of radio signals. The charging system of a modern automobile consists of a belt-driven, three-phase alternator and a solid-state voltage regulator. Interference from the charging system can affect receiver performance in two ways: RF radiation can be picked up by the antenna, and noise can be conducted directly into the circuitry through the power cable.

Alternator whine is a common form of conducted interference, and can affect both transmitting and receiving. This noise is characterized by a high-pitched buzz on the transmitted or received signal. The tone changes frequency as the engine speed changes.

VHF FM communications are the most affected by alternator whine, since synthesized carrier generators and local oscillators are easily frequency modulated by power-supply voltage fluctuation. The alternator ripple is most noticeable when transmitting, because the alternator is most heavily loaded in that condition.

Conducted noise can be minimized by connecting the radio power leads directly to the battery, as this point is the lowest impedance point in the system. Connect both the positive and negative leads directly to the battery, with a fuse rated to carry the transmit current installed in each lead. **Figure 4-22** shows a diagram of a typical mobile transceiver installation.

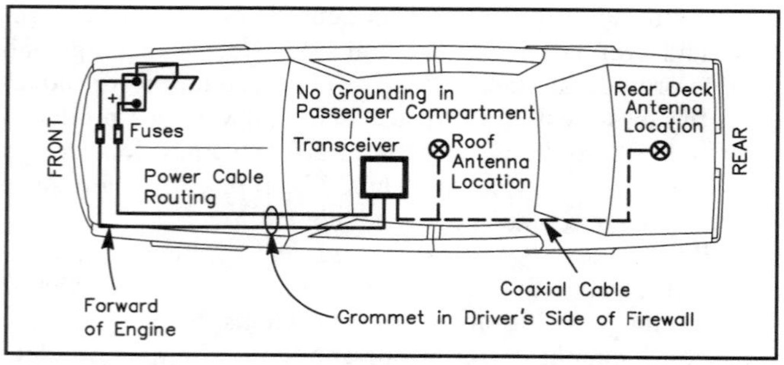

Figure 4-22 — To minimize alternator whine and conducted noise, connect both transceiver power-supply leads directly to the battery, as this wiring diagram shows. Include fuses in both power leads, near the battery.

If the voltage regulator is adjustable, set the voltage no higher than necessary to ensure complete battery charging. **Radiated noise** and conducted noise can be suppressed by filtering the alternator leads. **Coaxial capacitors** (about 0.5 µF) are suitable, but never connect a capacitor to the alternator field lead. The field lead can be shielded, or loaded with ferrite beads. Keep the alternator slip rings clean to prevent excess arcing. An increase in "hash" may indicate that the brushes need replacement.

Instrument Noise

Some automotive instruments can create noise. Among these gauges and senders are the engine-heat and fuel-level indicators. Ordinarily the installation of a 0.5-µF coaxial capacitor at the sender element will cure this problem.

Other noise-generating accessories include turn-signals, window-opener motors, heating-fan motors and electric windshield-wiper motors. The installation of a 0.25-µF capacitor across the motor winding will usually eliminate this type of interference.

Static Electricity Noise

When a charge of static electricity builds up on some object and then discharges to ground, it can produce a popping, crackling noise in receivers. The most common example of static noise is caused by a buildup of static electricity in the atmosphere. When there is a thunderstorm in your area you will definitely hear this type of noise on the lower-frequency HF bands, such as 160, 80 and 40 meters.

The noise from a thunderstorm can be heard in radio receivers for up to several hundred miles from the storm center. This static is louder on the lower HF frequencies, and the noise is more frequent in summer. Also, storms are more frequent in the lower latitudes than in the higher ones. Most static electricity noise in a receiver is produced by a thunderstorm somewhere in the area. Even when there is no thunderstorm you can be bothered by static noise, however. Any form of precipitation can pick up some static charge as it falls, so there can be high noise levels during any storm in your area. Noise produced by the falling precipitation is often called snow or rain static, depending on which type of precipitation is falling. Both mobile and fixed stations can be bothered by atmospheric static noise.

[Now turn to Chapter 10 and study exam questions E4D01 through E4D06. Also study question E4D08. Review this section if you have any difficulty with these questions.]

Electrical Line Noise

Electrical line noise can be particularly troublesome to operators working from a fixed location. The loud buzz or crackling sound of a line noise can cover all but the strongest signals. Most of this man-made interference is produced by some type of electrical arc. An electrical arc generates varying amounts of RF energy across the radio spectrum. (In the early days of radio, amateurs used spark-gap transmitters to generate their radio signals.)

When an electric current jumps a gap between two conductors, an arc is produced as the current travels through the air. See **Figure 4-23**. To produce such an arc, the voltage must be large enough to ionize the air between the conductors. Once an ionized path is established, there is a current through the gap. The electron flow through this gap is highly irregular compared to the rather smooth flow through a conductor. The resistance of the ionized air varies constantly, so the instantaneous current is also changing. This causes radio-frequency energy to be radiated and for the noise to be conducted along the power wires. There is a definite relationship between the length of

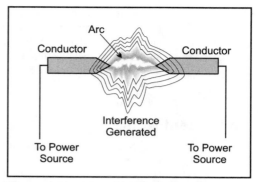

Figure 4-23 — An electrical arc can form through the air gap between two conductors.

the arc and the voltage needed to produce it. The longer the arc and the higher the voltage, the greater the interference it causes.

Electrical arcs are created in a variety of everyday appliances, especially those using brush-type motors. Electric shavers, sewing machines and vacuum cleaners are just a few examples. In addition, devices that control voltage or current by opening or closing a circuit can produce arcs that cause interference. Light dimmers and heater elements are a few examples of devices like this. Poor connections or defective insulators in the power distribution system can also produce line noise.

One effective way to reduce electrical noise produced by an electric motor is to use a "brute force" ac line filter in series with the power leads. This filter will block the noise from being conducted along the power wiring away from the motor. **Figure 4-24** shows the schematic diagram for such a filter. All components must be ac rated, and able to carry the current required by the motor or appliance connected to the filter. We recommend you purchase a UL-listed commercial filter for this application.

Tracking Down the Interference

Perhaps one of the most frustrating facts about a line noise is its intermittent nature. The noise will come and go without warning, usually with no apparent pattern or timing (except that it's in the middle of your favorite net or an exotic DX contact). This can make it extremely difficult to locate the source of the interference. There are two ways that noise interference can find its way into a receiver. If the interference source is located in the same building as the receiver, it's likely that the noise will flow along the house wiring. For example, if an oil-burner motor is at fault, the interference may ride the ac line from the furnace to wherever the receiver is located. The other way that this type of inter-

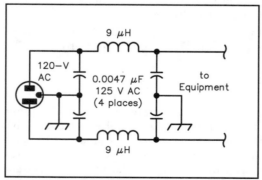

Figure 4-24 — Install a "brute-force" ac line filter in the power leads to an ac motor to reduce motor-noise interference.

ference is sometimes picked up is through an antenna and feed line.

The first step in tracking down the interference is to determine if it is being generated in your own house. Check this by pulling the main fuse or circuit breaker. You will need a battery-powered receiver for this test, but a portable AM receiver should work fine. Check to be sure you can hear the noise on your portable receiver. Tune to a clear frequency and listen for the interference.

If the noise goes away when you turn off all power to your house check to see if it comes back when you turn the power on again. It's possible that the offending device only produces the noise after it has been operating for a while, so the interference may not come back immediately when you restore power. If it doesn't, continue your investigation later.

After you are reasonably confident that the noise is being produced in your house, proceed with your investigation, one circuit at a time. When you narrow it down to a particular circuit, go around and unplug any appliances or other electrical devices one by one. Be persistent. You will probably have to continue your investigation over several days, weeks or even months.

If the interference is not being produced in your house, the search will become a bit more difficult. The problem may be an arc in the utility company's power distribution system, a neighbor's appliance or any of a number of other items. Your portable AM receiver can be used to "sniff" along the power lines, looking for stronger interference. A directional antenna may help you locate the noise source. As you will learn in the next section, when you get close to the noise source you may find that the null, or direction of weakest signal in the antenna pattern is more helpful in pinpointing the direction to the source. An FM receiver, with or without a directional antenna will not be helpful in tracking down the noise source, because FM receivers are not affected by line noise.

When you think you've located the source, contact the power company and explain the problem to them. Be as specific as possible about where you believe the interference is originating.

In addition to receiving interference from these various noise sources, your transmitter may also cause interference to some devices, such as a TV, radio or telephone. The ac and telephone wiring in your house may pick up signals from your transmitter and conduct them to the device. Such interference is often caused by *common-mode* signals. This means the signal flows in the same direction on both wires in the power or phone line, rather than having opposite directions along the wires as it would on a transmission line. You'll need a common-mode choke to cure this type of interference. Wind several turns of the power cord or phone line around a ferrite toroid core. Number 43 material is a good choice for most HF problems.

For more information about electrical power line noise and other types of interference, see *The ARRL RFI Book*. That book contains more detailed information about locating and curing this and other types of radio interference.

[Now turn to Chapter 10 and study examination question E4D07 and questions E4D09 through E4D11. Review this section as needed.]

COMPONENT MOUNTING TECHNIQUES

Sooner or later, most amateurs try building some piece of station equipment or accessory. They may use any of a variety of construction techniques to build this equipment.

Most electronics experimenters are familiar with printed circuit (PC) boards. This is often the technique used for a first project. An insulating board with copper circuit traces forms the basis for the project. Component leads are inserted through holes in the circuit-board material, and the leads are soldered to the copper traces. Some amateurs trace the circuit pattern onto the copper-clad circuit-board material and etch away the excess copper, while others prefer to purchase circuit boards prepared by commercial suppliers.

You can also build a project on "perf board," which is an insulating material with a grid of holes for component leads. Connecting wires may be soldered to the component leads or wire-wrap construction may be used. Others may prefer to build their projects on "universal" or "experimenters' circuit boards." These boards usually have holes and circuit pads for one or more dual in-line package (DIP) integrated circuits, with copper traces to additional pads for mounting other components.

Figure 4-25 — This photo shows how the "dead bug" construction method can be used to build a circuit that requires little or no advance preparation.

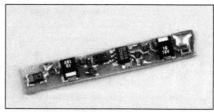

Figure 4-26 — Surface-mount components are used for circuits that require minimal stray inductance and capacitance between the component leads. Amateurs use such components especially for UHF and microwave circuits.

Still another circuit-construction technique that is popular with many hams is variously called "dead bug construction," "ugly construction" or "ground-plane construction." With this construction technique, a piece of copper-clad circuit-board material is used for a base. Components that connect to ground are soldered directly to the copper foil. Ungrounded connections are made directly to the component leads. Transistors and integrated circuits are often glued to the circuit board upside down, so their leads stick up for connections to other components. (This is where the name "dead bug" comes from.) **Figure 4-25** is a photo of a circuit built using this construction technique.

To reduce the effects of stray inductance and capacitance, and to aid robotic circuit construction, many components are available in **surface-mount packages**. These devices are soldered directly to circuit-board pads, and mount flush against the board. Amateurs can take advantage of these surface-mount packages, especially for UHF and microwave circuits, where stray capacitance and inductance must be considered. Surface-mount resistors and capacitors are small rectangles with metal end caps so they can be soldered to the circuit-board pads. These tiny components without wire leads are sometimes called "chips." **Figure 4-26** is a photo of a circuit that uses some surface-mounted components.

[Study question E4E01 in Chapter 10 before you go on to the next section. Review this material if you have difficulty with that question.]

DIRECTION-FINDING TECHNIQUES

Radio direction finding (RDF) is as old as radio itself. RDF is just what the name implies — finding the direction or location of a transmitted signal. The practical aspects of RDF include radio navigation, location of downed aircraft and identification of sources of signals that are jamming communications or causing radio-frequency interference. In many countries, the hunting of hidden transmitters has become a sport, often called *fox hunting*. Participants in automobiles or on foot use receivers and direction-finding techniques to locate a hidden transmitter.

The equipment required for an RDF system is a **directive antenna** and a device for detecting the radio signal. In Amateur Radio applications, the signal detector is usually a receiver with a meter to indicate signal strength. Some form of RF attenuation is desirable to allow proper operation of the receiver under high signal conditions, such as when "zeroing in" on the transmitter at close range. Otherwise the strong signals may overload the receiver. The directive antenna can take many forms.

In general, an antenna for direction-finding work should have good front-to-back and front-to-side ratios. Since the peak in an antenna pattern is often very broad, it can be difficult to identify the exact direction a signal is coming from. Antenna-pattern nulls are usually very narrow, however. The normal technique for RDF work is to use the pattern null to indicate the direction of the transmitted signal.

Loop Antennas

A simple antenna for RDF work is a small **loop antenna**, which consists of one or more turns of wire wound in the shape of a large open inductor. The loop is usually tuned to resonance with a capacitor. Several factors must be considered in the design of an RDF loop. The loop must be small compared to the wavelength — in a single-turn loop, the conductor should be less than 0.08 wavelength long. For 28 MHz, this represents a length of less than 34 inches (10-inch diameter). An ideal loop antenna has *maximum* response in the plane of the loop, as **Figure 4-27A** shows. An ideal loop has deep nulls at right angles to that plane. Because there are two nulls, we say the pattern is bidirectional. For RDF work, we often say the direction finding pattern is at right angles — or *broadside* — to the plane of the loop. The loop antenna is a simple one to construct, but the bidirectional pattern is a major drawback. You can't tell which of the two directions points to the signal source!

To obtain the most accurate bearings, the loop must be balanced electrostatically with respect to ground. Otherwise the loop will exhibit two modes of operation. One is the mode of the true loop, while the other is that of an essentially nondirectional vertical antenna of small dimensions. This second mode of operation is sometimes called **antenna effect**. The voltages introduced by the two modes are not in phase, and may add or subtract, depending on the direction from which the wave is coming.

Refer to Figure 4-27. The theoretical true loop pattern is shown in Figure 4-27A. When properly balanced, the loop exhibits two nulls that are 180° apart. Thus, a single null reading with a small loop antenna will not indicate the exact direction toward the transmitter — only the line along which it lies. When the antenna effect is appreciable and the loop is tuned to resonance, the loop may exhibit little directivity (Figure 4-27B). By detuning the loop to shift the phasing, a pattern similar to Figure 4-27C may be obtained. This pattern does exhibit a pair of nulls, although they are not symmetrical. The nulls may not be as sharp as that obtained with a well-balanced loop, and may not be at right angles to the plane of the loop.

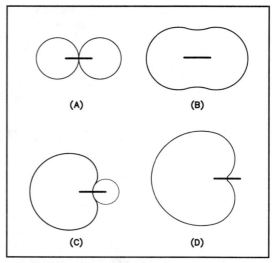

Figure 4-27 — Small loop-antenna field patterns. The heavy lines show the plane of the loop. Part A is for an ideal loop, B is the pattern with appreciable antenna effect present. Part C represents a loop that has been detuned to shift the phasing. Part D is the optimum detuning to produce a single null.

By suitable detuning, a unidirectional pattern may be approached (Figure 4-27D). There is no complete null in the pattern, but the loop is adjusted for the best null. An electrostatic balance can be obtained by shielding the loop, as shown in **Figure 4-28**. This eliminates the antenna effect, and the response of a well-constructed shielded loop is quite close to the ideal pattern of Figure 4-27A.

For the lower-frequency amateur bands, single-turn loops are generally not satisfactory for RDF work. Therefore, multiturn loops, such as shown in **Figure 4-29**, are generally used. This loop may also be shielded, and if the total conductor length remains below 0.08 wavelength, the pattern is that of Figure 4-27A.

A loop antenna responds to the magnetic field of the radio wave, not to the electric field. The voltage delivered by the loop is proportional to the amount of magnetic flux passing through the coil, and to the number of turns in the coil. The action is much the same as in the secondary winding of a transformer. You could increase the output voltage of the loop by increasing the number of turns in the loop or the loop area.

Sensing Antennas

Because there are two nulls 180° apart in the directional patterns of loop antennas, an ambiguity exists as to which one indicates the true direction of the signal. If there is more than one receiving station, or if the single receiving station takes bearings from more than one position, the ambiguity may be resolved through **triangu-**

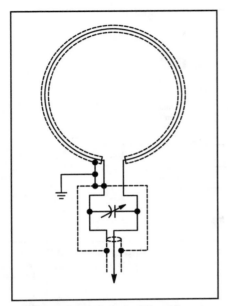

Figure 4-28 — This drawing shows a shielded loop for direction finding. The ends of the shielding turn are not connected. The shielding is effective against electric fields.

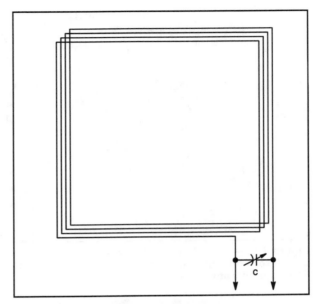

Figure 4-29 — A small loop consisting of several turns of wire makes an effective HF direction-finding antenna. The total conductor length is much less than a wavelength. Maximum response is in the plane of the loop, so the nulls are broadside to the loop.

lation. It may be more desirable for a pattern to have just one null, however, so there is no question about where the transmitter's true direction lies. A loop may be made to have a single null if a second antenna element, called a **sensing antenna**, is added. The second element must be omnidirectional, such as a short vertical. If the signals from the loop and the sensing antenna are combined with a 90° phase shift between the two, a **cardioid radiation pattern** results. This pattern has a single large lobe in one direction, with a deep, narrow null in the opposite direction. The deep null can help pinpoint the direction of the desired signal.

The development of the cardioid pattern is shown in **Figure 4-30A**. The loop and sensing-element patterns combine to form the cardioid pattern. In the top half of the graph the patterns add and in the bottom half they subtract. Figure 4-30B shows a circuit for adding a sensing antenna to a loop antenna. For the best null in the composite pattern, the signals from the loop and the sensing antenna must be of equal amplitude. R_1 is an internal adjustment, and is adjusted experimentally to control the signal level from the sensing antenna. The null of the cardioid is 90° away from the nulls of the loop, so it is customary to first use the loop alone to obtain a precise bearing line, then switch in the sensing antenna to resolve the ambiguity.

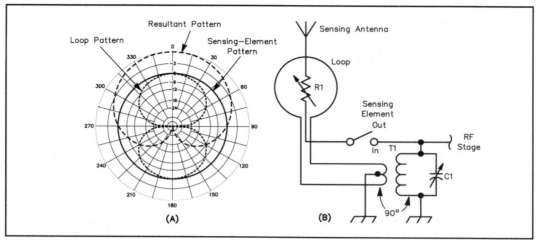

Figure 4-30 — At A, the directivity pattern of a loop antenna with sensing element. At B is a circuit for combining the signals from the two elements. C1 is adjusted for resonance with T1 at the operating frequency.

Phased Arrays

There are two general classifications of phased arrays — broadside and end-fire. Broadside arrays are inherently bidirectional — there are always at least two nulls in the pattern. Broadside arrays alone are seldom used for amateur RDF applications. Depending on the spacing and phasing of the elements, end-fire patterns may exhibit a null off one end of the axis of the elements, while at the same time the response is maximum off the other end of the axis, in the opposite direction from the null. A common arrangement uses two elements, spaced ¼ wavelength apart, and fed 90° out of phase. The result is a cardioid pattern, with the null in the direction of the leading element.

One of the most popular types of end-fire arrays was invented by F. Adcock, and patented in 1919. The **Adcock array** consists of two vertical elements, fed 180° apart, and mounted so the system can be rotated. Element spacing is not critical, and may be in the range ¹/₁₀ to ³/₄ wavelength. The two elements must be of the same length, but need not be self-resonant. Elements shorter than resonant are commonly used. Because neither spacing nor length is critical, the Adcock array may be operated over more than one amateur band.

The radiation pattern for the Adcock array is shown in **Figure 4-31A**. The nulls are in directions broadside to the array axis, and become sharper with greater element spacing. With an element spacing greater than ³/₄ wavelength, however, the pattern begins to have additional nulls off the ends of the array axis. At a spacing of one wavelength, the pattern is that of Figure 4-31B, and the array is unsuitable for RDF applications. The Adcock array, with its two nulls, has the same ambiguity as the loop antenna. Adding a sensing element to the Adcock array has not met with much success because of mutual coupling between the array elements and the sensing element, among other

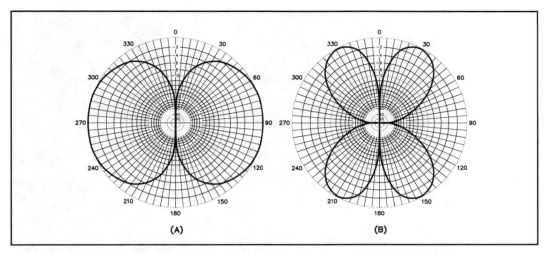

Figure 4-31 — At A, the pattern of the Adcock array with an element spacing of ¹/₂ wavelength. In these plots the elements are aligned with the horizontal axis. As element spacing is increased beyond ³/₄ wavelength, additional nulls develop off the ends of the array, and at a spacing of 1 wavelength the pattern at B is produced.

things. Because Adcock arrays are usually used for fixed-station operation, as part of a group of stations in an RDF network, the ambiguity presents no serious problems.

Most amateur RDF work is done within a few miles of the signal source, using **ground-wave signals**. At HF, **sky-wave signals** present some difficulties, mostly because of Faraday rotation as the signals travel through the ionosphere. This causes fading and makes it difficult to identify a true null direction. Loop antennas are generally unsatisfactory for sky-wave RDF, so the Adcock array has become the antenna of choice for this type of work. High radiation angles can still present problems, but the Adcock array offers several advantages for direction finding on sky-wave signals.

Triangulation

If two, or more, RDF bearing measurements are made at locations that are separated by a significant distance, the bearing lines can be drawn from those positions as represented on a map. See **Figure 4-32**. This technique is called **triangulation**. Notice that it is important that the two DF sites not be on the same straight line with the signal you are trying to find. The point where the lines cross (assuming the bearings are not the same nor 180° apart) will indicate a "fix" of the approximate transmitter location. The word "approximate" is used because there is always some uncertainty in the bearings obtained. Propagation effects may add to the uncertainty. In order to best indicate the probable location of the transmitter, the bearings from each position should be drawn as narrow sectors instead of single lines. Figure 4-32 shows the effect of drawing bearings in sectors — the location of the transmitter is likely to be found in the area bounded by the intersection of the various sectors.

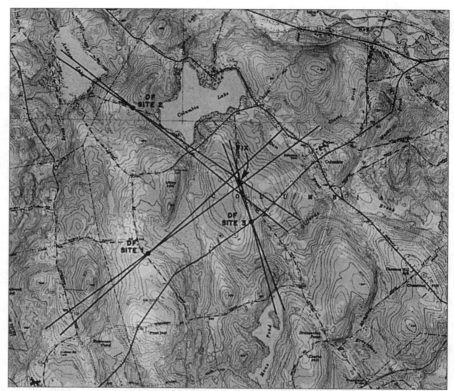

Figure 4-32 — Bearing sectors from three RDF positions are drawn on a map. The method is known as triangulation. Note that sensing antennas are not required at any of the RDF sites; antennas with two null indications 180° apart are quite acceptable when several separate bearings can be taken.

Terrain Effects

Most amateur RDF activity is conducted with ground-wave signals. The best accuracy in determining a bearing to a signal source is when the propagation path is over homogeneous terrain, and when only the vertically polarized component of the ground wave is present. (Homogeneous terrain means there are no hills, trees or buildings to block or reflect the signals.) If a boundary exists, such as between land and water, the different conductivities of the two mediums under the ground wave can cause bending (refraction) of the wave front. In addition, reflection of RF energy from vertical objects, such as mountains or buildings, can add to the direct wave and cause RDF errors.

The effects of refraction and reflection are shown in **Figure 4-33**. At A, the signal is actually arriving from a direction different than the true direction of the transmitter. This happens because the wave is refracted at the shoreline. Even the

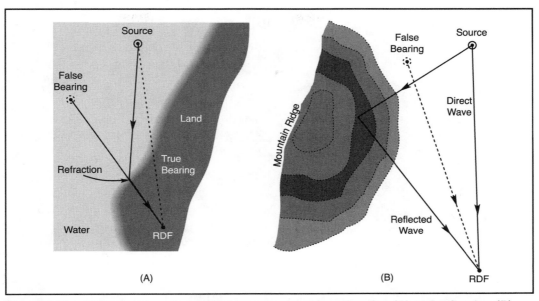

Figure 4-33 — This drawing shows RDF errors caused by refraction (A) and reflection (B). At A, a false reading is obtained because the signal actually arrives from a direction that is different from that to the source. At B, a direct signal from the source combines with a reflected signal from the mountain ridge. The two signals are averaged at the antenna, giving a false bearing somewhere between the two apparent sources.

most sophisticated RDF equipment will not indicate the true bearing in this instance, as the equipment can only show the direction from which the signal is arriving.

In Figure 4-33B, there are two apparent sources for the incoming signal — a direct wave from the source itself, and another wave that is reflected from the mountain ridge. In this case, the two signals add at the antenna of the RDF equipment. The uninitiated observer would probably obtain a false bearing in a direction somewhere between the directions to the two sources. The experienced RDF operator might notice that the null reading in this situation is not as sharp, or deep, as it usually is, but these indications would be subtle and easy to overlook.

Water towers, tall radio towers, and similar objects can also lead to false bearings. The effects of these objects become significant when they are large in terms of a wavelength. Local objects, such as buildings of concrete and steel construction, power lines, and the like, also tend to distort the field. It is important that the RDF antenna be in the clear, well away from surrounding objects.

[Before proceeding to Chapter 5, turn to Chapter 10 and study exam questions E4E02 through E4E12. Review this section if you have any difficulty with these questions.]

CHAPTER 5
KEYWORDS
KEYWORDS
KEYWORDS

Admittance — The reciprocal of impedance, often used to aid the solution of a parallel-circuit impedance calculation.

Apparent power — The product of the RMS current and voltage values in a circuit without consideration of the phase angle between them.

Average power — The product of the RMS current and voltage values associated with a purely resistive circuit, equal to one half the peak power when the applied voltage is a sine wave.

Back EMF — An opposing electromotive force (voltage) produced by a changing current in a coil. It can be equal to (or greater than) the applied EMF under some conditions.

Bandwidth — The frequency range (measured in hertz) over which a signal is stronger than some specified amount below the peak signal level. For example, if a certain signal is at least half as strong as the peak power level over a range of ± 3 kHz, the signal has a 3-dB bandwidth of 6 kHz.

Complex number — A number that includes both a real and an imaginary part. Complex numbers provide a convenient way to represent a quantity (like impedance) that is made up of two different quantities (like resistance and reactance).

Conductance — The reciprocal of resistance. This is the real part of a complex admittance.

Decibel (dB) — One tenth of a bel, denoting a logarithm of the ratio of two power levels— dB = 10 log (P2/P1). Power gains and losses are expressed in decibels.

Effective radiated power (ERP) — The relative amount of power radiated in a specific direction from an antenna, taking system gains and losses into account.

Electric field — A region through which an electric force will act on an electrically charged object.

Field — The region of space through which any of the invisible forces in nature, such as gravity, electric force or magnetic forces, act.

Half-power points — Those points on the response curve of a resonant circuit where the power is one half its value at resonance.

Imaginary number — A value that sometimes comes up in solving a mathematical problem, equal to the square root of a negative number. Since there is no real number that can be multiplied by itself to give a negative result, this quantity is imaginary. In electronics work, the symbol j is used to represent $\sqrt{-1}$. Other imaginary numbers are represented by $j\sqrt{x}$, where x is the positive part of the number. The reactance and susceptance of complex impedances and admittances are normally given in terms of j.

Joule — The unit of energy in the metric system of measure.

Magnetic field — A region through which a magnetic force will act on a magnetic object.

Optical shaft encoder — A device consisting of two pairs of photoemitters and photodetectors, used to sense the rotation speed and direction of a knob or dial. Optical shaft encoders are often used with the tuning knob on a modern radio to provide a tuning signal for the microprocessor controlling the frequency synthesizer.

Optocoupler (optoisolator) — A device consisting of a photoemitter and a photodetector used to transfer a signal between circuits using widely varying operating voltages.

Parallel-resonant circuit — A circuit including a capacitor, an inductor and sometimes a resistor, connected in parallel, and in which the inductive and capacitive reactances are equal at the applied-signal frequency. The circuit impedance is a maximum, and the current is a minimum at the resonant frequency.

Peak envelope power (PEP) — The average power of the RF envelope during a modulation peak. (Used for modulated RF signals.)

Peak power — The product of peak voltage and peak current in a resistive circuit. (Used with sine-wave signals.)

Phase — A representation of the relative time or space between two points on a waveform, or between related points on different waveforms. Also the time interval between two events in a regularly recurring cycle.

Phase angle — If one complete cycle of a waveform is divided into 360 equal parts, then the phase relationship between two points or two waves can be expressed as an angle.

Photoconductive effect — A result of the photoelectric effect that shows up as an increase in the electric conductivity of a material. Many semiconductor materials exhibit a significant increase in conductance when electromagnetic radiation strikes them.

Photoelectric effect — An interaction between electromagnetic radiation and matter resulting in photons of radiation being absorbed and electrons being knocked loose from the atom by this energy.

Polar-coordinate system — A method of representing the position of a point on a plane by specifying the radial distance from an origin, and an angle measured counterclockwise from the 0° line.

Potential energy — Stored energy. This stored energy can do some work when it is "released." For example, electrical energy can be stored as an electric field in a capacitor or as a magnetic field in an inductor. This stored energy can produce a current in a circuit when it is released.

Power — The time rate of transferring or transforming energy, or the rate at which work is done. In an electric circuit, power is calculated by multiplying the voltage applied to the circuit by the current through the circuit.

Power factor — The ratio of real power to apparent power in a circuit. Also calculated as the cosine of the phase angle between current and voltage in a circuit.

Q — A quality factor describing how closely a practical coil or capacitor approaches the characteristics of an ideal component.

Reactive power — The apparent power in an inductor or capacitor. The product of RMS current through a reactive component and the RMS voltage across it. Also called wattless power.

Real power — The actual power dissipated in a circuit, calculated to be the product of the apparent power times the phase angle between the voltage and current.

Rectangular-coordinate system — A method of representing the position of a point on a plane by specifying the distance from an origin in two perpendicular directions.

Resonant frequency — That frequency at which a circuit including capacitors and inductors presents a purely resistive impedance. The inductive reactance in the circuit is equal to the capacitive reactance.

Series-resonant circuit — A circuit including a capacitor, an inductor and sometimes a resistor, connected in series, and in which the inductive and capacitive reactances are equal at the applied-signal frequency. The circuit impedance is at a minimum, and the current is a maximum at the resonant frequency.

Skin effect — A condition in which ac flows in the outer portions of a conductor. The higher the signal frequency, the less the electric and magnetic fields penetrate the conductor and the smaller the effective area of a given wire for carrying the electrons.

Smith Chart — A coordinate system developed by Phillip Smith to represent complex impedances on a graph. This chart makes it easy to perform calculations involving antenna and transmission-line impedances and SWR.

Susceptance — The reciprocal of reactance. This is the imaginary part of a complex admittance.

Thevenin's Theorem — Any combination of voltage sources and impedances, no matter how complex, can be replaced by a single voltage source and a single impedance that will present the same voltage and current to a load circuit.

Time constant — The product of resistance and capacitance in a simple series or parallel RC circuit, or the inductance divided by the resistance in a simple series or parallel RL circuit. One time constant is the time required for a voltage across a capacitor or a current through an inductor to build up to 63.2% of its steady-state value, or to decay to 36.8% of the initial value. After a total of 5 time constants have elapsed, the voltage or current is considered to have reached its final value.

ELECTRICAL PRINCIPLES

While working your way up the ladder of Amateur Radio license classes, you have been learning some important electronics principles. You have studied both dc and ac circuit theory. To pass the Amateur Extra class exam, you will have to know about some more-complex topics.

There will be nine questions on your Extra class exam from the Electrical Principles subelement. The Element 4 question pool has nine groups of questions for this subelement, based on the syllabus points listed here.

E5A Characteristics of resonant circuits: Series resonance (capacitor and inductor to resonate at a specific frequency); Parallel resonance (capacitor and inductor to resonate at a specific frequency); half-power bandwidth

E5B Exponential charge/discharge curves (time constants): definition; time constants in RL and RC circuits

E5C Impedance diagrams: Basic principles of Smith charts; impedance of RLC networks at specified frequencies

E5D Phase angle between voltage and current; impedances and phase angles of series and parallel circuits; algebraic operations using complex numbers: rectangular coordinates (real and imaginary parts); polar coordinates (magnitude and angle)

E5E Skin effect; electrostatic and electromagnetic fields

E5F Circuit Q; reactive power; power factor

PHOTOELECTRIC EFFECT

In simple terms, the **photoelectric effect** refers to electrons being knocked loose from a material when light shines on the material. A thorough understanding of the photoelectric effect would require a course in the branch of physics known as quantum mechanics. Since that is beyond the scope of this book, we will simply describe some of the basic principles behind photoelectricity.

You are probably familiar with the basic structure of an atom, the building block for all matter. **Figure 5-1** is a simplified illustration of the parts of an atom. The nucleus contains protons (positively charged particles) and neutrons (with no electrical charge). The number of protons in the nucleus determines which type of element the atom will be. Carbon has 6 protons, oxygen has 8 and copper has 29, for example. The nucleus of the atom is surrounded by the same number of negatively charged electrons as there are protons in the nucleus. So an atom has zero net electrical charge.

The electrons surrounding the nucleus are found in specific energy levels, as shown in Figure 5-1. The increasing energy levels are shown as larger and larger spheres surrounding the nucleus. While this picture is not really accurate, it will help you get the idea of the atomic structure. For an electron to move to a different energy level it must either gain or lose a certain amount of energy. One way that an electron can gain the required energy is by being struck by a photon of light (or other electromagnetic radiation). The electron absorbs the energy from the light photon and jumps to a new energy level. Scientists say the electron is "excited" at this point. Since the electrons in an atom prefer to be at the lowest energy level possible, an excited electron will tend to radiate a photon of the required energy so that it can fall back into the lower level, if an opening exists at that level. (Only a certain number of electrons can exist at each energy level in an atom.)

This principle of electrons being excited to higher energy levels, then jumping back to lower levels and giving off a photon of light is the operating principle of "neon" signs. Electrical energy, in the form of a spark, passes through a gas inside a glass tube. Electrons are excited, then give off specific frequencies of light. Different gases inside the tubes produce different colors of light.

If an electromagnetic wave (usually in the range of visible light frequencies) with the right amount of energy strikes the surface of a material, electrons will be excited into higher energy levels. If the light photon has sufficient energy, then it may be able to knock an electron completely free of the atom. The result is a free electron and a positive ion created from what was an atom before the photon collided with it.

If a voltage is applied between two metal surfaces, and one of the surfaces is illuminated with light of the proper frequency (energy), then a spark can be made to jump across the gap at a lower voltage than if the surface were not illuminated. This property was first discovered by Heinrich Hertz in 1886. It wasn't until 1905

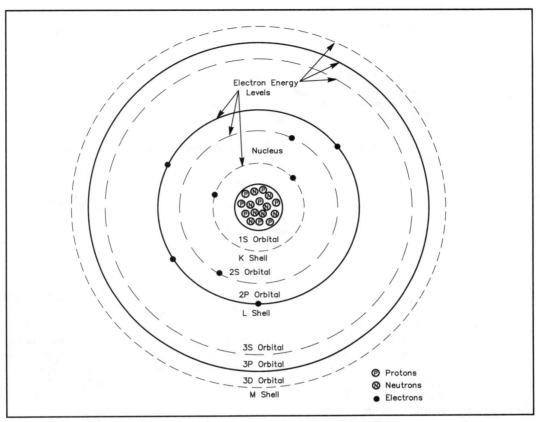

Figure 5-1 — This simple drawing illustrates the structure of an atom. The nucleus is made up of protons and neutrons. The electrons surrounding the nucleus are found in increasing energy levels as you go out from the center. This drawing represents an oxygen atom.

that Albert Einstein was able to offer a mathematical explanation of the photoelectric effect, however.

The Photoconductive Effect

With our simple model of the atom in mind, it is easy to see that an electric current through a wire or other material depends on electrons being pulled away from or knocked free of one atom and moving along to another. The rate of electrons moving past a certain point in the wire specifies the current. Every material presents some opposition to this flow of electrons, and that opposition is called the *resistivity* of the material. If you include the length and cross-sectional area of a specific object or piece of wire, then you know the resistance of the object:

$$R = \frac{\rho \ell}{A}$$

(Equation 5-1)

where:

ρ is the lower case Greek letter rho, representing the resistivity of the material.
ℓ is the length of the object.
A is the cross-sectional area of the object.
R is the resistance.

Conductivity is the reciprocal of resistivity, and conductance is the reciprocal of resistance:

$$\sigma = \frac{1}{\rho}$$

(Equation 5-2)

where σ is the lower case Greek letter sigma, which represents conductivity

and

$$G = \frac{1}{R}$$

(Equation 5-3)

where G is the conductance.

We learned earlier that with the photoelectric effect, electrons can be knocked loose from atoms when light strikes the surface of the material. With this principle in mind, you can see that those free electrons will make it easier for a current to flow through the material. But even if electrons are not knocked free of the atom, excited electrons in the higher-energy-level regions are more easily passed from one atom to another.

All of this discussion leads us to one simple fact: it is easier to produce a current when some of the electrons associated with an atom are excited. The conductivity of the material is increased, and the resistivity is decreased. The total conductance of a piece of wire may increase and the resistance decrease when light shines on the surface. This is called the **photoconductive effect**.

The **photoconductive effect** is more pronounced, and more important, for crystalline semiconductor materials than for ordinary metal conductors. With a piece of copper wire, for example, the conductance is normally high, so any slight increase because of light striking the wire surface will be almost unnoticeable. The conductivity of semiconductor crystals such as germanium, silicon, cadmium sulfide, cadmium selenide, gallium arsenide, lead sulfide and others is low when they are not illuminated, but the increase in conductivity is significant when light shines on their surfaces. (This also means that the resistance decreases.) Each material will show the biggest change in conductivity over a different range of light frequencies. For example, lead sulfide responds best to frequencies in the infrared region, while cadmium sulfide and cadmium selenide are both commonly used in visible light detectors, such as are found in cameras.

Of course, it's important to realize that most semiconductor devices are sealed in plastic or metal cases, so no light will reach the semiconductor junction. Light will not affect the conductivity, and hence the operating characteristics, of such a transistor or diode. But if the case is made with a window to allow light to pass through and reach the junction, then the device characteristics will depend on how much light is shining on it. Such specially made devices have a number of important applications in Amateur Radio electronics.

A phototransistor is a special device designed to allow light to reach the transistor junction. Light, then, acts as the control element for the transistor. In fact, in some phototransistors, the base lead is not even brought out of the package. In others, a base lead is provided, so you can control the output signal in the absence of light. You can also use the base lead to bias the transistor to respond to different light intensities. In general, the gain of the transistor is directly proportional to the amount of light shining on the transistor. A phototransistor can be used as a photodetector.

Electroluminescence in Semiconductors

If a semiconductor diode junction is forward biased, majority charge carriers from both the P- and N-type material will cross the junction, where they become minority carriers. As these minority carriers move away from the junction, they meet and combine with the opposite-type carriers. In this process, a photon of electromagnetic radiation will be produced. If the radiation frequency is in the range of visible light, and if the diode is constructed to allow the light to escape, we will have a light-emitting diode (LED)! The material used to dope the semiconductor to form the P- and N-type materials will determine the light color emitted. Some materials will emit light in the infrared range, rather than the visible range.

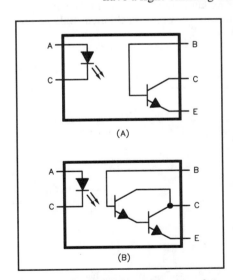

(A)

(B)

Figure 5-2 — Optocouplers consist of an LED and a phototransistor detector. The one shown at A uses a single-transistor detector, while the unit shown at B uses a Darlington phototransistor for improved transfer ratio.

Optocouplers

An **optocoupler**, or **optoisolator**, is an LED and a phototransistor in a common IC package. Applying current to the LED causes it to light, and the light from the LED causes the phototransistor to turn on. Because they use light instead of a direct electrical connection, optoisolators provide one of the safest ways to interface circuits using widely differing voltages. Signals from a high-voltage circuit can be fed into a low-voltage circuit without fear of damage to the low-voltage devices. Optoisolators have a very high impedance between the light source (input) and the phototransistor (output). There is no current between the input and output terminals. The LEDs in most optocouplers are infrared emitters, although some operate in the visible-light portion of the electromagnetic spectrum.

Figure 5-2A shows the schematic diagram of a typical optocoupler. In this example, the phototransistor base lead is brought outside the package. As shown at B, a Darlington phototransistor can be used to improve the current transfer ratio of the device.

In an IC optoisolator, the light is transmitted from the LED to the phototransistor detector by means of a plastic light pipe or small gap between the two sections. It is also possible to make an optoisolator by using discrete components. A separate LED or infrared emitter and matching phototransistor detector can be separated by some small distance to use a reflective path or other external gap. In this case, changing the path length or blocking the light will change the transistor output. This can be used to detect an object passing between the detector and light source, for example.

Other applications for a separate emitter and detector might be in a punched-paper-tape or computer-card reader, where you want to detect the position of holes to read a code. Such a system could also be used to send digital pulses over considerable distances using fiber optics. Amateur-Radio applications for this type of system might include the use of light pipes to connect various pieces of equipment in the presence of strong RF fields. Optical coupling can minimize interference between digital and analog equipment, such as computers and radios.

The Optical Shaft Encoder

An **optical shaft encoder** usually consists of two pair of emitters and detectors. A plastic disc with alternating clear and black radial bands rotates through a gap between the emitters and detectors. See **Figure 5-3**. By using two emitters and two detectors, a microprocessor can detect the rotation direction and speed of the wheel. Modern transceivers use a system like this to control the frequency of a synthesized VFO. To the operator, the tuning knob may feel like it is mechanically tuning the VFO, but there is no tuning capacitor or other mechanical linkage connected to the knob and light-chopping wheel.

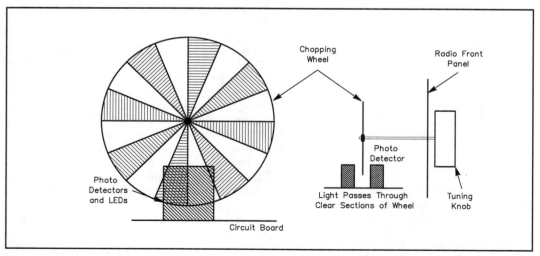

Figure 5-3 — This drawing illustrates the operation of an optical shaft encoder, often used as a tuning mechanism on modern transceivers.

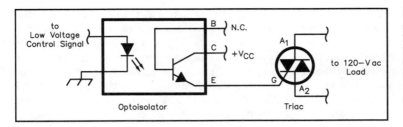

Figure 5-4 — An optoisolator and a triac can be connected to make the solid-state equivalent of a mechanical relay for 120-V ac household current.

A Solid-State Relay

An optoisolator and a triac, either packaged as one unit or as discrete components, form the solid-state equivalent of a mechanical relay. A solid-state relay can replace electromechanical units in many applications. Some of the advantages include freedom from contact bounce, no arcing, no mechanical wear and no noisy, clicking relay sounds. A solid-state relay can also operate much faster than a mechanical one. Some solid-state relays are rated to switch 10 A at 120 V using CMOS control signals. **Figure 5-4** illustrates how a solid-state relay can be connected in a circuit.

[Now turn to Chapter 10 and study examination questions E5I01 through E5I13. Review this section as needed.]

ELECTRICAL ENERGY

Before you can understand what electrical energy is, you must know some important definitions. Let's use some simple examples from your everyday experience to build those definitions. Pick up a stone, and carry it to an upstairs window of your house. You are doing work against the force of gravity exerted by the Earth as you move the stone farther away from the surface of the Earth. Gravity is an invisible force that the Earth exerts on the stone. There are many invisible forces in nature. These invisible forces work through space, so there is no physical contact required for the force to act. The space through which these invisible forces act is often referred to as a **field**. (No, this is not the same as a farmer's field!) What we mean here is a region of space through which a force acts without actual contact. When you pick up the stone, there is physical contact, so you are not exerting an invisible force — you are not a field, then. The gravitational field of the Earth pulls every object toward the center of the Earth.

Okay, so how much work did you do on the stone? Well, you have to multiply the distance you moved it through the gravitational field (let's say to a height of 10 feet above the Earth) times the force you had to exert, which was equal to the force of gravity on the stone (say 1 pound). So you have done 10 foot-pounds of work against the gravitational field. Now place the stone on a windowsill. By doing work on the stone, you have stored some energy in it. That energy is equal to the amount of work that you did, and is called **potential energy**. In effect, you are storing the energy by the position of the stone in the gravitational field of the Earth. If you push the stone out the window, it will fall back to the Earth, and while it is falling, the stored potential energy is being converted to kinetic energy, or energy of motion.

In electronics we are interested mainly in two types of invisible forces. Those are the *electric force* and the *magnetic force*. The space through which these forces act are the **electric field** and the **magnetic field**. These fields make up an electromagnetic wave. (A wave of this type is a field in motion.) We can store electrical **potential energy** as a voltage in an electric field and as a current in a magnetic field. In either case, that potential energy can be released in the form of an electric current in the circuit.

Storing Energy in an Electric Field

You can store energy in a capacitor by applying a dc voltage across the terminals. There will be an instantaneous inrush of current to charge the capacitor plates. The only thing limiting the current at the instant the voltage is connected is any resistance there may be in the circuit. (Of course there will always be some resistance in the wires connecting the components, and in the components themselves.) The capacitor builds up an electric charge as one set of plates accumulates an excess of electrons and the other set loses an equal number. The voltage across the capacitor rises as this charge builds up. Eventually, the voltage at the capacitor terminals is equal to the source voltage, and the current stops. If the voltage source is disconnected, the capacitor will remain charged to that voltage. The charge will stay on the capacitor plates as long as there is no path for the electrons to travel from one plate to the other.

At this point we should be careful to point out that our discussion in this section deals with ideal components. We are thinking of resistors that have no stray capacitance or inductance associated with the leads or composition of the resistor itself. Ideal capacitors exhibit no losses, and there is no resistance in the leads or capacitor plates. Ideal inductors are made of wire that has no resistance, and there is no stray capacitance between turns. Of course, in practice we do not have ideal components, so the conditions described here may be modified a bit in real-life circuits. Even so, components can come pretty close to the ideal conditions. For example, a capacitor with very low leakage will hold a charge for days or even weeks.

Stored electric potential energy produces an **electric field** in the capacitor. Since the charge is not moving, this field is sometimes called an *electrostatic field*. If a resistor or some other circuit is connected across the capacitor terminals, that field will return the stored energy by creating a current in the circuit.

Figure 5-5 illustrates a simple circuit for charging and discharging a capacitor, depending on the switch setting. The electric energy will be converted to heat energy in a resistor, or into other forms of energy, such as sound energy in a speaker. If the capacitance is high, and you connect a large-value resistance across the terminals, it may take a long time

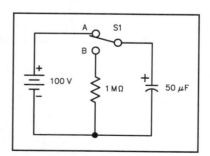

Figure 5-5 — A simple circuit for charging a capacitor and then discharging it through a resistor.

for all the energy to be dissipated and the capacitor voltage to drop to zero. On the other hand, by touching a wire across the terminals to short circuit the capacitor, you can discharge it very quickly. If you short a capacitor of several microfarads that has been charged to several hundred volts, you can produce quite a spark. If your skin touches the terminals instead of a piece of wire, you can get a dangerous shock. Large-value filter capacitors in a power supply have bleeder resistors connected across them to drain this charge when the supply is turned off.

The basic unit for expressing energy in the metric system, which is the system used to express all common electrical units, is the **joule** (pronounced with a long u sound, similar to jewel).

Joules measure the electrical energy stored in an electrostatic field across a capacitor. This electrical potential energy is returned to the circuit as a current when the capacitor discharges.

Storing Energy in a Magnetic Field

When electrons flow through a conductor, a **magnetic field** is produced. This magnetic field exists in the space around the conductor, and a magnetic force acts through this space. This can be demonstrated by bringing a compass near a current-carrying wire and watching the needle deflect. **Figure 5-6A** illustrates the magnetic field around a wire connected to a battery. If the wire is wound into a coil, so the fields from adjacent turns add together, then a much stronger magnetic field can be produced. The direction of the field, which points to the magnetic north pole, can be found using a "left hand rule." For a straight wire, point the thumb of your left hand in the direction of the *electron flow* and your fingers curl in the direction of the north pole of the magnetic field around the wire. In a similar manner, if you curl the fingers of your left hand around the coil in the direction of electron flow, your thumb points in the direction of the north pole of the field. Parts B and C of Figure 5-6 illustrate the fields around two coils wound in opposite directions.

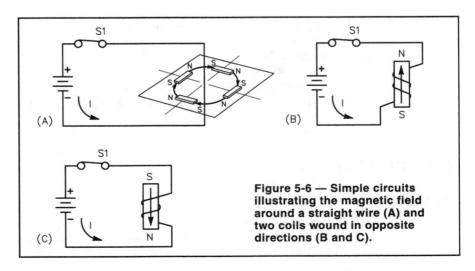

Figure 5-6 — Simple circuits illustrating the magnetic field around a straight wire (A) and two coils wound in opposite directions (B and C).

Notice that this *left-hand rule* describes the direction of the magnetic field in terms of the direction of *electron flow*. Older texts often describe the magnetic-field direction with a *right-hand rule*, using *conventional current*, which flows from positive to negative. Don't be confused by this difference if you read an older text.

The strength of the magnetic field depends on the amount of current, and is stronger when the current is larger. Electrical energy from the voltage source is transferred to the magnetic field in the process of creating the field. So we are storing energy by building up a magnetic field, and that means work must be done against some opposing force. That opposing force is the result of a voltage induced in the circuit whenever the magnetic field (or current) is changing. If the current remains constant, then the magnetic field remains a constant, and there is no more energy being stored.

When you first connect a dc source to a coil of wire, a current begins to flow, and a magnetic field begins to build up. The field is changing very rapidly at that time, so a large opposing voltage is created, preventing a large current from flowing. As a maximum amount of energy is stored and the magnetic field reaches its strongest value, the opposing voltage will decrease to zero, so the current increases gradually to a maximum value. That maximum current is limited only by the resistance of the wire in the coil, and by any internal resistance of the voltage source.

If the current decreases, then a voltage is induced in the wire that will try to prevent the decrease. The stored energy is being returned to the circuit in this case. As the magnetic field collapses, and the stored energy is returned to the circuit, current continues to flow.

This induced voltage or EMF is sometimes called a **back EMF**, since it is always in a direction to oppose any change in the amount of current. When the switch in the circuit of Figure 5-6B is first closed, this back EMF will prevent a sudden surge of current through the coil. Notice that this is just the opposite of the condition when a capacitor is charging. Likewise, if you open the switch to break the circuit, a back EMF will be produced in the opposite direction. This time the EMF tries to keep the current going, again preventing any sudden change in the magnetic-field strength.

The magnitude of the induced back EMF depends on how rapidly the current is changing. If you have a strong magnetic field built up in a coil, and then suddenly break the circuit at some point, a large voltage is induced in the coil, which tries to maintain the current. It is quite common to have a spark jump across the switch contacts as they open.

[Turn to Chapter 10 at this time, and study questions E5E06 through E5E11. Review this section if any of these questions give you difficulty.]

TIME CONSTANTS

RC Circuits

If you connect a dc voltage source directly to a capacitor, the capacitor will charge to the full voltage almost instantly. If the circuit contains resistance, however, the current will be limited, and it will take some time to charge the capacitor. The higher the resistance value, the longer it will take to charge the capacitor. Energy is stored in the form of an electric field between the capacitor plates as a capacitor charges.

Figure 5-7 shows a circuit that can be used to alternately charge and discharge a capacitor. With the switch in position A, there will be a current through the resistor to charge the capacitor to the battery voltage. When the switch is moved to position B, the capacitor will

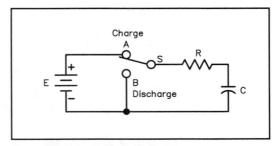

Figure 5-7 — A simple series circuit illustrates the principle of an RC time constant for charging and discharging a capacitor through a resistor.

give the energy stored in it back to the circuit as a current through the resistor. The amount of time it takes to charge or discharge the capacitor depends on the capacitor and resistor values. The product of the resistance and the capacitance is called a **time constant**:

$$\tau = RC \qquad \text{(Equation 5-4)}$$

where:

τ is the Greek letter tau, used to represent the time constant.
R is the total circuit resistance in ohms.
C is the capacitance in farads.

The simple circuit of Figure 5-7 is a series circuit. It is possible to have a circuit with several resistors and capacitors connected either in series or parallel. If the components are wired in series we can still use Equation 5-4, but we must first combine all of the resistors into one equivalent resistor, and all of the capacitors into one equivalent capacitor. Then calculate the time constant using Equation 5-4, as before. If the components are connected in parallel, there is an added complication when the circuit is charging, but for a discharging circuit you can still calculate a time constant. Again combine all of the resistors and all of the capacitors into equivalent values, and calculate the time constant using Equation 5-4. If you have forgotten how to combine resistors and capacitors in series and parallel, review the appropriate sections of *Now You're Talking!* or *The ARRL Handbook*. The required equations are:

$$R_T \text{ (series)} = R_1 + R_2 + R_3 + ... + R_n \qquad \text{(Equation 5-5)}$$

$$R_T \text{ (parallel)} = \cfrac{1}{\cfrac{1}{R_1} + \cfrac{1}{R_2} + \cfrac{1}{R_3} + ... + \cfrac{1}{R_n}} \qquad \text{(Equation 5-6)}$$

If you have only two resistors, you might prefer to use the simplified form of this equation:

$$R_T \text{ (parallel)} = \frac{R_1 \times R_2}{R_1 + R_2}$$

(Equation 5-6A)

$$C_T \text{ (series)} = \frac{1}{\dfrac{1}{C_1} + \dfrac{1}{C_2} + \dfrac{1}{C_3} + \cdots + \dfrac{1}{C_n}}$$

(Equation 5-7)

If you have only two capacitors, you can simplify this equation to be similar to Equation 5-6A.

$$C_T \text{ (series)} = \frac{C_1 \times C_2}{C_1 + C_2}$$

(Equation 5-7A)

$$C_T \text{ (parallel)} = C_1 + C_2 + C_3 + \ldots + C_n$$

(Equation 5-8)

Let's look at an example of calculating the time constant for a circuit like the one shown at Figure 5-7. We will pick values of 220 μF and 470 kΩ for C and R. To calculate the time constant, τ, we simply multiply the R and C values, in ohms and farads.

$\tau = RC = 470 \times 10^3$ ohms $\times 220 \times 10^{-6}$ farads $= 103.4$ seconds

You can calculate the time constant for any RC circuit in this manner.

If you have two 100-μF capacitors and two 470-kΩ resistors, all in series, we first combine the resistor values into a single resistance and the capacitor values into a single capacitance.

$R_T = R_1 + R_2 = 470 \text{ k}\Omega + 470 \text{ k}\Omega = 940 \text{ k}\Omega = 940 \times 10^3 \text{ }\Omega$

$$C_T \text{ (series)} = \frac{C_1 \times C_2}{C_1 + C_2} = \frac{100 \text{ μF} \times 100 \text{ μF}}{100 \text{ μF} + 100 \text{ μF}} = \frac{10{,}000 \text{ μF} \times \text{μF}}{200 \text{ μF}} = 50 \text{ μF} = 50 \times 10^{-6} \text{ F}$$

Then the time constant is:

$\tau = RC = 940 \times 10^3 \text{ }\Omega \times 50 \times 10^{-6} \text{ F} = 47$ seconds

Suppose you have two 220-μF capacitors and two 1-MΩ resistors all in parallel. Again, you must first combine the values into a single resistance and a single capacitance.

$$R_T \text{ (parallel)} = \frac{R_1 \times R_2}{R_1 + R_2} = \frac{1 \text{M}\Omega \times 1 \text{M}\Omega}{1 \text{M}\Omega + 1 \text{M}\Omega} = \frac{1 \text{M}\Omega \times \text{M}\Omega}{2 \text{M}\Omega} = 0.5 \text{M}\Omega = 5 \times 10^5 \text{ }\Omega$$

$C_T \text{ (parallel)} = C_1 + C_2 = 220 \text{ μF} + 220 \text{ μF} = 440 \text{ μF} = 440 \times 10^{-6} \text{ F}$

Then the time constant is:

$\tau = RC = 5 \times 10^5 \text{ }\Omega \times 440 \times 10^{-6} \text{ F} = 220$ seconds

Exponential Charge/Discharge Curve for Capacitors

The capacitor charge and discharge follows a pattern known as an exponential curve. **Figure 5-8** illustrates the charge and discharge curves, where the time axis is shown in terms of τ, and the vertical axis is expressed as a percentage of the battery voltage. These graphs hold true for any RC circuit, as long as you know the time constant and the maximum voltage the capacitor is charged to.

We can write equations to calculate the voltage on the capacitor at any instant of time, based on the exponential charge or discharge. For a charging capacitor:

$$V(t) = E\left(1 - e^{\frac{-t}{\tau}}\right)$$

(Equation 5-9)

where:

V(t) is the charge on the capacitor at time t.

E is the maximum charge on the capacitor, or the battery voltage.

t is the time in seconds that have elapsed since the capacitor began charging or discharging.

e is the base for natural logarithms, 2.718.

τ is the time constant for the circuit, in seconds.

If the capacitor is disharging, we have to write a slightly different equation:

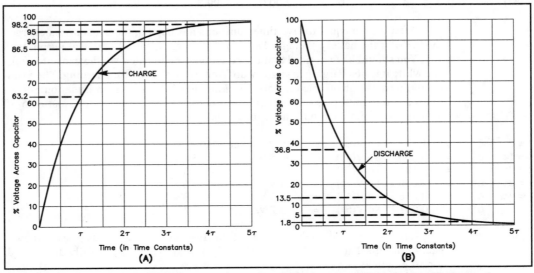

Figure 5-8 — The graph at A shows how the voltage across a capacitor rises, with time, when charged through a resistor. The curve at B shows the way in which the voltage decreases across the capacitor terminals while discharging through the same value of resistance. From a practical standpoint, a capacitor may be considered as charged (or discharged) after a time equal to 5τ.

$$V(t) = E\left(e^{\frac{-t}{\tau}} \right)$$

(Equation 5-10)

These exponential equations can be solved fairly easily with an inexpensive calculator that is able to work with natural logarithms (a key labeled LN or LN X). In that case you could calculate the value for $e^{-t/\tau}$ as the inverse natural log of $-t/\tau$, written as $\ln^{-1}(-t/\tau)$. (In mathematics, the abbreviation ln is normally used to mean the natural logarithm.) Actually, you do not have to know how to solve these equations if you are familiar with the results at a few important points. We'll show the solutions to the equations at these points. Practice those calculations and compare your results with the answers shown in this book.

As shown on the graphs of Figure 5-8, it is common practice to think of time in terms of a multiple of the circuit time constant when relating to the capacitor charge or discharge. If we select times equal to zero (starting time), one time constant (1τ), two time constants (2τ), and so on, then the exponential term in Equations 5-9 and 5-10 becomes simply e^0, e^{-1}, e^{-2}, e^{-3} and so forth. Then we can solve Equation 5-9 for those values of time. For simplicity, let's pick a value of E = 100 V. Then our answers will be in the form of a percentage of any battery voltage you are working with.

V(0)	=	100 V $(1 - e^{-0})$	=	100 V $(1 - 1)$	=	0 V,	or	0%
V(1τ)	=	100 V $(1 - e^{-1})$	=	100 V $(1 - 0.368)$	=	63.2 V,	or	63.2%
V(2τ)	=	100 V $(1 - e^{-2})$	=	100 V $(1 - 0.135)$	=	86.5 V,	or	86.5%
V(3τ)	=	100 V $(1 - e^{-3})$	=	100 V $(1 - 0.050)$	=	95.0 V,	or	95%
V(4τ)	=	100 V $(1 - e^{-4})$	=	100 V $(1 - 0.018)$	=	98.2 V,	or	98.2%
V(5τ)	=	100 V $(1 - e^{-5})$	=	100 V $(1 - 0.007)$	=	99.3 V,	or	99.3%

After a time equal to five time constants has passed, the capacitor is charged to 99.3% of the battery voltage. This is fully charged, for all practical purposes.

You should have noticed that the equation used to calculate the capacitor voltage while it is discharging is slightly different from the one for charging. The exponential term is not subtracted from 1 in Equation 5-10, as it is in Equation 5-9. At times equal to multiples of the circuit time constant, the solutions to Equation 5-10 have a close relationship to those for Equation 5-9.

t = 0,	e^{-0}	=	1,	so V(0)	=	100 V,	or	100%
t = 1τ,	e^{-1}	=	0.368,	so V(1τ)	=	36.8 V,	or	36.8%
t = 2τ,	e^{-2}	=	0.135,	so V(2τ)	=	13.5 V,	or	13.5%
t = 3τ,	e^{-3}	=	0.050,	so V(3τ)	=	5 V,	or	5%
t = 4τ,	e^{-4}	=	0.018,	so V(4τ)	=	1.8 V,	or	1.8%
t = 5τ,	e^{-5}	=	0.007,	so V(5τ)	=	0.7 V,	or	0.7%

Here we see that after a time equal to five time constants has passed, the capacitor has discharged to less than 1% of its initial value. This is fully discharged, for all practical purposes.

From the calculations of a charging capacitor we can define the **time constant** of an RC circuit as the time it takes to *charge* the capacitor to 63.2% of the supply voltage. From the calculations of a discharging capacitor we can also define the time constant as the time it takes to *discharge* the capacitor to 36.8% of its initial voltage.

Another way to think of these results is that the discharge values are the complements of the charging values. Subtract either set of percentages from 100 and you will get the other set. You may also notice another relationship between the discharging values. If you take 36.8% (0.368) as the value for one time constant, then the discharged value is $0.368^2 = 0.135$ after two time constants, $0.368^3 = 0.050$ after three time constants, $0.368^4 = 0.018$ after four time constants and $0.368^5 = 0.007$ after five time constants. You can change these values to percentages, or just remember that you have to multiply the decimal fraction times the battery voltage. If you subtract these decimal values from 1, you will get the values for the charging equation.

In many cases, you will want to know how long it will take a capacitor to charge or discharge to some particular voltage. Probably the easiest way to handle such problems is to first calculate what percentage of the maximum voltage you are charging or discharging to. Then compare that value to the percentages listed for either charging or discharging the capacitor. Often you will be able to approximate the time as some whole number of time constants.

Suppose you have a 0.01-μF capacitor and a 2-MΩ resistor wired in parallel with a battery. The capacitor is charged to 20 V, and then the battery is removed. How long will it take for the capacitor to discharge to 7.36 V? First, let's calculate the percentage decrease in voltage:

$$\frac{7.36 \text{ V}}{20 \text{ V}} = 0.368 = 36.8\%$$

You should recognize this as the value for the discharge voltage after 1 time constant. We can calculate the time constant for our circuit using Equation 5-4.

$$\tau = RC = 2 \times 10^6 \ \Omega \times 0.01 \times 10^{-6} \ F = 0.02 \text{ second}$$

It will take 0.02 seconds, or 20 milliseconds to discharge the capacitor to 7.36 V.

[Now turn to Chapter 10 and study examination questions E5B01 and E5B03 through E5B11. Be sure to work out all of the problems there, and review this section if you have any difficulty.]

RL Circuits

When resistance and inductance are connected in series there is a situation similar to what happens in an RC circuit. **Figure 5-9** shows a circuit for storing a magnetic field in an inductor. When the switch is closed, a current will try to flow immediately. The instantaneous transition from no current to the value that would flow in the circuit because of the voltage source and resistance represents a very large change in current, and a **back EMF** is developed by the inductance.

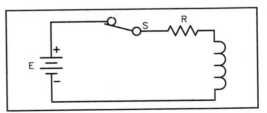

Figure 5-9 — This simple series circuit illustrates the principle of an RL time constant. When the switch closes, as the current begins to increase, there is a reverse voltage, or back emf, created by the inductor to oppose any change in current. Gradually the current increases to the maximum value, given by Ohm's Law from the circuit resistance and the applied voltage.

This **back EMF** is proportional to the rate of change of the current, and it tends to oppose the applied voltage. (This means the polarity of the back EMF is opposite to that of the applied voltage.) The result is that the initial current is very small, but it increases quickly at first and then gradually approaches the final current value.

Since the magnitude of the back EMF depends on the rate of change of the current, it decreases as the current stops increasing so fast. The current builds up to its final value as given by Ohm's Law (I = E / R), and the back EMF decreases toward zero.

Figure 5-10 shows how

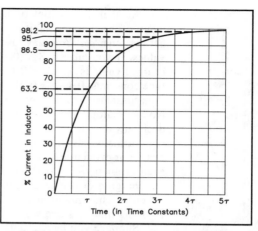

Figure 5-10 — **This graph shows the current build up in an RL circuit. Notice that the curve is identical to the voltage curve for a charging capacitor.**

the current through the inductor of Figure 5-9 increases as time passes. At any given instant, the back EMF will be equal to the difference between the voltage drop across the resistor and the battery voltage. You can see that when the switch is closed initially and there is no current, the back EMF is equal to the full battery voltage. Later on, the current will increase to a steady value and there will be no voltage drop across the inductor. The full battery voltage appears across the resistor and the back EMF goes to zero.

Theoretically, the back EMF will never quite disappear, so the current never quite reaches the value predicted by Ohm's Law when you ignore the inductance. In practice, the current is essentially equal to the final value after 5 time constants. The curve looks just like the one we found for a charging capacitor.

You can probably guess that this is another exponential curve, and that we can write an equation similar to Equation 5-10 to calculate the current in the circuit.

$$I(t) = \frac{E}{R}\left(1 - e^{\frac{-t}{\tau}}\right)$$

(Equation 5-11)

where:
  I(t) is the current in amperes at time t.
  E is the power-supply potential in volts.
  R is the circuit resistance in ohms.
  τ is the time in seconds after the switch is closed.

Here again, we define a time constant that depends on the circuit components, as we did with the RC circuit. For an RL circuit, the time constant is given by:

$$\tau = \frac{L}{R}$$
(Equation 5-12)

where τ is the Greek letter tau, used to represent a time constant for both RC and RL circuits.

If we choose values of time equal to multiples of the circuit time constant, as we did for the RC circuit, then we will find that the current will build up to its maximum value in the same fashion as the voltage does when a capacitor is being charged. This time let's pick a value of 100 A for the maximum current, so that our results will again come out as a percentage of the maximum current for any RL circuit.

t = 0,	e^{-0}	=	1,	so I(0)	=	100 A (1 − 1)	=	0 A,	or	0%
t = 1τ,	e^{-1}	=	0.368,	so I(1τ)	=	100 A (1 − 0.368)	=	63.2 A,	or	63.2%
t = 2τ,	e^{-2}	=	0.135,	so I(2τ)	=	100 A (1 − 0.135)	=	86.5 A,	or	86.5%
t = 3τ,	e^{-3}	=	0.050,	so I(3τ)	=	100 A (1 − 0.050)	=	95.0 A,	or	95%
t = 4τ,	e^{-4}	=	0.018,	so I(4τ)	=	100 A (1 − 0.018)	=	98.2 A,	or	98.2%
t = 5τ,	e^{-5}	=	0.007,	so I(5τ)	=	100 A (1 − 0.007)	=	99.3 A,	or	99.3%

Notice that the current through the inductor will increase to 63.2% of the maximum value during 1 time constant. After 5 time constants, the current has reached the maximum value.

[Turn to Chapter 10 now and study examination question E5B02. Review this section as needed.]

PHASE ANGLE BETWEEN CURRENT AND VOLTAGE

Now that you understand how capacitors and inductors store energy in electric and magnetic fields, we can learn how those devices react when an alternating voltage is applied to their terminals. While the resistance of a pure resistor does not vary with the frequency, you already know that the reactance of both a coil and a capacitor do change with frequency. That should tell us that coils and capacitors will behave differently with an ac voltage than a dc one. Remember that a pure inductor does not impede direct current, but does impede alternating current, and the higher the frequency the more it opposes the current. Also a pure capacitor will not allow dc to pass through it, but will hamper ac less and less as the frequency increases.

To understand the variations in voltage and current through inductors and capacitors, we must look at the amplitudes of these ac signals at certain instants of time. The relationship be-

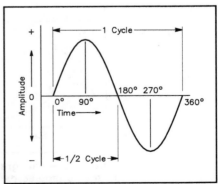

Figure 5-11 — An ac cycle is divided into 360° that are used as a measure of time or phase.

tween the current and voltage waveforms at a specific instant is called the **phase** of the waveforms. Phase essentially means time, or a time interval between when one event occurs and the instant when a second, related, event takes place. The event that occurs first is said to lead the second, while the second event lags the first.

Since each ac cycle takes exactly the same time as any other cycle of the same frequency, we can use the cycle as a basic time unit. This makes the phase measurement independent of the waveform frequency. If two or more different frequencies are being considered, phase measurements are usually made with respect to the lowest frequency.

It is convenient to relate one complete cycle of the wave to a circle, and to divide the cycle into 360 equal parts or degrees. So a phase measurement is usually specified as an angle. In fact, we often refer to the **phase angle** between two waveforms. **Figure 5-11** shows one complete cycle of a sine-wave voltage or current, with the wave broken into four quarters of 90° each.

AC Through a Capacitor

As soon as a voltage is applied across the plates of an ideal capacitor, there is a sudden inrush of current as the capacitor begins to charge. That current tapers off as the capacitor is charged to the full value of applied voltage. By the time the applied voltage is reaching a maximum, the capacitor is also reaching full charge, and so the current through the capacitor goes to zero. A maximum amount of energy has been stored in the electric field of the capacitor at this point. **Figure 5-12** shows the voltage across a capacitor as it charges and the charging current that flows into a capacitor with a dc voltage applied.

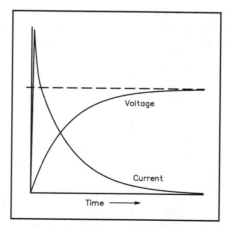

Figure 5-12 — This graph illustrates how the voltage across a capacitor changes as it charges with a dc voltage applied. The charging current is also shown.

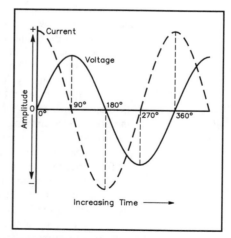

Figure 5-13 — Voltage and current phase relationships when an alternating voltage is applied to a capacitor.

The situation is a bit different when an ac voltage is applied. **Figure 5-13** graphs the relative current and voltage amplitudes over time, when an ac sine wave signal is applied. The two lines are shown differently to help you distinguish between them; the scale is not intended to show specific current or voltage values. When the applied voltage passes the peak and begins to decrease, the capacitor starts returning some of its stored energy to the circuit. Electrons are now flowing in a direction opposite to the direction they were flowing when the capacitor was charging. By the time the applied voltage reaches zero, the capacitor has returned all of its stored energy to the circuit, and the current is a maximum value in the reverse direction. Now the

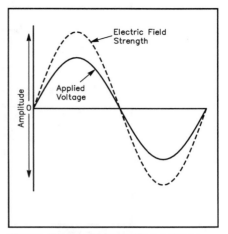

Figure 5-14 — The applied voltage and electric field in a capacitor are in phase.

applied voltage direction has also reversed, and a large charging current is applied to the capacitor, which decreases to zero by the time the applied voltage reaches a maximum value. The second half of the applied voltage cycle is exactly the same as the first half, except the relative directions of current and voltage are reversed.

Study Figure 5-13 to understand the current and voltage relationships for a capacitor over an entire cycle. Notice that the current reaches each point on a cycle 90° ahead of the applied-voltage waveform. We say that the current through a capacitor *leads* the applied voltage by 90°. You could also say that the voltage applied to a capacitor *lags* the current through it by 90°. To help you remember this relationship, think of the word ICE. This will remind you that the current (I) comes before (leads) the voltage (E) in a capacitor (C). Notice also that the applied-voltage waveform and the stored electric-field waveform are in phase — that is, similar points on those waveforms occur at the same instant (**Figure 5-14**).

AC Through an Inductor

The situation with ac through an inductor is a little more difficult to understand than the capacitor case. As we go through the conditions for a single current cycle, study the graph in **Figure 5-15**. That should make it easier to follow the changing conditions as they are described.

Let's apply an alternating current to an ideal inductor and observe what happens. At the instant when the current is zero and starts increasing in a positive direction, a magnetic field will start to build up around the coil, storing the applied energy. The current is increasing at a maximum rate, so the magnetic field strength is increasing at a maximum rate. As the current through the coil reaches a positive peak and begins to decrease, a maximum amount of energy has been stored in the magnetic field, and then the coil will begin to return energy to the circuit as the

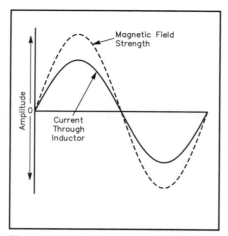

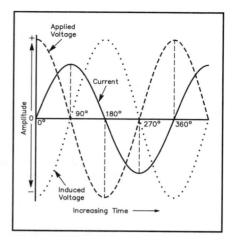

Figure 5-15 — The current through an inductor is in phase with the magnetic field strength.

Figure 5-16 — Phase relationships between voltage and current when an alternating voltage is applied to an inductance.

magnetic field collapses. When the current is crossing zero on the way down, it is again changing at a maximum rate, so the magnetic field is also changing at a maximum rate. As the current direction changes, it will begin to build a magnetic field in the direction opposite to the first field, and energy is being stored once again.

When the current reaches a maximum negative value, a maximum magnetic field has been built up, and then the stored energy begins to return to the circuit again as the current increases toward zero. The second half of the cycle is the same as the first, but with all polarities reversed. You should realize by now that the current and the magnetic field in an inductor are in phase — similar points on both waveforms occur at the same instant.

In the section on storing energy in an inductor, you learned about the back EMF that is induced in the coil. That EMF is greatest when the magnetic field is changing the fastest. Furthermore, it is in a direction that opposes the change in current or magnetic-field strength. So when the current is crossing zero on the way to a positive peak, the induced EMF is at its greatest negative value. When the current is at the positive peak, the back EMF is zero, and so on. **Figure 5-16** shows the phase relationship between the current through the inductor and the back EMF across it.

Since we know that the back EMF opposes the effects of an applied voltage, we can draw in the waveform for applied voltage as shown in Figure 5-16. The applied voltage is 180° out of phase with the induced voltage. From this fact, Figure 5-16 shows that the voltage across an inductor *leads* the current through it, or the current *lags* the voltage. The phase relationship between applied voltage and current through an inductor is just the opposite from the relationship for a capacitor. A useful memory technique to remember these relationships is the little saying, "ELI the ICE man." The L and C represent the inductor and capacitor, and the E and I stand for voltage and current. Right away you can see that E (voltage) comes before (leads) I (current) in an inductor and that I comes before E in a capacitor.

[To check your understanding of the voltage and current relationships for capacitors and inductors, turn to Chapter 10 and study questions E5D06 and E5D07. Review this section if you are uncertain about the answers to these questions.]

Phase Angle With Real Components

Up to this point we have been talking about ideal components in discussing the phase relationships between voltage and current in inductors and capacitors. Of course, any real components will have some resistance associated with the inductance or capacitance. Since the voltage across a resistor is in phase with the current through it, the overall effect is that the phase difference between voltage and current will be less than 90° in both cases.

Solving Problems Involving Inductors

Probably the easiest way to illustrate the effect of adding resistance to the circuit is by means of a calculation. **Figure 5-17A** shows a simple circuit with a resistor and an inductor in series with an ac signal source. Let's pick a frequency of 10 kHz for the signal generator, and connect it to a 20-mH inductor in series with a 1-kΩ resistor. The question is, what is the phase angle between the voltage and current in this circuit? To aid our solution, draw a set of **rectangular coordinates**, and label the X (horizontal) axis R (for resistance) and the Y (vertical) axis X (for reactance). The degree indications shown on Figure 5-17B show the standard way of relating the coordinate system to degrees around a circle. This coordinate system can represent either voltage or current, as needed. In this case, it represents voltage, since the current is the same in all parts of a series circuit. Next, we must calculate the inductive reactance of the coil:

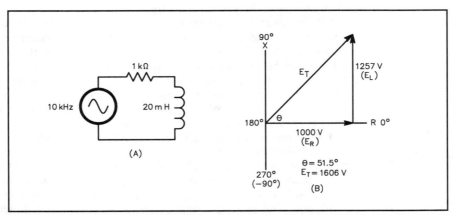

Figure 5-17 — A series RL circuit is shown at A. B shows the right triangle used to calculate the phase angle between the circuit current and voltage.

$X_L = 2\pi f L = 6.28 \times (10 \times 10^3 \text{ Hz}) \times (20 \times 10^{-3} \text{ H}) = 1257 \ \Omega$

Now calculate the voltage across the inductor using Ohm's Law. Since the actual current is not important, as long as we know it is the same through each part of the circuit, choose a simple value, such as 1 A:

$E_L = I X_L = 1 \text{ A} \times 1257 \ \Omega = 1257 \text{ V}$

The voltage across the resistor is also found easily using Ohm's Law:

$E_R = I R = 1 \text{ A} \times 1000 \ \Omega = 1000 \text{ V}$

Notice that the voltage is numerically equal to the reactance or resistance in these calculations. Wow! That leads to a nifty trick. By assuming a current of 1 A when the actual current value is not important to the problem, we can eliminate a step in the solution. There is no need to actually calculate voltages for a series-circuit problem. Just use the resistance and reactance values.

Okay, so now what? Well, we can't just add the two voltages to get the total voltage across the circuit, because they are not in phase. Remember that the voltage across an inductor *leads* the current, and the voltage across a resistor is *in phase* with the current. This means that the peak voltage across the inductor occurs 90° before the peak voltage across the resistor. On your graph, draw a line along the R (horizontal) axis to represent the 1000 V across the resistor. At the end of that line draw another line to represent the 1257 V across the inductor. Since this voltage leads the voltage in the resistor by 90°, it must be drawn vertically upward, along the 90° line. Now complete the figure by drawing a line from the origin (the point where the two axes cross) to the end of the E_L line. This right triangle represents the solution to our problem. Figure 5-17B shows a complete drawing of the triangle. The last line is the total voltage across the circuit, and the angle measured up from the E_R line to the E_T line is the phase angle between the voltage and current. If you use graph paper and make up a suitable scale of divisions per volt on the graph, you can actually solve the problem with no further calculation. Graphical solutions are usually not very accurate, however, and depend a great deal on how carefully you draw the lines.

Let's complete the mathematical solution to our problem. If you are not familiar with right triangles and their solutions (a branch of mathematics called trigonometry), just follow along. You should be able to pick up the techniques from a few examples, but if you continue to have trouble, there is an excellent discussion about the basics of trigonometry in *Understanding Basic Electronics*, published by ARRL. Alternatively, go to your library and check out a book on elementary high school trigonometry. You will find the tools of trigonometry to be very helpful for solving many electronics problems involving ac signals.

The various sides and angles of a right triangle are identified by their positions in relation to the right (90°) angle. The side opposite the right angle is always the longest side of a right triangle, and is called the hypotenuse. The other sides are either opposite or adjacent to the remaining angles. It is important to realize that these methods of trigonometry apply only to triangles that contain a 90° angle.

We will find the phase angle first, using the tangent function. Since the angle of interest is the one between the E_R line and the hypotenuse, E_L is the side opposite and E_R is the side adjacent to the angle. Tangent (often abbreviated as *tan*) is defined as:

$$\tan \theta = \frac{\text{side opposite}}{\text{side adjacent}}$$ (Equation 5-13)

Then for our problem,

$$\tan \theta = \frac{E_L}{E_R} = \frac{1257 \text{ V}}{1000 \text{ V}} = 1.257$$

where θ is the angle between the E_R and E_T lines.

Now that we know the tangent of the angle, it is a simple matter to refer to a table of trigonometric functions to find what angle has a tangent of that value. If you have an electronic calculator that is capable of doing trig functions, then you simply ask it to find the angle using the inverse tangent or arc tangent function, often written $\tan^{-1}$:

$$\tan^{-1}(1.257) = 51.5°$$

If the resistance and reactance of our components had been equal, then we would have found $\tan \theta$ to be 1, and the phase angle would be 45°. If the reactance were many times larger than the resistance, then the phase angle would be close to 90°, and if the resistance were many times larger than the reactance, then the phase angle would be close to 0°.

There is one more question we would like an answer to here, and that is the total assumed voltage across the circuit, given our assumed current of 1 ampere. We could find the actual voltage in the same manner, given an actual current through the circuit. There are two common methods available to find this, so let's look at both of them. The first method uses the sine function. (Sine is often abbreviated as *sin*.)

$$\sin \theta = \frac{\text{side opposite}}{\text{hypotenuse}}$$ (Equation 5-14)

We know that the phase angle is 51.5°, so we find the value of the sine function for that angle (using a trig table, calculator or slide rule). Sin (51.5°) = 0.7826. By solving Equation 5-14 for the hypotenuse, we can find the answer:

$$\text{hypotenuse} = E_r = \frac{\text{side opposite}}{\sin \theta} = \frac{E_L}{\sin (51.5°)} = \frac{1257 \text{ V}}{0.7826}$$

$$E_T = 1606 \text{ V}$$

The second method for finding E_T involves the use of an equation known as the Pythagorean Theorem (named after Pythagoras, a Greek mathematician who discovered this important relationship). The sides of a right triangle are related to each other by the equation:

$$C^2 = A^2 + B^2$$ (Equation 5-15)

where:
C = the length of the hypotenuse.
A and B = the lengths of the other two sides.

We can solve this equation for the length of the hypotenuse, and rewrite it:

$$C = \sqrt{A^2 + B^2}$$ (Equation 5-16)

Since the side we want to find is the hypotenuse of our triangle (side E_T), we can use Equation 5-16.

$$E_T = \sqrt{(1257 \text{ V})^2 + (1000 \text{ V})^2} = \sqrt{1580049 \text{ V}^2 + 1000000 \text{ V}^2}$$

$$E_T = \sqrt{2580049 \text{ V}^2} = 1606 \text{ V}$$

Notice that in this example when we squared the value of the voltage, we also show the units of volts being squared. We carried the units along with the calculations, and when we took the square root of the number, we also took the square root of the units. This technique, called *dimensional analysis*, can be useful for showing that the result has the units you expected. If the units don't work out, it may be because you selected an improper equation or followed an incorrect mathematical procedure!

It doesn't matter which method you use to find the total voltage, so pick whichever one seems easier, then stick with it. After you have worked a few of these problems, they will begin to seem much easier. It is very important to follow an organized, systematic approach, however, or you will become easily confused, and any slight change in the wording of the problem will throw you off.

Solving Problems Involving Capacitors

Let's try another problem, to see how well you understood that solution. **Figure 5-18A** shows a circuit with a capacitor instead of an inductor. For simplicity, we will keep the same signal-generator frequency and the same resistor in the circuit.

The 12660-pF capacitor is a strange value, but let's use it anyway, just for the sake of example.

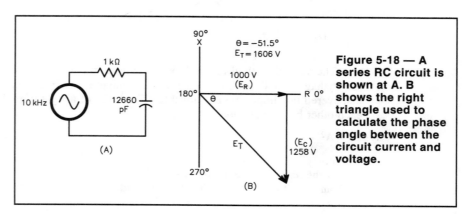

Figure 5-18 — A series RC circuit is shown at A. B shows the right triangle used to calculate the phase angle between the circuit current and voltage.

After drawing a set of coordinates on which to draw the triangle, we have to calculate the capacitive reactance.

$$X_C = \frac{1}{2\pi\, fc} = \frac{1}{2 \times 3.14 \times (10 \times 10^3 \text{ Hz}) \times (12660 \times 10^{-12} \text{ F})}$$

$$X_C = \frac{1}{7.95 \times 10^{-4}} = 1258\ \Omega$$

Assume a current of 1 A and use Ohm's Law to calculate the voltages across the resistor and capacitor. (Remember the trick we learned with the inductor problem. The voltages are numerically equivalent to the reactance and resistance when we assume a current of 1 A.) $E_R = 1000$ V and $E_C = 1258$ V. Draw the resistor-voltage line on your diagram, as before. Keep in mind that the voltage across a capacitor *lags* the current through it, so the capacitor voltage is 90° behind the resistor voltage. Show this on your graph by drawing the capacitor-voltage line vertically downward at the end of the E_R line. Our graph shows this as 270° but if you go clockwise from the resistance axis — the opposite direction from the way we normally measure angles — it is equivalent to –90°. Actually, we should refer to the capacitor voltage as a negative value, –1258 V, then. Complete the triangle by drawing the E_T line, and proceed with the solution as in the previous problem.

Tan $\theta = -\,1.258$, so $\tan^{-1}(-\,1.258) = -\,51.5°$ and $E_T = 1606$ V (1607 V by the Pythagorean Theorem solution). Notice that the phase angle came out negative this time, because the capacitor voltage was taken as a negative value. This indicates that the total voltage across the circuit is 51.5° behind the current — a result of the fact that the voltage across a capacitor *lags* the current through it. Of course we could also say that the current *leads* the voltage by 51.5°, and that would mean the same thing.

Solving Problems Involving Both Inductors and Capacitors

Series Circuits

The problem illustrated in **Figure 5-19** adds a slight complication, but don't get too confused. Start out the same way we did for the first two problems, by drawing a set of coordinates, labeling the axes, and then calculating the reactance values. Since this is still a series circuit, the current will be the same through all components, so we are still interested in calculating the voltages. The inductive reactance of a 1.60-µH inductor with a 10-MHz signal applied to it is 100 Ω, and the capacitive reactance of a 637-pF capacitor with the same signal is 25 Ω. Now write a value for the assumed voltages across each component: $E_R = 100$ V, $E_L = 100$ V and $E_C = -25$ V.

(Don't forget that the voltage across the capacitor is negative, 180° out of phase with the voltage across the inductor.) When we go to add these lines to make up our "triangle" you will notice that the lines for the inductor and capacitor voltages go in opposite directions. Just subtract them before drawing the line. In

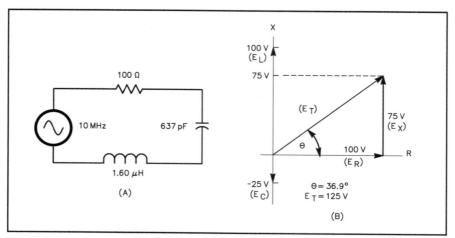

Figure 5-19 — A series RLC circuit is shown at A. B shows the right triangle used to calculate the phase angle between circuit current and voltage.

this example, the inductor voltage is the larger, so the total reactive voltage is 75 V, and is represented by a line drawn upward on the diagram. If the capacitor voltage were greater, then the total reactive voltage would be negative, and would be represented by a line drawn downward on the diagram. That just tells us if the phase angle between circuit voltage and current will be positive (upward) or negative (downward).

Okay, so now you have a triangle drawn to look like the one in Figure 5-19B, and the solution is straightforward. Tan θ = 0.75, $\tan^{-1}$ (0.75) = 36.9° and E_T = 125 V. The total voltage across our circuit leads the current by 36.9°.

[If you worked all of these problems along with the text, you should have no trouble with the phase-angle questions on the Extra class exam. Just to prove that to yourself, turn to Chapter 10 now. Study questions E5D01 through E5D05. Review the examples in this section if needed.]

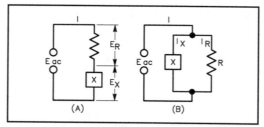

Figure 5-20 — Series and parallel circuits may contain resistance and reactance.

IMPEDANCE CALCULATIONS

When a circuit contains both resistance and reactance, the combined effect of the two is called impedance, symbolized by the letter Z. Impedance is a more general term than either resistance or reactance. The term is often used with circuits that have only resistance or reactance.

The reactance and resistance comprising an

impedance may be connected in series or in parallel, as shown in **Figure 5-20**. In these circuits, the reactance is shown as a box, to indicate that it can be either inductive or capacitive. In the series circuit shown at A, the current is the same through both elements, but with different voltages appearing across the resistance and reactance. In the parallel circuit shown at B, the same voltage is applied to both elements, but different currents may flow in the two branches.

You should remember that the current is in phase with the applied voltage through a resistance, but it is 90° out of phase with the voltage in a reactance. The mnemonic "ELI the ICE man" can help you remember that *the voltage leads the current in an inductor* and *the current leads the voltage in a capacitor.*

You can see, then, that the phase relationship between current and voltage for the whole circuit can be anything between zero and 90°. The phase angle depends on the relative amounts of resistance and reactance in the circuit.

It's important to realize that if there is more than one resistor in the circuit, you must combine them to get one equivalent resistance value. Likewise, if there is more than one reactive element, they must be combined to one equivalent reactance. If there are several inductors and several capacitors, combine all the like elements, then subtract the capacitive reactance value from the inductive reactance to calculate the total reactance.

Coordinate Systems

We commonly use coordinate systems to help us visualize or picture problems that we cannot see directly. For example, we can't actually see the electrons flowing in a circuit, or look at the voltage or impedance associated with the circuit. But we can draw pictures that will represent these quantities, and then use those pictures to aid our solution to the problems. There are a number of commonly used coordinate systems. Two commonly used coordinate systems are the **rectangular-coordinate system** shown

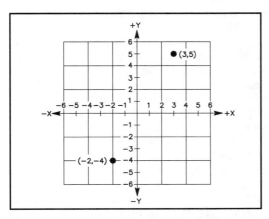

Figure 5-21 — A rectangular-coordinate system uses axes that are at right angles to each other. Any point on the plane can be expressed in terms of an X and a Y coordinate value.

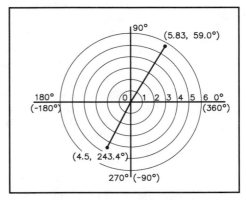

Figure 5-22 — A polar-coordinate system uses a radius, as measured from the center to the point and an angle measured counter-clockwise from the 0° line. Any point on the plane can be expressed in terms of a radius and an angle.

in **Figure 5-21** (sometimes called Cartesian coordinates) and the **polar-coordinate system** shown in **Figure 5-22**. Both of these systems are two dimensional, but they can be extended to three dimensional systems easily. If you add a third axis, usually labeled the Z axis (no relation to impedance), coming out of the page at right angles to the other two on the rectangular system, you have a three-dimensional system. If you rotate the polar-coordinate circle around a diameter, so it describes a sphere, you have a spherical coordinate system, which is handy for representing satellite orbits.

When using rectangular coordinates, a point anywhere on the plane formed by the X and Y axes can be described by a pair of numbers, (X, Y). For example, a point that is 3 units from the origin along the X axis and 5 units from the origin along the Y axis could be written as (3, 5). A point that is 2 units to the left of the origin along the X axis and 4 units below the origin along the Y axis would be described as (−2, −4). See Figure 5-21.

In the polar-coordinate system, points on the plane are also described by a number pair. In this case we use a length, or radius, measured from the origin and an angle, measured counterclockwise from the 0° line. So the two points described in the last paragraph could also be written as (5.83, /59.0°) and (4.5, /243.4°). If you measured this last angle clockwise from the 0° line, then it would be (4.5, /−116.6°). When the solution to a problem gives you a negative value for the angle, it indicates that the angle is measured clockwise from 0°, in the direction opposite to the standard way of measuring polar-coordinate angles.

In electronics, it's common to use both the rectangular and polar-coordinate systems when dealing with impedance problems. The examples in the next few pages of this book should help you become familiar with these coordinate systems and the techniques for changing between them.

Complex Numbers — Calculating Complex Impedances

Sometimes during the solution of a mathematical problem, you come up with a term that includes the square root of minus one ($\sqrt{-1}$). This number is a mathematical curiosity because it's impossible to find a number that you can multiply by itself and get a negative value ($-1 \times -1 = +1$). Since the square root of a negative number does not really exist, it's called an **imaginary number**. Any imaginary number can be expressed in terms of $\sqrt{-1}$ times the square root of the positive value of that number. Mathematicians use the symbol i to represent $\sqrt{-1}$, but in electronics we use j instead, to avoid confusion with the symbol for current. A number that includes an imaginary part is called a **complex number**. Solving a problem that includes an imaginary number involves the use of what is called complex-number algebra. (This name really has nothing to do with how hard — or easy — it is to understand and use the mathematics!)

We use these mathematical techniques to solve impedance problems, because we must distinguish between the resistance and reactance. The algebra of complex numbers provides a way to add, subtract, multiply and divide quantities that include both resistive and reactive components. You can best think of the j as an *operator* that produces a 90° rotation of the resistance line. (An operator is just a mathematical procedure applied to a quantity. For example, an exponent is an operator that tells you how many times to multiply a quantity by itself and the radical sign — $\sqrt{}$ — is an operator that tells you to take the square root.)

It's quite simple to learn how to calculate impedances using this method. We say that the X or horizontal axis (also called the real axis) represents the voltage or current associated with the resistances in our circuit. In the polar-coordinate system this is the line in the 0° - 180° direction. Then the Y or vertical axis (also called the reactance or imaginary axis) represents the voltage or current associated with the reactances. In the polar-coordinate system this is the line in the 90° - 270° direction. If you plot the impedance of a circuit and find the impedance point falls on the right side of the graph, on the horizontal or X axis, the impedance is equivalent to a pure resistance. If you find that the impedance point falls directly on the vertical, or Y axis of the graph, then the impedance is equivalent to either inductive reactance or capacitive reactance.

You actually began to learn about these ways to represent circuit values on graphs in the previous section. You used a rectangular coordinate system graph to find the phase angle between the voltage and current in circuits with resistors, inductors and capacitors.

Either the rectangular or polar-coordinate system can be used to describe a circuit impedance. You might choose to express an impedance value in rectangular coordinates if you want to visualize the resistive and reactive parts. You might choose to express an impedance in polar coordinates if you want to visualize the magnitude and the phase angle of the impedance.

Series Circuits

When the resistance and reactance are in series, the two values can be combined in a relatively straightforward manner. The current is the same in all parts of the circuit ($I_R = I_X$), and the voltage is different across each part. We can write an equation for the impedance in the form:

$$Z = \frac{E}{I} = \frac{E_R + E_X}{I}$$ (Equation 5-17)

Notice that this is really just Ohm's Law written for impedance instead of resistance, as we are used to seeing it. This equation also shows that we can consider the voltage and current associated with the resistive and reactive elements separately.

Since we are really interested in the impedance, and not the actual voltage or current in most cases, it is convenient to assume a current of 1 A, so the voltage and impedance have the same magnitude. If the reactance is inductive, the voltage will lead the current by 90°. We can use the voltage across the resistor as a reference, so the voltage across the reactance must be drawn in the $+j$ direction on our graph. If the reactance is capacitive, the voltage will lag the current by 90°. This time the voltage across the reactance is drawn in the $-j$ direction.

Common practice is to eliminate the voltage calculation and just plot the resistance and reactance values on the graph directly. It is helpful to remember, however, that the reason we label inductive reactance as $+j$ and capacitive reactance as $-j$ is because of the leading and lagging current-voltage relationship described here.

To specify an impedance on the rectangular-coordinate complex-number plane, you only need to know the resistance and the reactance value. If an inductance or capacitance is specified instead of a reactance, then you will first have to calculate the reactance.

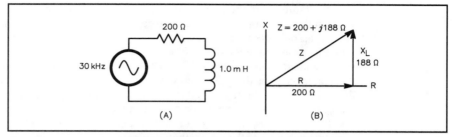

Figure 5-23 — Complex impedance values can be represented on a complex number plane using either rectangular or polar coordinates. The graph shown here uses rectangular coordinates. The horizontal axis represents the real (resistive) part of the impedance and the vertical axis represents the imaginary (reactive) part of the impedance. Inductive reactances are shown as positive values, and capacitive reactances are shown as negative values.

Suppose we have a circuit comprised of a 1.0-mH inductor in series with a 200-Ω resistor. What is the impedance of this circuit at 30 kHz? First, calculate the inductive reactance.

$$X_L = 2\pi \times 30 \times 10^3 \text{ Hz} \times 1.0 \times 10^{-3} \text{ H} = 188 \text{ } \Omega$$

So the new impedance is 200 + j188 ohms. **Figure 5-23** shows how you can plot this impedance value on a rectangular-coordinate system.

Now let's consider a circuit with a 40-Ω resistor in series with a 0.1-μF capacitor. What is the impedance of this combination when it is connected to a 50-kHz signal generator? First calculate the capacitive reactance.

$$X_C = \frac{1}{2\pi f C} = \frac{1}{2\pi \times 50 \times 10^3 \text{ Hz} \times 0.1 \times 10^{-6} \text{ F}} = \frac{1}{31.4 \times 10^{-3} \text{ Hz F}} = 31.8 \text{ } \Omega$$

Because the voltage across a capacitor lags the current, we place the X_C value on the $-j$ axis. The impedance value for this circuit, then, is 40 – j31.8 Ω. **Figure 5-24** shows how you can represent this impedance value on a rectangular-coordinate system.

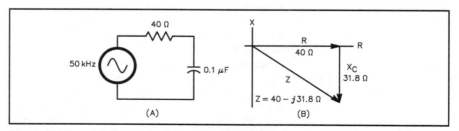

Figure 5-24 — A 0.1μF capacitor and a 40-Ω resistor are connected in series with a 50 kHz signal source. The graph at B shows how you can represent these values on a rectangular-coordinate system. The resulting "impedance triangle" helps you visualize the solution. The text includes a step-by-step solution.

Using Pi in Calculations

Pi (π) is a constant, equal to the circumference of a circle divided by its diameter. The value of π to ten places is 3.1415926535...; it is a nonrepeating decimal. A close approximation that is accurate enough for most Amateur Radio calculations is 22/7.

Many electronic calculators feature a programmed π key. Depending on the value programmed into your calculator, and depending on how you round off calculated values, your answers to problems requiring the use of this constant may vary slightly from those shown in this book. For example, your answer may be 3.508 MHz, while the answer given here may be 3.51 MHz. When you take your exam, the right answer should not be greatly different from the one you calculate, if you solved the problem correctly.

Sometimes you may want to express the impedance in polar-coordinate form. There are a number of methods that you could use to convert rectangular to polar coordinates. Probably the easiest way is with a calculator that is able to do the conversion for you! (Many inexpensive scientific calculators have this capability.) Enter the two rectangular-coordinate values, push a few other buttons, and read the polar-coordinate equivalent value. With this type of calculator you can also convert from polar to rectangular coordinates. (Consult the manual for your calculator to determine the exact method of performing these conversions — each brand of calculator is a little different.)

Lacking such a device, you will have to resort to some basic trigonometry. To help you visualize the solution, make a drawing on a piece of graph paper, or at least draw a set of axes, and label them R and X, for resistance and reactance. Then mark off a rough scale and plot the impedance point using the resistance and reactance values similar to the graphs shown in Figures 5-23 and 5-24. (Draw one line starting at the origin and going to the right to represent the resistance. Then draw another line starting at the end of that one and going either up or down depending on whether you have an inductive or capacitive reactance, to represent the reactance.) If you draw a line from the origin to the point at the end of the reactance line, you will form a right triangle, as shown on Figures 5-23 and 5-24. The Pythagorean Theorem equation is one way to calculate the impedance in polar coordinates:

$$|Z| = \sqrt{R^2 + X^2} \qquad \text{(Equation 5-18)}$$

We put vertical lines around the Z to indicate that we are calculating the total magnitude of the impedance.

To determine the phase angle between the current and voltage you will have to use the tangent function from trigonometry.

$$\tan \theta = \frac{\text{side opposite to the angle}}{\text{side adjacent to the angle}} \qquad \text{(Equation 5-19)}$$

As an example, suppose you have a circuit comprised of an inductor that has a reactance of 100 Ω at some frequency, and a 100-Ω resistor in series. You can look at these values and state the impedance in rectangular-coordinate form as

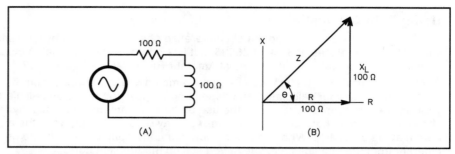

Figure 5-25 — A 100-Ω resistor is connected to an inductor that has a reactance of 100 Ω some frequency. The impedance triangle shown at B helps you visualize the solution, which is given in the text.

$100 + j100$ Ω. But suppose you wanted to know the impedance in polar-coordinate form? Make a sketch of the circuit and draw an "impedance triangle" as **Figure 5-25** shows. We can convert the impedance to polar form by finding the magnitude of the impedance using Equation 5-18.

$$|Z| = \sqrt{R^2 + X^2} = \sqrt{(100\,\Omega)^2 + (100\,\Omega)^2} = \sqrt{10{,}000\,\Omega^2 + 10{,}000\,\Omega^2} = \sqrt{20{,}000\,\Omega^2}$$

$$|Z| = 141\,\Omega$$

Notice that we included the units in this calculation. When you square the resistance and reactance values, the units of ohms were also squared. Then when you take the square root of the number, you also take the square root of the units, so the impedance value is expressed in ohms. Find the phase angle, marked θ on the drawing, using Equation 5-19:

$$\tan\theta = \frac{\text{side opposite to the angle}}{\text{side adjacent to the angle}} = \frac{100\,\Omega}{100\,\Omega} = 1$$

$$\tan^{-1}(1) = 45°$$

So we can also express this impedance in the form: 141 Ω at a phase angle of 45°. The positive phase angle implies that the voltage across the circuit leads the current through it, which is what we would expect from a circuit with inductive reactance. You will usually see this impedance written as 141 Ω, $\underline{/45°}$, where the $\underline{/}$ symbol indicates the angle of a polar-coordinate value.

The various combinations of series-circuit impedance problems (and questions you might be asked on your exam) all follow the techniques we've covered in this section. If you know the value of circuit inductance or capacitance, and the operating frequency, first calculate the inductive reactance or capacitive reactance. You can express any impedance either in rectangular coordinates or polar coordinates. If the circuit includes both inductance and capacitance, you can combine their reactance values to find a single total reactance value. For example, suppose you have a series circuit that consists of a 400-Ω resistance, an inductive reactance of

600 Ω and a capacitive reactance of 300 Ω. What is the impedance? The capacitive reactance of a series circuit is given a $-j$ operator and the inductive reactance has a $+j$ operator, so the total circuit reactance is $+j$ 600 Ω $-j$ 300 Ω $= +j$ 300 Ω. In rectangular-coordinate form, the impedance of this circuit is $400 + j$ 300 Ω. You can convert this value to polar-coordinate form as described earlier.

$$|Z| = \sqrt{R^2 + X^2} = \sqrt{(400\ \Omega)^2 + (300\ \Omega)^2} = \sqrt{160{,}000\ \Omega^2 + 90{,}000\ \Omega^2} = \sqrt{250{,}000\ \Omega^2}$$

$$|Z| = 500\ \Omega$$

Find the phase angle using Equation 5-18.

$$\tan\theta = \frac{\text{side opposite to the angle}}{\text{side adjacent to the angle}} = \frac{300\ \Omega}{400\ \Omega} = 0.75$$

$$\tan^{-1}(0.75) = 37°$$

So in polar-coordinate form, this impedance is 500 Ω $\underline{/37°}$.

[This is a good time to turn to the question pool and practice the series-circuit impedance calculations for your exam. You should be able to answer questions E5C10 and E5C12. Also study questions E5D08 through E5D13. If any of these questions give you difficulty, review this section of the text.]

Parallel Circuits

When a resistance and a reactance are connected in parallel, it is the applied voltage that is common to all parts of the circuit ($E_R = E_X$ this time), and the current through each part will be different. We still use Ohm's Law to calculate impedance, however.

$$Z = \frac{E}{I} = \frac{E}{I_R + I_X}$$ (Equation 5-20)

Keeping in mind our mnemonic, "ELI the ICE man," you can see that the current through an inductor lags the voltage, and that the current through a capacitor leads the voltage. Again, the current and voltage in a resistor are in phase, so we can use the resistive element as the reference point for a drawing to calculate the impedance and phase angle associated with the circuit. We can choose a convenient value for the voltage applied to the circuit, if all we want to know is the impedance and phase angle.

To find the current through each element, we consider the resistance and reactance elements separately.

$$I_R = \frac{E_R}{R} = E_R \times G$$ (Equation 5-21)

and

$$I_X = \frac{E_X}{X} = E_X \times B$$ (Equation 5-22)

The reciprocal of resistance (R) is **conductance** (G) and the reciprocal of reactance (X) is **susceptance** (B). The unit for conductance and susceptance is the siemens, abbreviated S. (The singular and plural name for this unit is the same.) For convenience, you can skip the current calculation, and just take the reciprocal of the resistance and reactance.

Let's look at an example to see how we can work through a parallel-impedance problem. Suppose you have a 100-pF capacitor in parallel with a 4000-ohm resistor and a 500-kHz signal generator. See **Figure 5-26**. If we assume E = 1 V for convenience, the current will be numerically equivalent to the reciprocal of the resistance or reactance. As indicated above, to calculate the impedance of a parallel circuit, first find the capacitive reactance, and then calculate the conductance and susceptance:

$$X_C = \frac{1}{2\pi f C} = \frac{1}{2\pi \times 500 \times 10^3 \text{ Hz} \times 100 \times 10^{-12} \text{ F}} = \frac{1}{3.14 \times 10^{-4}} = 3183 \ \Omega$$

$$G = \frac{1}{R} = \frac{1}{4000 \ \Omega} = 0.00025 \text{ S}$$

and

$$B = \frac{1}{X} = \frac{1}{-j3183 \ \Omega} = +j3.14 \times 10^{-4} \text{ S}$$

Don't be confused by the change of sign in front of the j here. That comes about because of a rule of algebra for complex numbers. When you have a j operator in the denominator, you must eliminate it before performing the division. That is accomplished using a technique called complex conjugates. If you multiply a complex number by the same number, but with an opposite-sign j operator, you get a $-j^2$ term. But since $j = \sqrt{-1}$, $-j^2 = -(\sqrt{-1})^2 = -(-1) = +1$! The only catch is that there is another rule of algebra that says if you multiply the denominator of a fraction by some number, you must multiply the numerator by the same value. Here are the steps involved:

$$B = \frac{1}{X} = \frac{1}{-j3183\Omega} \times \frac{+j3183\Omega}{+j3183\Omega} = \frac{+j3183\Omega}{-j^2(3183\Omega)(3183\Omega)} = \frac{+j}{-(-1)(3183\Omega)}$$

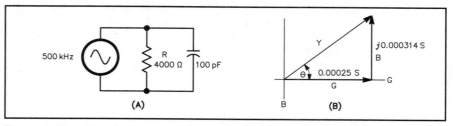

Figure 5-26 — Part A shows a 4000-Ω resistor connected in parallel with a 100-pF capacitor. Part B shows an "admittance triangle" to help you visualize the solution, which is described in the text.

$$B = + j3.14 \times 10^{-4} \text{ S}$$

From this example you can see that the only real effect of this process is that the sign in front of the j operator changes when you convert from reactance to susceptance. If you have a complex number that includes a real and an imaginary part, the process becomes just a little more involved.

Now we will turn to Equation 5-20, and write our impedance equation, choosing a value of 1 for the voltage:

$$Z = \frac{E}{I_R + I_X} = \frac{1 \text{ V}}{2.5 \times 10^{-4} + j3.14 \times 10^{-4} \text{ S}}$$

This is where we hit the first snag. We can either multiply the numerator and denominator of this fraction by the complex conjugate of the denominator, or we can convert the rectangular-coordinate form to polar-coordinate form. When you change to polar-coordinate form there is no j operator left in the expression. To divide by a number that is in polar-coordinate form, you just divide by the magnitude and subtract the angle from the numerator angle. (In this case the numerator angle is 0°.) Since the mathematics are a bit simpler using polar-coordinate form, let's use that method.

The numerator is especially easy. Since it is a value of 1, and there is no imaginary component, the angle will be zero; 1 V, $\underline{/0°}$. To calculate the magnitude of the denominator we use the Pythagorean Theorem (from Equation 5-18).

$$|I| = \sqrt{\left(2.5 \times 10^{-4} \text{ A}\right)^2 + \left(3.14 \times 10^{-4} \text{ A}\right)^2}$$

$$|I| = \sqrt{\left(6.25 \times 10^{-8} \text{ A}^2\right) + \left(9.86 \times 10^{-8} \text{ A}^2\right)} = \sqrt{16.11 \times 10^{-8} \text{ A}^2}$$

$$|I| = 4.01 \times 10^{-4} \text{ A}$$

To calculate the phase angle, you will have to use the tangent function from trigonometry:

$$\tan\theta = \frac{\text{side opposite to the angle}}{\text{side adjacent to the angle}} = \frac{3.14 \times 10^{-4}}{2.5 \times 10^{-4}} = 1.256$$

$$\tan^{-1}(1.256) = 51.5°$$

In polar-coordinate form, we can write the current of this circuit as 4.01×10^{-4} A, $\underline{/51.5°}$. Now we can substitute the calculated current value into the impedance equation.

$$Z = \frac{E}{I_R + I_X} = \frac{1 \text{ V}, \underline{/0°}}{4.01 \times 10^{-4} \text{ A}, \underline{/51.5°}} = 2494 \ \Omega, \underline{/-51.5°}$$

Notice that the sign of the angle changes, because the denominator angle is subtracted from the numerator angle.

Notice that since we are really just combining the conductance and susceptance values when we calculate the current value, the result is a value for the **admittance** (Y), which is the reciprocal of impedance. So $Y = 4.01 \times 10^{-4}$ S, $\underline{/-51.5°}$.

You would solve a problem involving an inductor and resistor in parallel by following the same method. You can find a circuit admittance if you know the impedance, and you can find the impedance if you know the admittance. These values are reciprocals of each other, so just divide the value into 1. Use polar-coordinate values when you have to divide. You can easily convert from rectangular coordinates to polar coordinates or from polar to rectangular using basic trigonometry.

There are many other ways to manipulate the mathematics of complex impedances for series and parallel circuits. Some books choose to present special equations that allow you to transform a parallel circuit into an equivalent series circuit, and then have you write the impedance of the equivalent circuit. That technique requires you to memorize several different equations, and then to be sure you know when to apply each equation. *The ARRL Handbook* includes such "circuit transformation" equations. The method described in this license manual requires only that you know some basic algebra of complex numbers, and the electronics principles that you are already familiar with, such as Ohm's Law. The syllabus for the Amateur Extra license includes algebraic operations using complex numbers as a topic that you may be tested on. To solve the actual impedance problems on your exam you can use any method that you choose. If you follow the step-by-step procedure described here, however, you won't go wrong.

[Turn to Chapter 10 and practice the calculations for examination question E5C11. Review this section as needed.]

THE SMITH CHART

Named after its inventor, Phillip H. Smith, this chart was originally described in the January 1939 issue of *Electronics*. The **Smith Chart** is an invaluable tool for calculating impedances and SWR values along transmission lines and many types of matching networks, both for antennas and other electronics circuitry.

At first glance it may seem formidable, but the Smith Chart is really nothing more than a specialized type of graph, with curved, rather than all straight lines. This coordinate system consists of two sets of circles, called the "resistance family" and the "reactance family."

Figure 5-27 illustrates how the resistance circles are drawn on the chart. These circles are centered on the *resistance axis*, which is the only straight line on an unused Smith Chart. The resistance circles are tangent to the outer circle at the bottom of the chart. Each circle is assigned a resistance value, which is indicated at the point where the circle crosses the resistance axis. All points along any one circle have the same resistance value.

Values assigned to these circles vary from zero at the top of the chart to infinity at the bottom, and actually represent a ratio with respect to the impedance value assigned to the center point on the chart, indicated by a 1.0 on Figure 5-27. This center point is called *prime center*. If prime center is assigned a value of 100 ohms, then 200 ohms of

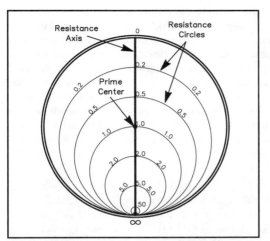

Figure 5-27 — This drawing shows the resistance circles of the Smith Chart coordinate system.

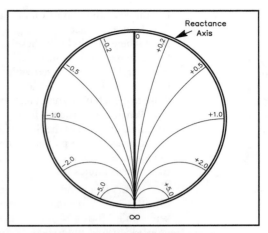

Figure 5-28 — This drawing shows the reactance circles (only segments are actually shown) on the Smith Chart coordinate system.

resistance is represented by the 2.0 circle, 50 ohms by the 0.5 circle, 20 ohms by the 0.2 circle and so on. If a value of 50 is assigned to prime center, then the 2.0 circle represents 100 ohms, the 0.5 circle 25 ohms and the 0.2 circle 10 ohms. In each case you can see that the value on the chart is determined by dividing the actual resistance by the number assigned to prime center. This process of assigning resistance values

with regard to the value at prime center is called *normalizing*. Conversely, values from the chart can be converted back to actual resistance values by multiplying the chart value times the value assigned to prime center. Normalization permits the Smith Chart to be used for any impedance value.

Now let's consider just the reactance circles, as shown on **Figure 5-28**. These circles appear as curved lines on the chart because only segments of the complete circles are drawn. These circles are tangent to the resistance axis, which itself is a member of the reactance family (it has a radius of infinity). The centers of the reactance circles are displaced to the right or left of the resistance axis, along a line that is perpendicular to the resistance axis at the bottom of the chart. The larger outer circle bound-

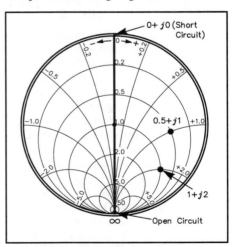

Figure 5-29 — This drawing shows the complete Smith Chart coordinate system. For simplicity, only a few divisions are shown for the resistance and reactance values.

ing the coordinate portion of the chart is called the *reactance axis*.

Each reactance-circle segment is assigned a value of reactance, indicated near the point where the circle touches the reactance axis. All points along any one segment have the same reactance value. As with the resistance circles, the values assigned to each reactance circle are normalized with respect to the value assigned to prime center. Values to the right of the resistance axis are considered to be positive (inductive reactance) and those to the left of the resistance axis are considered to be negative (capacitive reactance).

When the resistance family and the reactance family of circles are combined, the coordinate system of the Smith Chart results, as shown in **Figure 5-29**. Complex series impedances can be plotted on this coordinate system. The Smith Chart normally has more circles than are shown here, and as a result there will be less interpolation between chart circles for any value you have to plot.

[This is a good time to turn to Chapter 10 and study the questions about the Smith Chart coordinate system. Study questions E5C01 through E5C08, and review this section if you have any difficulty with these questions. After you are confident you understand the definitions, go on to the next sections, where you will learn more about how to use the Smith Chart.]

Impedance Plotting

Suppose we have an impedance consisting of 50 ohms resistance and 100 ohms inductive reactance ($Z = 50 + j100 \ \Omega$). If a value of 100 ohms is assigned to prime center, you would normalize this impedance by dividing each component by 100. The normalized impedance would then be:

$$\frac{50}{100} + \frac{j100}{100} \ \Omega = 0.5 + j1.0 \ \Omega$$

This impedance would be plotted on the Smith Chart at the intersection of the 0.5 resistance circle and the +1.0 reactance circle, as indicated on the chart shown in Figure 5-29. If a value of 50 ohms had been assigned to prime center, as for 50-Ω coaxial line, the same impedance would be plotted at the intersection of the 1.0 resistance circle and the + 2.0 reactance circle (also indicated on the chart shown in Figure 5-29).

From these examples, you can see that the same impedance may be plotted at different points on the chart, depending on the value assigned to prime center. When solving transmission-line problems, it is customary to assign a value to prime center that is equal to the characteristic impedance, or Z_0, of the line being used.

Short and Open Circuits

On the subject of plotting impedances, two special cases deserve consideration. These are short circuits and open circuits. A true short circuit has zero resistance and zero reactance, or an impedance of $0 + j0 \ \Omega$. This impedance is plotted at the top of the Smith Chart, at the intersection of the resistance and reactance axes. An open circuit has infinite resistance, and would therefore be plotted at the bottom of the chart, at the intersection of the resistance and reactance axes. These two points will be the same regardless of the value assigned to prime center. They are special cases sometimes used to calculate matching stubs.

Standing-Wave-Ratio Circles

Members of a third family of circles, which are not printed on the chart but which are added during the problem-solving process, are *standing-wave-ratio* (SWR) *circles.* This family is centered on prime center, and appears as concentric circles inside the reactance axis. During calculations, one or more of these circles may be added with a drawing compass. Each circle represents an SWR value, and that value for a given circle may be determined directly from the chart coordinate system by reading the resistance value where the SWR circle crosses the resis-

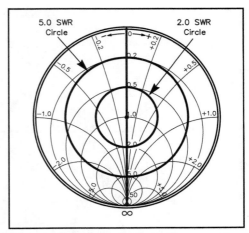

Figure 5-30 — Here is a Smith Chart with SWR circles added.

tance axis, below prime center. (The reading where the circle crosses the resistance axis above prime center indicates the inverse ratio.) **Figure 5-30** illustrates a Smith Chart with SWR circles added.

Consider the situation where a load mismatch in a length of transmission line causes an SWR of 3 to 1. If we temporarily disregard line losses, we may state that the SWR remains constant throughout the entire length of this line. This is represented on the Smith Chart by drawing a 3:1 constant SWR circle (a circle with a radius of 3 on the resistance axis) as shown in **Figure 5-31**. The chart design is such that an impedance encountered anywhere along the length of this mismatched line will fall on the SWR circle, and may be read directly from the coordinates merely by moving around the SWR circle by an amount corresponding to the length of the line involved.

This brings us to using the *wavelength scales*, which appear around the perimeter of the Smith Chart circle, as shown in Figure 5-31. These scales are calibrated in terms of electrical wavelength along a transmission line. One scale, running counterclockwise, starts with the generator (input) end of the line at the top of the chart, and progresses toward the load (output) end of the line. The other scale starts with the load end of the line at the top of the scale, and proceeds toward the generator in a clockwise direction. The complete circle represents one half wavelength. Progressing once around the perimeter of these scales corresponds to progressing along a transmission line for a half wavelength. Because impedances will repeat every half wavelength along a piece of line, the chart may be used for any line length by subtracting an integer (whole number) multiple of half wavelengths from the total line length. A set of external scales are also shown along the left edge of the chart in Figure 5-31. These scales can be thought of as replacements for the resistance axis. For example, there is a scale labeled Standing Wave Voltage Ratio. This scale starts with an SWR of 1 at prime center, and increases toward the top of the scale. You could measure the radius for your 3:1 SWR circle along this scale, then transfer it to the resistance axis and draw the circle. The advantage of using this external scale is that finer calibration marks can be included than are

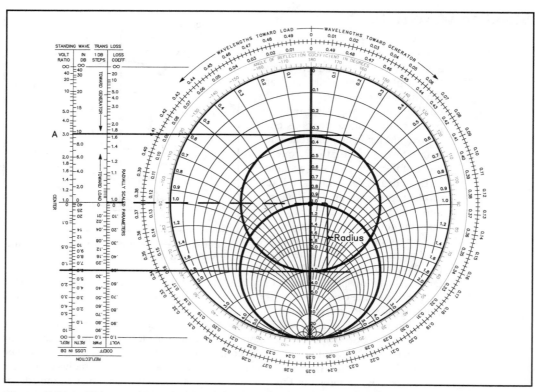

Figure 5-31 — This complete Smith Chart shows an example of impedance calculations. The text explains the steps involved.

practical on the resistance scale. Circle radii for SWR values less than 2:1, for example, would have to be estimated on the resistance scale, but could be measured more accurately on the SWR scale.

[At this point, you should turn to Chapter 10 and study examination question E5C09. Review this section as needed.]

RESONANT CIRCUITS

With all of the problems so far, we have used inductor and capacitor values that give different inductive and capacitive reactances. So there was always a bigger voltage across one of the series components or a larger current through one of the parallel ones. But you may have wondered about the condition if both reactances turned out to be equal, with equal (but opposite) voltages or currents. You have probably realized that in a series circuit with an inductor and a capacitor, if the inductive reactance is equal to the capacitive reactance, then the voltage drop across each component is the same, but they are 180° out of phase. The two values cancel, and the only remaining voltage drop is across any resistance in the circuit.

It is a common practice to say that the two reactances cancel in this case, and to talk about inductive reactance as a positive value and capacitive reactance as a negative value. Just keep in mind that this terminology is a simplification that applies only to series circuits, and comes about because the voltage across the inductor is positive or leads the current and that the voltage across the capacitor is negative or lags the current.

With a parallel circuit, if the inductive and capacitive reactances are equal, then the currents through the components will be equal, but 180° out of phase. The two currents cancel, and the only current

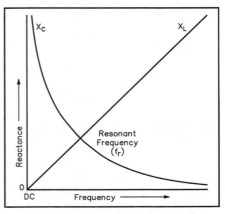

Figure 5-32 — A graph showing the relative change in inductive reactance and capacitive reactance as the frequency increases. For a specific inductor and a specific capacitor, the point where the two lines cross ($X_L = X_C$) is called the resonant frequency.

in the circuit is a result of a parallel resistance. Here again, it is common practice to say that the reactances cancel. In the parallel circuit, however, it is the inductive component that is considered to be negative and the capacitive component is considered to be positive. Of course that terminology comes from the fact that the current through a parallel inductor lags the applied voltage and the current through a parallel capacitor leads the applied voltage.

Whether the components are connected in series or parallel, we say the circuit **resonant frequency** occurs when the inductive-reactance value is the same as the capacitive-reactance value. Remember that inductive reactance increases as the frequency increases and that capacitive reactance decreases as frequency increases. **Figure 5-32** is a graph of inductive and capacitive reactances for two general components.

Neither the exact frequency scale nor the exact reactance scale is important. The two lines cross at only one point, and that point represents the resonant frequency of a circuit using those two components. Every combination of a capacitor and an inductor will be resonant at some frequency.

Since resonance occurs when the reactances are equal, we can derive an equation to calculate the resonant frequency of any capacitor-inductor pair:

$$X_L = 2\pi f_r L = X_C = \frac{1}{2\pi f_r C}$$

$$1 = \frac{1}{(2\pi f_r L)(2\pi f_r C)}$$

$$f_r^2 = \frac{1}{(2\pi)^2 \, L \, C}$$

$$f_r = \frac{1}{2\pi \sqrt{L \, C}} \qquad \text{(Equation 5-23)}$$

Series Resonant Circuits

Figure 5-33 shows a signal generator connected to a series RLC circuit. The signal generator produces a current through the circuit, which will cause a voltage drop to appear across each component. The voltage drops across the inductor and capacitor are always 180° out of phase. When the signal generator produces an output signal at the resonant frequency of the circuit, the voltage drops across the inductor and capacitor can be many times larger than the voltage applied to the circuit. In fact, those voltages are sometimes at least 10 times as large, and may be as much as a few hundred times as large as the applied voltage in a practical circuit. With perfect components and no resistance in the circuit, there would be nothing to restrict the current in the circuit. An ideal **series-resonant circuit**, then, "looks like" a short circuit to the signal generator. There is always some resistance in a circuit, but if the total resistance is small, the current will be large, by Ohm's Law. See Figure 5-33B. The large current that results from this condition produces large, but equal, voltage drops across each reactance. The phase relationship between the voltages across each reactance means that the voltage across the coil reaches a positive peak at the same time that the voltage across the capacitor reaches a

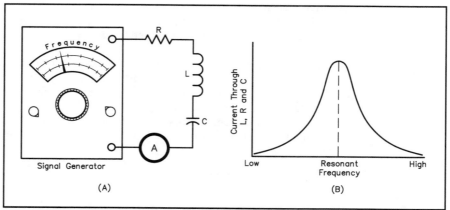

Figure 5-33 — A series-connected LC or RLC circuit behaves like a very low resistance at the resonant frequency. Therefore, at resonance, the current passing through the components reaches a peak.

negative peak. We can say that the impedance of an ideal series-resonant circuit is zero, and is approximately equal to the circuit resistance in an RLC circuit. The main purpose of the series resistor is to prevent the circuit from overloading the generator. The current at the input to a series RLC circuit is a maximum at resonance.

We can think of the resonance condition in another way, which may help you understand these conditions a little better. The large voltages across the reactances develop because of energy stored in the electric and magnetic fields associated with the components. The energy going into the magnetic field on one half cycle is coming out of the electric field of the capacitor. Then, when all of the energy has been transferred to the magnetic field, it is returned to the circuit and is stored in the electric field of the capacitor again. A large amount of energy can be handed back and forth between the inductor and the capacitor without the source supplying any additional amount. The source only has to supply the actual power dissipated in the resistance of imperfect inductors and capacitors plus what is used by the resistor.

Let's see how you do with a series-resonant-circuit problem. What frequency should the signal generator in Figure 5-33 be tuned to for resonance if the resistor is 47 Ω, the coil is a 50-μH inductor and the capacitor has a value of 40 pF? Probably the biggest stumbling block on these calculations will be remembering to convert the inductor value to henrys and the capacitor value to farads. After you have done that, use Equation 5-23 to calculate the resonant frequency.

$$50 \ \mu H = 50 \times 10^{-6} \, H = 0.000050 \, H$$

$$40 \ pF = 40 \times 10^{-12} \, F = 0.000000000040$$

$$f_r = \frac{1}{2\pi \sqrt{LC}} = \frac{1}{2\pi \sqrt{(50 \times 10^{-6})(40 \times 10^{-12})}}$$

$$f_r = \frac{1}{2\pi \sqrt{(2000 \times 10^{-18})}} = \frac{1}{(6.28)(44.7 \times 10^{-9})} = \frac{1}{2.81 \times 10^{-7}}$$

$$f_r = 3.56 \times 10^6 \, Hz = 3.56 \, MHz$$

[Turn to Chapter 10 and study questions E5A01 through E5A05 and question E5A07. Review this section if any of these questions give you trouble.]

Parallel-Resonant Circuits

With a **parallel-resonant circuit** there are several current paths, but the same voltage is applied to the components. **Figure 5-34** shows a parallel LC circuit connected to a signal generator. The series resistor is just a precaution to prevent the circuit from overloading the generator. The applied voltage will force some current through the branches. The current through the coil will be 180° out of phase with the current in the capacitor, and again they add up to zero. So the total current into or out of the generator is very small. Part B is a graph of the relative generator current. It is a mistake to assume, however, that because the generator current is small there is a small current flowing through the capacitor and inductor. In a par-

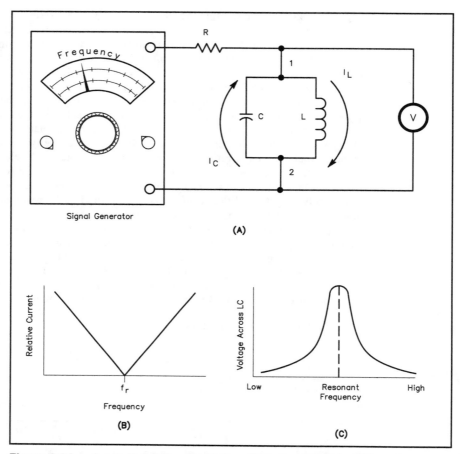

Figure 5-34 — A parallel-connected LC or RLC circuit behaves like a very high resistance at the resonant frequency. Therefore, the voltage measured across the circuit reaches a peak at the resonant frequency.

allel resonant circuit, the current through the inductor and capacitor is very large. Part C is a graph of the voltage across the inductor and capacitor. Because the voltage can be very large near resonance, the current through the components can also be very large. We call this current the *circulating current*, or *tank current*.

The current at the input of a parallel RLC circuit at resonance is at a minimum. The circulating current within the components, however, is at a maximum.

The parallel resonant circuit has a high impedance, and can appear to be an open circuit to the signal generator, because the current from the signal generator is quite small. A resistor placed in parallel with the inductor and capacitor will draw current from the signal generator that depends on the generator output voltage and the resistance value. At resonance, the magnitude of the impedance of a circuit with a resistor, inductor and capacitor all connected in parallel will be approximately equal to the circuit resistance.

If we use points 1 and 2 in Figure 5-34A as reference points and examine the branch currents, we observe an interesting phenomenon. Because the current through the capacitor and inductor are 180° out of phase, the branch currents appear to flow in opposite directions. For example, in the figure the current through the inductor may appear to flow in the direction from 1 to 2, while the current through the capacitor is flowing from 2 to 1. When the applied current from the generator reverses polarity, the branch currents reverse direction. From the point of view of the circulating current, the two reactive components are actually in series. Current flows from point 1 through the inductor to point 2, and from point 2 through the capacitor to point 1. At resonance, this circulating current will only be limited by resistive losses in the components and in the wire connecting them.

We can again think about the energy in the circuit being handed back and forth between the magnetic field of the inductor and the electric field of the capacitor. Large amounts of energy are being transferred, but the generator only has to supply a small amount to make up for the losses in imperfect components. While the total current from the generator is small at resonance, the voltage measured across the tank reaches a maximum value at resonance.

It is also interesting to consider the phase relationship between the voltage across a resonant circuit and the current through that circuit. Because the inductive reactance and the capacitive reactance are equal but opposite, their effects cancel each other. The current and voltage associated with a resonant circuit are in phase, then. This is true whether we are considering a series resonant circuit or a parallel resonant circuit. Calculating the resonant frequency of a parallel circuit is exactly the same as for a series circuit.

[Now turn to Chapter 10 and study questions E5A06 and E5A08 through E5A11. Review this section if any of these questions give you trouble.]

Calculating Component Values for Resonance

Using the same technique that we used to derive Equation 5-23, we can easily derive equations to calculate either the inductance or capacitance to resonate with a certain component at a specific frequency:

$$L = \frac{1}{\left(2\pi f_r\right)^2 C} \qquad \text{(Equation 5-23A)}$$

$$C = \frac{1}{\left(2\pi f_r\right)^2 L} \qquad \text{(Equation 5-23B)}$$

You should always be sure to change the frequency to hertz, the capacitance to farads and the inductance to henrys when using these equations. Let's try solving a problem, just to be sure you know how to handle the equations.

What value capacitor is needed to make a circuit that is resonant in the 80-meter band if you have a 20-μH coil? Choose a frequency in the 80-meter band to work with. Let's pick 3.6 MHz. Then convert to fundamental units: $f_r = 3.6 \times 10^6$ Hz and $L = 20 \times 10^{-6}$ H. Select Equation 5-23B, since that one is written to find capacitance, the quantity we are looking for:

$$C = \frac{1}{\left(2\pi\right)^2 \left(3.6 \times 10^6\right)^2 \left(20 \times 10^{-6}\right)}$$

$$C = \frac{1}{\left(39.48\right) \left(1.30 \times 10^{13}\right) \left(20 \times 10^{-6}\right)}$$

$$C = \frac{1}{1.03 \times 10^{10}} = 9.7 \times 10^{-11} \text{ F} = 97 \times 10^{-12} \text{ F} \approx 100 \text{ pF}$$

You will need a 100-pF capacitor. If you try solving this problem for both ends of the 80-meter band, you will find that you need a 103-pF capacitor at 3.5 MHz and a 79-pF unit at 4 MHz. So any capacitor value within this range will resonate in the 80-meter band with the 20-μH inductor.

[For more practice with resonant circuits, study question E7C06 in Chapter 10. Review the procedures described in this section if you have difficulty with this question.]

Q — THE QUALITY FACTOR OF REAL COMPONENTS

We have talked about ideal resistors, capacitors and inductors, and how they behave in ac circuits. We have shown that resistance in a circuit causes some departure from the ideal conditions. But how can we determine how close to the ideal a certain component comes? Or how much of an effect it will have on the designed circuit conditions? We can assign a number to the coil or capacitor that will tell us the relative merits of that component — a quality factor of sorts. We call that number **Q**. We can also assign a Q value to an entire circuit, and that is a measure of how close to the ideal that circuit performs — at least in terms of its resonance properties.

One definition of Q is that it is the ratio of reactance to resistance. **Figure 5-35** shows that a capacitor can be thought of as an ideal capacitor in series with a resistor and a coil can be considered as an ideal inductor in series with a resistor. This internal resistance can't actually be separated from the coil or capacitor, of course, but it acts just the same as if it were in series with an ideal, lossless component. The Q of a real inductor, L, is equal to the inductive reactance divided by the resistance and the Q of a real capacitor, C, is equal to the capacitive reactance divided by the resistance:

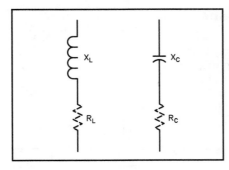

Figure 5-35 — A practical coil can be considered as an ideal inductor in series with a resistor, and a practical capacitor can be considered as an ideal capacitor in series with a resistor.

$$Q = \frac{X}{R}$$ (Equation 5-24)

If you want to know the Q of a circuit containing both internal and external resistance, both resistances must be added together to find the value of R used in the equation. Since added external resistance can only raise the total resistance, the Q always goes down when resistance is added in series. There is no way to reduce the internal resistance of a coil or capacitor and raise the Q, therefore, except by building a better component. The internal resistance of a capacitor is usually much less than that for a coil, so we often ignore the resistance of a capacitor and consider only that associated with the coil. **Figure 5-36A** shows a series RLC circuit with a Q of 10. To calculate that value, select either value of reactance and divide it by the resistor value.

At Figure 5-36B, we have increased the input frequency 5 times. The reactance of our inductor has increased 5 times, and we have selected a new capacitor to provide a resonant circuit. This time the components are arranged to provide a parallel resonant circuit. The circuit Q is still found using Equation 5-24. You will notice that the Q for this circuit is 50. Increasing the frequency increased the inductive reactance, so as long as the internal resistance stays the same, the Q increases by the same factor.

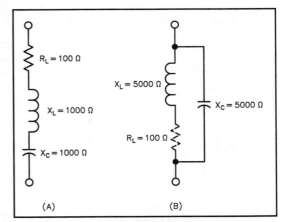

Figure 5-36 — The Q in a series-resonant circuit such as is shown at A and a parallel-resonant circuit such as is shown at B is found by dividing the inductive reactance by the resistance.

Skin Effect Increases Resistance

Unfortunately, the internal resistance of the coil (due mainly to the resistance of the wire used to wind it) increases somewhat as the frequency increases. In fact, the coil Q will increase with increasing frequency up to a point, but then the internal resistance becomes greater and the coil Q degrades. **Figure 5-37** illustrates how coil Q changes with increasing frequency.

The major cause of this increased resistance at higher frequencies is something known as **skin effect**. As the frequency increases, the electric and magnetic fields of the signal do not penetrate as deeply into the conductor. At dc, the entire thickness of the wire is used to carry currents but as the frequency increases, the effective area gets smaller and smaller. In the HF range, the current all flows in the outer few thousandths of an inch of the conductor, and at VHF and UHF, the depth is on the range of a few ten thousandths of an inch. This makes the wire less able to carry the electron flow, and increases the effective resistance.

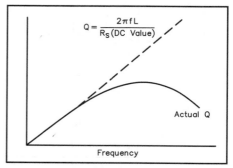

Figure 5-37 — For low frequencies, the Q of an inductor is proportional to frequency. At high frequencies, increased losses in the coil cause the Q to be degraded from the expected value.

[Before going on to the next section, you should turn to Chapter 10 and study questions E5E01 through E5E05. Review this section if you are uncertain about the importance of skin effect.]

Q in Parallel-Resonant Circuits

Often a parallel resistor is added to a parallel-resonant circuit to decrease the Q and increase the **bandwidth** of the circuit, as shown in **Figure 5-38**. Such a resistor should have a value more than 10 times the reactance of the coil (or capacitor) at resonance or the resonance conditions will change. With the added parallel resistor, circuit Q is found by dividing the resistor value by the reactance value. So for a parallel resonant circuit:

$$Q = \frac{R}{X} \qquad \text{(Equation 5-25)}$$

Figure 5-38 shows a parallel circuit made with a 2.7-μH inductor and a 47-pF capacitor. The circuit has an 18-kΩ resistor in parallel with the inductor and capacitor. The resonant frequency of this circuit is 14.128 MHz. What is the circuit Q?

First we must calculate the reactance of either the inductor or capacitor. For this example, we will use the inductor.

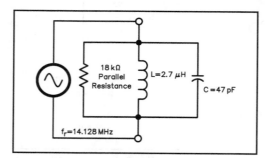

Figure 5-38 — The Q in a parallel resonant circuit with parallel resistance is found by dividing the parallel-resistance value by the inductive reactance.

$$X_L = 2 \pi f_r L = 6.28 \times 14.128 \times 10^6 \text{ Hz} \times 2.7 \times 10^{-6} \text{ H} = 239.6 \ \Omega$$

Next we will divide the circuit resistance by this reactance value to find the Q.

$$Q = \frac{R}{X} = \frac{18 \times 10^3 \ \Omega}{239.6 \ \Omega} = 75.1$$

As in the case of a series circuit, any added resistance degrades the circuit Q and increases the **bandwidth**. (Bandwidth is discussed in the next section.) When a resistor is added in parallel, the circuit Q is always less than if no resistor were included. If no additional resistor were included, the parallel resistance would be infinitely large. Adding a resistor to the circuit always decreases that value, and so the Q is also reduced. The larger the value of added resistance, the higher the Q.

[Now study questions E5F01 through E5F06 in Chapter 10. Review this section as needed.]

Resonant-Circuit Bandwidth

The higher the circuit Q, the sharper the frequency response of a resonant circuit will be. **Figure 5-39** shows the relative **bandwidth** of a circuit with two different Q values. Bandwidth refers to the frequency range over which the circuit response is no more than 3 dB below the peak response. The –3-dB points are shown on Figure 5-39, and the bandwidths are indicated. (The **decibel**, or **dB**, represents the logarithm of a ratio of two power levels. If you are not familiar with the use of decibels, see ARRL's *Understanding Basic Electronics*, *The ARRL Handbook*, or a math text on the subject.) Since this 3-dB decrease in signal represents the points where the circuit power is one half of the resonant power, the –3-dB points are also called **half-power points**. The voltage and current have been reduced to 0.707 times their peak values.

The half-power points are called f_1 and f_2; Δf is the difference between these two frequencies, and represents the half-power (or 3-dB) bandwidth. It is possible to calculate the bandwidth of a resonant circuit based on

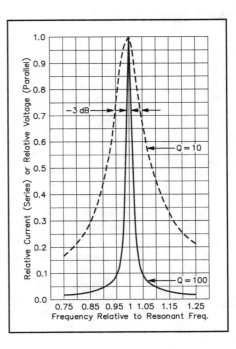

Figure 5-39 — The relative –3-dB bandwidth of two resonant circuits is shown. The circuit with the higher Q has a steeper response, and a narrower bandwidth. Notice that the vertical scale represents current for a series circuit and voltage for a parallel one. The two circuits are considered to have equal peak response for either case.

the circuit Q and the resonant frequency.

$$\Delta f = \frac{f_r}{Q}$$ (Equation 5-26)

where:

Δf = the half-power bandwidth.

f_r = the resonant frequency of the circuit.

Q = the circuit Q, as given by Equation 5-24 or 5-25 as appropriate.

Let's calculate the half-power bandwidth of a parallel circuit that has a resonant frequency of 1.8 MHz and a Q of 95. The half-power bandwidth is found by Equation 5-26:

$$\Delta f = \frac{f_r}{Q} = \frac{1.8 \times 10^6 \text{ Hz}}{95} = 18.9 \times 10^3 \text{ Hz} = 18.9 \text{ kHz}$$

Notice that the response of this circuit will be at least half of the peak signal power for signals in the range 1.79055 to 1.80945 MHz. To find the upper and lower frequency limits you have to subtract *half* the total bandwidth from the center frequency for the lower limit and add half the bandwidth to get the upper frequency limit.

[This is a good time to practice the half-power bandwidth calculations from the Extra class question pool. Study questions E5A12 through E5A17. Review the example in this section if you have difficulty with any of these questions.]

POWER IN REACTIVE CIRCUITS

Earlier in this chapter, we learned that energy is stored in the magnetic field of an inductor when current increases through it, and in the electric field of a capacitor when the voltage across it increases. That energy is returned to the circuit when the current through the inductor decreases or when the voltage across the capacitor decreases. We also learned that the voltages across and currents through these components are 90° out of phase with each other. One way to think of this situation is that in one half of the cycle the power source gives some energy to the inductor or capacitor, only to have the same amount of energy handed back on the next half cycle. A perfect capacitor or coil does not consume any energy, but current does flow in the circuit when a voltage is applied to it.

Power is the time rate of doing work, or the time rate of using energy. Going back to our example of work and energy at the beginning of this chapter, if you did the 10 foot-pounds of work in 5 seconds, then you have developed a power of 2 foot-pounds per second. If you could develop 550 foot-pounds per second of power, then we would say you developed 1 horsepower. So power is a way to express not only how much work you are doing (or how much energy is being stored); it also tells how fast you are doing it. In the metric system of measure, which is used to express all of our common electrical units, power is expressed in terms of the watt, which means energy is being stored at the rate of 1 joule per second, or work is being done at the rate of 1 joule per second. To pass the General class license exam, you learned that electrical power is equal to the current times the voltage:

$$P = I E \qquad \text{(Equation 5-27)}$$

But there is one catch. That equation is only true when the current is through a resistor and the voltage is across that resistor. In other words, the current and voltage must be in phase.

Peak and Average Power

In a dc circuit, it is easy to calculate the power. Simply measure or calculate the current and voltage, and multiply the values. But in an ac circuit, you must specify whether you are using the peak or effective (root-mean-square — RMS) values. Peak voltage times peak current will give the **peak power**, and RMS voltage times RMS current gives the **average power** value. Notice that we do not call this RMS power! It is interesting to note that with pure sine waves, the peak power is just two times the average power. With a modulated RF wave, which does not result in a pure sine wave, the relationship between peak and average power is not quite that simple. To avoid any possible confusion, we usually refer to the average power output of an RF amplifier. So when you say the output power from your transmitter is 100 W, you mean that the average output power is 100 W.

Another term that is often used in relation to RF power amplifiers is **peak-envelope power (PEP)**. This is the power specification used by the FCC to determine the maximum permissible output power. PEP is defined as the average power of the RF envelope during a modulation peak. So PEP refers to the *maximum average power* of a modulated wave, and not the *peak power* as defined in the previous paragraph! It is interesting to note that in the case of a CW transmitter with the key held down, the PEP and average power will be the same, and can be measured on a peak-reading wattmeter.

Power Factor

An ammeter and a voltmeter connected in an ac circuit to measure voltage across and current through an inductor or capacitor will read the correct RMS values, but multiplying them together does not give a true indication of the power being dissipated in the component. If you multiply the RMS values of voltage and current read from these meters, you will get a quantity that is referred to as **apparent power**. This term should tell you that the power found in that way is not quite correct! The apparent power should be expressed in units of volt-amperes (VA) rather than watts. The apparent power in an inductor or capacitor is called **reactive power** or nonproductive, wattless power. Reactive power is expressed in volt-amperes-reactive (VARs).

When there are inductors and capacitors in the circuit, the voltage is not in phase with the current. As we have already said, there is no power developed in a circuit containing a pure capacitor or a pure inductor. With ideal inductors and capacitors, the power is handed back and forth between the magnetic field in the inductor and the electric field in the capacitor, but no power is dissipated. Only the resistive part of the circuit will dissipate power.

There is something about the phase angle between the voltage and current that must be taken into account. That something is called the **power factor**. Power factor is a quantity that relates the apparent power in a circuit to the real power. You can find the **real power** in a circuit by:

$$P = I^2 R$$ (Equation 5-28)

for a series circuit, or:

$$P = \frac{E^2}{R}$$ (Equation 5-29)

for a parallel circuit. Notice that both of these equations are easily derived by using Ohm's Law to solve for either voltage or current, ($E = I \times R$ and $I = E / R$) and replacing that term with the Ohm's Law equivalent.

One way to calculate the power factor is to simply divide the real power by the apparent power:

$$\text{Power factor} = \frac{P_{\text{REAL}}}{P_{\text{APPARENT}}}$$ (Equation 5-30)

Figure 5-40 shows a series circuit containing a 75-Ω resistor and a coil with an inductive reactance of 100 Ω at the signal frequency. The voltmeter reads 250-V RMS and the ammeter indicates a current of 2-A RMS. This is an apparent power of 500 VA. Use Equation 5-28 to calculate the power dissipated in the resistor, $P_{\text{REAL}} = (2\,\text{A})^2 \times 75\,\Omega = 4\,\text{A}^2 \times 75\,\Omega = 300\,\text{W}$. Now by using Equation 5-30, we can calculate the power factor:

$$\text{Power factor} = \frac{300\,\text{W}}{500\ \text{VA}} = 0.6$$

Another way to calculate the real power, if you know the power factor, is given by:

$$P_{\text{REAL}} = P_{\text{APPARENT}} \times \text{Power factor}$$ (Equation 5-31)

In our example,

$$P_{\text{REAL}} = 500\ \text{VA} \times 0.6 = 300\ \text{W}$$

Of course the value found using Equation 5-31 must agree with the value found by either Equation 5-28 or 5-29, depending on whether the circuit is a series or a parallel one.

Figure 5-40 — Only the resistance actually consumes power. The voltmeter and ammeter read the proper RMS value for the circuit, but their product is apparent power, not real average power.

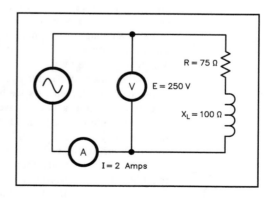

What if you don't have the benefit of a voltmeter and an ammeter in a circuit? How can you calculate the real power or power factor? Well, it turns out that the phase angle between the total applied voltage and the circuit current can be used. We already learned how to calculate the phase angle of either a series or a parallel circuit, so that should be no problem. If you don't remember how to calculate the phase angle, go back and review that section of this chapter.

The power factor can be calculated from the phase angle by finding the cosine value of the phase angle:

Power factor = cos θ   (Equation 5-32)

where θ is the phase angle between voltage and current in the circuit and cosine is the trigonometric function with values that vary between 0 for an angle of − 90° (such as for the voltage across a capacitor lagging the current) to 1 for an angle of 0° (such as for the voltage and current being in phase for a resistor) and back to 0 for an angle of 90° (such as for the voltage across an inductor leading the current).

From this discussion, we can see that for a circuit containing only resistance, where the voltage and current are in phase, the power factor is 1, and the real power is equal to the apparent power. For a circuit containing only pure capacitance or pure inductance, the power factor is 0, so there is no real power! For most practical circuits, which contain resistance, inductance and capacitance, and the phase angle is some value greater than or less than 0°, the power factor will be something less than one. In such a circuit, the real power will always be something less than the apparent power. This is an important point to remember.

Let's try a sample problem, just to be sure you understand all this. We will assume you can calculate the phase angle between voltage and current, given the component values and generator frequency. After that, it doesn't matter if the circuit is a series or parallel one. The procedure is the same. What is the power factor for an R-L circuit having a phase angle of 60°? Use Equation 5-32 to answer this question. Find the cosine value using your calculator or a table of trigonometric values.

Power factor = cos 60° = 0.500

Let's try another example. Suppose you have a circuit that draws 4 amperes of current when 100 V ac is applied. The power factor for this circuit is 0.2. What is the real power (how many watts are consumed) for this circuit? The apparent power is calculated using Equation 5-27:

$P_{APPARENT}$ = 100 V × 4 A = 400 VA

Real power is then found using Equation 5-31:

P_{REAL} = 400 VA × 0.2 = 80 W

[If you have practiced these example calculations with the text you should have no difficulty with questions E5F07 through E5F12 in Chapter 10. Try those questions now, to be sure you understand the answers and the calculations. Review this section as needed.]

EFFECTIVE RADIATED POWER (ERP)

Knowing the output power from your transmitter is important to ensure that you stay within the limits set by FCC rules for your Amateur Radio Station. Sometimes it is more helpful, in evaluating your total station performance, to know how much power is actually being radiated. This is especially true with regard to repeater systems. The effective power radiated from the antenna helps establish the coverage area of the repeater. In addition, the height of the repeater antenna as compared to buildings and mountains in the surrounding area (height above average terrain, or HAAT) has a large effect on the repeater coverage. In general, for a given coverage area, with a greater antenna HAAT, less **effective radiated power (ERP)** is needed. A frequency coordinator may even specify a maximum ERP for a repeater, to help reduce interference between stations using the same repeater frequencies.

You may be wondering why the transmitter power output is not the same as the power radiated from the antenna. Well, there is always some power lost in the feed line, and often there are other devices inserted in the line, such as a watt-meter, SWR bridge or an impedance-matching network. In the case of a repeater system, there is usually a duplexer so the transmitter and receiver can use the same antenna, and perhaps a circulator to reduce the possibility of intermodulation interference. These devices also introduce some loss to the system. Antennas are compared to a reference, and they may exhibit gain or loss as compared to that reference.

The two types of antennas commonly used for reference are a half-wave dipole and a theoretical isotropic radiator. For our discussion and calculations here, we will assume the antenna gain is with reference to a half-wave dipole. A beam antenna will have some gain over a dipole, at least in the desired radiation direction. The exact amount of gain will depend on the design and installation.

These system gains and losses are usually expressed in **decibels (dB)**. The decibel is a logarithm of the ratio of two power levels, and the gain in dB is calculated by:

$$dB = 10 \log \left(\frac{P_2}{P_1} \right)$$

(Equation 5-33)

where:

P_1 is the reference power.
P_2 is the power being compared to the reference.

In the case of calculating ERP, the transmitter output power is considered as the reference, and the power at any other point is P_2.

The main advantage of using decibels is that system gains and losses expressed in these units can simply be added, with losses written as negative values. Suppose we have a repeater station that uses a 50-W transmitter and a feed line with 4 dB of loss. There is a duplexer in the line that exhibits 2 dB of loss and a circulator that adds another 1 dB of loss. This repeater uses an antenna that has a gain of 6-dBd (the antenna gain is compared to a dipole). Our total system gain looks like:

System gain = −4 dB + −2 dB + −1 dB + 6 dBd = −1 dB

Note that this is a loss of 1 dB total for the system. Using Equation 5-33:

$$-1 \text{ dB} = 10 \log \left(\frac{P_2}{50 \text{ W}} \right)$$

$$\frac{-1 \text{ dB}}{10} = \log \left(\frac{P_2}{50 \text{ W}} \right)$$

$$\log^{-1} \left(\frac{-1 \text{ dB}}{10} \right) = \log^{-1} \left(\log \left(\frac{P_2}{50 \text{ W}} \right) \right)$$

$\left(\log^{-1} \text{ means the antilog, or inverse log function} \right)$

$$\log^{-1} \left(-0.1 \text{ dB} \right) = \frac{P_2}{50 \text{ W}}$$

$$P_2 = \log^{-1} (-0.1) \times 50 \text{ W} = 0.79 \times 50 \text{ W} = 39.7 \text{ W}$$

This is consistent with our expectation that with a 1-dB system loss we would have somewhat less ERP than transmitter output power.

As another example, suppose we have a transmitter that feeds a 100-W output signal into a feed line that has 1-dB of loss. The feed line connects to an antenna that has a gain of 6-dBd. What is the effective radiated power from the antenna? To calculate the total system gain (or loss) we add the decibel values given:

System gain = – 1 dB + 6 dBd = 5 dB

Then we can use Equation 5-21 to find the ERP:

$$5 \text{ dB} = 10 \log \left(\frac{P_2}{100 \text{ W}} \right)$$

$$\frac{5 \text{ dB}}{10} = \log \left(\frac{P_2}{100 \text{ W}} \right)$$

$$\log^{-1} \left(\frac{5 \text{ dB}}{10} \right) = \log^{-1} \left(\log \left(\frac{P_2}{100 \text{ W}} \right) \right)$$

$$\log^{-1} \left(0.5 \text{ dB} \right) = \frac{P_2}{100 \text{ W}}$$

$$P_2 = \log^{-1} (0.5 \text{ dB}) \times 100 \text{ W} = 3.16 \times 100 \text{ W} = 316 \text{ W}$$

The total system has positive gain, so we should have expected a larger value for ERP than the transmitter power. Keep in mind that the gain antenna concentrates more of the signal in a desired direction, with less signal in undesired directions. So the antenna doesn't really increase the total available power.

[Turn to Chapter 10 now and study examination questions E5G01 through E5G11. If you have any difficulty with those questions, review this section.]

THEVENIN'S THEOREM

Thevenin's Theorem is a useful tool for simplifying complex networks of resistors (or other components). Our discussion here is limited to resistors, to make it easier for you to understand. (Besides, you don't have to know how to handle this technique with reactive components for the Element 4 examination!) The theorem states that any two-terminal network of resistors and voltage sources, no matter how complex, can be replaced by a circuit consisting of a single voltage source and a single series resistor. Doing this conversion for a part of a complex circuit greatly simplifies calculations involving other parts of the circuit.

We will use a simple example to illustrate the technique provided by

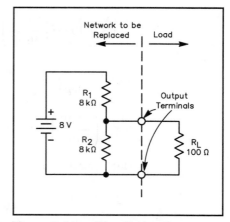

Figure 5-41 — Circuit showing a combination of a voltage source and a resistive network that can be simplified using Thevenin's Theorem to find the load current.

Thevenin's Theorem. While it may seem easier to solve this problem using other methods that you are familiar with, the advantages become obvious with more complex problems. **Figure 5-41** shows a simple circuit involving a voltage source, two parallel resistors and a series resistor. We want to know the current through R_L, so that is considered as the load for the Thevenin-equivalent circuit.

We could solve this problem by calculating the resistance of the parallel-resistor combination, adding the series resistance, then using Ohm's Law to find the total circuit current. By calculating the voltage drops across the series resistor and the parallel pair, we could finally solve for the current through R_L. A lot of work, even for this simple problem.

To apply Thevenin's Theorem, first remove the load from the network to be simplified. The new Thevenin-equivalent voltage source has a voltage equal to the open-circuit (no load) voltage at the output terminals. **Figure 5-42A** shows the network to be replaced, with the output voltage. You can find that output voltage by using Ohm's Law to calculate the total circuit current and the voltage drop across R_2:

$$V_{OUT} = I \times R_2$$

But the circuit current is equal to the source voltage divided by the total resistance:

$$I = \frac{E}{R_1 + R_2}$$

Then:

$$V_{OUT} = \frac{E \times R_2}{R_1 + R_2} = \frac{8 \text{ V} \times 8 \times 10^3 \text{ }\Omega}{8 \times 10^3 \text{ }\Omega + 8 \times 10^3 \text{ }\Omega} = \frac{64 \times 10^3 \text{ V}}{16 \times 10^3} = 4 \text{ V}$$

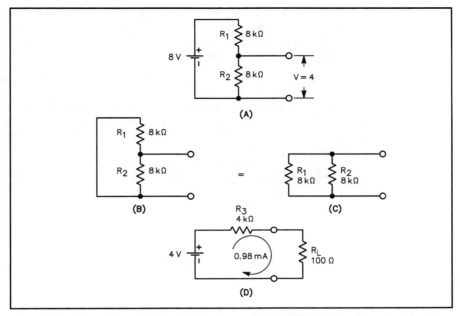

Figure 5-42 — Steps showing how a two-terminal network can be simplified to find a Thevenin-equivalent voltage source (A) and the Thevenin-equivalent resistance (B and C) to calculate the load current (D).

Now replace the voltage source in the original network with its "internal resistance." (Most voltage sources can be assumed to have an internal resistance of zero ohms, so we will use a piece of wire in place of the voltage source.) See Figure 5-42B. The Thevenin-equivalent series resistor is equal to that resistance which would be measured by an ohmmeter connected to the load terminals of the circuit with the voltage source shorted out. We can redraw the circuit of Figure 5-42B as shown in part C, which clearly shows that these two resistors are simply in parallel. Resistors in parallel are combined by adding their reciprocals, then taking the reciprocal of that sum. You probably remember the equation to combine two resistors in parallel from your studies for the Technician license exam:

$$R_T = \frac{R_1 \times R_2}{R_1 + R_2} \qquad \text{(Equation 5-34)}$$

$$R_T = \frac{8 \text{ k}\Omega \times 8 \text{ k}\Omega}{8 \text{ k}\Omega + 8 \text{ k}\Omega} = \frac{64 \text{ k}\Omega^2}{16 \text{ k}\Omega} = 4 \text{ k}\Omega$$

Now we can replace the original voltage-divider network with our new Thevenin equivalent circuit, consisting of a 4-V source in series with a 4-kilohm resistor. Adding the load resistor back in, as shown in Figure 5-42D, enables us to calculate the load current easily using Ohm's Law:

$$I = \frac{E}{R_3 + R_L} = \frac{4\,V}{4000\,\Omega + 100\,\Omega} = \frac{4\,V}{4100\,\Omega} = 9.8 \times 10^{-4}\,A = 0.98\,mA$$

The whole procedure is much more difficult to describe than it is to carry out, and after practicing a few problems, you should be ready to tackle problems with much more complicated networks, and maybe even a few inductors and capacitors thrown in with an ac voltage source!

[To get some of the practice that you need, turn to Chapter 10, and solve questions E5H01 through E5H11. Review this section as needed.]

Alpha (α) — The ratio of transistor collector current to emitter current. It is between 0.92 and 0.98 for a junction transistor.

Alpha cutoff frequency — A term used to express the useful upper frequency limit of a transistor. The point at which the gain of a common-base amplifier is 0.707 times the gain at 1 kHz.

Anode — The terminal that connects to the positive supply lead for current to flow through a device.

Avalanche point — That point on a diode characteristic curve where the amount of reverse current increases greatly for small increases in reverse bias voltage.

Bipolar junction transistor — A transistor made of two PN semiconductor junctions using two layers of similar-type material (N or P) with a third layer of the opposite type between them.

Cathode — The terminal that connects to the negative supply lead for current to flow through a device.

Cathode-ray tube (CRT) — An electron beam tube in which the beam can be focused on a luminescent screen. The spot position can be varied to produce a pattern or picture on the screen.

Charge-coupled device (CCD) — An integrated circuit that uses a combination of analog and digital circuitry to sample and store analog signal voltage levels, passing the voltages through a capacitor string to the circuit output.

Complementary metal-oxide semiconductor (CMOS) — A type of construction used to make digital integrated circuits. CMOS is composed of both N-channel and P-channel MOS devices on the same chip.

Crystal-lattice filter — A filter that employs piezoelectric crystals (usually quartz) as the reactive elements. They are most often used in the IF stages of a receiver or transmitter.

Depletion mode — Type of operation in a JFET or MOSFET where current is reduced by reverse bias on the gate.

Digital IC — An integrated circuit whose output is either on (1) or off (0).

Doping — The addition of impurities to a semiconductor material, with the intent to provide either excess electrons or positive charge carriers (holes) in the material.

Drain — The point at which the charge carriers exit an FET. Corresponds to the plate of a vacuum tube.

Enhancement mode — Type of operation in a MOSFET where current is increased by forward bias on the gate.

Field-effect transistor (FET) — A voltage-controlled semiconductor device. Output current can be varied by varying the input voltage. The input impedance of an FET is very high.

Forward bias — A voltage applied across a semiconductor junction so that it will tend to produce current.

Gate — Control terminal of an FET. Corresponds to the grid of a vacuum tube. **Gate** also refers to a combinational logic element with two or more inputs and one output. The output state depends upon the state of the inputs.

Hot-carrier diode — A type of diode in which a small metal dot is placed on a single semiconductor layer. It is superior to a point-contact diode in most respects.

Integrated circuit — A device composed of many bipolar or field-effect transistors manufactured on the same chip, or wafer, of silicon.

Junction field-effect transistor (JFET) — A field-effect transistor created by diffusing a gate of one type of semiconductor material into a channel of the opposite type of semiconductor material.

Light-emitting diode — A device that uses a semiconductor junction to produce light when current flows through it.

Linear IC — An integrated circuit whose output voltage is a linear (straight line) representation of its input voltage.

Maximum average forward current — The highest average current that can flow through the diode in the forward direction for a specified junction temperature.

Metal-oxide semiconductor FET (MOSFET) — A field-effect transistor that has its gate insulated from the channel material. Also called an IGFET, or insulated gate FET.

Monolithic microwave integrated circuit (MMIC) — A small pill-sized amplifying device that simplifies amplifier designs for microwave-frequency circuits. An MMIC has an input lead, an output lead and two ground leads.

N-type material — Semiconductor material that has been treated with impurities to give it an excess of electrons. We call this a "donor material."

Offset voltage — As related to op amps, the differential amplifier output voltage when the inputs are shorted. It can also be measured as the voltage between the amplifier input terminals in a closed-loop configuration.

Operational amplifier (op amp) — A linear IC that can amplify dc as well as ac. Op amps have very high input impedance, very low output impedance and very high gain.

Peak inverse voltage (PIV) — The maximum instantaneous anode-to-cathode reverse voltage that is to be applied to a diode.

Persistence — A property of a cathode-ray tube (CRT) that describes how long an image will remain visible on the face of the tube after the electron beam has been turned off.

Phase-locked loop (PLL) — A servo loop consisting of a phase detector, low-pass filter, dc amplifier and voltage-controlled oscillator.

Piezoelectric effect — The physical deformation of a crystal when a voltage is applied across the crystal surfaces.

PIN diode — A diode consisting of a relatively thick layer of nearly pure semiconductor material (intrinsic semiconductor) with a layer of P-type material on one side and a layer of N-type material on the other.

PN junction — The contact area between two layers of opposite-type semiconductor material.

Point-contact diode — A diode that is made by a pressure contact between a semiconductor material and a metal point.

P-type material — A semiconductor material that has been treated with impurities to give it an electron shortage. This creates excess positive charge carriers, or "holes," so it becomes an "acceptor material."

Reverse bias — A voltage applied across a semiconductor junction so that it will tend to prevent current.

Semiconductor material — A material with resistivity between that of metals and insulators. Pure semiconductor materials are usually doped with impurities to control the electrical properties.

Source — The point at which the charge carriers enter an FET. Corresponds to the cathode of a vacuum tube.

Toroid — A coil wound on a donut-shaped ferrite or powdered-iron form.

Transistor-transistor logic (TTL) — Digital integrated circuits composed of bipolar transistors, possibly as discrete components, but usually part of a single IC. Power supply voltage should be 5 V.

Tunnel diode — A diode with an especially thin depletion region, so that it exhibits a negative resistance characteristic.

Unijunction transistor (UJT) — A three-terminal, single-junction device that exhibits negative resistance and switching characteristics unlike bipolar transistors.

Varactor diode — A component whose capacitance varies as the reverse-bias voltage is changed. This diode has a voltage-variable capacitance.

Vidicon tube — A type of photosensitive vacuum tube widely used in TV cameras.

Zener diode — A diode that is designed to be operated in the reverse-breakdown region of its characteristic curve.

Zener voltage — A reverse-bias voltage that produces a sudden change in apparent resistance across the diode junction, from a large value to a small value.

CIRCUIT COMPONENTS

Before you can understand the operation of most complex electronic circuits, you must know some basic information about the parts that make up those circuits. This chapter presents the information about circuit components that you need to know in order to pass your Amateur Extra class license exam. You will find descriptions of several types of diodes; transistors; linear and digital integrated circuits (ICs) including phase-locked loops, operational amplifiers and basic logic gates; vidicon tubes and cathode-ray and liquid crystal displays. These components are combined with other devices to build practical electronic circuits, some of which are described in Chapter 7.

The Extra class exam contains five questions from the Circuit Components subelement. There are five groups of questions in this subelement of the Element 4 question pool, and the five questions on your exam will be taken from these five groups. The syllabus topics for this subelement are:

E6A   Semiconductor material: Germanium, Silicon, P-type, N-type; Transistor types: NPN, PNP, junction, unijunction, power; filed-effect transistors (FETs): enhancement mode; depletion mode; MOS; CMOS; N-channel; P-channel

E6B   Diodes: Zener, tunnel, varactor, hot-carrier, junction, point contact, PIN and light emitting; operational amplifiers (inverting amplifiers, noninverting amplifiers, voltage gain, frequency response, FET amplifier circuits, single-stage amplifier applications); phase-locked loops

E6C TTL digital integrated circuits; CMOS digital integrated circuits; gates

E6D Vidicon and cathode-ray tube devices; charge-coupled devices (CCDs); liquid crystal displays (LCDs); toroids: permeability, core material, selecting, winding

E6E Quartz crystal (frequency determining properties as used in oscillators and filters); monolithic amplifiers (MMICs)

As you study the characteristics of the components described in this chapter, be sure to turn to the questions in Chapter 10 when you are directed to do so. That will show you where you need to do some extra studying. If you thoroughly understand how these components work, you should have no problem learning how they can be connected to make a circuit perform a specific task.

SEMICONDUCTOR MATERIALS

Silicon and germanium are the materials normally used to make **semiconductor materials**. (The element silicon is not the same as the household lubricants and rubber-like sealers called silicone.) Silicon has 14 protons and 14 electrons, while germanium has 32 of each. Silicon and germanium each have an electron structure with four electrons in the outer energy layers. The silicon and germanium atoms will share these four electrons with other atoms around them.

Some atoms share their electrons so the atoms arrange themselves into a regular pattern. We say these atoms form crystals. **Figure 6-1** shows how silicon and germanium atoms produce crystals. (Different kinds of atoms might arrange themselves into other patterns.) The crystals made by silicon or germanium atoms do not make good electrical conductors. They aren't good insulators either, however. That's why we call them semiconductor materials. Sometimes they act like conductors and sometimes they act like insulators. Semiconductor material exhibits properties of both metallic and nonmetallic substances.

Manufacturers add other atoms to these crystals through a carefully controlled process, called **doping**. The atoms added in this way produce a material that is no longer pure silicon or pure germanium. We call the added atoms impurity atoms.

As an example, the manufacturer might add some atoms of arsenic or antimony to the silicon or germanium while making the crystals. Arsenic and antimony atoms each have five electrons to share.

Figure 6-2 shows how an atom with five electrons in its outer layer fits into the crystal structure. In such a case, there is an extra electron in the crystal. We refer to this as a *free electron*, and we call the semiconductor material made in this way **N-type material**. (This name comes from the extra negative charge in the crystal structure.)

The impurity atoms are electrically neutral, just as the silicon or germanium atoms are. The free electrons result from the crystal structure itself. These impurity atoms create (donate) free electrons in the crystal structure. That is why we call them *donor* impurity atoms.

Now let's suppose the manufacturer adds some gallium or indium atoms instead of arsenic or antimony. Gallium and indium atoms only have three electrons

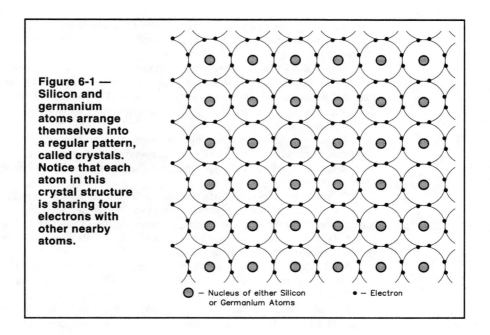

Figure 6-1 — Silicon and germanium atoms arrange themselves into a regular pattern, called crystals. Notice that each atom in this crystal structure is sharing four electrons with other nearby atoms.

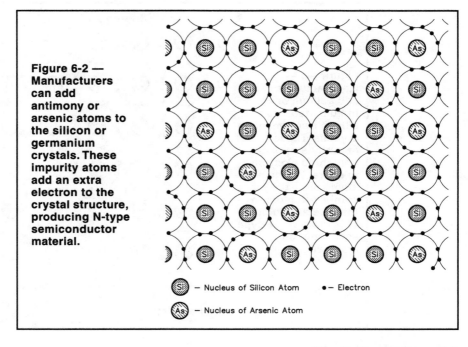

Figure 6-2 — Manufacturers can add antimony or arsenic atoms to the silicon or germanium crystals. These impurity atoms add an extra electron to the crystal structure, producing N-type semiconductor material.

that they can share with other nearby atoms. When there are gallium or indium atoms in the crystal there is an extra space where an electron could fit into the structure.

Figure 6-3 shows an example of a crystal structure with an extra space where an electron could fit. We call this space for an electron a *hole*. The semiconductor material produced in this way is **P-type material**. These impurity atoms produce holes that will accept extra electrons in the semiconductor material. That is why we call them *acceptor* impurity atoms.

Again, you should realize that the impurity atoms have the same number of electrons as protons. The material is still electrically neutral. The *crystal structure* is missing an electron in P-type material. Similarly, the *crystal structure* has an extra electron in N-type material.

Suppose we apply a voltage across a crystal of N-type semiconductor. The positive side of the voltage attracts electrons. The free electrons in the structure move through the crystal toward the positive side. Since most of the current through this N-type semiconductor material is produced by these free electrons, we call them the *majority charge carriers*.

Next suppose we apply a voltage to a crystal of P-type material. The negative voltage attracts the holes, and they move through the material toward the negative side. The *majority charge carriers* in P-type semiconductor material are holes.

Free electrons and holes move in opposite directions through a crystal. Manufacturers can control the electrical properties of semiconductor materials by

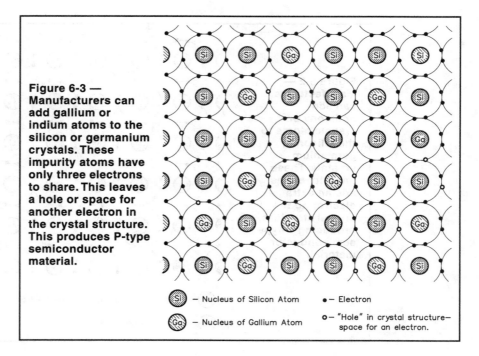

Figure 6-3 —
Manufacturers can add gallium or indium atoms to the silicon or germanium crystals. These impurity atoms have only three electrons to share. This leaves a hole or space for another electron in the crystal structure. This produces P-type semiconductor material.

⊘ – Nucleus of Silicon Atom   • – Electron

⊘ – Nucleus of Gallium Atom   ○ – "Hole" in crystal structure—
                                              space for an electron.

carefully controlling the amount and type of impurities that they add to the silicon or germanium crystals.

Semiconductors are solid crystals. They are strong and not easily damaged by vibration or rough handling. We refer to electronic parts made with semiconductor materials as solid-state devices.

Other materials are used to make semiconductor materials for special-purpose applications. For example, gallium arsenide semiconductor material has performance advantages for use at microwave frequencies, and is often used to make solid-state devices for operation on those frequencies.

[Turn to Chapter 10 and study questions E6A01 through E6A05 before you go on to the next section. Review this material if you have any difficulty with those questions.]

DIODES

To qualify for the Extra class license, you must be familiar with some specialized types of diodes. Rectifier circuits were covered on the General class examination; for the Extra class test, you must be familiar with the two main structural categories of diodes: junction and point-contact types. PIN, Zener, light-emitting, tunnel, varactor and hot-carrier diodes are special-purpose devices that come under those main headings.

Junction Diodes

The junction diode, also called the **PN-junction** diode, is made from two layers of semiconductor material joined together. One layer is made from P-type (positive) material. The other layer is made from N-type (negative) material. The name PN junction comes from the way the P and N layers are joined to form a semiconductor diode. **Figure 6-4** illustrates the basic concept of a junction diode.

The P-type side of the diode is called the **anode**, which is the lead that normally connects to the positive supply lead. The N-type side is called the **cathode**, and that lead normally connects to the negative supply lead. When voltage is applied to a junction diode as shown at A in **Figure 6-5**, carriers flow across the barrier and the diode conducts (that is, electrons will flow through it). When the diode anode is positive with respect to the cathode, electrons are attracted across the junction from the N-type material, through the P-type material and on through the circuit to the positive battery terminal. Holes are attracted in the opposite direction by the

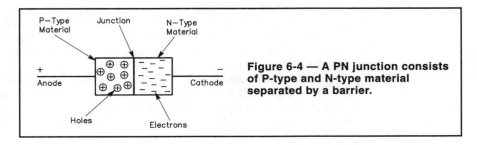

Figure 6-4 — A PN junction consists of P-type and N-type material separated by a barrier.

negative potential of the battery. When the diode anode is connected in this manner it is said to be **forward biased**.

Conventional current (which flows from positive to negative) in a diode flows from the anode to the cathode. The electrons flow in the opposite direction. Electrons and holes are also called carriers because they are the means by which current is carried from one side of the junction to the other. When no voltage is applied to a diode, the junction between the P-type and N-type material acts as a barrier that prevents carriers from flowing between the layers.

Figure 6-5B shows the schematic symbol for a diode, drawn as it would be used instead of the semiconductor blocks used in part A. Here we can see one of the advantages of considering conventional current instead of electron flow through the diode. The arrow on the schematic symbol points in the direction of conventional current.

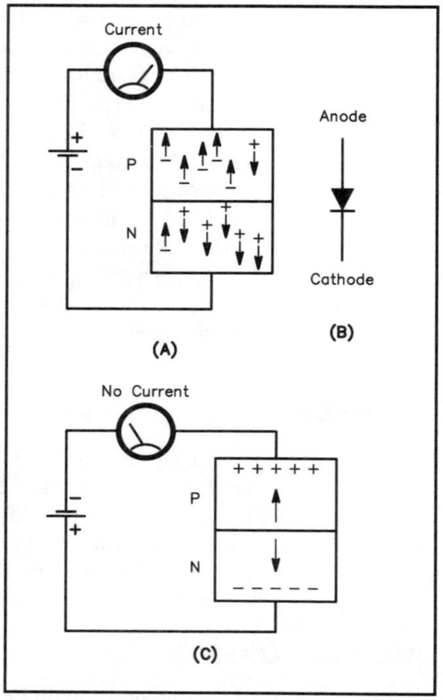

If the battery polarity is reversed, as shown at C in Figure 6-5, the excess electrons in the N-type material are attracted away from the junction by the positive battery terminal. Similarly, the holes in the P-type material are attracted away from the junction by the negative battery side. When this happens, the area around the junction has no current carriers; electrons do not flow across the junction to the P-type material, and the diode does not conduct. When the anode is connected to a negative voltage source and the cathode is connected to a positive voltage source, the diode does not conduct, and the device is said to be **reverse biased**.

Figure 6-5 — At A, the PN junction is forward biased and conducting. B shows the schematic symbol used to represent a diode, drawn as it would be used instead of the semiconductor blocks of part A. *Conventional current* flows in the direction indicated by the arrow in this symbol. At C, the PN junction is reverse biased, so it does not conduct.

Junction diodes are used as rectifiers because they allow current in one direction only. See **Figure 6-6**. When an ac signal is applied to a diode, it will be forward biased during one half of the cycle, so it will conduct, and there will be current to the load. During the other half of the cycle, the diode is reverse biased, and there will be no current. The diode output is pulsed dc, and current always flows in the same direction.

In a junction diode, the P and N layers are sandwiched together, separated by the junction. Although the spacing between the layers is extremely small, there is some capacitance at the junction. The structure can be thought of in much the same way as a simple capacitor: two charged plates separated by a thin dielectric.

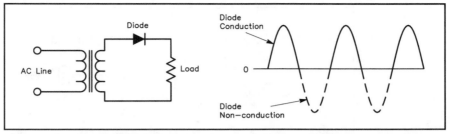

Figure 6-6 — PN-junction diodes are used as rectifiers because they allow current in one direction only.

Although the internal capacitance of a PN junction diode may be only a few picofarads, this capacitance can cause problems in RF circuits, especially at VHF and above. Junction diodes may be used from dc to the microwave region, but there is a special type of diode with low internal capacitance that is specially designed for RF applications. This device, called the **point-contact diode**, is discussed later in this chapter.

Diode Ratings

Junction diodes have maximum voltage and current ratings that must be observed, or damage to the diode could result. The voltage rating is called **peak inverse voltage (PIV)**, and the rectified current rating is called **maximum average forward current**. With present technology, diodes are commonly available with ratings up to 1000 PIV and 100 A.

Peak inverse voltage is the voltage that a diode must withstand when it isn't conducting. Although a diode is normally used in the forward direction, it will conduct in the reverse direction if enough voltage is applied. A few hole/electron pairs are thermally generated at the junction when a diode is reverse biased. These pairs cause a very small reverse current, called leakage current. Semiconductor diodes can withstand some leakage current. If the inverse voltage reaches a high enough value, however, the leakage current rises abruptly, resulting in a large reverse current. The point where the leakage current rises abruptly is called the **avalanche point**. A large reverse current usually damages or destroys the diode.

The maximum average forward current is the highest average current that can flow through the diode in the forward direction for a specified junction temperature. This specification varies from device to device, and it depends on the maximum allowable junction temperature and on the amount of heat the device can dissipate. As the forward current increases, the junction temperature will increase. If allowed to get too hot, the diode will be destroyed.

Impurities at the PN junction cause some resistance in the diode. This resistance results in a voltage drop across the junction. For silicon diodes, this drop is approximately 0.6 to 0.7 V; it is 0.2 to 0.3 V for germanium diodes. When current flows through the junction, some power is dissipated in the form of heat. The amount of power depends on the current through the diode. For example, it would be approximately 6 W for a silicon rectifier with 10 A flowing through it

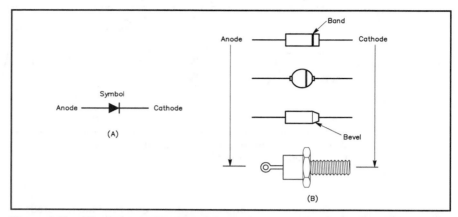

Figure 6-7 — The schematic symbol for a diode is shown at A. Diodes typically are packaged in one of the case styles shown at B.

$(P = I \times E; P = 10 \text{ A} \times 0.6 \text{ V})$. If the junction temperature exceeds the safe level specified by the manufacturer the diode is likely to be damaged or destroyed.

Diodes designed to safely handle forward currents in excess of 6 A generally are packaged so they may be mounted on a heat sink. These diodes are often referred to as stud-mount devices. The heat sink helps the diode package dissipate heat more rapidly, thereby keeping the diode junction temperature at a safe level. The metal case of a stud-mount diode is usually one of the contact points, so it must be insulated from ground.

Figure 6-7 shows some of the more common diode-case styles, as well as the general schematic symbol for a diode. The line, or spot, on a diode case indicates the cathode lead. On a high-power, stud-mount diode, the stud may be either the anode or cathode. Check the case or the manufacturer's data sheet for the correct polarity.

Varactor and Varicap Diodes

As mentioned before, junction diodes exhibit an appreciable internal capacitance. It is possible to change the internal capacitance of a diode by varying the amount of reverse bias applied to it. Manufacturers have designed certain kinds of diodes, called voltage-variable capacitors or variable-capacitance diodes (Varicaps) and **varactor diodes** (variable reactance diodes) to take advantage of this property.

Varactors are designed to provide various capacitance ranges from a few picofarads to more than 100 pF. Each style has a specific minimum and maximum capacitance, and the higher the maximum amount, the greater the minimum amount. A typical varactor can provide capacitance changes over a 10:1 range

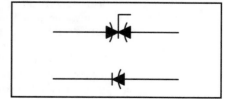

Figure 6-8 — These schematic symbols are commonly used to represent varactor diodes.

with bias voltages in the 0- to 100-V range.

Varactors are similar in appearance to junction diodes. Common schematic symbols for a varactor diode are given in **Figure 6-8**. These devices are used in frequency multipliers at power levels as great as 25 W, remotely tuned circuits and simple frequency modulators.

PIN Diodes

A **PIN** (positive/intrinsic/negative) **diode** is formed by diffusing P-type and N-type layers onto opposite sides of an almost pure silicon layer, called the I re-

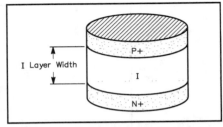

Figure 6-9 — This diagram illustrates the inner structure of a PIN diode. The top and bottom layers are designated P+ and N+ to indicate heavy doping in those layers.

gion. See **Figure 6-9**. This layer is not "doped" with P-type or N-type charge carriers, as are the other layers. Any charge carriers found in this layer are a result of the natural properties of the pure semiconductor material. In the case of silicon, there are relatively few free charge carriers. PIN-diode characteristics are determined primarily by the thickness and area of the I region. The outside layers are designated P+ and N+ to indicate heavier than normal doping of these layers.

PIN diodes have a forward resistance that varies inversely with the amount of forward bias applied. When a PIN diode is at zero or reverse bias, there is essentially no charge, and the intrinsic region can be considered as a low-loss dielectric. Under reverse-bias conditions, the charge carriers move very slowly. This slow response time causes the PIN diode to look like a resistor, blocking RF currents.

When forward bias is applied, holes and electrons are injected into the I region from the P and N regions. These charges do not recombine immediately. Rather, a finite quantity of charge always remains stored, resulting in a lower I-region resistivity. The amount of resistivity that a PIN diode exhibits to RF can be controlled by changing the amount of forward bias applied.

PIN diodes are commonly used as RF switches, variable attenuators and phase shifters. PIN diodes are faster, smaller, more rugged and more reliable than relays or other electromechanical switching devices.

Figure 6-10 shows how PIN diodes can be used to build an RF switch. This diagram shows a transmit/receive switch for use between a 2-meter transceiver and a UHF or microwave transverter. With no bias, or with reverse bias applied to the diode, the PIN diode exhibits a high resistance to RF, so no signal will flow from the generator to the load. When forward bias is applied, the diode resistance will decrease, allowing the RF signal to pass. The amount of insertion loss (resistance to RF current) is determined primarily by the amount of forward bias applied; the greater the forward bias current, the lower the RF resistance.

PIN diodes are packaged in case styles similar to conventional diodes. The package size depends on the intended application. PIN diodes intended for low-power UHF and microwave work are packaged in small epoxy or glass cases to minimize internal capacitance. Others, intended for high-power switching, are of the stud-mount variety so they can be attached to heat sinks. PIN diodes are shown

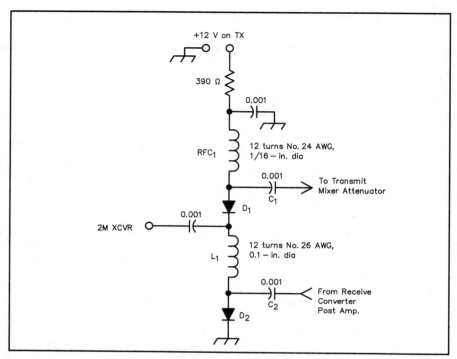

Figure 6-10 — PIN diodes may be used as RF switches. This diagram shows a PIN-diode T/R switch for use between a 2-meter transceiver and a UHF or microwave transverter.

D1, D2 — Phillips BA182, Motorola MPN3401 or equivalent.

by the schematic symbol shown in Figure 6-7.

Zener Diodes

Zener diodes are a special class of junction diode used as voltage references and as voltage regulators. When they are used as voltage regulators, Zener diodes provide a nearly constant dc output voltage, even though there may be large changes in load resistance or input voltage. As voltage references, they provide an extremely stable voltage that remains constant over a wide temperature range.

As discussed earlier, leakage current rises as reverse (inverse) voltage is applied to a diode. At first, this leakage current is very small and changes very little with increasing reverse voltage. There is a point, however, where the leakage current rises suddenly. Beyond this point, the current increases very rapidly for a small increase in voltage; this is called the **avalanche point**. **Zener voltage** is the voltage necessary to cause avalanche. Normal junction diodes would be destroyed immediately if they were operated in this region, but Zener diodes are specially manufactured to safely withstand the avalanche current.

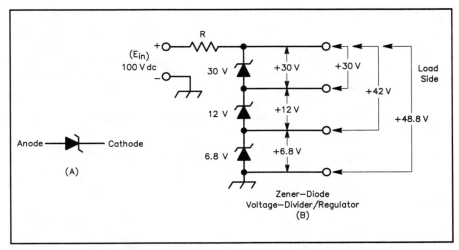

Figure 6-11 — The schematic symbol for a Zener diode is shown at A. B is an example of how Zener diodes are used as voltage regulators.

Since the current in the avalanche region can change over a wide range while the voltage stays practically constant, this kind of diode can be used as a voltage regulator. The voltage at which avalanche occurs can be controlled precisely in the manufacturing process. Zener diodes are calibrated in terms of avalanche voltage.

Zener diodes are currently available with ratings between 1.8 and 200 V. The power ratings range from 250 mW to 50 W. They are packaged in the same case styles as junction diodes. Usually, Zener diodes rated for 10-W dissipation or more are made in stud-mount cases. The schematic symbol for a Zener diode is shown in **Figure 6-11**, along with an example of how such a device is used as a voltage regulator.

Tunnel Diodes

The **tunnel diode** is a special type of device that has no rectifying properties. When properly biased, it possesses an unusual characteristic: negative resistance. Negative resistance means that when the voltage across the diode increases, the current decreases. This property makes the tunnel diode capable of *amplification* and *oscillation*.

At one time, tunnel diodes were expected to dominate in microwave applications, but other devices with better performance soon replaced them. The tunnel diode is seldom used today. The schematic symbol for a tunnel diode is shown in **Figure 6-12**.

Figure 6-12 — This is the schematic symbol for a tunnel diode.

Light-Emitting Diodes

Light-emitting diodes (LEDs) are designed to emit light when they are forward biased, and current passes through their PN junctions. The junction of an LED is made from gallium arsenide, gallium phosphide or a combination of these two materials. The color and intensity of the LED depends on the material, or combination of materials, used for the junction. LEDs are available in many colors.

LEDs are packaged in plastic cases, or in metal cases with a transparent end. LEDs are useful as replacements for incandescent panel and indicator lamps. In this application they offer long life, low current drain and small size. One of their most important applications is in numeric displays, in which arrays of tiny LEDs are arranged to provide illuminated segments that form the numbers. Schematic symbols and typical case styles for the LED are shown in **Figure 6-13**.

Figure 6-13 — The schematic symbol for an LED is shown at A. At B is a drawing of a typical LED case style.

A typical red LED has a voltage drop of 1.6 V. Yellow and green LEDs have higher voltage drops (2 V for yellow and 4 V for green). The forward-bias current for a typical LED ranges between 10 and 20 mA for maximum brilliance. Bias currents of about 10 mA are recommended for longest device life. As with other diodes, the current through an LED can be varied with series resistors. Varying the current through an LED will affect its intensity; the voltage drop, however, will remain fairly constant.

Point-Contact Diodes

Figure 6-14 illustrates the internal structural differences between a junction diode and a **point-contact diode**. As you can see by this diagram, the point-contact

Figure 6-14 — The internal structure of a point-contact diode is shown at A. B shows the internal structure of a PN-junction diode.

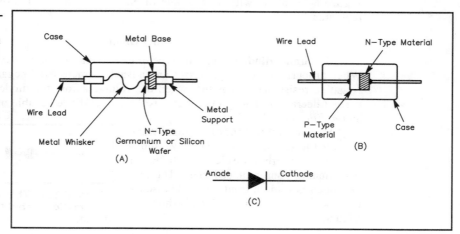

diode has a much smaller surface area at the junction than does a PN-junction diode. When a point-contact diode is manufactured, the main portion of the device is made from N-type material, and a thin aluminum wire, often called a *whisker*, is placed in contact with the semiconductor surface. This forms a metal-semiconductor junction. The result is a diode that exhibits much less internal capacitance than PN-junction diodes, typically 1 pF or less. This means point-contact diodes are better suited for VHF and UHF work than are PN-junction diodes.

Point-contact diodes are packaged in a variety of cases, as are junction diodes. The schematic symbol is the same as is used for junction diodes, and is included in Figure 6-14. Point-contact diodes are generally used as UHF mixers and as RF detectors at VHF and below.

Hot-Carrier Diodes

Another type of diode with low internal capacitance and good high-frequency characteristics is the **hot-carrier diode**. This device is very similar in construction to the point-contact diode, but with an important difference. Compare the inner structure of the hot-carrier diode depicted in **Figure 6-15** to the point-contact diode shown in Figure 6-14.

The point-contact device relies on the touch of a metal whisker to make contact with the active element. In contrast, the whisker in a hot-carrier diode is physically attached to a metal dot that is deposited on the element. The hot-carrier diode is mechanically and electrically superior to the point-contact diode. Some of the advantages of the hot-carrier type are improved power-handling characteristics, lower contact resistance and improved immunity to burnout caused by transient noise pulses.

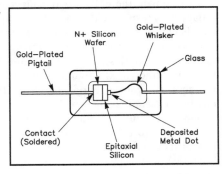

Figure 6-15 — This drawing represents the internal structure of a hot-carrier diode.

Hot-carrier diodes are similar in appearance to point-contact and junction diodes and share the same schematic symbol. They are often used in mixers and detectors at VHF and UHF. In this application, hot-carrier diodes are superior to point-contact diodes because they exhibit greater conversion efficiency and lower noise figure.

[This completes your study of diode types and characteristics for your Extra class license exam. You should turn to Chapter 10 now and study examination questions E6B01 through E6B10. Review this section as needed.]

TRANSISTORS

The **bipolar junction transistor** is a type of three-terminal, PN-junction device that is able to amplify signal energy (current). It is made up of two layers of like semiconductor material with a layer of the opposite-type material sandwiched in between. See **Figure 6-16**. If the outer layers are P-type material, and the middle layer is N-type material, the device is called a PNP transistor because of the layer

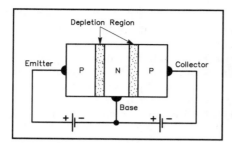

Figure 6-16 — A bipolar junction transistor consists of two layers of like semiconductor material separated by a layer of the opposite material. This drawing represents the internal structure of a PNP transistor.

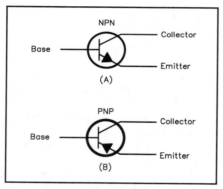

Figure 6-17 — Part A shows the schematic symbol for an NPN bipolar transistor. B is the schematic symbol for a PNP bipolar transistor.

arrangement. If the outer layers are N-type material, the device is called an NPN transistor. A transistor is, in effect, two PN-junction diodes back-to-back. **Figure 6-17** shows the schematic symbols for PNP and NPN bipolar transistors.

The three layers of the transistor sandwich are called the emitter, base and collector. These are functionally analogous to the cathode, grid and plate of a vacuum tube. A diagram of the construction of a typical PNP transistor is given in Figure 6-16. In an actual bipolar transistor, the center layer (in this case, N-type material) is much thinner than the outer layers. As shown in the diagram, forward-bias voltage across the emitter-base section of the sandwich causes electrons to flow through it from the base to the emitter. As the free electrons from the N-type material flow into the holes of the P-type material, the holes in effect travel into the base. Some of the holes will be neutralized in the base by free electrons, but because the base layer is so thin, some will move right on through into the P-type material in the collector.

As shown, the collector is connected to a negative voltage with respect to the base. Normally, no current will flow because the base-collector junction is reverse biased. The collector, however, now contains an excess of holes because of those from the emitter that overshot the base. Since the voltage source connected to the collector produces a negative charge, the holes from the emitter will be attracted to the power-supply connection. The amount of emitter-to-collector current is approximately proportional to the base-to-emitter current. Because of the transistor construction, however, the current through the collector will be considerably larger than that flowing through the base.

When a transistor is forward biased, collector current increases in proportion to the amount of bias applied. The transistor is saturated when the collector current reaches its maximum value, and the transistor is said to be fully on. Further increases in bias voltage do not increase the collector current when the transistor is saturated. At the other end of the curve, when the transistor is reverse-biased, the transistor is turned off. There is no current from the emitter to the collector, and the transistor is at cutoff.

A load line is a graphical representation of the range of the transistor resistance for collector-current points between cutoff and saturation. At one end of the load line, the transistor has an infinite resistance to current (cutoff); at the other, it has zero resistance (saturation). Normally, a transistor is operated at some point on the load line between these two extremes. For best efficiency and stability, the transistor in a solid-state power amplifier is operated at a point on the load line that is just below saturation.

Figure 6-16 shows an area around each junction that is called the depletion region. The depletion region, sometimes called the transition region, is an area near a PN junction that is devoid of holes and excess electrons. This region is caused by the repelling forces of the ions on opposite sides of the junction. When the PN junction is reverse biased, the depletion region becomes larger because the electrons and holes are attracted away from the junction. When the PN junction is forward biased, the depletion region becomes smaller because the electrons and holes move toward each other.

Transistor Characteristics

As mentioned before, the current through the collector of a bipolar transistor is approximately proportional to the current through the base. The ratio of collector current to base current is called the current gain, or beta. Beta is expressed by the Greek symbol β. It can be calculated from the equation:

$$\beta = \frac{I_c}{I_b} \qquad \text{(Equation 6-1)}$$

where:
I_c = collector current
I_b = base current

For example, if a 1-mA base current results in a collector current of 100 mA, the beta is 100. Typical betas for junction transistors range from as low as 10 to as high as several hundred. Manufacturers' data sheets specify a range of values for β: Individual transistors of a given type will have widely varying betas.

Another important transistor characteristic is **alpha**, expressed by the Greek letter α. Alpha is the ratio of collector current to emitter current, given by:

$$\alpha = \frac{I_c}{I_e} \qquad \text{(Equation 6-2)}$$

where:
I_c = collector current
I_e = emitter current

The smaller the base current, the closer the collector current comes to being equal to that of the emitter, and the closer alpha comes to being 1. For a junction transistor, alpha is usually between 0.92 and 0.98.

Transistors have important frequency characteristics. The **alpha cutoff frequency** is the frequency at which the current gain of a transistor in the common-base configuration decreases to 0.707 times its gain at 1 kHz. Alpha cutoff frequency is considered to be the practical upper frequency limit of a transistor configured as a common-base amplifier.

Figure 6-18 — Transistors are packaged in a wide variety of case styles, depending on their intended application.

Beta cutoff frequency is similar to alpha cutoff frequency, but it applies to transistors connected as common-emitter amplifiers. Beta cutoff frequency is the frequency at which the current gain of a transistor in the common-emitter configuration decreases to 0.707 times its gain at 1 kHz. (These amplifier configurations are explained in Chapter 7.)

Bipolar junction transistors are used in a wide variety of applications, including amplifiers (from very low level to very high power), oscillators and power supplies. They are used at all frequency ranges from dc through the UHF and microwave range.

Transistors are packaged in a wide variety of case styles. Some of the more common case styles are depicted in **Figure 6-18**.

Unijunction Transistors

Another three-terminal semiconductor device is the **unijunction transistor (UJT)**, sometimes called a double-base diode. The internal structure of a UJT is shown in **Figure 6-19**. The elements of a UJT are base 1, base 2 and emitter. There is only one PN junction, and this is between the emitter and the silicon substrate. The base terminals are ohmic contacts; that is, the current is a linear function of the applied voltage. Current flowing between the bases sets up a voltage gradient along the substrate. In operation, the direction of current flow causes the emitter junction to be reverse biased.

The most common application for the UJT is in relaxation oscillator circuits. UJTs are packaged in cases similar to small-signal bipolar transistors. Two schematic symbols for UJTs are given in **Figure 6-20**. The substrate is made from one type of semiconductor material, and the emitter is made from the other type of material. The large block of material that the two bases connect to in Figure 6-20 is called the substrate. When the substrate is N-type material, we call the UJT an N-channel device. Similarly, the substrate of a P-channel UJT is formed from P-type semiconductor material.

[Proceed to Chapter 10 and study questions E6A06 through E6A10. Review this section as needed.]

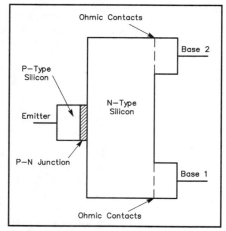

Figure 6-19 — This drawing represents the internal structure of a unijunction transistor.

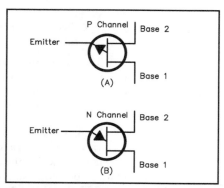

Figure 6-20 — These schematic symbols are used to represent P-channel and N-channel unijunction transistors.

FIELD-EFFECT TRANSISTORS (FETS)

Field-effect transistors (FETs) are given that name because the current through them is controlled by a varying electric field created by the *voltage* applied to the **gate** lead. By contrast, in a bipolar transistor, output current is controlled by the *current* applied to the *base*.

There are two types of field-effect transistors in use today: the **junction FET (JFET)** and the **metal-oxide semiconductor FET (MOSFET)**. Like bipolar transistors, the JFET has no insulation between its elements. The MOSFET has a thin layer of oxide between the gate or gates and the drain/source junction.

The basic characteristic of both FET types is a very high input impedance — typically 1 megohm or greater. This is considerably higher than the input impedance of a bipolar transistor. Although some FETs have only one gate, others have two.

JFETs

The basic **junction field-effect transistor (JFET)** construction is shown in **Figure 6-21**. The JFET can be thought of simply as a bar of silicon semiconductor material that acts like a resistor. The terminal into which the charge carriers enter is called the **source**. The opposite terminal is called the **drain**. There are two types of JFET (N-channel and P-channel), so named for the type of material used to form the drain/source channel. The schematic symbols for the two JFET types are illustrated in **Figure 6-22**.

Two gate regions, made of the semiconductor material opposite to that used for the channel, are diffused into the JFET channel. When a reverse voltage is applied to the gates, an electric field is set up perpendicular to the channel. The electric field interferes with the normal electron flow through the channel. As the gate voltage changes, this electric field varies, and that causes a variation in source-to-drain current. Thus, the FET is an example of a voltage-controlled current source, similar to a triode vacuum tube. The gate terminal is always reverse biased, so the JFET has a very high input impedance, much like a vacuum tube with its high grid input impedance.

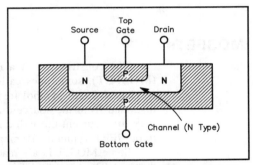

Figure 6-21 — This drawing illustrates the construction of a junction field-effect transistor (JFET).

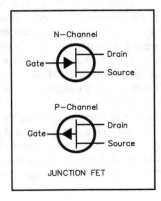

Figure 6-22 — The schematic symbol for an N-channel JFET has an arrow pointing toward the center line. The symbol for a P-channel JFET has an arrow pointing away from the center line.

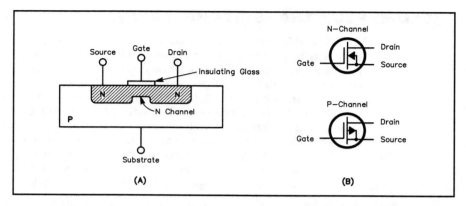

Figure 6-23 — Part A shows the construction of a MOSFET. Part B shows the schematic symbols of N-channel and P-channel MOSFETs. Notice that the arrow on the N-channel device points in to the center line while the arrow on the P-channel device points away from the line.

MOSFETs

The construction of a **metal-oxide semiconductor field-effect transistor (MOSFET)**, sometimes called an insulated gate field-effect transistor (IGFET), and its schematic symbol are illustrated in **Figure 6-23**. In the MOSFET, the gate is insulated from the source/drain channel by a thin dielectric layer. Since there is very little current through this dielectric, the input impedance is even higher than in the JFET, typically 10 megohms or greater.

Some MOSFETs are made with two gates rather than one. This type of FET is widely used as an RF amplifier or mixer in receivers and converters. The schematic symbols for N-channel and P-channel dual-gate MOSFETs are shown in **Figure 6-24**.

Sometimes both P- and N-channel MOSFETs are placed on the same wafer. The resulting transistor arrays can be interconnected on the wafer and are designed to perform a variety of special functions. This construction is called **complementary metal-oxide semiconductor (CMOS)** because the P- and N-channel transistors complement each other.

Nearly all the MOSFETs manufactured today have built-in gate-protective Zener diodes. Without this provision the gate insulation can be perforated easily by small static charges on the user's hand or by the application of excessive voltages to the device. The protective diodes are connected between the gate (or gates) and the source lead of the FET.

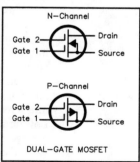

Figure 6-24 — This drawing shows the schematic symbols for N-channel, and P-channel dual-gate MOSFETs. Notice the directions of the arrows in relation to the center line.

Enhancement- and Depletion-Mode FETs

Field-effect transistors are classified into two main groupings for applications in circuits — **enhancement mode** and **depletion mode**. A depletion-mode device corresponds to Figure 6-21, where a channel exists with no bias applied. The gate of a depletion-mode device is reverse biased in operation. When the reverse bias is applied to the gate, the channel is *depleted* of charge carriers, and current decreases.

Enhancement-mode devices are those specifically constructed so they have no channel when there is no voltage on the gate. They become useful only when a gate voltage is applied, which causes a channel to be formed. When the gate of an enhancement-mode device is forward biased, current begins to flow through the source/drain channel. The more forward bias on the gate, the more current through the channel. JFETs cannot be used as enhancement-mode devices, because if the gate is forward biased it will conduct, like a forward-biased diode. MOSFETs have their gates insulated from the channel region, so they may be used as enhancement-mode devices, since both polarities may be applied to the gate without the gate becoming forward biased and conducting. Some MOSFETs are designed to be used with no bias on the gate. In this type of operation, the control signal applied to the gate swings the bias forward part of the time and reverse part of the time. The MOSFET operates in the enhancement mode when the gate is forward biased, and in the depletion mode when the gate is reverse biased.

To sum up, a depletion-mode FET is one that has a channel constructed; thus, a current will flow through the channel with zero gate voltage applied. Enhancement-mode FETs are those that have no channel, so there is no current when there is zero gate voltage applied.

Designing with FETs is much like designing with vacuum tubes. They are used in RF amplifiers and oscillators, and they are ideal for use in voltmeters, where their high input impedance will not load down the circuit being tested.

FETs are made in the same types of packages as bipolar transistors. Some different case styles are shown in **Figure 6-25**.

Figure 6-25 — FETs are packaged in cases much like those used for bipolar transistors.

FET Amplifier Characteristics

The circuit of **Figure 6-26** is a simple common-source, two-stage FET amplifier. In this circuit, R_1 and R_4 are gate bias resistors. The input impedance of a common-source FET amplifier is approximately equal to the value of the gate bias resistor. R_2 and R_5 are source resistors that are used to bias the individual FETs at a specific point. R_3 is the load resistor for the first stage and R_6 is the load resistor for the second stage. The output impedance of a common-source FET amplifier is approximately equal to the value of the drain load resistor. C_1 and C_3 are source

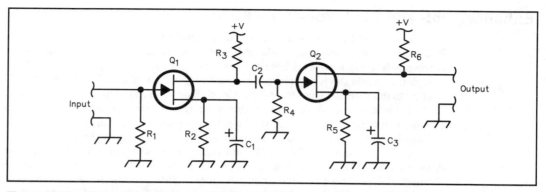

Figure 6-26 — This simple two-stage FET amplifier circuit illustrates the use of coupling and bypass capacitors. C_2 is a coupling capacitor; C_1 and C_3 are bypass capacitors. The gate bias resistors, R_1 and R_4 determine the input impedance of each stage and the drain load resistors, R_3 and R_6 determine the output impedance of each stage.

bypass capacitors and C_2 is a coupling capacitor. The overall frequency response of this circuit is affected strongly by the choice of capacitance values for C_1, C_2 and C_3.

[Now turn to Chapter 10 and study exam questions E6A11 through E6A17. Also study question E6B14. Review this section if you have any difficulty with these questions.]

LINEAR INTEGRATED CIRCUITS

Integrated circuits (ICs) comprise many transistors on a single wafer, or chip, of silicon. Integrated circuits may be made up of bipolar or field-effect transistors, and may be linear (smoothly varying output) or digital (on/off output) in operation. Most ICs today are packaged in a plastic dual-in-line package, or DIP. Some are packaged in a metal can similar to a transistor, with many leads. The DIP and metal-can packages are illustrated in **Figure 6-27**.

Linear ICs have this name because in their usual operating mode, the output voltage is a linear function of the input voltage. This does not mean they cannot be operated in a nonlinear mode, such as for a class-C amplifier. The bias will determine the operating mode, class A through class C. You will need to un-

Figure 6-27 — Integrated circuits may be packaged in 8-pin and 14-pin (or more) dual-in-line packages (DIP) or TO-204 and TO-205 case styles similar to a transistor, with many leads.

derstand two types of linear ICs for the Amateur Extra exam: operational amplifiers (op amps) and phase-locked loops (PLL).

Operational Amplifiers

The **operational amplifier (op amp)** is a high-gain, direct-coupled, differential amplifier that will amplify dc signals as well as ac signals. An op amp is characterized by its high input impedance and low output impedance. To amplify dc and ac signals without frequency limitations, the amplifier input and output signals must be coupled directly rather than having dc blocking capacitors or ac bypass capacitors connected. With capacitors connected to the input and output leads, the amplifier frequency response can be tailored to meet specific needs. This means the characteristics of the amplifier circuit in which the op amp is used will be determined by external components.

The first op amps were designed for use in analog computers, where they performed such mathematical operations as multiplying numbers and extracting square roots; hence the name *operational* amplifier. Op amps are some of the most versatile ICs available.

The most obvious use for operational amplifiers is as a low-distortion audio amplifier, but it has many other uses as well. Op amps can be made into oscillators to generate sine, square and even sawtooth waves. Used with negative feedback, their high input impedance and linear characteristics make them ideal for use as instrumentation amplifiers. It would take an entire book to illustrate the many uses to which op amps can be put.

A theoretically perfect (ideal) op amp would have the following characteristics: infinite input impedance, zero output impedance, infinite voltage gain, flat frequency response within its frequency range and zero output when the input is zero. These criteria can be approached in a practical situation, but not realized entirely.

Operational amplifiers have two inputs, an inverting input and a noninverting input. The circuit will amplify the difference between the input signals, so an op amp is basically a differential amplifier. By connecting the inputs in a variety of configurations, op amps can be used to obtain output signals other than just the difference between two input signals. Three of the more common op amp configurations are the inverting amplifier, noninverting amplifier and difference amplifier. These modes are compared in **Figure 6-28**, which also shows the phase relationship of the input and output for each configuration.

Op amps can be assembled from discrete transistors, but better thermal stability results from fabricating the circuit on a single silicon chip. IC op amps are manufactured with bipolar, JFET and MOSFET devices, either exclusively or in combination.

The gain of a practical op amp without feedback (open-loop gain) is often as high as 500,000. Op amps are rarely used as amplifiers in the open-loop configuration, however. Usually, some of the output is fed back to the inverting input, where it acts to reduce the stage gain. The more negative feedback, the more stable the amplifier circuit will be. **Figure 6-29** illustrates a basic op-amp circuit using a feedback resistor to reduce the circuit gain.

The gain of the circuit with negative feedback is called the closed-loop gain. The higher the open-loop gain, the more negative feedback that can be used and still provide enough closed-loop gain. If the open-loop gain is many times greater than the closed-

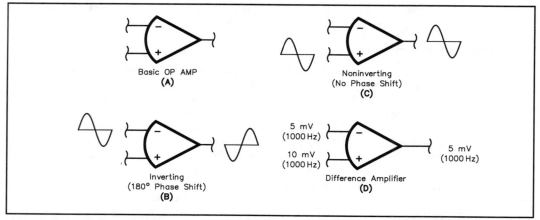

Figure 6-28 — Part A shows the basic schematic symbol for an operational amplifier (op amp). Parts B through D show various ways that op amps can be used.

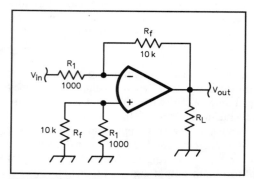

Figure 6-29 — This drawing shows the configuration of an inverting op-amp circuit. Resistance values determine the gain of the circuit.

loop gain, the stage gain is determined by the external feedback network components rather than by the gain of the op amp itself.

The most common application for op amps is in negative-feedback circuits operating from dc to perhaps a few hundred kilohertz. Provided the op amp has sufficient open-loop gain, the amplifier transfer function is determined almost entirely by the external feedback network. (The terms open loop and closed loop refer to whether or not there is a feedback path.) The differential inputs on an op amp allow for both inverting and noninverting circuits.

Voltage gain and voltage output for the inverting op-amp circuit can be determined easily. Figure 6-29 illustrates the basic circuit arrangement. R_1 and R_f determine the circuit gain.

$$V_{out} = -\frac{R_f}{R_1} V_{in}$$

(Equation 6-3)

$$V_{gain} = \frac{V_{out}}{V_{in}}$$

(Equation 6-4)

Then substituting Equation 6-3 into Equation 6-4, we get:

$$V_{gain} = \frac{\dfrac{R_f}{R_1} V_{in}}{V_{in}} = \frac{R_f}{R_1}$$

(Equation 6-5)

We can also express this gain in decibels if we remember that V_{gain} is a voltage ratio, and that we have to multiply the log of a voltage ratio by 20 when converting to decibels.

$$\text{Gain (dB)} = 20 \log (V_{gain}) \qquad \text{(Equation 6-6)}$$

These equations illustrate that the circuit gain is determined by the resistors rather than by the op-amp characteristics and the power-supply voltage. In the inverting configuration, the op amp can act as a summing amplifier. For example, two or more separate audio signals may be brought to the input; the circuit then acts as an active audio mixer.

Op Amp Specifications

If the input terminals of an op amp are shorted together, the output voltage should be zero. With most inexpensive op amps there will be a small output voltage, however. The **offset voltage** specification indicates the potential between the amplifier input terminals that will produce a zero output voltage in the closed-loop condition. Offset results from imbalance between the differential input transistors in the IC. Offset-voltages range from millivolts in ordinary consumer-grade devices to only nanovolts or microvolts in premium units designed to meet military specifications.

The temperature coefficient of offset voltage is called *drift*. Drift is usually considered in relation to time. Heat generated by the op amp itself or by associated circuitry will cause the offset voltage to change over time. A few microvolts per degree Celsius (at the input) is a typical drift specification.

All op amps will generate some noise. If your application requires the lowest possible amount of noise in the circuit, you can select one of the more expensive low-noise op amps. For some analytical purposes, drift is considered as a very low frequency noise component. Op amps that have been optimized for offset, drift and noise are called instrumentation amplifiers.

The small-signal bandwidth of an op amp is the frequency range over which the open-loop voltage gain is at least unity (0 dB). This specification depends mostly on the frequency compensation scheme used. **Figure 6-30** shows how the maximum

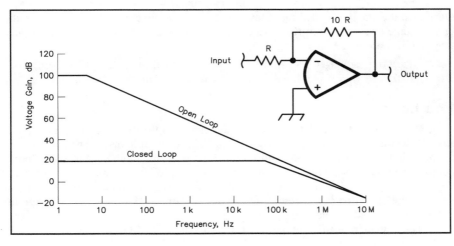

Figure 6-30 — The open-loop gain and closed-loop gain are shown as a function of frequency. The vertical distance between the curves is the feedback or gain margin.

closed-loop gain varies with frequency. This figure also shows how the external resistors reduce gain and stabilize the amplifier in a closed-loop configuration.

The gain of any op-amp circuit is set by the value of a few resistors. The output, or load, resistance does not appear in the equations, implying that the output impedance is zero. This condition results from the application of heavy negative feedback. Most IC op amps have built-in current limiting. This feature not only protects the IC from damage caused by short circuits, but also limits the values of load resistance for which the output impedance is zero. Most op amps work best with load resistances of at least 2 kΩ.

Since the op amp magnifies the difference between the voltages applied to its inputs, applying negative feedback has the effect of equalizing the input voltages. Kirchhoff's current law establishes a zero impedance, or virtual ground at the junction of R_f and R_1. The circuit input impedance is just R_1. Negative feedback applied to the noninverting configuration causes the input impedance to approach infinity.

Voltage gain and voltage output for an op amp can be determined easily by means of the equations given earlier. For example, assume that we have an inverting op-amp circuit similar to the one shown in Figure 6-29. Calculate the circuit gain if R_1 is 1 kΩ and R_f is 10 kΩ. Substitute those values into Equation 6-5:

$$V_{gain} = \frac{R_f}{R_1} = \frac{10 \ k\Omega}{1 \ k\Omega} = 10$$

If a −100-mV signal were applied to the input, the output voltage would be 1000 mV or 1 V. (Notice that the output signal polarity is inverted from the input polarity.)

You should remember that the gain of an ideal op amp does not vary with changing frequency. The open loop gain of a practical IC op amp does decrease linearly with increasing frequency, however. By connecting the op amp in a closed-loop circuit (one that uses negative feedback to stabilize the amplifier), the amplifier gain does remain constant over a wide frequency range. Figure 6-30 shows a graph of gain for both the open- and closed-loop circuits.

The power bandwidth of an op amp is a function of *slew rate*, and is always less than the small-signal value. Slew rate is a measurement of the maximum output voltage swing per unit time. Values from 0.8 to 13 volts per microsecond are typical of modern devices.

[Turn to Chapter 10 and study questions E6B11 through E6B13 and E6B15. Review this section if you have difficulty with any of those questions.]

Phase-Locked Loops

Phase-locked loop (PLL) circuits have many applications in Amateur Radio: as FM demodulators, frequency synthesizers, FSK demodulators and a host of other applications. A basic phase-locked loop circuit is really an electronic servo loop, consisting of a phase detector, a loop filter, a dc amplifier and a voltage-controlled oscillator. (A servo is a remote control device that makes automatic adjustments to correct the operation of the device it controls.) **Figure 6-31** shows a block diagram of a phase-locked-loop circuit.

The signal from the voltage-controlled oscillator (VCO) and the input signal

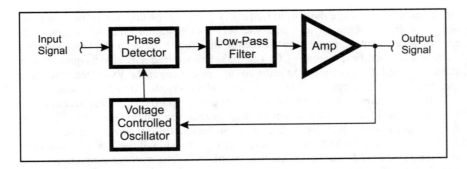

Figure 6-31 — This simple block diagram represents a phase-locked-loop (PLL) system. This type of circuit could be used to demodulate an FM signal.

are fed to the phase detector, which produces an error voltage corresponding to the frequency difference between the VCO and the input signals. This error voltage is filtered, amplified and sent back to the VCO so the oscillator adjusts to the same frequency as the input signal. When the input signal to the phase detector and the output from the VCO approach the same frequency, the error voltage coming from the phase detector decreases toward zero, and we say that the VCO is locked onto the incoming-signal frequency. Any changes in the phase of the input signal, indicating a change in its frequency, are sensed at the phase detector and the error voltage readjusts the VCO so that it remains locked to that signal.

One important use of a phase-locked-loop circuit is as an FM demodulator or detector. If the incoming signal is frequency modulated, then the error voltage coming out of the dc amplifier is a copy of the audio variations used to modulate the FM transmitter. Thus, taking the output from the dc amplifier allows the phase-locked loop to be used directly as a demodulator or detector in an FM receiver.

Many frequency synthesizers used in amateur transceivers involve phase-locked-loop ICs. **Figure 6-32** illustrates the parts of such a circuit. The frequency-divider block allows a wide range of output frequencies to be generated using a single, stable reference oscillator. In most circuits the division ratio is controlled by elec-

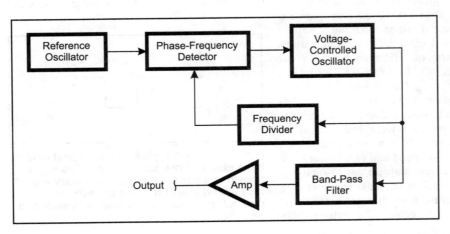

Figure 6-32 — An indirect frequency synthesizer uses a phase-locked loop and a variable-ratio frequency divider.

tronic means. To change the output frequency you vary the division ratio. Again, we are using the principle of an electronic servo, or negative feedback system. The phase-locked loop compares the frequency difference between the output from a voltage-controlled oscillator (VCO) and the reference frequency (or frequency standard). The PLL then produces an error voltage that changes the VCO frequency. The average frequency error is reduced to zero, and we say the circuit is locked.

If the difference between the VCO frequency and the reference frequency is too great, the PLL circuit will not be able to lock the output frequency. The frequency range over which the PLL circuit can lock is called the *capture range*.

Notice that if the VCO is not stable when it is outside of the phase-locked loop, it will still change frequency when it is in the loop. The average frequency is correct, but there is a frequency variation that produces a phase modulation of the output signal. This "phase noise" is directly audible in some receivers, and the noise is also present on the transmitted signal of some rigs that use a PLL synthesizer.

The phase detector, filter and amplifier functions shown in the block diagrams of Figures 6-31 and 6-32 can be contained in a single IC. A few external components can set the VCO frequency and loop filter characteristics. PLL integrated circuits are also used for signal conditioning, AM demodulation, FSK demodulation and a host of other applications.

[Before you go on to the next section, turn to Chapter 10 and study questions E6B16 and E6B17. Review this section as needed.]

DIGITAL INTEGRATED CIRCUITS

Digital electronics is an important aspect of Amateur Radio. Everything from simple digital circuits to sophisticated microcomputer systems are used in modern Amateur Radio systems. The applications include digital communications, code conversion, signal processing, station control, frequency synthesis, amateur satellite telemetry, message handling, word processing and other information-handling operations.

The fundamental principle of digital electronics is that a device can have only a finite number of states. In binary digital systems, there are two discrete states, represented in base-2 arithmetic by the numerals 0 and 1. The binary states described as 0 and 1 may represent an off and on condition or a space and mark in a communications transmission such as CW or RTTY. **Figure 6-33** illustrates a typical binary signal. The simplest digital devices are switches and relays. Computers built before 1950 were made almost entirely with mechanical relays. Low speed and rapid wear were the main objections to such mechanical devices. The next generation of digital instruments used electron tubes as the switch-

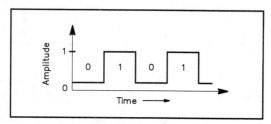

Figure 6-33 — A typical binary signal may have either of two signal levels. The signal shown is a square wave, but a binary signal may spend unequal times at each level, depending on how the signal is intended to be used.

ing elements. Physical size and power consumption then limited the complexity of possible circuits. Modern semiconductor technology allows digital systems with tremendous complexity to be built at a small fraction of the cost (and size) of previous methods. A **digital IC**, then, is an integrated circuit chip that generates, detects or in some way processes digital signals.

There are several types of digital-logic integrated circuits. You should be familiar with the characteristics of two types for the Amateur Extra exam: TTL (transistor-transistor logic) and CMOS (complementary metal-oxide semiconductor).

Transistor-Transistor-Logic Characteristics

Transistor-transistor logic (TTL) is a bipolar logic family, so called because the gates are made with bipolar transistors. (A digital-logic **gate** is a combinational logic element that has two or more inputs and one output. The output state — 1 or 0 — depends on the state of the inputs.) Discrete-component logic circuits could be built using TTL, but modern IC technology builds complete circuits containing many transistors on a single semiconductor wafer.

Most TTL ICs are identified by 7400/5400 series numbers. For example, the 7490 is a decade counter IC. This IC, as its name implies, can count to 10 (one decade). Decade counters produce one output pulse for every 10 input pulses.

Chapter 7 describes the operation of various digital-logic circuits. All of those circuits have TTL IC implementations. Some examples are the 7400 quad NAND gate, the 7432 quad OR gate and the 7408 quad AND gate. (The *quad* in these names refers to the fact that there are four individual gate circuits on the single IC chip.) Other examples of 7400 series ICs are the 7404 hex inverter, and the 7476 dual flip-flop. The 7404 contains six separate inverters, each with one input and one output, in a single 14-pin DIP. (Hex refers to the six inverters on a single IC.) An inverter is also called a NOT gate, because the output is the opposite of the input, so the output is *not* the input. A diagram of the 7404 is shown in **Figure 6-34**. The 7476 includes two J-K flip-flops on one IC. The flip-flop circuit is sometimes known as a latch, because the flip-flop can be set, or latched, either high or low to store one bit of information. Chapter 7 will give more details about flip-flops and how they work.

We have mentioned several types of digital-logic circuits, or gates in these examples. There are three main types of gates. An AND gate will have a HI output only if all the inputs are HI. An OR gate will have a HI output if one or more inputs are HI. The inverter, or NOT gate provides an output that is the opposite of the input. If we combine an AND gate with a NOT gate, the result is a NAND gate, which means *not and*. Likewise, if we combine an OR gate with a NOT gate, the result is a

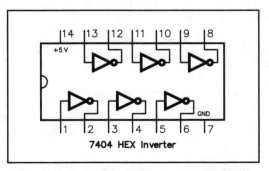

7404 HEX Inverter

Figure 6-34 — This symbol is the schematic representation of a 7404 hex inverter.

NOR gate, which is short for *not or*. **Figure 6-35** shows the schematic symbols used to represent these various logic functions. You should be familiar with these schematic symbols for your Extra class exam. You can distinguish between the schematic symbols for the AND and NAND gates because the small circle on the output represents the NOT function, or inversion. The same is true for the OR and NOR gates. So the symbol with the circle on the output is the NAND or NOR gate, depending on the shape of the basic symbol. You will learn more about these digital-logic gates in Chapter 7.

Figure 6-35 — These symbols represent digital-logic gates. The AND and OR gates each have a distinctive shape. When the small circle from the NOT symbol is added to the output lead of an AND gate, the resulting symbol represents a NAND gate. Adding the NOT circle to the output lead of an OR gate produces the symbol for a NOR gate.

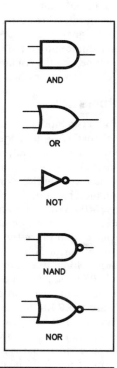

TTL ICs require a +5-V power supply. The supply voltage can vary between 4.7 and 5.3, but 5 is optimum. There are also limits on the input-signal voltages. To ensure proper logic operation, a HI, or 1 input must be at least 2 V and a LO, or 0 input must be no greater than 0.8 V. To prevent permanent damage to a TTL IC, HI inputs must be no greater than 5.5 V, and LO inputs no more negative than −0.6 V. TTL HI outputs will fall somewhere between 2.4 V and 5.0 V, depending on the individual chip. TTL LO outputs will range from 0 V to 0.4 V. The ranges of input and output levels are shown in **Figure 6-36**. Note that the guaranteed output levels fall conveniently within the input limits. This ensures reliable operation when TTL ICs are interconnected. While a LO input can be as much as 0.6 V negative and the device will still operate properly, it is poor engineering practice to allow a LO input to be negative. Sometimes the LO input range is specified as 0 to 0.8 V for that reason.

TTL inputs that are left open, or allowed to "float," will assume a HI or 1 state, but operation of the gate may be unreliable. If an input should be HI, it is better to tie the input to the positive supply through a pull-up resistor (usually a 10-kΩ resistor).

When a TTL gate changes state, the amount of current that it draws changes rapidly. These current changes, called switching transients, appear on the power-supply line and can cause false triggering of other devices. Good engineering practice suggests being generous with decoupling (or bypass) capacitors. Typical decoupling schemes for the +5-V lines include: a 20-μF electrolytic capacitor at the input to the circuit board, a 6.8-μF electrolytic capacitor in parallel with a 0.1 to 1.5-μF tantalum capacitor

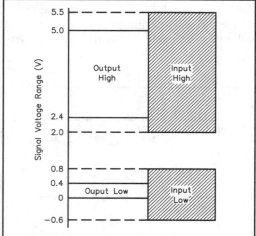

Figure 6-36 — This diagram illustrates the typical TTL-device input and output signal-voltage ranges.

near every 1 to 8 devices and a 0.01 to 0.5-µF ceramic capacitor for every 1 or 2 adjacent devices.

CMOS Logic Characteristics

Complementary metal-oxide semiconductor (CMOS) gates are composed of N-channel and P-channel field-effect transistors (FETs) combined on the same substrate. The major advantages of using the CMOS family are low current (and power) consumption and high noise immunity. Many CMOS ICs are identified by 4000 series numbers. For example, a 4001 IC is a quad, two-input NOR gate. The 4001 contains four separate NOR gates, each with two inputs and one output. Some other examples are the 4011 quad NAND gate, the 4081 quad AND gate and the 4069 hex inverter. Manufacturers are also developing CMOS ICs that are pin-compatible with the 7400 TTL family. If you come across a 74HC00 or a 74HC04 device you should be aware that a C in the part number probably indicates that it is a CMOS device.

CMOS ICs will operate over a much larger power-supply range than TTL ICs. The power-supply voltage can vary from 3 V to as much as 18 V. CMOS output voltages depend on the power-supply voltage. A HI output is generally within 0.1 V of the positive supply connection, and a LO output is within 0.1 V of the negative supply connection (ground in most applications). For example, if you are operating CMOS gates from a 9-volt battery, a logic 1 output will be somewhere between 8.9 and 9 volts, and a logic 0 output will be between 0 and 0.1 volts.

The switching threshold for CMOS inputs is approximately half the supply voltage. **Figure 6-37** shows these input and output voltage characteristics. The wide range of input voltages gives the CMOS family great immunity to noise, since noise spikes will generally not cause a transition in the input state.

CMOS ICs require special handling because of the thin layer of insulation between the gate and substrate of the MOS transistors. Even small static charges can cause this insulation to be punctured, destroying the gate. CMOS ICs should be stored with their pins pressed into special conductive foam. They should be installed in a socket or else a soldering iron with a grounded tip should be used to solder them on a circuit board. Wear a grounded wrist strap when handling CMOS ICs to ensure that your body is at ground potential. Any static electricity on the IC will probably destroy it.

[Now turn to Chapter 10 and study exam questions E6C01 through E6C11. Review this section as needed.]

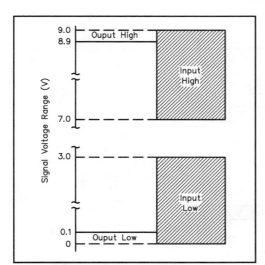

Figure 6-37 — This diagram shows CMOS-device input and output signal-voltage ranges with a 9-V supply.

VIDICON AND CATHODE-RAY TUBE DEVICES

The **vidicon tube** is a relatively simple, inexpensive TV-camera pickup tube. Vidicons are used in closed-circuit applications in banks and factories because of their small size and low cost. Amateurs use vidicon tubes for both fast and slow-scan television. **Figure 6-38** shows the physical construction of a vidicon tube.

The photoconductive layer and signal electrode can best be thought of as an array of leaky capacitors. As the electron beam scans the photoconductor, it charges each miniature capacitor to the cathode voltage (usually about −20 V with respect to the signal electrode). Horizontal and vertical deflection of the electron beam in a vidicon is accomplished with magnetic fields generated by coils on the outside of the tube. The varying electromagnetic fields control the beam as it scans the tube face.

As soon as each capacitor is charged, it starts discharging through its leakage resistance, the rate of discharge depending on the amount of light reaching it. On the next scan of the area, the electron beam will deposit enough electrons to recharge the capacitor to cathode potential. In the instant that it does this, a net current flows through the cathode/signal electrode circuit with an amplitude proportional to the amount of capacitor discharge. Since this discharge depends on the amount of light hitting that portion of the screen, the beam current, as the beam sweeps by that area, is proportional to the light intensity. The output beam current is very low (a fraction of a microamp), and the output impedance of the vidicon is very high, so the video preamp must be designed with care to minimize hum and noise problems.

A **cathode-ray tube (CRT)** is a display device that converts an electrical signal into a visible image. The CRTs used in TV sets and oscilloscopes differ quite a bit, but both operate on the same basic principles. **Figure 6-39** illustrates the operation of a simple CRT. An electron beam is produced in the tube and directed toward a phospho-

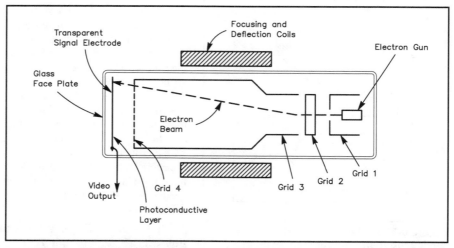

Figure 6-38 — This diagram shows the construction of a vidicon tube, used in many amateur TV cameras.

rescent material on the inside face of the tube. When the electrons in the beam strike this "phosphor," the phosphor begins to glow. By scanning the electron beam across the tube face, an image is produced. The relative brightness of each part of the image depends on the number of electrons striking the surface at that point.

In a TV picture tube, horizontal and vertical deflection of the electron beam is produced with magnetic fields generated by external coils. While sufficient for television purposes, this electromagnetic deflection is not suitable for measurement purposes. To display high-frequency signals on a lab-type oscilloscope, *electrostatic deflection* must be used. In this case, a charge is stored on the plates of a capacitor, with the CRT between the capacitor plates. As the charge on the plates varies, it causes the electron beam to be deflected.

Persistence describes the length of time that an image remains on a CRT screen after the electron beam is turned off. In the case of a broadcast television receiver, the image should remain until the entire scan is completed, but not much longer than that. As the second scan begins, $1/30$ of a second later, you don't want the previous image to interfere with the new one. On the other hand, for some oscilloscope displays, you may want the image to remain much longer, so you can examine the waveform that results from a single sweep across the screen. In the early days of slow-scan television experimentation, amateurs needed the image to remain during the entire 8 seconds it took to scan the complete picture. Otherwise they would not be able to see the entire picture on the screen at the same time. The choice of materials used as the phosphor on the inside surface of the display screen will determine the persistence that a tube will exhibit.

The anode voltage in a CRT pulls the electrons away from the cathode, and accelerates them down the length of the tube. Higher anode voltages pull more electrons away from the cathode, and move them at faster speeds. You should be certain that you operate a CRT within its specified anode voltage. If you operate the tube at a higher anode voltage, the electrons will strike the phosphor on the

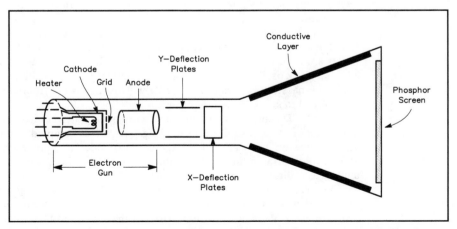

Figure 6-39 — This diagram illustrates the basic con-struction and operation of a cathode-ray tube (CRT).

front of the tube with more force, and that could produce X-rays. The X-rays would probably leave the tube through the front of the glass, right toward you. So operating a CRT with a higher than rated anode voltage presents a safety hazard from the production of X-rays. The increased operating voltage would also make the electrons move at a higher speed, and that would result in less deflection of the electron beam. Therefore, the display image size would be decreased.

For some display applications, liquid-crystal display (LCD) technology has advantages. When a voltage is applied across the liquid crystal material it changes the way light is refracted through the material. With no voltage applied, the crystal material is virtually transparent, but with the applied voltage, the light is blocked, making the crystal appear black. LCDs have much lower operating voltages than CRTs. Since there is no need for an electron gun and deflection circuitry, LCDs require no depth behind the display screen. The principal advantage of these display devices is that they consume very little power. Of course there are also disadvantages. LCDs are slow to operate at low temperatures, and may be damaged by high temperatures. With the present technology they do not have the sharp resolution available from a CRT, and color displays are not as vivid.

CHARGE-COUPLED DEVICES

A **charge-coupled device (CCD)** is a special type of integrated circuit. It combines analog and digital signal-handling properties in a single IC. The CCD is made from a string of metal-oxide semiconductor (MOS) capacitors with a MOSFET on the input and output sides. The first capacitor stores a sample of the input-signal voltage. When a control pulse biases the MOSFETs to conduct, the first capacitor passes its sampled voltage on to the second capacitor, and it takes another sample. With successive control pulses, each input sample is passed to the next capacitor in the string. When the MOSFETs are biased off, each capacitor stores its charge. This process is sometimes described as a "bucket brigade," because the analog signal is sampled and then passed in stages through the CCD to the output.

It is easy to imagine that one application of a CCD is to form an audio delay line. Each signal sample taken at the input will appear at the output some time later. CCDs are used to build very accurate filters, called switched-capacitor filters. An external clock oscillator and resistors control the filter bandwidth and function (high-pass, low-pass, band-pass or notch filter). This means the filter characteristics can be changed to suit current conditions.

The CCD samples the input signal voltage at times controlled by a clock signal, so this is a form of digital sampling. The actual sampled voltage is an analog value, however. The sampled voltage is not rounded to a predetermined step size, as it would be in an analog-to-digital converter. So a CCD cannot be used as an analog-to-digital converter.

In recent years a two-dimensional array of charge-coupled device elements was developed using light-sensitive materials. The charge entering each capacitor is proportional to the amount of light striking the surface at that point. The capacitors store the charge as signals corresponding to the pixels of the array surface. The signal is shifted out of the CCD array a line at a time in a pattern that matches a vidicon scan. This CCD array forms the basis for a modern video camera.

[To check your understanding of this section, turn to Chapter 10 and study examination questions E6D01 through E6D07. If you have difficulty with any of these questions, review the material in this section.]

TOROIDS

A donut-shaped coil form is called a **toroid**. When a wire is wound on such a coil form, a *toroidal inductor* is produced. Toroidal inductors are one of the most popular inductor types in RF circuits. See **Figure 6-40** for a photo of a variety of toroidal inductors. Toroidal inductors are also called ferromagnetic inductors, because the coil forms are made with ferrite and powdered-iron materials. The chemical names for iron compounds are based on the Latin word for iron, *ferrum*, so this is how these materials get the name ferrite.

Figure 6-40 — This photo shows a variety of inductors wound on powdered-iron and ferrite toroids.

A primary advantage of using a toroidal core to wind an inductor rather than a linear core is that nearly all the magnetic field is contained within the core of a toroid. With a linear core, the magnetic field extends through the space surrounding the inductor. The magnetic field of one linear-core inductor will interact with other nearby inductors, so external shields or other isolation methods must be used. Toroidal inductors can be located close to each other on the circuit board and there will be almost no interaction, however.

Manufacturers offer a wide variety of materials, or mixes, to provide cores that will perform well over a desired frequency range. By careful selection of core material, it is possible to produce toroidal inductors that can be used from dc to at least 1000 MHz. Cores made by mixing various amounts of powdered iron with binder materials are called *powdered-iron* toroids. If other materials, such as nickel-zinc and manganese-zinc compounds are mixed with the iron, *ferrite* toroids are produced.

The inductance of a toroidal core is determined by the number of turns of wire on the core, and on the core *permeability*. Permeability refers to the strength of a magnetic field in the core as compared to the strength of the field if no core were used. Cores with higher values of permeability will produce larger inductance values for the same number of turns on the coil. In other words, if you make two inductors with 10 turns on the coil forms, the core with a higher permeability will have more inductance.

The choice of core materials for a particular inductor presents a compromise of features. The powdered-iron cores generally have better temperature stability. Ferrite toroids generally have higher permeability values, however, so coils made with ferrite toroids usually require fewer turns to produce a given inductance value.

Calculating the inductance of a particular toroidal coil is simple. First you must know the *inductance index* value for the particular core you will use. This value, known as A_L, is found in the manufacturer's data. For powdered-iron toroids, A_L values are given in microhenrys per 100 turns. **Table 6-1** gives an example of the data for several core types. The information for this table is taken from *The ARRL Handbook*, and is courtesy of Amidon Associates (one of the major distributors of toroidal cores in small quantities to the amateur market) and Micrometals (the manufacturer of the cores distributed by Amidon). See *The ARRL Handbook* for more complete information about these cores and their applications.

To calculate the inductance of a powdered-iron toroidal coil, when the number of turns and the core material are known, use Equation 6-7.

$$L = \frac{A_L \times N^2}{10,000}$$ (Equation 6-7)

where:

L = inductance in μH.
A_L = inductance index, in μH per 100 turns.
N = number of turns.

For example, suppose you have a T-50 sized core made from the number 6 mix, which is good for inductors from about 10 to 50 MHz. From Table 6-1 we find that this core has an A_L value of 40. What is the inductance of a coil that has 10 turns on this core?

$$L = \frac{A_L \times N^2}{10,000} = \frac{40 \times 10^2}{10,000} = \frac{40 \times 100}{10,000} = \frac{4000}{10,000} = 0.4 \text{ μH}$$

Table 6-1
A_L Values for Selected Powdered-Iron and Ferrite Toroids

A_L Values for Powered-Iron Cores (μH per 100 turns)

Size	Mix				
	2	3	6	10	12
T-12	20	60	17	12	7.5
T-20	27	76	22	16	10.0
T-30	43	140	36	25	16.0
T-50	49	175	40	31	18.0
T-200	120	425	100	na	na

A_L Values for Ferrite Cores (mH per 1000 turns)

Size	Mix			
	43	61	63	77
FT-23	188.0	24.8	7.9	396
FT-37	420.0	55.3	19.7	884
FT-50	523.0	68.0	22.0	1100
FT-114	603.0	79.3	25.4	1270

Data from *The ARRL Handbook*, 2001, courtesy of Amidon Associates and Micrometals.

Often you want to know how many turns to wind on the core to produce an inductor with a specific value. In that case, you simply solve Equation 6-7 for N.

$$N = 100 \sqrt{\frac{L}{A_L}}$$ (Equation 6-8)

Suppose you want to know how many turns to wind on the T-50-6 core used in the previous example to produce a 5-μH inductor? (The A_L value = 40.)

$$N = 100 \sqrt{\frac{L}{A_L}} = 100 \sqrt{\frac{5}{40}} = 100 \sqrt{0.125} = 100 \times 0.35 = 35$$

So we will have to wind 35 turns of wire on this core to produce a 5-μH inductor. Perhaps the most common error made when winding an inductor involves counting the correct number of turns. Keep in mind that if the wire simply passes through the center of the core, you have a 1-turn inductor. Each time the wire passes through the center of the core, it counts as another turn. The common error is to count one complete wrap around the core ring as one turn. That produces a 2-turn inductor, however. See **Figure 6-41**.

The calculations with ferrite toroids are nearly identical, but the A_L values are given in millihenrys per 1000 turns instead of microhenrys per 100 turns. This requires a change of the constant in Equation 6-7 from 10,000 to 1,000,000. Use Equation 6-9 to calculate the inductance of a ferrite toroidal inductor.

$$L = \frac{A_L \times N^2}{1,000,000}$$ (Equation 6-9)

where:
L = inductance in mH.
A_L = inductance index, in mH per 1000 turns.
N = number of turns.

Suppose we have an FT-50-sized core, made from 43-mix material. What is

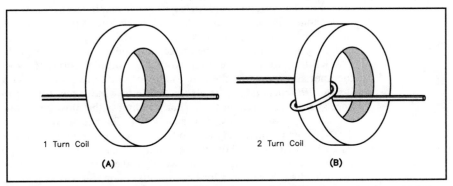

1 Turn Coil
(A)

2 Turn Coil
(B)

Figure 6-41 — Proper turns counting is important when you wind a toroidal inductor. Each pass through the center of the core must be counted. Part A shows a 1-turn inductor and Part B shows a 2-turn inductor.

the inductance of a 10-turn coil? Table 6-1 shows that the $A_L = 523$ for this core.

$$L = \frac{A_L \times N^2}{1,000,000} = \frac{523 \times 10^2}{1,000,000} = \frac{523 \times 100}{1,000,000} \quad \frac{52300}{1,000,000} = 0.0523 \text{ mH} = 52.3 \text{ } \mu\text{H}$$

Again, it is a simple matter to solve this equation for N, so you can calculate the number of turns required to produce a specific inductance value for a particular ferrite core.

$$N = 1000 \sqrt{\frac{L}{A_L}}$$
(Equation 6-10)

How many turns must we wind on a T-50-43 core to produce a 1-mH inductor? ($A_L = 523$.)

$$N = 1000 \sqrt{\frac{L}{A_L}} = 1000 \sqrt{\frac{1}{523}} = 1000 \sqrt{1.91 \times 10^{-3}} = 1000 \times 0.0437 = 43.7$$

Winding 43 or 44 turns on this core will produce an inductor of about 1 mH.

Toroidal cores are available in a wide variety of sizes. It is important to select a core size large enough to be able to hold the required number of turns to produce a particular inductance value. For a high-current application you will have to use a large wire size, so a larger core size is required. To wind an inductor for use in a high-power antenna tuner, for example, you may want to use number 10 or 12 wire. If your inductor requires 30 turns of this wire, you would probably select a 200-size core, which has an outside diameter of 2 inches and an inside diameter of $1\frac{1}{4}$ inches. You might even want to select a larger core for this application. A *ferrite bead* is a very small core with a hole designed to slip over a component lead. These are often used as parasitic suppressors at the input and output terminals of VHF and UHF amplifiers.

The use of ferrite beads as parasitic suppressors points out another interesting property of these core materials. While we normally want to select a core material that will have low loss at a particular frequency or over a certain range for winding our inductors, at times we want to select a core material that will have high loss. Toroid cores are very useful for solving a variety of radio-frequency interference (RFI) problems. For example, you might select a *type 43 mix* ferrite core and wind several turns of a telephone wire or speaker leads through the core to produce a *common-mode choke*. Such a choke is designed to suppress any RF energy flowing on these wires. So the audio signals flow through the choke unimpeded, but the RF signals are blocked.

RF transformers are often wound on toroidal cores. If two wires are twisted together and wound on the core as a pair to place two windings on the core, we say it is a *bifilar winding*. It is also possible to wind three, four or more wires on the core simultaneously, but the bifilar winding is the most common.

You may wonder how you can tell the difference between, say a T-50-6 core and a T-50-10 core if you found them both in a piece of surplus equipment or in a grab bag of parts. For that matter, how can you tell if a particular core is a powdered-iron or a ferrite core. Unfortunately, the answer is, you can't! There is no standard way of marking or coding these cores for later identification. So it is important to purchase your cores from a reliable source, and store them separate in marked containers.

[Turn to Chapter 10 now and study questions E6D08 through E6D16. Review this section if you have any difficulty with these questions about toroids.]

CRYSTAL-LATTICE FILTERS

Crystal-lattice filters are used in SSB transmitters and receivers where high-Q, narrow-bandwidth filtering is required. Such filters typically are used at intermediate frequencies above 500 kHz in receivers. In SSB transmitters, crystal-lattice filters frequently are used after the balanced modulator to attenuate the unwanted sideband. A quartz crystal acts as an extremely high-Q circuit. The equivalent electrical circuit is depicted in **Figure 6-42A**. Figure 6-42B shows a graph of the reactance versus frequency for the crystal.

A quartz crystal is known as a *piezoelectric device*. That means if a voltage is applied across a crystal, the crystal will be physically deformed. This physical deformation is known as the **piezoelectric effect**. This physical deformation results in crystal vibrations at a particular frequency, and these vibrations can be used to control the operating frequency of a circuit. This is the operating principle for a crystal oscillator. Crystal filters use the piezoelectric effect to pass desired frequencies while blocking undesired ones.

Although single crystals can be used as filtering devices, the normal practice is to wire two or more together in various configurations to provide a desired response curve. **Figure 6-43** depicts a configuration known as the half-lattice filter. In this arrangement, crystals Y1 and Y2 are on different frequencies. The bandwidth and response shape of the filter depends on the relative frequencies of these two crystals. The overall filter bandwidth is equal to approximately 1 to 1.5 times the frequency separation of the crystals. The closer the crystal frequencies, the narrower the bandwidth of the filter. In general, a crystal lattice filter has narrow bandwidth and steep response skirts, as shown in Figure 6-43B.

A good crystal-lattice filter for double-sideband (DSB) voice use would have a bandwidth of approximately 6 kHz at the − 6 dB points on the response curve. A good crystal filter for single-sideband (SSB) phone service is significantly narrower;

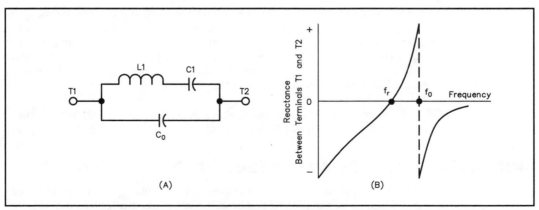

Figure 6-42 — The equivalent circuit for a piezoelectric quartz crystal is shown at A. At B is a graph of reactance versus frequency for the crystal.

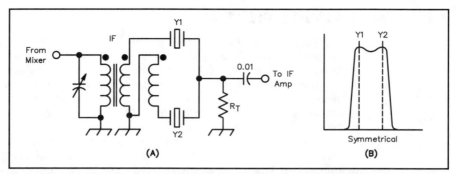

Figure 6-43 — Part A shows a schematic diagram of a half-lattice crystal filter. B shows a typical response curve for this type of filter. Note the steep skirts on the response, representing good rejection of signals outside the passband.

typical bandwidth is 2.1 kHz at the – 6 dB points. For CW use, crystal filters typically have 250 to 500-Hz bandwidths.

The home construction of crystal filters can be time consuming but can be relatively inexpensive. For home builders, crystal ladder filters may be easier to design than the lattice variety. **Figure 6-44** illustrates a simple crystal ladder filter for CW, using three crystals.

Start with a collection of crystals that have approximately the

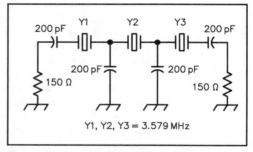

Figure 6-44 — This diagram shows a simple crystal ladder filter that uses three crystals and is suitable for CW use.

same frequency and characteristics. TV color-burst-oscillator crystals are inexpensive, easy to obtain and work well for such filters. Use an oscillator circuit to measure the actual operating frequency of each of your crystals. Carefully select the crystals to build your filter. The frequency difference between the selected crystals should be less than 10% of the desired filter *bandwidth*. In other words, if you want to build a 500-Hz bandwidth filter, select crystals that are all within 50 Hz of each other.

[To check your understanding of this section, turn to Chapter 10 and study questions E6E01 through E6E06. Review this section as needed.]

MONOLITHIC MICROWAVE INTEGRATED CIRCUITS

The **monolithic microwave integrated circuit (MMIC)** is unlike most other ICs that you may be familiar with. These ICs are quite small, often classified as "pill sized" devices, perhaps because they look like a small pill, with four leads coming out of the device at 90° to each other.

Figure 6-45 — This simple utility amplifier is suitable for transmitting and receiving on 903, 1296, 2304, 3456 and 5760 MHz. It consists of three MMICs, three ordinary resistors and four chip (surface mount) capacitors. In addition, a feed-through capacitor is used to bring the bias-supply voltage into the project case box.

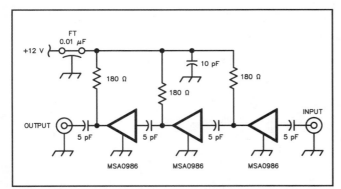

Figure 6-46 — This is the schematic diagram of the utility amplifier pictured in Figure 6-45. Note that the operating bias voltage is applied to the MMICs through a resistor connected to the device output leads. An RF choke is also often included in series with the supply lead.

MMIC devices typically have predefined operating characteristics, and require only a few external components for proper operation. This can greatly simplify an amplifier design for circuits at UHF and microwave frequencies. For example, an MSA-0135 MMIC would be an excellent choice to build a receive preamplifier for a 1296-MHz receiver. This device provides 18.5 dB of gain for signals up to 1300 MHz, with a noise figure of 5.5 dB.

An MSA-0735 MMIC might be used to construct a 3456-MHz receiver amplifier. This device can be expected to provide 13 dB of gain for signals up to about 2500 MHz, and slightly less gain for higher-frequency signals. This amplifier would have a noise figure of about 4.5 dB.

Circuits built using MMICs generally employ *microstrip* construction techniques. Double-sided circuit board material is used, and one side serves to form a ground plane for the circuit. Circuit traces connecting the MMICs or other active devices to the signal input and output connectors form sections of feed line. The line widths, along with the circuit-board thickness and dielectric constant of the insulating material form sections of feed line with the desired characteristic impedance. Components are soldered to these feed line sections. **Figure 6-45** is a photo of a general-purpose amplifier constructed using this technique with three MMICs, three resistors, four chip (surface-mount) capacitors and a feed-through capacitor to bring the supply voltage into the box.

The operating bias voltage is supplied to an MMIC through a resistor connected to the device output lead. **Figure 6-46** is the schematic diagram of the general-purpose amplifier shown in the photo of Figure 6-45. Although this example does not include one, an RF choke is also often included in series with the bias voltage supply.

[Before you go on to the next chapter, you should turn to Chapter 10 and study questions E6E07 through E6E11. Review this section if you have difficulty with any of these questions.]

Amplifier transfer function — A graph or equation that relates the input and output of an amplifier under various conditions.

AND gate — A logic circuit whose output is 1 only when both of its inputs are 1.

Astable (free-running) multivibrator — A circuit that alternates between two unstable states. This circuit could be considered as an oscillator that produces square waves.

Balanced modulator — A circuit used in a single-sideband suppressed-carrier transmitter to combine a voice signal and an RF signal. The balanced modulator isolates the input signals from each other and the output, so that only the difference of the two input signals reaches the output.

Band-pass filter — A circuit that allows signals to go through it only if they are within a certain range of frequencies. It attenuates signals above and below this range.

Bistable multivibrator — Another name for a flip-flop circuit that has two stable output states.

Butterworth filter — A filter whose passband frequency response is as flat as possible. The design is based on a Butterworth polynomial to calculate the input/output characteristics.

Chebyshev filter — A filter whose passband and stopband frequency response has an equal-amplitude ripple, and a sharper transition to the stop band than does a Butterworth filter. The design is based on a Chebyshev polynomial to calculate the input/output characteristics.

Counter (divider, divide-by-n counter) — A circuit that is able to change from one state to the next each time it receives an input signal. A counter produces an output signal every time a predetermined number of input signals have been received.

Crystal-controlled marker generator — An oscillator circuit that uses a quartz crystal to set the frequency, and which has an output rich in harmonics that can be used to determine band edges on a receiver. An output every 100 kHz or less is normally produced.

Detector — A circuit used in a receiver to recover the modulation (voice or other information) signal from the RF signal.

Double-balanced mixer (DBM) — A mixer circuit that is balanced for both inputs, so that only the sum and the difference frequencies, but neither of the input frequencies, appear at the output. There will be no output unless both input signals are present.

Elliptical filter — A filter with equal-amplitude passband ripple and points of infinite attenuation in the stop band. The design is based on an elliptical function to calculate the input/output characteristics.

Exclusive OR gate — A logic circuit whose output is 1 when any single input is 1 and whose output is 0 when no input is 1 or when more than one input is 1.

Flip-flop (bistable multivibrator) — A circuit that has two stable output states, and which can change from one state to the other when the proper input signals are detected.

High-pass filter — A filter that allows signals above the cutoff frequency to pass through. It attenuates signals below the cutoff frequency.

Inverter — A logic circuit with one input and one output. The output is 1 when the input is 0, and the output is 0 when the input is 1.

Latch — Another name for a bistable multivibrator flip-flop circuit. The term **latch** reminds us that this circuit serves as a memory unit, storing a bit of information.

Linear electronic voltage regulator — A type of voltage-regulator circuit that varies either the current through a fixed dropping resistor or the resistance of the dropping element itself. The conduction of the control element varies in direct proportion to the line voltage or load current.

L network — A combination of a capacitor and an inductor, one of which is connected in series with the signal lead while the other is shunted to ground.

Low-pass filter — A filter that allows signals below the cutoff frequency to pass through. It attenuates signals above the cutoff frequency.

Mixer — A circuit that takes two or more input signals, and produces an output that includes the sum and difference of those signal frequencies.

Modulator — A circuit designed to superimpose an information signal on an RF carrier wave.

Monostable multivibrator (one shot) — A circuit that has one stable state. It can be forced into an unstable state for a time determined by external components, but it will revert to the stable state after that time.

NAND (NOT AND) gate — A logic circuit whose output is 0 only when both inputs are 1.

Neutralization — Feeding part of the output signal from an amplifier back to the input so it arrives out of phase with the input signal. This negative feedback neutralizes the effect of positive feedback caused by coupling between the input and output circuits in the amplifier. The negative-feedback signal is usually supplied by connecting a capacitor from the output to the input circuit.

Noninverting buffer — A logic circuit with one input and one output, and whose output level is the same as the input level.

NOR (NOT OR) gate — A logic circuit whose output is 0 if either input is 1.

OR gate — A logic circuit whose output is 1 when either input is 1.

Oscillator — A circuit built by adding positive feedback to an amplifier. It produces an alternating current signal with no input signal except the dc operating voltages.

Parasitics — Undesired oscillations or other responses in an amplifier.

Phase modulator — A device capable of modulating an ac signal by varying the reactance of an amplifier circuit in response to the modulating signal. (The modulating signal may be voice, data, video or some other kind.) The circuit capacitance or inductance changes in response to an audio input signal. Used in PM (or FM) systems, this circuit acts as a variable reactance in an amplifier tank circuit.

Pi network output-coupling circuits — A combination of two like reactances (coil or capacitor) and one of the opposite type. The single component is connected in series with the signal lead and the two others are shunted to ground, one on either side of the series element.

Prescaler — A divider circuit used to increase the useful range of a frequency counter.

Reactance modulator — A device capable of modulating an ac signal by varying the reactance of an oscillator circuit in response to the modulating signal. (The modulating signal may be voice, data, video or some other kind.) The circuit capacitance or inductance changes in response to an audio input signal. Used in FM systems, this circuit acts as a variable reactance in an oscillator tank circuit.

Selectivity — A measure of the ability of a receiver to distinguish between a desired signal and an undesired one at some different frequency. Selectivity can be applied to the RF, IF and AF stages.

Sequential logic — A type of circuit element that has at least one output and one or more input channels, and in which the output state depends on the previous input states. A flip-flop is one sequential-logic element.

Switching regulator — A voltage-regulator circuit in which the output voltage is controlled by turning the pass element on and off at a high rate, often several kilohertz. The control-element duty cycle is proportional to the line or load conditions.

Truth table — A chart showing the outputs for all possible input combinations to a digital circuit.

Zener diode — A diode that is designed to be operated in the reverse-breakdown region of its characteristic curve.

PRACTICAL CIRCUITS

Now that you have studied some dc and ac electronics principles, and have learned about the basic properties of some modern solid-state components, you are ready to learn how to apply those ideas to practical Amateur Radio circuits. This chapter will lead you through examples and explanations to help you gain experience with those circuits. At various places throughout your study you will be directed to turn to Chapter 10 to use the exam questions as a review exercise.

There will be seven questions from the Practical Circuits subelement on your Extra class exam, so the questions for this chapter are divided into seven groups. The topics covered in this chapter are:

E7A   Digital logic circuits: Flip flops; Astable and monostable multivibrators; Gates (AND, NAND, OR, NOR); Positive and negative logic

E7B   Amplifier circuits: Class A, Class AB, Class B, Class C, amplifier operating efficiency (ie, DC input versus PEP), transmitter final amplifiers; amplifier circuits: tube, bipolar transistor, FET

E7C   Impedance-matching networks: Pi, L, Pi-L; filter circuits: constant K, M-derived, band-stop, notch, crystal lattice, pi-section, T-section, L-section, Butterworth, Chebyshev, elliptical; filter applications (audio, IF, digital signal processing {DSP})

E7D   Oscillators: types, applications, stability; voltage-regulator circuits: discrete, integrated and switched mode

E7E   Modulators: reactance, phase, balanced; detectors; mixer stages; frequency synthesizers

E7F   Digital frequency divider circuits; frequency marker generators; frequency counters

E7G   Active audio filters: characteristics; basic circuit design; preselector applications

You should keep in mind that entire books have been written about every topic covered in this chapter. So if you do not understand some of the circuits from our brief discussion, or if you feel we have left out some of the details, you may want to consult some other reference books. *The ARRL Handbook* is a good starting point, but even that won't tell you everything there is to know about each topic. Our discussion in this chapter should help you understand the circuits well enough to pass your Amateur Extra class exam, however.

DIGITAL ELECTRONICS BASICS

In electronic digital-logic systems, binary information is represented as voltage levels. For example, 0 V may represent a binary 0, and + 5 V may equal a binary 1. Because you can't always achieve those exact voltages in practical circuits, digital circuits consider the signal to be a 1 or 0 if the voltage comes within certain bounds. The voltage ranges used for HI and LO logic levels in TTL and CMOS ICs are discussed in Chapter 6.

Combinational Logic — Boolean Algebra

In binary digital-logic circuits, a combination of inputs results in a specific output or combination of outputs. Except during switching transitions, the state of the output is determined by the simultaneous state(s) of the input channel(s). A combinational logic function has one and only one output state corresponding to each combination of input states. Combinational logic networks have no feedback loops.

Networks made up of many combinational logic elements may perform arithmetic or logical manipulations. Regardless of the purpose, these operations are usually expressed in arithmetic terms. Digital networks add, subtract, multiply and divide, but normally do it in binary form using two states that we represent with the numerals 0 and 1.

Binary digital circuit functions may be represented by equations using *Boolean algebra*. The symbols and laws of Boolean algebra are somewhat different from those of ordinary algebra. The symbol for each logical function is shown here in the descriptions of the individual logical elements. The logical function of a particular element may be described by listing all possible combinations of input and output values in a **truth table**, also called a state table. Such a list of all input combinations and their corresponding outputs characterize, or describe, the function of any digital device. Standard logic circuit symbols and their corresponding Boolean expressions and truth tables are shown in Figures 7-1 through 7-7.

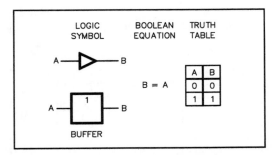

Figure 7-1 — Schematic symbols for a noninverting buffer are shown. The distinctive (triangular) shape is used by ARRL and in most US publications. The square symbol is used in some other countries. The Boolean equation for the buffer and a truth table for the operation are also given.

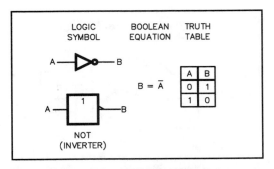

Figure 7-2 — Schematic symbols for an inverter (NOT) are shown. The distinctive (triangular) shape is used by ARRL and in most US publications. The square symbol is used in some other countries. The Boolean equation and truth table for the operation are also given.

One-Input Elements

There are two logic elements that have only one input and one output: the **noninverting buffer** and the **inverter** or NOT circuit (**Figures 7-1** and **7-2**). The noninverting buffer simply passes the same state (0 or 1) from its input to its output. In an inverter or NOT circuit, a 1 at the input produces a 0 at the output, and vice versa. NOT indicates inversion, negation or complementation. Notice that the only difference between the noninverting buffer and the inverter is the small circle on the output lead. This small circle is used to indicate an inverted output on any digital-logic circuit symbol. The Boolean algebra notation for NOT is a bar over the variable or expression.

The AND Operation

A gate is usually defined as a combinational logic element with two or more inputs and one output state that depends on the state of the inputs. (Emitter-coupled logic — ECL — devices violate this definition by having two outputs.) Gates perform simple logical operations and can be combined to form complex switching functions. So as we talk about the logical operations used in Boolean algebra, you should keep in mind that each function is implemented by using a gate with the same name. So an **AND gate** implements the AND operation.

The AND operation results in a 1 only when all operands are 1. That is, if the inputs are called A and B, the output is 1 only if A and B are both 1. In Boolean notation, the logical operator AND is usually represented by a dot between the variables, centered on the line (•). The AND function may also be signified by no space between the variables. Both forms are shown in **Figure 7-3**, along with the schematic symbol for an AND gate.

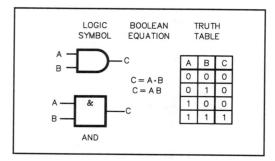

Figure 7-3 — Schematic symbols for a two-input AND gate are shown. The distinctive (round-nosed) shape is used by ARRL and in most US publications. The square symbol is used in some other countries. The Boolean equation and truth table for the operation are also given.

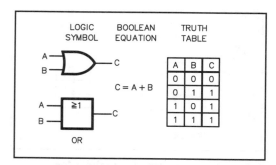

Figure 7-4 — Schematic symbols for a two-input OR gate are shown. The distinctive (pointed bullet) shape is used by ARRL and in most US publications. The square symbol is used in some other countries. The Boolean equation and truth table for the operation are also given.

The OR Operation

The OR operation results in a 1 at the output if any or all inputs are 1. In Boolean notation, the + symbol is used to indicate the OR function. The **OR gate** shown in **Figure 7-4** is sometimes called an *Inclusive* OR. In Boolean algebra notation, a + sign is used between the variables to represent the OR function.

Study the truth table for the OR function in Figure 7-4. You should notice that the OR gate will have a 0 output only when all inputs are 0.

The EXCLUSIVE OR Operation

The EXCLUSIVE OR (XOR) operation results in an output of 1 if only one of the inputs is 1, but if both inputs are 1, then the output is 0. The Boolean expression ⊕

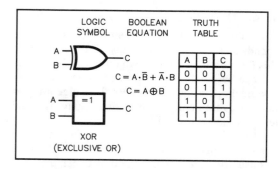

Figure 7-5 — Schematic symbols for a two-input EXCLUSIVE OR (XOR) gate are shown. The Boolean equation and truth table for the operation are included.

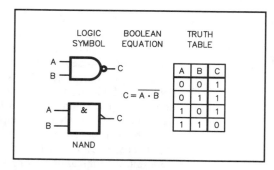

Figure 7-6 — Schematic symbols for a two-input NAND gate are shown. The Boolean equation and truth table for the operation are also given.

represents the EXCLUSIVE OR function. **Figure 7-5** shows the schematic symbol for an **EXCLUSIVE OR gate** and its truth table.

The NAND Operation

The NAND operation means NOT AND. A **NAND gate** (**Figure 7-6**) is an AND gate with an inverted output. A NAND gate produces a 0 at its output only when all inputs are 1. In Boolean notation, NAND is usually represented by a dot between the variables and a bar over the combination, as shown in Figure 7-6.

The NOR Operation

The NOR operation means NOT OR. A **NOR gate** (**Figure 7-7**) produces a 0 output if any or all of its inputs are 1. In Boolean notation, the variables have a + symbol between them and a bar over the entire expression to indicate the NOR function. When you study the truth table shown in Figure 7-7, you will notice that a nor gate produces a 1 output only when all of the inputs are 0.

Logic Polarity

Logic systems can be designed to use two types of polarity. If the highest voltage level (HI) represents a binary 1, and the lowest level (LO) represents a 0, the logic is *positive*. If the opposite representation is used (HI = 0 and LO = 1), the logic is *negative*. In the gate descriptions discussed so far, positive logic was assumed.

Positive and negative logic symbols are compared in **Figure 7-8**. Small circles (state indicators) on the input side of a gate signify negative logic. The use of negative logic sometimes simplifies the Boolean algebra associated with logic networks. Consider a circuit having two inputs and one output, and suppose you desire a high output only when both inputs are low. A search through the truth tables shows the NOR gate has the proper characteristics. The way the problem is posed (the words *only* and *both*) suggests the AND (or NAND) function, however. A negative-logic

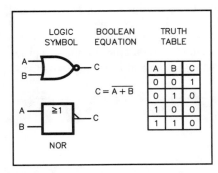

Figure 7-7 — Schematic symbols for a two-input NOR gate are shown. The Boolean equation and truth table for the operation are included.

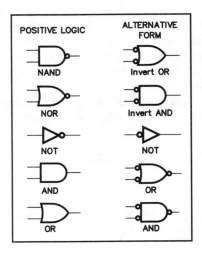

Figure 7-8 — This drawing compares the positive and negative-true logic symbols for the common logic functions.

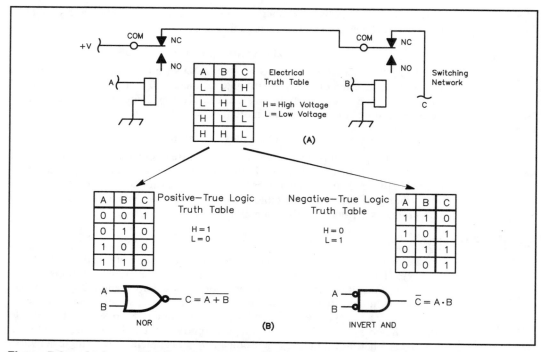

Figure 7-9 — At A, combinational logic is implemented with relays, shown with an electrical truth table. Assigning values of 1 and 0 to the electrical states as shown in B leads to two schematic symbols, one for positive-true logic and one for negative-true logic. The two symbols are electrically equivalent; depending on the application, one symbol may represent the logical operation being performed better than the other.

NAND is functionally equivalent to a positive-logic NOR gate. The NAND symbol better expresses the circuit function in the application just described. **Figure 7-9** traces the evolution of an electromechanical switching circuit into a NOR or NAND gate, depending on the logic convention chosen. Notice that the truth tables prove the circuits perform identical functions. Compare the lists of input and output conditions to verify this.

[Now turn to Chapter 10 and study examination questions E7A07 through E7A13. Review this section as needed.]

DIGITAL LOGIC CIRCUITS

Flip-Flops

The output state of a **sequential-logic** circuit is a function of its present inputs and past output states. The dependence on previous output states implies a capability

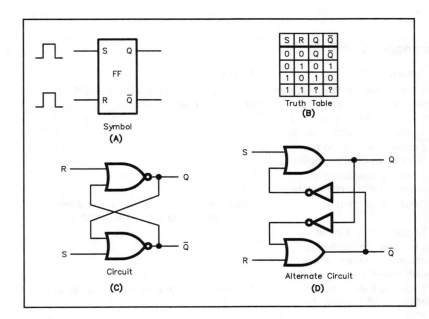

Figure 7-10 — A positive-logic, unclocked R-S flip-flop is used to illustrate the operation of flip-flops in general. Where Q and $\overline{Q}$ are shown in the truth table, the previous output states are retained. A question mark (?) indicates an invalid state, and you cannot be sure what the output will be. C shows an R-S flip-flop made from two NOR gates. The circuit shown at D is another implementation using two OR gates and two inverters.

S	R	Q	$\overline{Q}$
0	0	Q	$\overline{Q}$
0	1	0	1
1	0	1	0
1	1	?	?

Truth Table
(B)

Symbol
(A)

Circuit
(C)

Alternate Circuit
(D)

of, and requirement for, some type of memory.

A **flip-flop** (also known as a **bistable multivibrator**) is a binary sequential-logic element with two stable states: the set state (1 state) and the reset state (0 state). Thus, a flip-flop can store one bit (from *bi*nary digi*t*) of information. This type of circuit is sometimes called a **latch**. The schematic symbol for a flip-flop is a rectangle containing the letters FF, as shown in **Figure 7-10A**. (These letters may be omitted if no ambiguity results.)

Flip-flop inputs and outputs are normally identified by a single letter, as outlined in **Tables 7-1** and **7-2**. A letter followed by a subscripted letter (such as D_C), means *that* input is dependent on the input of the subscripted letter (input D_C is dependent on input C). Note that simultaneous 1 states for the R and S inputs are not allowed. The truth table of Figure 7-10B shows that you can't be sure what the outputs (Q and $\overline{Q}$) will be if both inputs are high at the same time. There are normally two output lines, which are complements of each other, designated Q and $\overline{Q}$ (read as Q NOT). If Q = 1 then $\overline{Q}$ = 0 and vice versa.

Synchronous and Asynchronous Flip-Flops

The terms synchronous and asynchronous are used to characterize a flip-flop or individual inputs to an IC. In *synchronous flip-flops* (also called clocked, clock-driven or gated flip-flops), the output follows the input only at prescribed times determined by the clock input. *Asynchronous flip-flops* are sometimes called unclocked or data-driven flip-flops because the output is determined solely by the inputs. Asynchronous inputs are those that can affect the output state independently of the clock. Synchronous inputs affect the output state only on command of the clock.

Table 7-1

Flip-Flop Input Designations

Input	Action	Restriction
R (Reset)	1 resets the flip-flop. A return to 0 causes no further action.	Simultaneous 1 states for R and S inputs are not allowed.
S (Set)	1 sets the flip-flop. A return to 0 causes no further action.	
R_D (Direct Reset)	1 causes flip-flop to reset regardless of other inputs.	
S_D (Direct Set)	1 causes flip-flop to set regardless of other inputs.	
J	Similar to S input.	Simultaneous 1 states for J and K inputs cause the flip-flop to change states.
K	Similar to the R input.	
G (Gating)	1 causes the flip-flop to assume the state of G's associated input.	
C (Control)	1 causes the flip-flop to assume the state of the D input.	A return to 0 produces no further action.
T (Toggle)	1 causes the flip-flop to change states.	A return to 0 produces no further action.
D (Data)	A D input is always dependent on another input, usually C. C = 1, D = 1 causes the flip-flop to set. C = 1, D = 0 causes the flip-flop to reset.	A return of C to 0 causes the flip-flop to remain in the existing state (set or reset).
H1 (Hold for 1 state)	A 1 input will prevent the flip-flop from being reset after it has been set.	The signal has no effect on the flip-flop if it is in the reset state.
HO (Hold for 1 state)	A 1 input will prevent the flip-flop from being set after it has been reset.	The signal has no effect on the flip-flop if it is in the set state.

Dynamic versus Static Inputs

Dynamic (edge-triggered) inputs are sampled only when the clock changes state. This type of input is indicated on logic symbols by a small isosceles triangle (called a dynamic indicator) inside the rectangle where the input line enters. Unless there is a negation indicator (a small circle outside the rectangle), the 0-to-1 transition is recognized. This is called *positive-edge triggering*. The negation indicator means that the input is *negative-edge triggered*, and is responsive to 1-to-0 transitions.

Static (level-triggered) inputs are recognizable by the absence of the dynamic indicator on the logic symbol. Input states (1 or 0) cause the flip-flop to act.

Table 7-2
Flip-Flop Output Designations

Output	Action	Restrictions
Q (Set)	Normal output	Only two output states are possible: Q = 1 and Q = 0
$\overline{Q}$ (Reset)	Inverted output	Output states are the opposite of Q: $\overline{Q}$ = 0 and $\overline{Q}$ = 1

Notes
1) $\overline{Q}$ is the complement of Q.
2) The normal output is normally marked Q or unmarked.
3) The inverted output is normally marked $\overline{Q}$. If there is a 1 state at Q, there will be a 0 state at $\overline{Q}$.
4) Alternatively the inverted output may have a (negative) polarity indicated (a small right triangle on the outside of the flip-flop rectangle at the inverted output line). For lines with polarity indicators, be aware that a 1 state in negative logic is the same as a 0 state in positive logic. This is the convention followed by the International Electrotechnical Commission.

Many different types of flip-flops exist. These include the clocked and unclocked R-S, D, T, J-K and master/slave (M/S) types. These names come from the type of input lines that the flip-flop has. See Tables 7-1 and 7-2 for a summary of the operation of these lines.

R-S, D and T Flip-Flops

One simple circuit for storing a bit of information is the R-S (or S-R) flip-flop. The inputs for this circuit are set (S) and reset (R). Figure 7-10 shows the schematic symbol for the flip-flop along with a truth table to help you determine the outputs for given input conditions. Two implementations of this circuit using discrete digital logic gates are also shown. When S = 0 and R = 0 the output will stay the same as it was at the last input pulse. This is indicated by a Q in the truth table.

If S = 1 and R = 0, the Q output will change to 1. If S = 0 and R = 1, the Q output will change to 0. If both inputs became 1 simultaneously, the output states would be indeterminate, meaning there is no way to predict how they may change. A clocked R-S flip-flop also has a clock input, in which case no change in the output state can occur until a clock pulse is received.

When a D flip-flop is wired with the Q NOT ($\overline{Q}$) output to the D input, it forms a *toggle* or T flip-flop, also known as a complementing flip-flop. The timing diagram of **Figure 7-11** shows that the flip-flop output changes state with each positive clock pulse. So if the output is 0 initially, it will

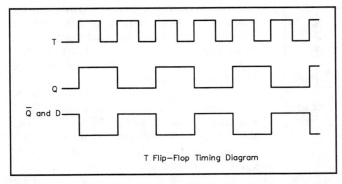

T Flip-Flop Timing Diagram

Figure 7-11 — The timing diagram for a T flip-flop is shown. As you can see, the T flip-flop serves as a divide-by-two counter.

change to a 1 on the leading edge of the first positive clock pulse and it will change back to 0 on the leading edge of the next positive clock pulse. The output of a bistable T flip-flop changes state two times for every two trigger pulses applied to the input. Another way to say this is that the T flip-flop provides one output pulse for every two input pulses. The result is that a bistable multivibrator, such as this flip-flop circuit, electronically divides the input signal by two. Two such flip-flops could be connected to divide the input signal by four, and so on. (There is more about digital frequency-divider circuits later in this chapter.)

There are other types of flip-flop circuits, each with different input and clocking arrangements. They too can be implemented using discrete digital logic elements, although all types of flip-flops are available as single IC packages. Some ICs include several flip-flop circuits in a single package.

One-Shot Multivibrator

A **monostable multivibrator** (or **one-shot**) has one stable state and an unstable (or quasi-stable) state that exists for a time determined by RC circuit components connected to the one-shot. When the time constant has expired, the one-shot reverts to its stable state until retriggered.

In **Figure 7-12**, a 555 timer IC is shown connected as a one-shot multivibrator. The action is started by a negative-going trigger pulse applied between the trigger input and ground. The trigger pulse causes the output (Q) to go positive and capacitor C to charge to two-thirds of V_{CC} through resistor R. At the end of the trigger pulse, the capacitor is quickly discharged to ground. The output remains at logic 1 for a time determined by:

$$T = 1.1\,RC \qquad \text{(Equation 7-1)}$$

where:
R is resistance in ohms.
C is capacitance in farads.
T is time in seconds.

Astable Multivibrator

An **astable** or **free-running multivibrator** is a circuit that alternates between two unstable states. It can be synchronized by an input signal of a frequency that is slightly higher than the astable multivibrator free-running frequency.

An astable multivibrator circuit using the 555 timer IC is shown in **Figure 7-13**. Capacitor C_1 charges to two-thirds V_{CC} through R_1 and R_2, and discharges to one-third V_{CC} through R_2. The ratio (R_1:R_2) sets the duty cycle. The frequency is determined by:

$$f = \frac{1.46}{\left(R_1 + \left(2 \times R_2\right)\right)C_1} \qquad \text{(Equation 7-2)}$$

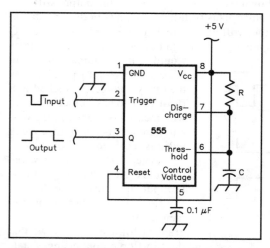

Figure 7-12 — A 555 timer IC can be connected as a one-shot multivibrator. See text for the formula to calculate values for R and C.

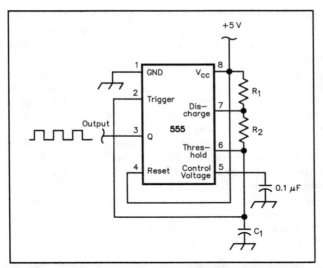

Figure 7-13 — A 555 timer IC can be connected as an astable multivibrator. See text for the formula to calculate values of R_1, R_2 and C_1.

where:
R is resistance in ohms.
C is capacitance in farads.

[Before you go on to the next section, turn to Chapter 10 and study examination questions E7A01 through E7A06. Review this section as needed.]

DIGITAL FREQUENCY-DIVIDER CIRCUITS

A **counter**, **divider** or **divide-by-n counter** is a circuit that stores pulses and produces an output pulse when a specified number (n) of pulses has been stored. In a counter consisting of flip-flops connected in series, when the first stage changes state it affects the second stage and so on.

A ripple, ripple-carry or asynchronous counter passes the count from stage to stage; each stage is clocked by the preceding stage. In a synchronous counter, each stage is controlled by a common clock.

Most counters have the ability to clear the count to 0. Some counters can also be preset to a desired count. Some counters may count up (increment) and some down (decrement). Up/down counters are also available. They are able to count in either direction, depending on the status of a control line.

Internally, a decade counter IC has 10 distinct states. Each input pulse toggles the counter to the next state. Some counters have a separate output pin for each of these 10 states, while others have only one output connected to the last bit of the counter. The last state produces one output pulse for every 10 input pulses.

Circuit Applications

Counter or divider circuits find application in various forms. Common uses for these circuits are in marker generators and frequency counters.

The regulations governing amateur operation require that your transmitted signal be maintained inside the limits of certain frequency bands. The exact frequency need not be known, as long as it is not outside the limits. Staying inside the limits is not difficult to do, and requires only a marker generator or frequency counter, and some care.

Marker Generator

Many receivers and transceivers include a **crystal-controlled marker generator**. This circuit employs a high-stability crystal-controlled oscillator that generates a series of reference signals at known frequency intervals. When these signals are detected in a receiver, they mark the exact edges of the amateur frequency assignments. Most US amateur band limits are exact multiples of 25 kHz, whether at the band extremes or at points marking the subdivisions between types of emission and license-class restrictions. A 25-kHz fundamental frequency will produce the desired marker signals, provided that the oscillator harmonics are strong enough to be heard throughout the desired range. But if the receiver calibration is not accurate enough to positively identify which harmonic you are hearing, there may still be a problem in determining how close to the band edge you are operating.

Rather than using a 25-kHz oscillator, an oscillator frequency of 100 kHz is often used. A divider circuit coupled to the oscillator provides markers at increments of other than 100 kHz. In the circuit of **Figure 7-14**, two divide-by-2 stages are switch selected to produce markers at 50 and 25-kHz points. U3 is a 4013 CMOS dual D-type flip-flop. A 4001 CMOS quad NOR gate and a diode matrix provide the required switching and signal routing.

These D-type flip-flops are wired to form T flip-flops by connecting the $\overline{Q}$

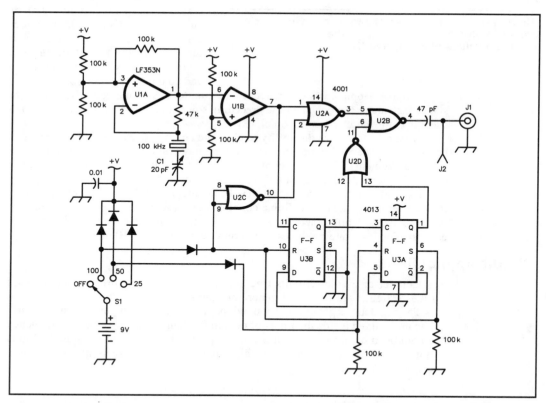

Figure 7-14 — This schematic diagram shows a simple 100, 50 and 25-kHz marker generator. Two switch-selected divide-by-two stages produce the 50 and 25-kHz markers.

output of each to its D input. These flip-flops form two divide-by-two counters to provide 50-kHz and 25-kHz outputs.

A marker-generator oscillator must produce lots of harmonic signals, so you can hear the output across the amateur bands. A sinusoidal crystal oscillator by itself would *not* make a good marker generator. A class C amplifier following the crystal oscillator will generate the required harmonics. You can also build an oscillator from TTL digital ICs that will generate lots of harmonics. The most popular method is to follow the crystal oscillator with a frequency divider circuit, because the output pulses from the frequency divider will contain lots of harmonic energy. For example, you could use a 1-MHz crystal oscillator and a divide-by-ten circuit to produce a 100-kHz signal. You should be able to hear the harmonics of this circuit every 100 kHz with your receiver.

Frequency Counters

One of the most accurate means of measuring frequency is the frequency counter. This instrument actually counts the pulses of the input signal over a specified time, and displays the frequency of the signal, usually on a digital readout. For example, if an oscillator operating at 14.230 MHz is connected to the counter input, 14.230 would be displayed. Some counters are usable well up into the gigahertz range. Most counters that are used at high frequencies make use of a **prescaler** ahead of a basic low-frequency counter. The prescaler divides the high-frequency signal by increments of 10, 100, 1000 or some other amount so that the low-frequency counter can display the input frequency.

Frequency-counter accuracy depends on an internal crystal-controlled reference oscillator. The more accurate the crystal reference, the more accurate the counter readings will be. A crystal frequency of 1 MHz has become more or less standard for use in the reference oscillator. The crystal should have excellent temperature stability so the oscillator frequency won't change appreciably with temperature changes. That is also a very important consideration for a crystal-controlled marker generator. Using a 1-MHz crystal also makes it relatively easy to compare harmonics of the crystal-oscillator signal to WWV or WWVH signals at 5, 10 or 15 MHz, for example, and to adjust the oscillator to zero beat (that is, the frequencies of the beating signals or their harmonics are equal).

A frequency counter will indicate the frequency of the strongest signal on its antenna or input terminal. Most counters receive their input through an antenna placed close to a transmitter, rather than having a direct connection to the transmitter. For low-level signals, however, a probe or other input connection may be used. The measured signal must have a frequency within the measurement range of the frequency counter or prescaler circuit.

[At this time you should turn to Chapter 10 and study examination questions E7F01 through E7F11. Review this section as needed.]

OPERATIONAL AMPLIFIER CIRCUITS

In Chapter 6 we discussed the basic fundamentals of integrated-circuit operational-amplifiers. We also learned how a basic operational amplifier circuit is wired and how to calculate the gain of the circuit. In this chapter we will study some other circuits that use op-amp ICs. As you remember, an op amp is a high-gain, direct-coupled differential amplifier whose characteristics are chiefly determined by components external to the IC.

Op Amps as Active Audio Filters

An active filter is defined as one that has some gain, rather than insertion loss. Op amps are often used to design an active filter because the gain and frequency response of the filter are controlled by a few resistors and capacitors connected external to the op amp. Filters of this type designed for use at audio frequencies are often called RC active audio filters.

Although there are numerous applications for RC active filters, their principal use in amateur work is as audio filters used with receivers. Such filters establish selectivity at audio frequencies, and are useful for receivers that need more selectivity than their IF filters provide. Op amps have the distinct advantage of providing gain and variable parameters when used as audio filters. Passive filters that contain inductive and capacitive elements are generally designed for some fixed frequency and exhibit an insertion loss. Also, LC filters are usually physically larger and heavier than their op-amp counterparts. The design of any filter, or a preselector used with a receiver, involves a trade-off between bandwidth and insertion loss. To achieve narrower bandwidth, an LC filter usually has more insertion loss.

The use of an op-amp IC, such as a type TL081, results in a compact filter section that provides stable operation. Only five connections are made to the IC. The gain of the filter sections and the frequency characteristics are determined by the choice of resistors and capacitors external to the IC.

Figure 7-15 shows a single-section band-pass filter. To select the component values for a specific filter, you must first select the band-pass characteristics: desired filter Q, voltage gain (A_v) and operating frequency (f_o). Next you should choose a value for C_1 and C_2. They have an equal value and should be high-Q, temperature-stable components. Polystyrene capacitors are excellent for such use. Disc-ceramic capacitors are not recommended. Choose standard capacitor values because capacitors with unusual values are not generally available, and it is more difficult to combine capacitors in series or parallel to achieve a desired value than with resistors. R_4 and R_5 are equal in value and are used to establish the op-amp reference voltage, which is $V_{CC} / 2$. Then:

$$R_1 = \frac{Q}{2\pi f_0 C_1}$$
(Equation 7-3)

$$R_2 = \frac{Q}{\left(2Q^2 - A_V\right)\left(2\pi f_0 C_1\right)}$$
(Equation 7-4)

$$R_3 = \frac{2Q}{2\pi f_0 C_1}$$
(Equation 7-5)

$$R_4 = R_5 \approx 0.02 \times R_3$$
(Equation 7-6)

Single filter sections can be cascaded for greater **selectivity**. One or two sections may be used as a band-pass or low-pass section for improving the audio-channel passband characteristics during SSB or AM reception. Up to four filter sections are frequently used to obtain selectivity for CW or RTTY reception. The greater the number of filter sections, up to a practical limit, the sharper the filter skirt response will be. Not only does a well-designed RC filter help to reduce QRM, but it also improves the signal-to-noise ratio in some receiving systems.

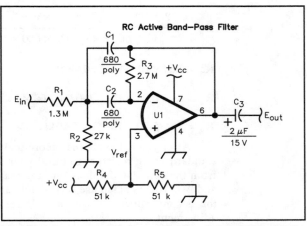

Figure 7-15 — This diagram shows a basic circuit for a single-section, RC active band-pass filter.

The component values shown in Figure 7-15 illustrate the design of a single-section band-pass filter. The value of f_o was chosen as 900 Hz for the calculation, but for CW reception you may prefer frequencies between 200 and 700 Hz. An A_V (gain) of 1 and a Q of 5 were chosen for this example. Both the gain and the Q can be increased for a single-section filter if desired, but for a multisection RC active filter, it is best to restrict the gain to 1 or 2 and limit the Q to no more than 5. This helps prevent unwanted filter "ringing" and audio instability. Filter ringing occurs when the filter shape, as measured in the frequency domain (bandwidth), is too narrow for the signal being received.

Standard-value 680-pF capacitors are chosen for C_1 and C_2. For certain design parameters and C_1-C_2 values, unwieldy resistance values may result. If this happens, select a new value for C_1 and C_2. Use Equations 7-3 through 7-6 to calculate the required resistance values.

$$R_1 = \frac{Q}{2\pi f_0 A_V C_1} = \frac{5}{6.28 \times 900\,\text{Hz} \times 1 \times 680 \times 10^{-12}\,\text{F}} = \frac{5}{3.84 \times 10^{-6}}$$

$$R_1 = 1.30 \times 10^6\,\Omega = 1.3\,\text{M}\Omega$$

$$R_2 = \frac{Q}{\left(2Q^2 - A_V\right)\left(2\pi f_0 C_1\right)} = \frac{5}{\left(2 \times 5^2 - 1\right)\left(6.28 \times 900\,\text{Hz} \times 680 \times 10^{-12}\,\text{F}\right)}$$

$$R_2 = \frac{5}{\left(2 \times 25 - 1\right)\left(3.84 \times 10^{-6}\right)} = \frac{5}{49 \times 3.84 \times 10^{-6}} = \frac{5}{1.88 \times 10^{-4}}$$

$$R_2 = 2.65 \times 10^4\,\Omega = 26.5\,\text{k}\Omega$$

$$R_3 = \frac{2Q}{2\pi f_0 C_1} = \frac{2 \times 5}{6.28 \times 900\,\text{Hz} \times 680 \times 10^{-12}\,\text{F}} = \frac{10}{3.84 \times 10^{-6}}$$

$$R_3 = 2.60 \times 10^6\,\Omega = 2.60\,\text{M}\Omega$$

$$R_4 = R_5 \approx 0.02 \times R_3 = 0.02 \times 2.60 \times 10^6\,\Omega$$

$$R_4 = R_5 \approx 5.20 \times 10^4\,\Omega = 52.0\,\text{k}\Omega$$

After you calculate the theoretical values for the resistors, select the nearest standard-value resistors to build your filter. The actual values can vary slightly from the calculated values, but the main effect of this is a slight alteration of f_o and A_V. You can also use series and parallel combinations of resistors to achieve the design values, or use variable resistors in the proper range to allow filter adjustment.

In use, an RC active filter should be inserted in the low-level audio stages. This prevents overloading the filter during strong-signal reception, such as would occur if the filter were placed at the audio output, just before the speaker or phone jack. The receiver AF gain control should be placed between the audio preamplifier and the input of the RC active filter for best results. If audio-derived AGC is used in the receiver, the RC active filter will give its best performance when it is contained within the AGC loop.

[Before proceeding to the next section, turn to Chapter 10 and study examination questions E7G01 through E7G11. If you have trouble with any of these questions, review this section.]

AMPLIFIERS

The energy picked up by a receiving antenna is extremely small. Therefore, the strength of the signals must be increased by a large factor in a receiver — often a million or more times. You might think that this amplification process could be carried on indefinitely, so that even the weakest radio signals could be brought up to usable strength. Unfortunately, there are limitations on how much you can amplify a signal.

Noise across the radio spectrum limits the amount of useful amplification. Unlimited amplification of a particular frequency cannot reduce the noise mask covering a desired signal, since both are RF emissions. Where does this noise come from? Electrical currents generated in nature (static) and in electrical circuits and devices all produce noise. Occurring at all frequencies, this noise is inescapable.

Amateurs don't give up in their battle against noise, though, because there are means to make it less bothersome. One of the big objectives in designing good receivers is to improve the signal-to-noise ratio. This means amplifying the signal as much as possible while amplifying the noise as little as possible.

Amplifier Operating Class

The basic operation of an amplifier is selected by choosing a bias or dc input when there is no signal applied to the amplifier. In the most common amplifier configuration, tubes and PNP transistors are operated with a slightly negative bias voltage applied to the grid or base, and NPN transistors with a slightly positive bias voltage applied to the base. The actual bias voltage depends on the amplifying device used and the type of amplification desired.

The amplifier operating characteristics define several classes of operation. The three basic operating classes are Class A, Class B and Class C. Another class of operation is sometimes added between the A and B categories, and this is called Class-AB operation.

To understand what we mean by operating class, let's consider something called an **amplifier transfer function**. This is simply a way of expressing the amplifier output in terms of the input. All amplifiers have their own transfer function, but they all have certain characteristics in common.

There is usually a point at which a bigger input signal will not produce a bigger output signal. This is called the *saturation region*. There is also a point at which the output will not decrease if a weaker (or no) input signal is supplied. This is called the *cutoff region*.

Between saturation and cutoff is an area where the input and output signals vary in a linear fashion. Of course, this is called the *linear region*. In practice, the saturation region is considered to be that area where the output signal no longer increases linearly with increasing input signals. Likewise, the cutoff region is the area below the linear-response region. **Figure 7-16** illustrates a typical transfer function, showing the important regions on the curve.

Two factors about the input are extremely important for establishing the type of operation required of a specific amplifier. The first, called the bias, is the average signal input level, or the dc input level. The second is the size of the input signal. These levels must be set carefully to produce the desired type of operation.

With a Class-A amplifier, the bias level and input-signal amplitude are set so that all of the input signals appear between the saturation and cutoff regions. This means the amplifier is operating in the linear region of its transfer function, and the output is a linear (but larger) reproduction of the input signal. There is an output signal for the full 360° of input signal. At its weakest point, the input signal is large enough to produce an output signal, and at its strongest point the input signal does not drive the output signal into the saturation region.

That portion of the input cycle for which there is an output is called the conduction angle. For a Class-A amplifier, the conduction angle is 360°. **Figure 7-17A** is a graph of the output current from a Class-A amplifier. The saturation and cutoff levels are indicated. The *quiescent point* sets the operating bias point, and represents the

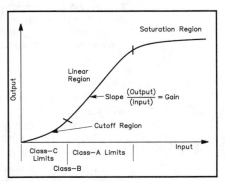

Figure 7-16 — A typical transfer-function graph displaying output versus input for different amplifier classes.

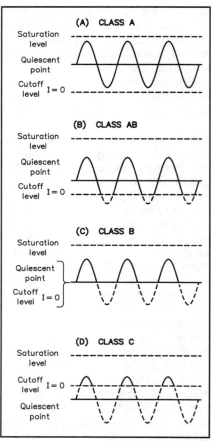

Figure 7-17 — Amplifying-device output current for various classes of operation. All graphs assume a sine-wave input signal.

amplifier voltage or current conditions with no input signal.

The efficiency of a Class-A amplifier is low, because there is always a significant amount of current drawn from the power supply — even with no input signal. This no-signal current is called the quiescent current of the amplifier. The maximum theoretical efficiency of a Class-A amplifier is 50%, but in practice it is more like 25 to 30%.

For a Class-AB amplifier, the drive level and dc bias are adjusted so output current flows for more than half the input cycle, but not for the entire cycle. The conduction angle is between 180° and 360°, and the operating efficiency is often more than 50%. Figure 7-17B shows the output signal for a Class-AB amplifier.

Class-B operation sets the bias right at the cutoff level. In this case, there is output current only during half the input sine wave. This represents a conduction angle of 180°. See Figure 7-17C. The output is not as linear as with a Class-A amplifier, but is still acceptable for many applications. The advantage is increased efficiency; up to 65% efficiency is theoretically possible with a Class-B amplifier, and practical amplifiers often attain 60% efficiency.

Class-C amplification requires that the bias be well below cutoff on the transfer-characteristic curve, and that the signal be large enough to bring some part of the top half into the conduction region.

A Class-C amplifier has a conduction angle of less than 180°. The output current will just be pulses at the signal frequency, as shown in Figure 7-17D. The amplifier is cut off for considerably more than half the cycle, so the operating efficiency can be quite high — up to 80% with proper design. Linearity is very poor, however.

The linearity of the amplifier stage is important, because it describes how faithfully the input signal will be reproduced at the output. Any nonlinearity results in a distorted output. So you can see that a Class-A amplifier will have the least amount of distortion, while a Class-C amplifier produces a severely distorted output. A side effect of this distortion or nonlinearity is that the output will contain harmonics of the input signal. Odd-order intermodulation products are formed, which are close to the desired frequency, and so will not be filtered out by a resonant tank circuit. So a pure sine wave input signal becomes a complex combination of sine waves at the output.

By now you are probably asking, "But why would anyone want to have an amplifier that generates a distorted output signal?" That certainly is a good question. At first thought, it sure doesn't seem like a very good idea. You must remember, however, that every circuit design consists of compromises between fundamentally opposing ideals. We would like to have perfect linearity for our amplifiers, but we would also like them to have 100% efficiency. You have just learned that those two ideals are exclusive. The closer you get to one of them, the further you get from the other. So you will have to compromise the ideals a bit to achieve a workable design. Your particular application and circuit conditions help you determine which compromises to make.

An important point to keep in mind is that most RF amplifiers will have a tuned-output tank circuit. That tuned circuit stores electrical energy like a mechanical flywheel, which is used to store mechanical energy. A heavy wheel, with its mass concentrated as close to the outer rim as possible, will have a large moment of inertia. That means that it has a lot of angular momentum when it is spinning, so it will continue to spin unless there is something to make it stop. If you turn off the motor that was used to set the flywheel in motion, the wheel will continue to spin for a long time. (In the ideal condition, where there is no friction, and the wheel is spinning on perfect bearings, it would never stop.) Your car engine uses a flywheel to store some of the energy from the gasoline exploding in the engine, and keep the system spinning smoothly. Without a flywheel, there would be a noticeable "bump" every time a cylinder fires.

A parallel tuned circuit is the electrical equivalent of a flywheel. The electrons that make up the current in the tank circuit oscillate, or circulate, back and forth in the inductance (L) and capacitance (C) of the parallel-resonant circuit. (This type of circuit is often called a tank circuit because you can also think of it as a storage tank for electrical energy.) By placing such a circuit in series with the amplifier output, you can use it to smooth out the "bumps" that will occur when there is no output because the amplifier is turned off. This is especially useful if you are amplifying a pure sine-wave signal, such as for CW, and want to take advantage of the increased efficiency offered by a Class-C amplifier.

The electric current flowing in the circuit is passing the energy back and forth between the electric field of the capacitor and the magnetic field of the inductor. The signal source only has to supply small amounts of energy to keep the current flowing. The tuned tank circuit will filter out the unwanted harmonics generated by a nonlinear amplifier stage. Usually, the tank circuit should have a Q of at least 12 to reduce these harmonics to an acceptable level.

If you are amplifying an audio signal, linearity may be the most important consideration. Use a Class-A amplifier for audio stages. For an AM or SSB signal, which is an RF signal envelope that varies at an audio rate, you would want to use a linear amplifier. You may be willing to accept a bit of nonlinearity to obtain increased efficiency, in which case Class-AB operation would be indicated. In some instances, you may even be willing to use a Class-B amplifier, and let the flywheel effect of a tank circuit fill in the waveform voids for you.

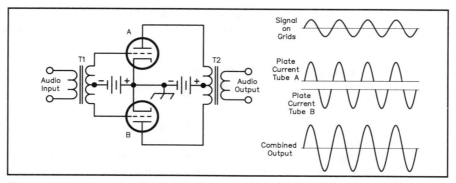

Figure 7-18 — Two Class-B amplifiers connected to operate as a push-pull amplifier. The graphs show current waveforms at the points indicated.

You can take advantage of the nonlinearity of a Class-C amplifier by using it as a frequency-multiplier stage. We mentioned earlier that one consequence of a nonlinear amplifier is that it would generate harmonic signals. If you want to multiply the frequency for operation on another band, sometimes you can do that by using a tank circuit tuned to a harmonic of the input frequency, and selecting the appropriate harmonic from the output of a Class-C amplifier. This is especially useful for generating an FM signal at VHF or UHF, where you may start with a signal in the HF range, apply modulation through a reactance modulator, and then multiply the signal frequency to the desired range on 2 meters or higher. The exact bias point for the Class-C amplifier will determine which harmonic frequency will be the strongest in the output, so careful selection of bias point is important.

Class-B amplifiers are often used for audio frequencies by connecting two of them back to back in push-pull fashion. **Figure 7-18** illustrates a simple triode-tube push-pull amplifier and the waveforms associated with this type of operation. A push-pull amplifier can also be used at RF. While one tube is cut off, the other is conducting, so both halves of the signal waveform are present in the output. This reduces the amount of distortion in the output, and will reduce or eliminate even-order harmonics.

Voltage, Current and Power Amplification

When we talk about amplifiers, it is common to think of a *power* amplifier. An amateur amplifier is said to be "1 W in, 10 W out," for example. We can also build circuits to amplify voltage or current, however. An instrumentation amplifier may be designed to amplify a very small voltage so that it can be measured with a voltmeter; in a multistage amplifier, we may wish to amplify the output current from one stage to drive the next stage. *Stage*, by the way, is the name we give to one of a number of signal-handling sections used one after another in an electronic device. The instrumentation amplifier, for example, may increase the voltage level but not the available current, so the available power is not increased significantly.

The input and output impedance of an amplifier circuit will vary, depending on what type of amplification the circuit is designed to provide. Generally speaking, voltage amplifiers have a very low input impedance and a high output impedance. If too much current is drawn from the output of a voltage amplifier, its operation can be affected. Current and power amplifiers have a lower output impedance, and can be used to supply both voltage and current to a load.

Amplifier Gain

The *gain* of an amplifier is the ratio of the output signal to the input signal. Voltage amplifier gain is based on the ratio of output and input voltages, current amplifier gain on the ratio of output and input current levels and power amplifier gain is determined by the ratio of output and input power levels.

This ratio can get to be a very large number when several stages are combined or when the gain is very large. The decibel expresses the ratio in terms of a logarithm, making the number smaller and easier to work with. We often state the gain of a stage as a voltage gain of 16 or a power gain of 250, which are ratios, or refer to a gain of 24 dB, for example.

[Turn to Chapter 10 and study questions E7B01, E7B02, E7B05, E7B06 and E7B08. Review this section as needed.]

Transistor Amplifiers

We are limiting our discussion here to the operation of bipolar-transistor-amplifier circuits, but many of the techniques and general circuit configurations also apply to tube-type amplifiers. One major difference between tube-type amplifiers and transistor amplifiers is that tubes are voltage-operated devices, but bipolar transistors are current operated. A transistor amplifier is essentially a current amplifier. To use it as a voltage amplifier, the current is drawn through a resistor, and the resulting voltage drop provides the amplifier output. FET amplifiers operate as voltage amplifiers. To learn more about how tube-type amplifiers function, we recommend that you turn to appropriate sections of *The ARRL Handbook*. At some point in your Amateur Radio career, you may want to learn about tube amplifiers, but for the purpose of helping you pass your Extra class exam, we will concentrate on transistor circuits.

Bipolar-transistor diode junctions must be forward biased in order to conduct significant current. If your circuit includes an NPN transistor, the collector and base must be positive with respect to the emitter, and the collector must be more positive than the base. When working with a PNP transistor, the base and collector must be negative with respect to the emitter and the collector more negative than the base. The required bias is provided by the collector-to-emitter voltage, and by the emitter-to-base voltage. These bias voltages cause two currents to flow: Emitter-to-collector current and emitter-to-base current. Either type of transistor, PNP or NPN, can be used with a negative- or positive-ground power supply. Forward bias must still be maintained, however. Remember that the amount of bias current sets the class of amplifier operation.

The lower the forward bias, the less collector current will flow. As the forward bias is increased, the collector current rises and the junction temperature rises. If the bias is continuously increased, the transistor eventually overloads and burns out. This condition is called thermal runaway. To prevent damage to the transistor, some form of bias stabilization should be included in a transistor amplifier design.

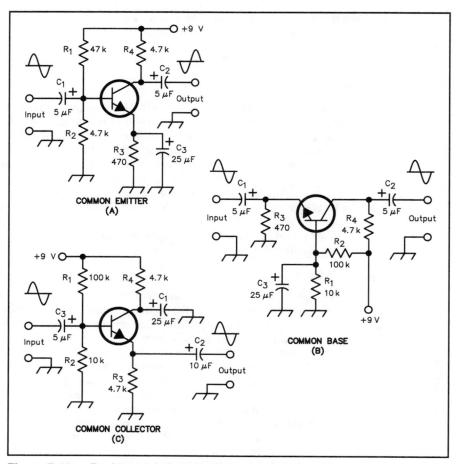

Figure 7-19 — Basic transistor-amplifier circuits. Typical component values are given for use at audio frequencies where these circuits are often used. The input and output phase relationship of each amplifier is shown.

Even if the bias is not increased, however, thermal runaway can occur. As the transistor heats up, its beta increases, causing more collector current to flow. This causes more heating, even higher beta and even more current, until eventually the transistor burns out.

Amplifier circuits used with bipolar transistors fall into one of three types, known as the common-base, common-emitter and common-collector circuits. These are shown in **Figure 7-19** in elementary form. The three circuits correspond approximately to the grounded-grid, grounded-cathode and cathode-follower vacuum-tube circuits, respectively.

Common-Emitter Circuit

The common-emitter circuit is shown in Figure 7-19A. The base current is small and the input impedance is fairly high — several thousand ohms on average. The collector resistance is some tens of thousands of ohms, depending on the signal-circuit source impedance. The common-emitter circuit has a lower cutoff frequency than does the common-base-circuit, but it gives the highest power gain of the three configurations.

In this circuit, the output (collector) current phase is opposite to that of the input (base) current. So any feedback through the small emitter resistance is negative and that will stabilize the amplifier, as we will show by an example.

Common-emitter amplifiers are probably the most common bipolar-transistor-amplifier type, so we will use this circuit to illustrate some of the design procedures. R_1 and R_2 make up a divider network to provide base bias. These resistors provide a fixed, stable operating point, and tend to prevent thermal runaway. R_1 and R_2 are sometimes called *fixed-bias* resistors. R_3 provides the proper value of bias voltage, when normal dc emitter current flows through it. This biasing technique is often referred to as *self bias*. C_3 is used to bypass the ac signal current around the emitter resistor. This *emitter-bypass* capacitor should have low reactance at the signal frequency. C_1 and C_2 are coupling capacitors, used to allow the desired signals to pass into and out of the amplifier, while blocking the dc bias voltages. Their values should be chosen to provide a low reactance at the signal frequency.

The emitter-to-base resistance is approximately:

$$R_{e-b} = \frac{26}{I_e} \qquad \text{(Equation 7-7)}$$

where I_e is the emitter current in milliamperes. The voltage gain is the ratio of collector load resistance to (internal) emitter resistance. The letter A represents amplifier gain, which you can think of as *amplification factor*, to help you identify the meaning of the symbol in the equation.-

$$A_V = \frac{R_L}{R_{e-b}} \qquad \text{(Equation 7-8)}$$

For the example of Figure 7-19A, if the emitter current is 1.6 mA, then R_{e-b} is 16.25 Ω, and

$$A_V = \frac{R_L}{R_{e-b}} = \frac{4.7 \text{ k}\Omega}{16.25 \ \Omega} = \frac{4.7 \times 10^3 \ \Omega}{16.25 \ \Omega} = 289$$

If you would like to express this voltage-gain ratio in decibels, just take the logarithm of 289, and multiply by 20:

Gain in dB = $20 \times \log (289) = 20 \times 2.46 = 49$ dB.

The base input impedance is given by:

$$R_b = \beta \ R_{e-b} \qquad \text{(Equation 7-9)}$$

For our example, $R_b = 1625 \ \Omega$, assuming beta = 100. The actual amplifier input impedance is found by considering R_1 and R_2 to be in parallel with this resistance, so the input impedance for our amplifier is about 1177 Ω.

If you omit the emitter bypass capacitor, then all of the emitter signal current must flow through R_3. This resistor dominates the emitter impedance in the voltage gain equation, so we have:

$$A_V = \frac{R_L}{R_E}$$

(Equation 7-10)

Now we obtain a gain of 10 from our amplifier (20 dB). The base impedance of the unbypassed-emitter circuit becomes $\beta R_E = 47 \text{ k}\Omega$, which when combined with the bias resistors, gives an amplifier input impedance of 3.92 kΩ. Notice that the unbypassed emitter resistor introduced 29 dB of negative feedback. This has the effect of stabilizing the gain and impedance values over a wide frequency range. A frequently used amplifier design splits the emitter resistor into two series resistors. The first, connected to the emitter, is unbypassed, and the second, which connects to ground, is bypassed with a capacitor.

Figure 7-20 shows a set of characteristic curves for a typical NPN transistor amplifier. The lines on this graph show how the collector current changes as the collector to emitter voltage varies. Since the base current controls the collector current for this amplifier, there are individual lines to represent various values of base current. The collector supply voltage and bias resistors establish the quiescent point, or operating point for the amplifier. The amplifier load lines represent the amplifier operation. The changing conditions of an ac input signal move the load line along the bias curve. For best efficiency and stability, you usually want to operate the amplifier at a point on the load line that is just below saturation.

Common-Base Circuit

The input circuit of a common-base amplifier must be designed for low impedance. Equation 7-7 can be used to calculate the base-emitter junction resistance. The optimum output load impedance, R_L, may range from a few thousand ohms to

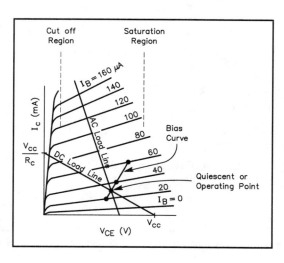

Figure 7-20 — This transistor characteristic curve shows the Q (quiescent) point or operating point for a particular amplifier. The base bias current varies with the input signal, moving the load lines along the bias curve.

100,000 Ω, depending on the circuit. In the common-base circuit (Figure 7-19B), the phase of the output (collector) current is the same as that of the input (emitter) current. The parts of these currents that flow through the base resistance are likewise in phase, so the circuit tends to be regenerative and will oscillate if the current amplification factor is greater than one. You will notice that the bias resistors have much the same configuration as with the common-emitter circuit, and that input and output-coupling capacitors are used. C_3 bypasses the base bias resistor, placing the base at ac-ground potential.

Common-Collector Circuit

The common-collector transistor amplifier, sometimes called an emitter-follower amplifier, has high input impedance and low output impedance. The input resistance depends on the load resistance, being approximately equal to the load resistance divided by $(1 - \alpha)$. The fact that input resistance is directly related to the load resistance is a disadvantage of this type of amplifier, especially if the load is one whose resistance or impedance varies with frequency.

The cutoff frequency of the common-collector circuit is the same as in the common-emitter amplifier. The input and output currents are in phase, as shown in Figure 7-19C. C_1 is a collector bypass capacitor. This amplifier also uses input- and output-coupling capacitors, C_2 and C_3. R_3 is a feedback resistor, and is also used to develop an output voltage for the next stage. Because of this use to develop the output voltage, R_3 is often called the *emitter load* resistor for the common collector circuit.

[Study the examination question E7B03 and questions E7B09 through E7B13 in Chapter 10 now. Review this section as needed.]

Amplifier Stability

Excessive gain or undesired feedback may cause amplifier instability. Oscillation may occur in unstable amplifiers under certain conditions. Damage to the active device from over-dissipation is only the most obvious effect of oscillation. Deterioration of noise figure, spurious signals generated by the oscillation and reradiation of the oscillation through the antenna, causing RFI to other services, can also occur from amplifier instability. Negative feedback will stabilize an RF amplifier. Care in terminating both the amplifier input and output can produce stable results from an otherwise unstable amplifier. Attention to proper grounding and proper isolation of the input from the output by means of shielding can also yield stable operating conditions.

Neutralization

A certain amount of capacitance exists between the input and output circuits in any active device. In the bipolar transistor it is the capacitance between the collector and the base. In an FET it's the capacitance between the drain and gate. In vacuum tubes it is the capacitance between the plate and grid circuit. So far we have simply ignored the effect that this capacitance has on the amplifier operation. In fact, it doesn't have much effect at the lower frequencies. Above 10 MHz or so, however, the capacitive reactance may be low enough to cause complications. Os-

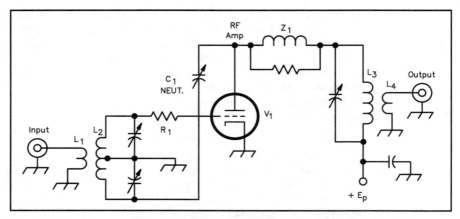

Figure 7-21 — An example of a neutralization technique used with a tube-type RF amplifier.

cillations can occur when some of the output signal is fed back in phase, so that it adds to the input (positive feedback). As the output voltage increases so will the feedback voltage: The circuit adds fuel to its own fire and the amplifier is now an oscillator. The output signal is no longer dependent on the input signal, and the circuit is useless as an amplifier.

In order to rid the amplifier of this positive feedback, it is necessary to provide a second feedback path, which will supply a signal that is 180° out of phase with the positive feedback voltage. This is called *negative feedback.* This path should supply a voltage that is equal to that causing the oscillation, but of opposite polarity.

One **neutralization** technique for vacuum-tube amplifiers is shown in **Figure 7-21**. In this circuit, the neutralization capacitor is adjusted to have the same value as the interelectrode capacitance that is causing the oscillation.

With solid-state amplifiers, a similar technique could be used, although the interelement capacitances tend to be much smaller. It is more common to include a small value of resistance in either the base or collector lead of a low-power amplifier. Values between 10 and 20 Ω are typical. For higher power levels (above about 0.5 W), one or two ferrite beads are often used on the base or collector leads.

Parasitic Oscillations

Oscillations can occur in an amplifier on frequencies that have no relation to those intended to be amplified. Oscillations of this sort are called **parasitics** mainly because they absorb power from the circuits in which they occur. Parasitics are brought on by resonances that exist in either the input or output circuits. They can also occur below the operating frequency, which is usually the result of an improper choice of RF chokes and bypass capacitors. High-Q RF chokes should be avoided, because they are most likely to cause a problem.

Parasitics are more likely to occur above the operating frequency as a result of stray capacitance and lead inductance along with interelectrode capacitances. In some cases it is possible to eliminate such oscillations by changing lead lengths or the position of leads so as to change the capacitance and inductance values. An effective method with vacuum tubes is to insert a parallel combination of a small coil and a resistor in series with the grid or plate lead. The coil serves to couple the VHF energy to the resistor, and the resistor value is chosen so that it loads the VHF circuit so heavily that the oscillation is prevented. Values for the coil and resistor have to be found experimentally as each different layout will probably require different suppressor networks.

With transistor circuits, ferrite beads are often used on the device leads, or on a short connecting wire placed near the transistor. These beads act as a high impedance to the VHF or UHF oscillation, and block the parasitic current flow. In general, proper neutralization will help prevent parasitic oscillations.

[Before proceeding, turn to Chapter 10 and study examination questions E7B04 and E7B07. Review this section as needed.]

IMPEDANCE MATCHING NETWORKS

When most hams talk about impedance matching networks, they probably think of a circuit used between a transmitter or transceiver and an antenna system. Such an *antenna-coupling circuit* has two basic purposes: (1) to match the output impedance of a power-amplifier tube or transistor to the input impedance of the antenna feed line, so the amplifier has a proper resistive load, and (2) to reduce unwanted emissions (mainly harmonics) to a very low value.

Most tube-type transmitters or amplifiers use **pi-network output-coupling circuits** (**Figure 7-22**). A pi-network consists of one inductor and two capacitors or two inductors and one capacitor. As used to match a transmitter output impedance with an antenna system impedance, a pi-network consists of one capacitor in parallel with the input and another capacitor in parallel with the output. An inductor is in series between the two capacitors. The circuit is called a pi network because it resembles the Greek letter pi (π) — if you use your imagination a bit while you look at the two capacitors drawn down from the ends of the horizontally drawn inductor. Because of the series coil and parallel capacitors, this circuit acts as a low-pass filter to reduce harmonics, as well as acting as an impedance-matching device. (A pi network with two coils shunted to ground and a series capacitor would make a high-pass filter, and is virtually never used as an amateur output-coupling circuit.) The circuit Q will be equal to the plate load impedance divided by the reactance of C_1. Coupling is adjusted by varying C_2, which generally has a reactance somewhat less than the load resistance (usually 50 Ω). Circuit design information for pi networks appears in *The ARRL Handbook*.

Harmonic radiation can be re-

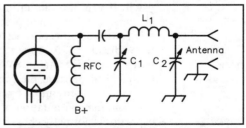

Figure 7-22 — A pi-network output-coupling circuit. C_2 is the coupling (loading) control and C_1 adjusts the tuning.

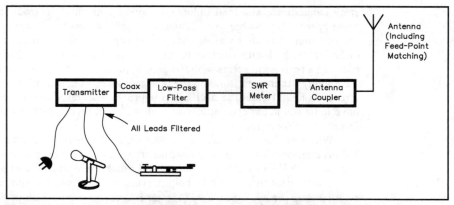

Figure 7-23 — Proper connecting arrangement for a low-pass filter and antenna coupler. The SWR meter is included for use as a tuning indicator for the antenna coupler.

duced to any desired level by sufficient shielding of the transmitter, filtering of all external power and control leads, and inclusion of a low-pass filter (of the proper cutoff frequency) connected with shielded cable to the transmitter antenna terminals (see **Figure 7-23**). Unfortunately, low-pass filters must be operated into a load of close to their design impedance or their filtering properties will be impaired, and damage may occur to the filter if high power is used. For this reason, if the filter load impedance is not within limits, a device must be used to transform the load impedance of the antenna system (as seen at the transmitter end of the feed line) into the proper value. For this discussion we will assume this impedance to be 50 Ω. This device is usually an external antenna-matching network.

Impedance-matching devices are variously referred to as Transmatches, matchboxes or antenna couplers. Whatever you call it, an impedance-matching unit transforms one impedance to be equivalent to another. To accomplish this it must be able to cancel reactances (provide an equal-magnitude reactance of the opposite type) and change the value of the resistive part of a complex impedance.

The **L network**, which consists of an inductor and a capacitor, (**Figure 7-24A**) will match any unbalanced load with a series resistance higher than 50 Ω. (At least it will if you have an unlimited choice of values for L and C.) Most unbalanced antennas will have an impedance that can be matched with an L network. To adjust this L network for a proper match, the coil tap is moved one turn at a time, each time adjusting C for lowest SWR. Eventually a combination should be found that will give an acceptable SWR value. If, however, no combination of L and C is available to perform the proper impedance transformation, the network may be reversed input-to-output by moving the capacitor to the transmitter side of the coil.

The major limitation of an L network is that a combination of inductor and capacitor is normally chosen to operate on only one frequency band because a given LC combination has a relatively small impedance-matching range. If the operating frequency varies too greatly, a different set of components will be needed.

If you are using balanced feed lines to your antenna, they may be tuned by means of the circuit shown in Figure 7-24B. A capacitor may be added in series with the input to tune out link inductance. As with the L network, the coil taps and tuning-capacitor settings are adjusted for lowest SWR, with higher impedance loads being tapped farther out from the coil center. For very low load impedances, it may be necessary to put C_1 and C_2 in series with the antenna leads (with the coil taps at the extreme ends of the coil).

You can convert an L network into a pi network by adding a variable capacitor to the transmitter side of the coil. Using this circuit, any value of load impedance (greater or less than 50 Ω) can be matched using some values of inductance and capacitance, so it provides a greater impedance-transformation range. Harmonic suppression with a pi network depends on the impedance-transformation ratio and the circuit Q.

If you need more attenuation of the harmonics from your transmitter, you can add an L network in series with a pi network, to build a pi-L network. **Figure 7-25A** shows a pi network and an L network connected in series. It is common to include the value of C_2 and C_3 in one variable

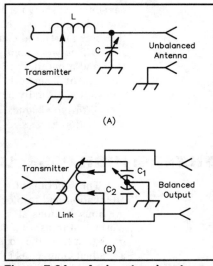

Figure 7-24 — An L-network antenna coupler, useful for an unbalanced feed system is shown at A. B shows an inductively coupled circuit for use with a balanced feed line.

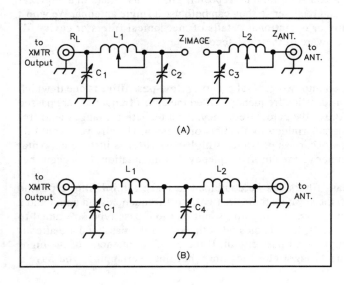

Figure 7-25 — The pi-L network uses a pi network to transform the transmitter output impedance (R_L) to the image impedance (Z_{IMAGE}). An L network transforms Z_{IMAGE} to the antenna impedance, Z_{ANT}.

capacitor, as shown at Figure 7-25B, so the pi-L network consists of two inductors and two capacitors. The pi-L network provides the greatest harmonic attenuation of the three most-used matching networks — the L, pi and pi-L networks.

A T-network configuration consists of two capacitors in series with the signal lead and a parallel, or shunt-connected inductor between them to ground. While this type of T-network will transform impedances, it also acts as a high-pass filter, and so it is virtually never used in an amateur antenna-matching network.

[Turn to Chapter 10 and study the questions E7C01 through E7C05. Review this section as needed.]

FILTERS: HIGH PASS, LOW PASS AND BAND PASS

The function of a filter is to transmit a desired band of frequencies without attenuation and to block all other frequencies. The resonant circuits discussed in previous sections all do this. The term *filter* is frequently reserved for those networks that transmit a desired band with little variation in output, and in which the transition from the "pass" band to the "stop" band is very sharp, rather than being a gradual change, as is the case with simple resonant circuits.

In this section you will learn about *passive filters* and *active filters*. Passive filters always result in some loss of signal strength at the desired signal frequencies. This is called *insertion loss*. Active filters include an amplifying device, to overcome the filter insertion loss, and sometimes even provide signal gain.

The passive filters discussed in this section use inductors and capacitors to set the filter frequency response. There are other types of passive filters, however. For example, *mechanical filters* used to be very popular as receiver IF filters, although they are seldom used in modern receivers. A mechanical filter uses mechanically resonant disks at the design frequency, and a pair of electromagnetic transducers to change the electrical signal into a mechanical wave and back again. *Cavity filters* are used to build a duplexer for a 2-meter repeater. The cavity filters in a duplexer isolate the transmitter and receiver so they can both use a single antenna. We won't go into the construction or operational details of mechanical filters or cavity filters, but you should be aware of their existence.

Filter Classification

Filters are classified into two general groups. A **low-pass filter** is one in which all frequencies below a specified frequency (called the cutoff frequency) are passed without attenuation. Above the cutoff frequency, the attenuation changes with frequency in a way that is determined by the network design. Usually you want this transition region to be as sharp as possible. A **high-pass filter** is just the opposite; there is no attenuation above the cutoff frequency, but attenuation does occur below that point.

High- and low-pass filters can be combined to make a third filter type, the **band-pass filter**. With these filters there are two cutoff frequencies, an upper and a lower one. If a pair of filters, one high-pass and one low-pass type, are tuned to have overlapping passbands, frequencies on both sides of the passband are attenuated. This makes a simple band-pass circuit. If the cutoff frequencies of the high- and low-pass filters are brought close together, but not overlapping, you have a

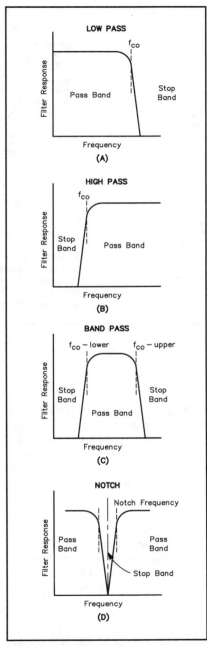

LOW PASS

(A)

HIGH PASS

(B)

BAND PASS

(C)

NOTCH

(D)

Figure 7-26 — Ideal filter-response curves for low-pass, high-pass, band-pass and notch filters.

notch filter. In this case, signals on either side of a certain band are passed, while signals in the middle are attenuated. **Figure 7-26** illustrates idealized response curves for the four types of filters.

Modern Network Design

Modern filter network designs use techniques based on exact mathematical equations that can be applied to filter characteristics. Some examples are filters based on equations called Butterworth and Chebyshev polynomials and elliptical functions. With these mathematical techniques, it is possible to build a catalog of filter characteristics, with appropriate component values, and to select a design from the tabulated data. Tables summarizing these computations can be found in *The ARRL Handbook,* and other reference books.

There are many kinds of so-called "modern filters," and they are usually referred to by the name of the mathematical function used to calculate the design. **Butterworth**, **Chebyshev** and **elliptical filters** are three kinds that have many applications in Amateur Radio. A Butterworth filter is used when you want a response that is as flat as possible in the passband, with no ripple. (Ripple is a variation of attenuation, and you could get these "ups and downs" inside the passband and/or outside the passband.) Unfortunately, the transition from passband to stopband is not very sharp with a Butterworth filter. The Chebyshev filter has a sharp cutoff, but with some ripple in the passband. Higher SWR increases the ripple. The elliptical filter has the sharpest cutoff, but it has ripple in the passband and stopband. The elliptical filter also has one or more infinitely deep notches in the stopband, which can be positioned at specific frequencies that you want to attenuate. **Figure 7-27** illustrates the filter-response curves for these three types.

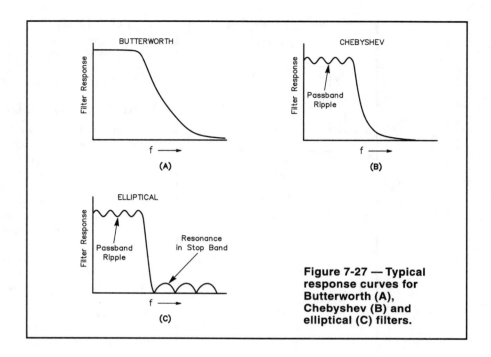

Figure 7-27 — Typical response curves for Butterworth (A), Chebyshev (B) and elliptical (C) filters.

Crystal Filters

The filters we have been talking about so far all use electronic circuits to provide the filtering action. The IF section of a radio requires very good band-pass filters to provide the narrow bandwidth needed for a top-performance rig. Filters that do not use inductors and capacitors as the primary circuit elements are often used in the IF section of a radio. In Chapter 6, we described crystal filters and their operation. Filters of this type, with piezoelectric quartz crystals to provide high-Q, narrow-bandwidth characteristics are often used in modern receivers and transmitters.

Active Filters

Operational amplifiers (op amps) are often used to build *active filters* for the audio range because the gain and frequency response of the filter are controlled by a few resistors (R) and capacitors (C) connected to the op amp. Such filters are often called RC active audio filters. **Figure 7-28** is a simple RC active band-pass filter suitable for CW use.

You can also build an RC active notch filter using an op amp. A notch filter is used to reject a narrow band of frequencies. This type of filter is particularly useful to attenuate an interfering carrier signal while receiving an SSB transmission.

Active filters have a number of advantages over passive (LC) filters at audio frequencies. They provide gain and good frequency-selection characteristics. They do not require the use of inductors, and they can be accurately tuned to a specific design frequency using a potentiometer.

There are a few disadvantages to using active filters as well. The useful upper frequency is limited to a few hundred kilohertz with low-cost op amps. The output voltage swing must be less than the dc supply voltage. Strong out-of-band input signals may overload the active device and distort the output signal. The op amp may generate some noise and add that to the signals, resulting in a lower signal-to-noise ratio than you would have with an LC filter.

Digital Signal Processing (DSP) Filters

Digital signal processing (DSP) is one of the great technological innovations of recent time in electronics. The basic idea of DSP is to represent a signal waveform by a series of numbers, perform some manipulation on those numbers (usually with a computer) and then convert the new series of numbers to a modified signal waveform.

We won't go into detail here about how DSP works. If you want to learn more about this fascinating new technology, start with the Digital Signal Processing chapter of the latest edition of *The ARRL Handbook*. From there you may want to check out some of the other books and articles referenced in that chapter.

There are numerous advantages to processing signals digitally. Since the processing takes place in a computer, the "circuit" never needs tuning because the computer program doesn't change characteristics with age or temperature. Since

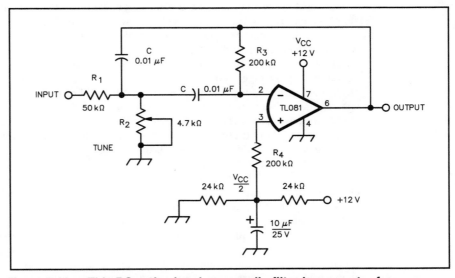

Figure 7-28 — This RC active band-pass audio filter has a center frequency of 800 Hz, and would be suitable for receiving CW signals. R_2 is used to adjust the filter operating frequency. This adjustment compensates for the tolerance variations of the other resistors and capacitors in the circuit.

the signal processing is controlled by the computer, it is simply a matter of changing the program to process the signal in a new way. In fact, this leads to some interesting applications, because the computer can be programmed to respond differently to different types of signals or conditions. This is called *adaptive processing*.

An adaptive DSP filter can be useful for removing unwanted noise from a received SSB signal, for example. A notch filter might automatically identify an interfering heterodyne, lock onto that signal and remove it from the received audio. Such a filter can even track the interfering signal as it moves through the receiver passband!

Digital communications modes require a filter response that is the same to all signals in the passband. This means the filter must have a linear phase response. Typical SSB IF filters introduce phase distortion to signals because the inductive and capacitive reactance of the elements changes with frequency. A DSP finite impulse response (FIR) filter provides a constant delay to signals regardless of their frequency, so this type of filter has a linear phase response.

A Hilbert-transform filter introduces a 90° phase shift to all frequency components of a signal. This phase shift can be used to generate a single sideband (SSB) signal by the phasing method. (You will learn more about the phasing method of generating SSB signals in Chapter 8.)

[Before proceeding to the next section, turn to Chapter 10 and study examination questions E7C07 through E7C13. Review this section as needed.]

ELECTRONIC VOLTAGE-REGULATION PRINCIPLES

Almost every electronic device requires some type of power supply. The power supply must provide the required voltages when the device is operating and drawing a certain current. The output voltage of most power supplies varies inversely with the load current. If the device starts to draw more current, the applied voltage will be pulled down. In addition, the operation of most circuits will change as the power-supply voltage changes. Modern solid-state devices are more sensitive to slight voltage changes than many tube circuits are. For this reason, there is a voltage-regulator circuit included in the power supply of almost every electronic device that uses transistors or integrated circuits. The purpose of this circuit is to stabilize the power-supply output voltage and/or current under changing load conditions.

Linear electronic voltage regulators make up one major category of regulator. With these, the regulation is accomplished by varying the conduction of a control element in direct proportion to the line voltage or load current. The control element can be a fixed dropping resistance through which the current changes as input voltage or load currents change, or it can be the resistance of the dropping element that changes.

Zener-diode regulator circuits and gaseous-regulator-tube circuits control the current through a fixed dropping resistance. Electronic regulators use a tube or transistor as the voltage-dropping rather than a resistor. By varying the dc voltage at the grid or the current at the base of these elements, the conductivity of the device may be varied as necessary to hold the output voltage constant. In solid-state regulators, the series dropping element is called a pass transistor. Power transistors are available which will handle several amperes of current at several hundred volts, but solid-state regulators of this type are usually operated at potentials below 100 V.

The second major regulator category is the **switching regulator**, where the dc source voltage is switched on and off electronically, with the duty cycle proportional to the line or load conditions. The average dc voltage available from the regulator is proportional to the duty cycle of the switching waveform, or the ratio of the on time to the total switching-cycle period. Switching frequencies of several kilohertz are normally used, to avoid the need for extensive filtering to smooth the switching frequency from the dc output. We won't go into the operating details of switching regulators in this manual, but you should at least know they exist.

Discrete-Component Regulators

Zener-Diode Shunt Regulators

A **Zener diode** can be used to stabilize a voltage source, as shown in **Figure 7-29**. Note that the cathode side of the diode is connected to the positive supply side. This places a reverse bias voltage across the diode. A Zener diode limits the voltage drop across its junction when a specified current passes through it in the reverse-breakdown direction. The diode is connected in parallel with, or shunted across, the load. This type of linear voltage regulator is often referred to as a shunt regulator.

Zener diodes are available in a wide variety of voltage and power ratings. Voltage ratings range from less than two to a few hundred volts. Power ratings specify the power the diode can dissipate, and run from less than 0.25 W to 50 W. The ability of the Zener diode to stabilize a voltage depends on the diode conducting impedance, which can be as low as one ohm or less in a low-voltage, high-power diode, to as high as a thousand ohms in a low-power, high-voltage diode.

Zener diodes of a particular voltage rating have varied maximum current capabilities, depending on the diode power ratings. The Ohm's Law relationships you are familiar with can be used to calculate power dissipation, current rating and conducting impedance of a Zener diode.

$$P = I \times E \tag{Equation 7-11}$$

$$I = \frac{P}{E} \tag{Equation 7-12}$$

and

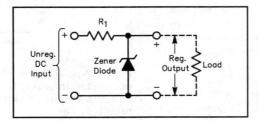

Figure 7-29 — A voltage-regulator circuit using a Zener diode.

$$Z = \frac{E}{I} \qquad \qquad \text{(Equation 7-13)}$$

where:

P = diode maximum-safe-power-dissipation rating.
E = Zener voltage.
I = maximum current that can safely flow through the diode.
Z = diode conducting impedance.

The power-handling capability of most Zener diodes is rated at 25°C; approximately room temperature. If the diode is operated at a higher ambient temperature, its power-handling capability must be derated. A typical 1-W diode can safely dissipate only $\frac{1}{2}$ W at 100°C. The breakdown voltage of Zener diodes also varies with temperature. Those rated for operation at 5 or 6 V have the smallest variation with temperature changes, and so are most often used as a voltage reference where temperature stability is a consideration.

Obtaining Other Voltages

Figure 7-30 shows how two Zener diodes may be connected in series to obtain regulated voltages that you could not achieve otherwise. This is especially useful when you want to use a 6-V Zener diode for maximum temperature stability, but you need a reference voltage other than 6 V. Two 6-V diodes in series, for example, will provide a 12-V reference. Another advantage with a circuit of this type is that you have two regulated output voltages. The diodes need not have equal breakdown voltages. You must pay attention to the current-handling capability of each diode, however.

Series Regulators

The previous section outlines some of the limitations when Zener diodes are used as regulators. Greater currents can be accommodated if the Zener diode is used as a voltage reference at low current, permitting the bulk of the load current to flow through a series pass transistor (Q_1 of **Figure 7-31**). Q_1, then, increases the current-handling capability of this linear voltage regulator circuit. An added benefit in using a pass transistor is that ripple on the output waveform is reduced. This technique is commonly referred to as "electronic filtering."

The pass transistor serves as a simple emitter-follower dc amplifier. It increases the load resistance seen by the Zener diode by a factor of beta (β). In this circuit arrangement, the Zener diode, D_1, provides a voltage reference. It is only required to supply the base current for Q_1. The net result is that the load regulation and ripple characteristics are

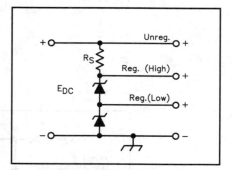

Figure 7-30 — Zener diodes can be connected in series to obtain multiple output voltages from one regulator circuit.

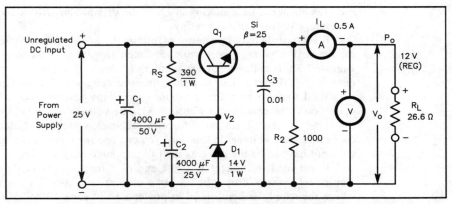

Figure 7-31 — Illustration of a power-supply regular including a pass transistor to provide more current than is available from a circuit using only a Zener diode.

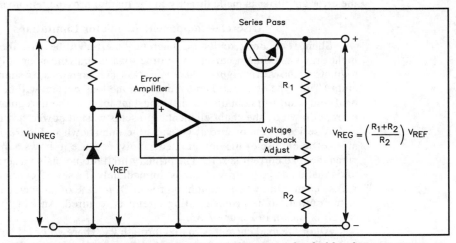

Figure 7-32 — A sample of the output voltage can be fed back to an error amplifier to obtain better regulation.

improved by a factor of beta. C_1 charges to the peak value of the input voltage ripple, smoothing, or filtering, the dc input supply voltage to the regulator. R_S supplies operating current to D_1. The addition of C_2 bypasses any remaining hum or ripple around the reference element, D_1, reducing the output ripple even more. Many simple supplies do not make use of a capacitor in this part of the circuit. C_3 provides some final filtering and helps to stabilize the transistor circuit, preventing it from oscillating. R_2 provides a constant minimum load for the pass transistor, Q_1.

The greater the value of transformer secondary voltage, the higher the power dissipation in Q_1. This not only reduces the overall power-supply efficiency, but requires stringent heat sinking at Q_1 if the dissipation will be more than a small percentage of the transistor rating.

It is possible to obtain better regulation by adding a few components to monitor the output voltage from your regulated supply. This technique, illustrated in **Figure 7-32**, is called feedback regulation. You will notice that an error amplifier detects the difference between the reference voltage and a feedback voltage. This amplified signal is then applied to the base of the pass transistor. The basic circuit operation is the same as for Figure 7-31. The voltage-divider circuit, connected at the supply output, is adjusted to set a feedback voltage that matches the Zener-diode reference voltage. In this way, you can use any convenient reference diode. It does not have to be the same as the full regulated output voltage.

If the load is connected at the end of a long supply cable, there may be a significant voltage drop when a large current is drawn. This would be the case where the supply is being used to power a solid-state 100-W HF transceiver, for example. By monitoring the voltage at the load rather than at the regulated-supply output, the regulating element can respond to the voltage drop in the cable. This technique is called remote-sensed feedback regulation. The feedback connection to the error amplifier is made directly to the load to provide remote sensing.

Discrete-Component-Regulator Limitations

Shunt-regulator circuits are inherently inefficient because the regulating element draws maximum current when the load is drawing none. (The Zener diode regulators shown in Figures 7-29 and 7-30 are examples of shunt-regulator circuits.) This type of circuit is most useful if the load current will be fairly constant, or if you want to maintain a nearly constant load on the unregulated supply. The current drawn by the shunt element represents wasted power, however.

A series-regulator circuit (such as the one shown in Figure 7-31) will make more efficient use of the unregulated supply, because it draws a minimum current when the load current is zero. The primary limitation of using a series pass transistor is that it can be destroyed almost immediately if a severe overload occurs at R_L. A fuse cannot blow fast enough to protect Q_1 in case of an accidental short circuit at the output, so a current-limiting circuit is required. An example of a suitable circuit is shown in **Figure 7-33**.

All of the load current is routed through R_2. There will be a voltage difference across R_2 that depends on the exact load current at a given instant. When the load current exceeds a predetermined safe value, the voltage drop across R_2 will forward bias Q_2, causing it to conduct. If you select a silicon transistor for Q_2 and a silicon diode for D_2, the combined voltage drops through them (roughly 0.6 V each) will be 1.2 V. Therefore, the voltage drop across R_2 must exceed 1.2 V before Q_2 can turn on. Choose a value for R_2 that provides a drop of 1.2 V when the maximum safe load current is drawn. In this example, there will be a 1.2-V drop across R_2 when I_L reaches 0.43 A.

When Q_2 turns on, some of the current through R_S flows through Q_2, thereby depriving Q_1 of some of its base current. Depending on the amount of Q_1 base current at a precise moment, this action cuts off Q_1 conduction to some degree, thus limiting the current through it.

If the collector-emitter junction of the pass transistor becomes shorted, the

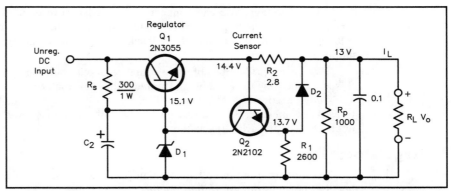

Figure 7-33 — Overload protection for a regulated supply can be effected by addition of a current-sense transistor to turn off the regulator in case of a short circuit on the output.

full unregulated rectifier voltage can be applied to the load. This can lead to a disaster, especially if the load is an expensive piece of electronic equipment! So you see, the load can draw too much current and damage the supply, or the supply can provide too high a voltage, and damage the load. The current-sense transistor protects the supply, and a "crowbar circuit" is often used to protect the load.

Figure 7-34 shows a simple crowbar circuit. A silicon-controlled rectifier (SCR) is selected that will turn on when a set voltage is applied to its gate terminal. Zener diode D_2, and resistors R_1 and R_2, are used to sense a voltage for the SCR. When the set-point voltage is exceeded, the SCR turns on, creating a short circuit across the power-supply output terminals. This sustained short circuit will blow a fuse, turning the supply off. The key is that the SCR "fires" quickly, shorting the output voltage before it can damage the load equipment.

Just as it is possible to add several Zener diodes in series to increase the regu-

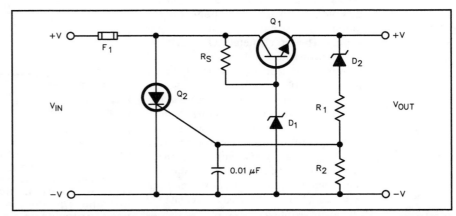

Figure 7-34 — An SCR provides a "crowbar" feature to short the unregulated supply and blow the fuse if the pass transistor fails.

lated output voltage, you can connect pass transistors in parallel to increase the current handling capability of the regulator. You can include current limiting, overvoltage protection, remote-sensed feedback regulation and other features in a single supply.

[We've covered quite a bit of information about voltage-regulator circuits, so this is a good time to take a break from learning new material. Turn to Chapter 10 and study questions E7B14 through E7B16. Also study questions E7D08 through E7D13 and question E7D15. Review this section about discrete-component regulators if you have difficulty with any of these questions.]

IC Regulators

The modern trend in voltage regulators is toward the use of integrated-circuit devices known as *three-terminal regulators*. Inside these tiny packages is a voltage reference, a high-gain error amplifier, current-sense resistors and transistors, and a series pass element. Some of the more sophisticated units have thermal shutdown, overvoltage protection and foldback current limiting.

Three-terminal regulators have a connection for unregulated dc input, one for regulated dc output and one for ground. They are available in a wide range of voltage and current ratings. It is easy to see why regulators of this sort are so popular when you consider the low price and the number of individual components they can replace. The regulators are available in several different package styles. The package and mounting methods you choose will depend on the amount of current required from your supply. The larger metal TO-3 package, mounted on a heat sink, for example, will handle quite a bit more current than a plastic DIP IC.

Three-terminal regulators are available as positive or negative types. In most cases, a positive regulator is used to regulate a positive voltage and a negative regulator for a negative voltage (with respect to ground). Depending on the system ground requirements, however, each regulator type may be used to regulate the "opposite" voltage.

Parts A and B of **Figure 7-35**

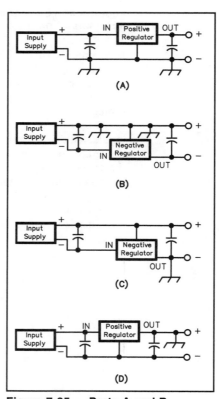

Figure 7-35 — Parts A and B illustrate the conventional manner of connecting three-terminal regulators. Parts C and D show how one regulator polarity can be used to provide an output voltage of the opposite polarity.

illustrate the conventional method of connecting an IC regulator. Several regulators can be used with a common input supply to deliver a variety of voltages with a common ground. Negative regulators may be used in the same manner, to provide several negative voltages, or with positive regulators to provide supplies with both positive and negative polarities.

Parts C and D of Figure 7-35 show how a regulator can be connected to provide opposite-polarity voltages, as long as no other supplies operate from the unregulated input source. In these configurations the input supply is floated; neither side of the input is tied to the system ground.

When choosing a three-terminal regulator for a given application, there are several important specifications to look for. Be sure the maximum and minimum input voltage and the maximum output current and voltage ratings aren't exceeded. In addition, you should consider the line regulation, load regulation and power dissipation of the device you choose, to be sure the device will handle your circuit requirements.

In use, most of these regulators require an adequate heat sink, since they may be called on to dissipate a fair amount of power. Also, since the chip contains a high-gain error amplifier, bypass capacitors on the input and output leads are essential for stable operation. Most manufacturers recommend bypassing the input and output directly at the IC leads. Tantalum capacitors are usually recommended because of their excellent bypass capabilities up into the VHF range.

Adjustable-Voltage IC Regulators

Adjustable-voltage regulators are similar to the fixed-voltage devices just described. The main difference is the lead provided to connect the error-amplifier sense voltage to a voltage divider. See **Figure 7-36**. This allows you to select the portion of the output voltage that is fed back to the amplifier, thus providing an output voltage to suit your needs, at least within certain limits. The ratio of reference voltage to total output voltage will be the same as the ratio $R_1 / (R_2 + R_1)$. For example, the regulator shown in Figure 7-36 (the LM-317 is a 1.2 to 37-V adjustable regulator) requires a reference voltage of 1.2 V. To use this regulator to build a 12-V supply, the voltage ratio is 1.2 V / 12 V = 0.1. If we select $R_1 = 1$ kΩ, then:

$R_2 + R_1 = 1000\ \Omega\ /0.1 = 10,000\ \Omega$,

so

$R_2 = 10,000\ \Omega - 1000\ \Omega = 9000\ \Omega$.

Select a 9.1-kΩ standard-value component for R_2.

IC regulators are readily available to provide an output voltage of from 5 to 24 V at up to 5 A. The same precautions should be taken with these types of regulators as with the fixed-voltage units. Proper heat sinking and lead bypassing is essential for proper circuit operation.

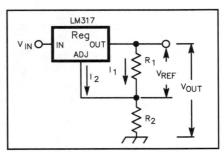

Figure 7-36 — By varying the ratio of R_2 to R_1 in this simple regulator circuit, a wide range of output voltages is possible.

It is a rather simple task to design a regulated power supply around an IC regulator. Multiple voltages can be obtained by simply adding other ICs in parallel. The main consideration is to provide an input voltage within the range specified for the device you select, and to be sure your load will not draw more current than the IC can safely supply.

[Now turn to Chapter 10 and study the examination questions E7D14 and E7D16. Review this section as needed.]

OSCILLATORS

When we discussed amplifiers earlier in this chapter, we described an amplifier-circuit problem that occurs if some of the output signal makes its way back to the input in a manner that creates positive feedback. This circuit instability turns the amplifier into an **oscillator**. When we want the circuit to behave as an amplifier, this is a problem. But sometimes we want a circuit that will generate a signal (often at radio frequencies) without any input signal, and in that case we can take advantage of the instability of an amplifier.

To start a circuit oscillating, we need to feed power from the plate back to the grid of a tube or from the collector to the base of a transistor. This is called feedback. Feedback is what happens when you are using a public address system and you get the microphone too close to the loudspeaker. When you speak, your voice is amplified in the public address system and comes out over the loudspeaker. If the sound that leaves the speaker enters the microphone and goes through the whole process again, the amplifier begins to squeal, and the sound keeps getting louder until the "circuit" is broken, even though you are not talking into the microphone anymore. Usually, the microphone has to be moved away from the speaker or the volume must be turned down. This is a good example of an amplifier becoming unstable through positive feedback, and breaking into oscillation.

When the output signal applied to the input reinforces the signal at the input, we call it positive feedback. Negative feedback opposes the regular input signal.

Positive feedback can be created in many ways — so many ways, in fact, that we can't possibly cover them all here. There are three major oscillator circuits used in Amateur Radio. They are shown in **Figure 7-37**. These oscillators can be built using vacuum tubes or FETs, and the feedback circuits will be much the same as the ones shown here for bipolar transistor circuits. The amount of feedback required to make the circuit oscillate is determined by the circuit losses. There must be at least as much energy fed back to the input as is lost in heating the components, or the oscillations will die out.

The Hartley oscillator uses inductive feedback. Alternating current flowing through the lower part of the tapped coil induces a voltage in the upper part, which is connected to the circuit input. A Hartley oscillator is the least stable of the three major oscillator types. (You might find it helpful to remember the H for henrys and Hartley as a mnemonic to remind you that this oscillator uses a tapped coil.)

The second general type is the Colpitts oscillator circuit. It uses capacitive feedback. (The Cs give a good mnemonic clue to remind you that the Colpitts oscillator gets the feedback energy from the capacitors.) The collector-circuit energy is fed back by introducing it across a capacitive voltage divider, which is part of

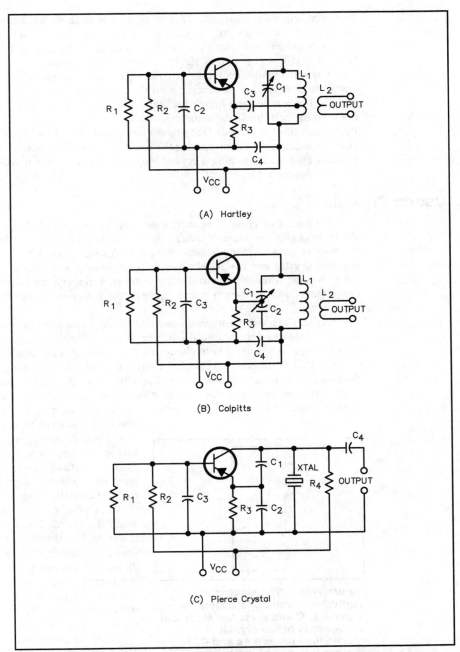

Figure 7-37 — Three common types of transistor oscillator circuits. Similar circuits can be built using tubes or FETs.

the tuning-circuit capacitance. This coupling sets up an RF voltage across the whole circuit, and is consequently applied to the transistor base. The large values of capacitance used in the tuning circuit tend to stabilize the circuit, but the oscillation frequency still will not be as stable as a crystal oscillator.

The most stable oscillator circuit is the Pierce crystal oscillator. The Pierce circuit uses capacitive feedback, with the necessary capacitances supplied by C_1 and C_2. For a tube-type Pierce oscillator the feedback capacitance is supplied by the tube interelectrode capacitances. Besides its stability, another reason why the Pierce circuit is popular is that you do not have to build and tune an LC tank circuit. Adjusting the tank circuit for the exact resonant frequency can be touchy, but it is a simple matter to plug a crystal into a circuit and know it will oscillate at the desired frequency!

Quartz Crystals

A number of crystalline substances can be found in nature. Some have the ability to change mechanical energy into an electrical potential and vice versa. This property is known as the *piezoelectric effect*. A small plate or bar properly cut from a quartz crystal and placed between two conducting electrodes will be mechanically strained when the electrodes are connected to a voltage source. The opposite can happen, too. If the crystal is squeezed, a voltage will develop between the electrodes.

Crystals are used in microphones and phonograph pickups, where mechanical vibrations are transformed into alternating voltages of corresponding frequency. They are also used in headphones to change electrical energy into mechanical vibration.

Crystalline plates have natural frequencies of vibration ranging from a few thousand hertz to tens of megahertz. The vibration frequency depends on the kind of crystal, and the dimensions of the plate. What makes the crystal resonator (vibrator) valuable is that it has an extremely high Q, ranging from a minimum of about 20,000 to as high as 1,000,000.

The mechanical properties of a crystal are very similar to the electrical properties of a tuned circuit. We therefore have an "equivalent circuit" for the crystal. The electrical coupling to the crystal is through the holder plates, which "sandwich" the crystal. These plates form, with the crystal as the dielectric, a small capacitor constructed of two plates with a dielectric between them. The crystal itself is equivalent to a series-resonant circuit and, together with the capacitance of the holder, forms the equivalent circuit shown in **Figure 7-38**.

Can we change the crystal in some way so that it will resonate at a different frequency? Sure. If we cut a new crystal, and make it longer or thicker, the reso-

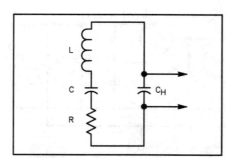

Figure 7-38 — The electrical equivalent circuit of a quartz crystal. L, C and R are the electrical equivalents of the crystal mechanical properties and C_H is the capacitance of the holder plates, with the crystal serving as the dielectric.

nant frequency will go down. On the other hand, if we want the crystal to vibrate at a higher frequency, we would make it thinner and shorter.

There are two major limitations to the use of crystals. First, we can't have more than two terminals to the circuit, since there are only two crystal electrodes. In other words, a crystal can't be tapped as we might tap a coil in a circuit. Second, the crystal is an open circuit for direct current, so you can't feed operating voltages through the crystal to the circuit.

The major advantage of a crystal used in an oscillator circuit is its frequency stability, especially with mechanical vibration. The spacing of coil turns can change with vibration, and the plates of a variable capacitor can move. On the other hand, the frequency of a crystal is much less apt to change when the equipment is bounced around (short of actually cracking the crystal, anyway).

One disadvantage of using crystals to control an oscillator frequency is that they can be easily affected by temperature changes. Manufacturers can cut a crystal at various angles across the plane of the quartz structure, and are thus able to control the temperature coefficient and other parameters, however. If better frequency stability is required, the crystal can also be placed in an "oven" to maintain a constant temperature over varying external environmental conditions.

Variable Frequency Oscillators

The major advantage of a variable-frequency oscillator (VFO) is that a single oscillator can be tuned over a frequency range. Since the frequency is determined by the coil and capacitors in the tuned circuit, the frequency is often not as stable as in a crystal-controlled circuit. The Colpitts oscillator shown in **Figure 7-39** is one of the more common types of VFO because it is a stable oscillator. Hartley oscillators are also often used in VFO circuits. The FET in the diagram is connected like a source follower (similar to the bipolar-transistor emitter follower) with the input (gate) connected to the top of the tuned circuit and the output (source) tapped down on the tuned circuit with a capacitive divider. Although the voltage gain of a source follower can never be greater than one, the voltage step-up that results from tapping down on the tuned circuit ensures that the signal at the gate will be large enough to sustain oscillation. Although the FET voltage gain is less than one, the power gain must be great enough to overcome losses in the tank circuit and the load.

The amplitude and frequency stability of a variable-frequency Colpitts oscillator is quite good, although it doesn't approach that of a crystal-controlled oscillator. The large capacitances used in the tank circuit minimize

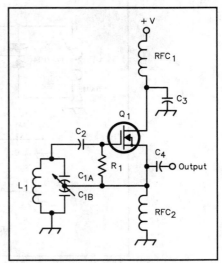

Figure 7-39 — A Colpitts VFO circuit.

frequency shifts that result from small capacitance variations caused by vibration or variations in tube or transistor characteristics.

Most oscillators provide an output rich in harmonic content. The relative amplitude of any given harmonic can be enhanced by choosing the optimum bias voltage. A Colpitts oscillator that has a second tank circuit tuned to the desired harmonic and connected to the output makes an excellent harmonic generator.

Variable capacitors can be large and relatively expensive. Mounting the capacitor so the control shaft reaches through the chassis front panel is sometimes inconvenient. In addition, providing a smoothly operating mechanism with fine tuning adjustments may require the use of an expensive vernier drive. There is an alternative to these problems, however. Varactor diodes have a junction capacitance that is controlled by the reverse bias voltage applied to the diode.

Figure 7-40 is a diagram of a Colpitts oscillator that uses a varactor diode to change the oscillator frequency. The variable resistor, R_2, adjusts the reverse bias voltage applied to the varactor diode. This resistor can be a multiturn potentiometer for fine tuning control. It can be mounted away from the actual VFO circuit, at a convenient location on the front panel. The varactor diode and variable resistor have the added advantage of small size, making this the tuning method of choice for small, battery-operated low power (QRP) equipment.

[Turn to Chapter 10 and study questions E7D01 through E7D06. Review this section as needed.]

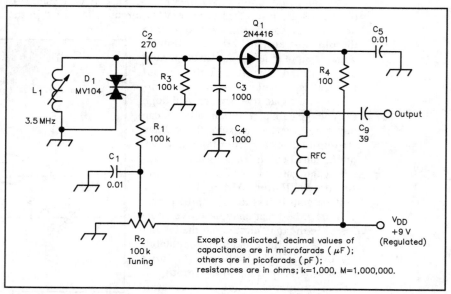

Figure 7-40 — This schematic diagram shows a Colpitts oscillator circuit with varactor diode tuning.

FREQUENCY SYNTHESIZERS

Frequency synthesizers serve much the same purpose as VFOs. That is, they are used to provide a stable, variable tuning range, generally to control the operating frequency of a radio. They are much more stable with changes in temperature and vibration than are Hartley or Colpitts VFOs. There are two major methods of frequency synthesis commonly used in modern commercially built HF radios: phase-locked loop (PLL) synthesis and direct digital synthesis (DDS).

Phase-locked loop synthesizers are the type most commonly used in modern equipment, although more and more radios are beginning to include some form of direct digital synthesizer. The phase-locked-loop (PLL) configuration uses some specialized integrated-circuit chips. **Figure 7-41** shows the major blocks of a PLL synthesizer. The signal frequency from a voltage-controlled oscillator (VCO) is divided by some integer value in a programmable-divider IC. The divider output is compared with a stable reference frequency in a phase-detector circuit. The phase detector produces an output that indicates the phase difference between the VCO and the reference signal. This signal is then fed back to the VCO through a filter. The loop provides a feedback circuit that tends to adjust the phase-detector output to zero. That means the divider output is always on the same frequency as the reference oscillator. By changing the division factor, you can change the synthesizer output. Mathematically:

$$F = N f_r$$
(Equation 7-14)

where:

F = synthesizer output frequency.
N = the division factor.
f_r = reference-oscillator frequency.

A PLL synthesizer feedback loop is constantly "correcting" the output signal frequency. Any variations in the frequency of the reference oscillator will cause variations in the phase of the output signal, from one cycle to the next. These unwanted variations in the oscillator signal phase are called phase noise. This is a

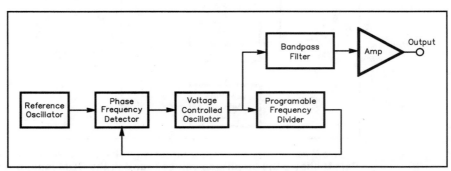

Figure 7-41 — Block diagram of an indirect frequency synthesizer, which uses a phase-locked loop and a variable-ratio divider.

broadband noise around the desired output frequency, and is the major spectral impurity component of a phase-locked loop synthesizer. A very stable reference oscillator must be used for a phase-locked loop frequency synthesizer, to minimize this phase noise on the output signal.

Figure 7-42A is a block diagram of a direct digital synthesizer. This type of synthesizer is based on the concept that we can define a sine wave of any frequency by specifying a series of values (sine or cosine) taken at equal time intervals (or phase angles). The crystal oscillator sets the *sampling rate* for these values. The *phase increment* input to the adder block sets the number of samples for one cycle. The oscillator *clock* signal tells the phase accumulator to sample the data input from the adder, and then increment the adder value by the phase increment. The phase accumulator value varies between 0 and 360. The ROM lookup table contains the amplitude values for the sine (or cosine) of each angle represented by the phase accumulator.

For example, suppose our synthesizer uses a 10-kHz crystal oscillator. This means there will be one sample every 0.1 ms. If the phase increment is set to 36°,

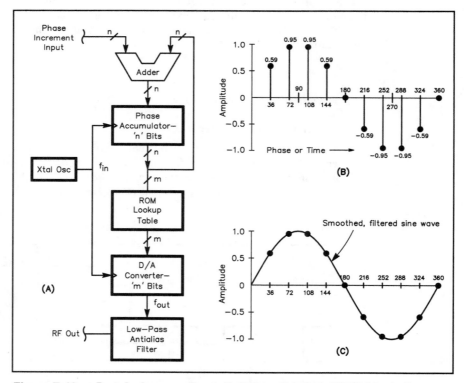

Figure 7-42 — Part A shows a direct digital synthesizer (DDS) block diagram. Part B shows the amplitude values found in the ROM lookup table for a particular sine wave being generated. Part C shows the smoothed output signal from the DDS, after it goes through the low-pass antialias filter.

there will be ten samples in each cycle: 0°, 36°, 72°, 108°, 144°, 180°, 216°, 252°, 288° and 324°. The next sample, at 360° starts the second cycle. The total time for these ten samples is 1 ms, which means the sine wave defined by these samples has a frequency of 1 kHz. Figure 7-42B shows a representation of the sine values found in the lookup table for these phase angles.

The sine values are fed to a digital to analog converter (DAC), and the analog output signal goes through the low-pass antialias filter. A smoothed sine-wave signal results. (See Figure 7-42C.)

We can change the frequency of the signal produced by our direct digital synthesizer by changing the value of the phase increment input to the adder. For example, if the new phase increment is 72° there will be five samples per cycle. Each cycle will take 0.5 ms, so the frequency of this new signal is 2 kHz.

DDS has the disadvantage of requiring a more sophisticated control circuit to set the proper phase increment to achieve a desired output frequency. A computer or microprocessor controller circuit is normally used. Phase noise is not a problem with DDS circuits, however. The major spectral impurity components produced by a direct digital synthesizer are spurious (unwanted) signals — *spurs* — at specific discrete frequencies.

Careful design can place those spurs outside of the amateur bands. Commercially built HF radios are beginning to use hybrid synthesizers, which combine some of the best features of PLL and DDS circuits.

[Study question E7D07 and questions E7E10 through E7E15 in Chapter 10. Review this section as needed.]

MIXERS, DETECTORS AND MODULATORS

Mixer circuits are used to change the frequency of a desired signal. In a receiver, this means converting all received signals to a single frequency (called the intermediate frequency or IF) so they can be processed more efficiently. In this way the amplifier chain can be adjusted for maximum efficiency and the best signal-handling characteristics without the need to retune many circuit elements every time you change bands or turn the radio dial. By converting all received signals to a single IF for processing, the selectivity of the receiver is improved greatly.

Mixers are also used to change the frequency of a signal as it progresses through a transmitter. In fact, a mixer circuit is even used in most transmitters to produce the modulated radio wave being sent out across the airwaves. If the circuit introduces a signal loss during processing, it is called a passive mixer. If some gain is provided (or at least no signal loss) because an amplifier is included, then the circuit is called an active mixer,

A **detector** circuit is used to "reclaim" the information that has been superimposed on a radio wave at a transmitter. That information may be the operator's voice, or it may be Morse code or RTTY signals. It could even be the information that will allow you to reproduce a slow-scan TV or facsimile picture that has been transmitted across the miles. There are a variety of detector methods described in this section, but they are all associated with receivers. Like mixer circuits, detectors can be classified as either passive or active.

A **modulator** is the circuit that combines an information signal with a radio-wave signal. The type of modulator circuit used at the transmitter will determine the type of signal that is transmitted, and the particular detection methods that can be used in a receiver for that signal. For example, different modulator circuits are used to produce AM phone, single-sideband (SSB) phone and FM phone signals.

The principles of operation are much the same for mixers, detectors and modulators, so when you become familiar with the operation of one type, you will find the others easier to understand.

Mixers

When two sine waves are combined in a nonlinear circuit, the result is a complex waveform that includes the two original signals, a sine wave that is the sum of the two frequencies and a sine wave that is the difference between the two. Also included are combinations of the harmonics from the input signals, although these are usually weak enough to be ignored. One of the product signals can be selected at the output by using a filter. Of course, the better the filter, the less of the unwanted products or the two input signals that will appear in the final output signal. By using balanced-mixer techniques, the mixer circuit provides isolation of the various ports so the signals at the radio frequency (RF), local oscillator (LO) and intermediate frequency (IF) ports will not appear at any other. This prevents the two input signals from reaching the output. In that case the filter needs to remove only the unwanted mixer products.

In a typical amateur receiving application, you want to mix the RF energy of a desired signal with the output from an oscillator in the receiver so you can produce a specific output frequency (IF) for further processing in the radio. The local oscillator (LO) frequency determines the frequency that the input signal is converted to. By using this frequency-conversion technique, the IF stages can be designed to operate over a relatively narrow frequency range, and filters can be designed to provide a high degree of selectivity in the IF stages.

The mixer stages in a high-performance receiver must be given careful consideration. These stages will have a great impact on the dynamic range of the receiver. The RF signal should be amplified only enough to overcome mixer noise. Otherwise, strong signals will cause desensitization, cross-modulation and IMD products in the mixer. Spurious mixer products will be produced if an excessive amount of input-signal energy reaches the mixer circuit. The level of these spurious mixer products may be increased to the point that they appear in the output. One result of these effects is that the receiver may be useless in the presence of extremely strong signals. A mixer should be able to handle strong signals without being affected adversely.

Passive Mixers

One simple mixer that has good strong-signal characteristics is the diode mixer shown in **Figure 7-43A**. This circuit makes use of a trifilar-wound broadband toroidal transformer to balance the mixer, effectively canceling the RF and LO input signals, so only the sum and difference frequencies appear at the IF output terminal. Sometimes this circuit is referred to as a single balanced mixer. Figure 7-43B shows an improved circuit that includes two balance transformers and four diodes arranged in a circular, or ring, pattern. This circuit is usually called

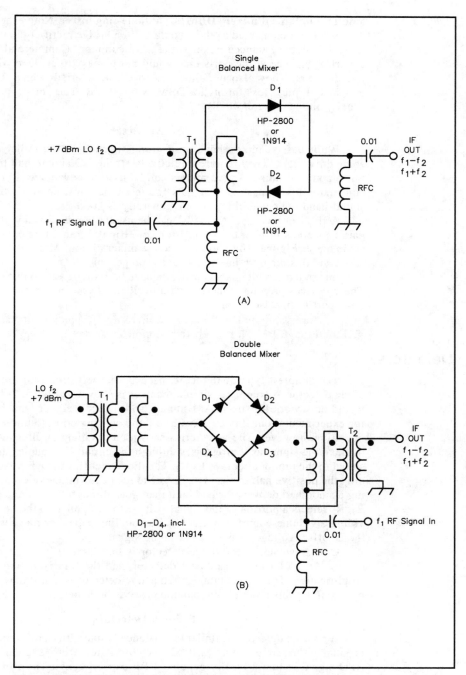

Figure 7-43 — Single and double balanced diode mixers.

a **double balanced mixer (DBM)** or a diode-ring mixer. Note especially that the diodes are not connected as they would be in a bridge rectifier circuit.

The double balanced mixer is the most common. Commercial modules offer electrical balance at the ports that would not be easy to achieve with homemade transformers. They also use diodes whose characteristics have been carefully matched. Typical loss through a DBM is 6 to 9 dB. The port-to-port isolation is usually on the order of 40 dB.

Active Mixers

While passive mixers have good strong-signal-handling ability, they also have some drawbacks. They require a relatively strong LO signal, and they generate a fair amount of noise. Active mixers can be used to advantage if you require less conversion loss, weaker LO signals and less noise. Just be aware that the strong-signal-handling capabilities are not generally as good.

A JFET or dual-gate MOSFET can be used as a mixer, and will provide some gain as well as mixing the signals for you. Bipolar transistors could be used, but seldom are. **Figure 7-44** shows an active mixer circuit. Many variations are possible, and this diagram just shows one arrangement.

Integrated-circuit mixers are available in both single and double balanced types. These devices provide at least several decibels of conversion gain, low noise and good port-to-port isolation.

[Before going on to the next section, turn to Chapter 10 and study questions E7E07 through E7E09. Review this section as needed.]

Detectors

The simplest type of **detector**, used in the very first radio receivers, is the diode detector. A complete, simple receiver is shown in **Figure 7-45**. This circuit would only work for strong AM signals, so it is not used very much today, except for experimentation. It does serve as a good starting point to understand detector operation, however. The waveforms shown on the diagram illustrate the changes made to the signal as it progresses through the circuit. L_1 couples the received RF signal to the tuning circuit of L_2-C_1. The diode rectifies the RF waveform, passing only the positive half cycles. C_2 charges to the peak voltage of each pulse, producing a smoothed dc waveform, which then goes through the voltage divider R_1 and R_2. R_2 selects a portion of the signal voltage to be applied to the headphones, providing a volume-control feature. C_4 is a coupling capacitor that serves to remove the dc offset voltage, leaving an ac audio signal.

One drawback of the diode detector is that there is some signal loss in the circuit. An FET can be used as a detector, and the transistor will provide some amplification. The disadvantage of an active detector is that it may be overloaded by strong signals more easily than the passive diode detector.

Product Detectors

A *product detector* is similar to a balanced mixer. It is a detector whose output is equal to the product of a beat-frequency oscillator (BFO) and the RF signal. The BFO signal is like a locally generated RF carrier signal. The BFO frequency is chosen to provide detector output at audio frequencies. A double balanced diode-

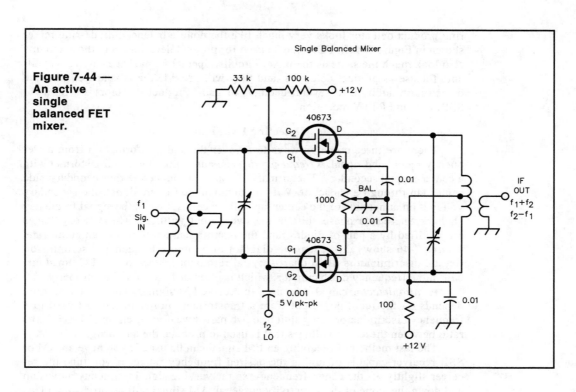

Figure 7-44 — An active single balanced FET mixer.

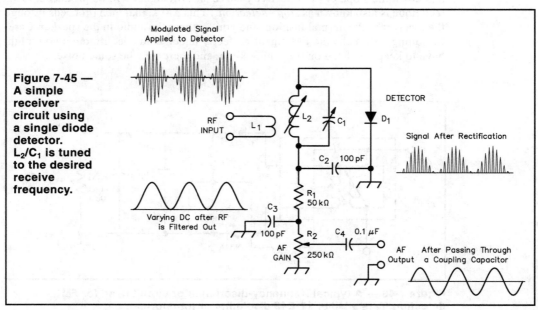

Figure 7-45 — A simple receiver circuit using a single diode detector. L_2/C_1 is tuned to the desired receive frequency.

ring product detector looks very much like the double balanced diode-ring mixer shown in Figure 7-43B. FETs can be used for product detectors, and those circuits also look much the same as the mixer circuits. Special IC packages are also available for use as product detectors, and they will provide the advantages of diode detectors in addition to several decibels of gain. Product detectors are used for SSB, CW and RTTY reception.

Detecting FM Signals

There are three common ways to recover the audio information from a frequency-modulated signal. A *frequency-discriminator* circuit uses a transformer with a center-tapped secondary. The primary signal is introduced to the secondary-side center tap through a capacitor. With an unmodulated input signal, the secondary voltages on either side of the center tap will cancel. But when the signal frequency changes, there is a phase shift in the two output voltages. These two voltages are rectified by a pair of diodes, and the resulting signal varies at an audio rate. **Figure 7-46** shows the schematic diagram of a simple frequency discriminator. Crystal discriminators use a quartz-crystal resonator instead of the LC tuned circuit in the frequency discriminator, which is often difficult to adjust properly.

A *ratio detector* can also be used to receive FM signals. The basic operation depends on the rectified output from a transformer similar to the one used in a frequency discriminator being split into two parts by a divider circuit. Then it is the ratio between these two voltages that is used to produce the audio signal.

The last method of receiving an FM signal can be used if you have an AM or SSB receiver capable of tuning the desired frequency range. If you tune the receiver slightly off the center frequency, so the varying signal frequency moves up and down the slope of the selectivity curve, an AM signal will be produced. (This technique is also known as *slope detection*.) This AM signal then proceeds through the receiver in the normal fashion, and you can hear the audio in the speaker. Careful tuning will make the FM signal perfectly understandable, although you might have to keep one hand on the tuning knob, and there may be some noise.

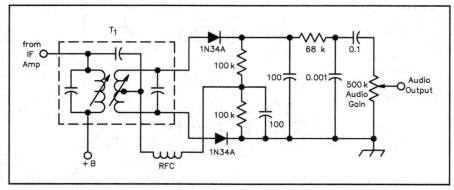

Figure 7-46 — A typical frequency-discriminator circuit used for FM detection. T$_1$ is a Miller 12-C45 disciminator transformer.

Modulator Circuits

Modulation is really a mixing process whereby information is imposed upon a carrier; any mixer or converter circuit could be used for generating a modulated signal. Instead of introducing two radio frequencies into a mixer circuit, we simply introduce one radio frequency (the carrier frequency) and the voice-band audio frequencies.

Mixer circuits used in receivers are designed to handle a small signal and a large local-oscillator voltage. This means that the percentage of modulation of the IF signal is low. In a transmitter we want to get as close as possible to 100% modulation, and we also want more power output. For these reasons **modulator** circuits differ in detail from receiving mixers, although they are much the same in principle.

Amplitude Modulation

Although double-sideband, full-carrier, amplitude modulation (DSB AM) is seldom used on the amateur bands anymore, we will discuss a simple system just to help you understand how a single-sideband, suppressed-carrier (SSB) signal is generated. When RF and AF signals are combined in a DSB AM transmitter, four principle output signals are generated. First, there is the original audio signal, which is easily rejected by the tuned RF circuits in the stages following the modulator. Then there is the RF carrier, also unchanged from its original form. Of primary importance are the other two signals: one being the sum of the carrier frequency and the audio, the other being the difference between the original signals. These two new signals are called sideband signals. The amplitude of these signals at any given instant depends on the amplitude of the original audio signal at that instant. The greater the audio-signal strength, the greater the amplitude of the sideband signals.

The sum component is called the upper sideband. As the audio-signal frequency increases, so does the frequency of the upper-sideband signal. The difference component is called the lower sideband. This sideband is inverted, which means that as the audio-signal frequency increases, the sideband frequency decreases. **Figure 7-47** illustrates this principle.

The result of all this is that the RF envelope, as viewed on an oscilloscope, has the general shape of the modulating waveform. The envelope varies in ampli-

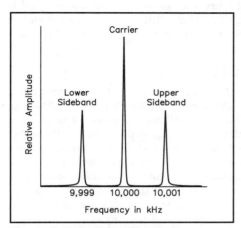

Figure 7-47 — A 10-MHz carrier, modulated by a 1-kHz sine wave, produces an output as shown.

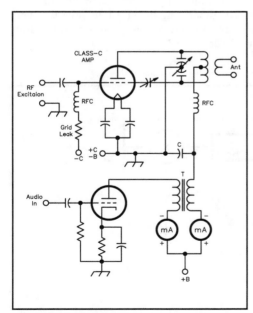

Figure 7-48 — Plate modulation of a Class-C RF amplifier.

tude because it is the vector sum of the carrier and the sidebands. Note that the *carrier* amplitude is not changed in amplitude modulation. This is a common misunderstanding; one that often leads to confusion. It is the RF *envelope* that varies in amplitude.

You can produce amplitude modulation by applying the AF signal to the plate or collector of an RF amplifier stage. **Figure 7-48** shows how the audio signal is used to modulate the plate voltage of a Class-C amplifier. You can also modulate the control element of the amplifier (grid, base or gate). A wide variety of modulator circuits have been used over the years.

SSB: The Filter Method

One way to generate an SSB signal is to remove the carrier and one of the sidebands from an ordinary DSB AM signal. The block diagram shown in **Figure 7-49** shows how this is done. The RF oscillator generates a carrier wave that is injected into a **balanced modulator**. With this system, the oscillator frequency is changed, depending on which sideband you want to use. Another system maintains the same oscillator frequency, but switches filters to remove the opposite sideband. The audio information is amplified by the speech amplifier and is then applied to the modulator. A balanced modulator takes these two inputs and supplies, as its output, both sidebands without the carrier. This meets the first requirement for the generation of an SSB signal — removal of the carrier.

Let's see how the balanced modulator accomplished this. There are many different types of balanced modulators and it would be impossible to show them all here. One of the more popular types is illustrated in **Figure 7-50**. This particular circuit is called a diode-ring balanced modulator. If you get the feeling that you have seen all this before, don't worry. The diode-ring balanced modulator is very similar to the double balanced diode-ring mixer. Actually, you could redraw Figure 7-50 to look almost identical to Figure 7-43B.

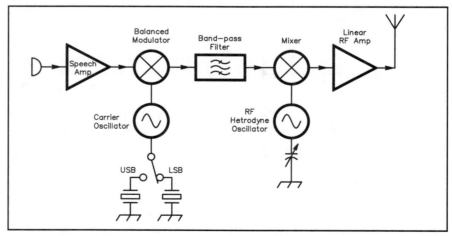

Figure 7-49 — A block diagram showing the filter method of generating an SSB signal.

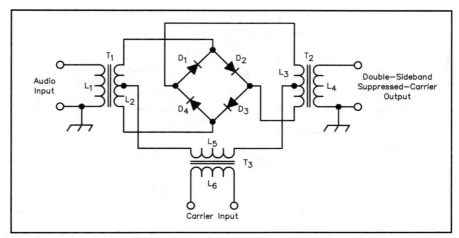

Figure 7-50 — One example of a diode-ring double balanced modulator.

Audio information is coupled into the circuit through transformer T_1. The carrier is injected through coils L_5 and L_6 and the double-sideband, suppressed-carrier output is taken through L_3 and L_4. To better understand the circuit operation, let's first analyze the circuit with only the carrier applied. See **Figure 7-51A**. The polarity of voltage shown across L_5 will cause electrons to flow in the direction indicated by the arrows. D_1 and D_3 will conduct. The current that flows through each half of L_3 is equal and opposite, causing a canceling effect. Output at L_4 will be zero. During the next half cycle, the polarity of voltage across L_5 will reverse, D_2 and D_4 will conduct, and again the output at L_4 will be zero.

Now for a moment, let's remove the carrier signal and connect an audio source to the audio-input terminals. Figure 7-51B illustrates this condition. During one half of the audio cycle, D_2 and D_3 will conduct and the output will be zero. On the other half cycle of the audio signal, D_1 and D_4 will conduct and again the output at L_4 will be zero. Notice that in this case, there is no signal through L_3 at all. We can see that if either the carrier or the audio is applied without the other, there will be no output.

Both signals must be applied if the circuit is to work as intended. In practice, the carrier level is made much larger than the audio input. Diode conduction is, therefore, determined by the carrier. There are four conditions that can exist as far as the voltage polarity of the carrier and audio signals are concerned. Both waves can be positive, both can be negative or they can be opposites. We must keep in mind that the carrier is going through many cycles while the audio sine wave goes through only one.

Let's look again at the drawing at Figure 7-51C. With the polarities indicated, the major electron flow will be through D_3 since the audio and carrier voltages are aiding (adding together) in this path. Since the balance through L_3 has been upset

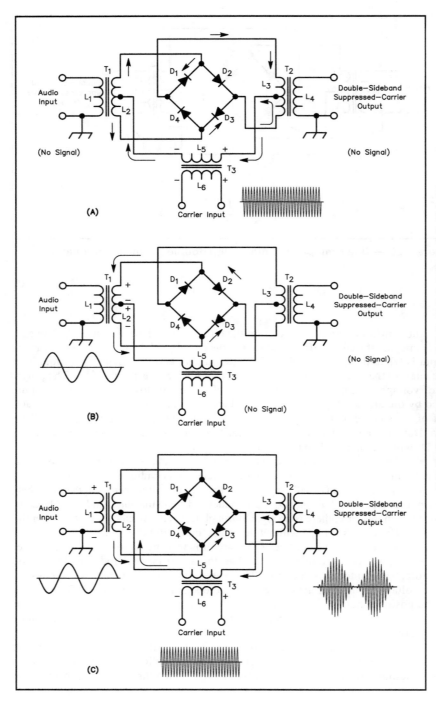

Figure 7-51 — Arrows indicate the direction of electron flow in a diode-ring mixer circuit. Part A shows the condition when only a carrier signal is applied. B shows the conditions with audio applied but no carrier, and C shows the conditions for one polarity relationship of carrier- and audio-input signals. See the text for more information about the other three possible polarity relationships.

(the bottom half of L_3 has more current flowing through it than does the top half), there will be an output present at L_4. Next, the carrier polarity will reverse while the audio polarity remains the same. This is the order in which the polarities would change, since the carrier is reversing polarity at a much faster rate than the audio signal. Later, the carrier polarity will be back to what it was at C; however, now the audio will be on the negative portion of its sine wave and so its polarity will be reversed. Maximum current under these conditions is through D_1 and again an output signal appears at L_4. The fourth condition that will exist is shown when the audio-signal polarity is the same as in C, but the carrier is reversed. This time, D_4 will be the main current path. As you have probably guessed, there is output at L_4 under these conditions.

Figure 7-52 shows a composite drawing of the audio and carrier waveforms. The shaded areas represent the double-sideband, suppressed-carrier output from the balanced modulator.

Another type of diode-ring balanced modulator is shown in **Figure 7-53**. Two balance controls are provided so that the circuit can be adjusted for optimum carrier suppression (50 dB is a practical amount). This circuit operates basically the same way the modulator shown in Figures 7-50 and 7-51 works. With either the audio or carrier applied separately, the circuit is balanced and there will be no output. With both the audio and carrier applied, the balance is upset and output will be present at T_1. We won't go into the detail that we did in the previous circuit. You should be able to determine which diodes are conducting for the different voltage polarities presented by the audio and carrier signals.

The two circuits we have examined can be classified as passive balanced modulators That is, they do not provide gain (amplify) but actually cause a small amount of signal loss (insertion loss). Balanced modulators can also be built using active devices.

One such modulator is shown in **Figure 7-54**. This circuit makes use of two FETs as the active devices. As was true with the two other modulators, the circuit is in a balanced condition if either the audio or carrier signals are applied separately. In this circuit the tuned output network (C_5/L_1) is adjusted for resonance at the RF carrier frequency. At audio frequencies, this circuit represents a very low imped-

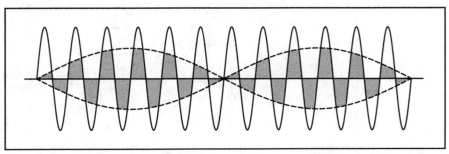

Figure 7-52 — Superimposed audio and RF waveforms. The shaded area represents the double-sideband, suppressed-carrier output from a balanced modulator.

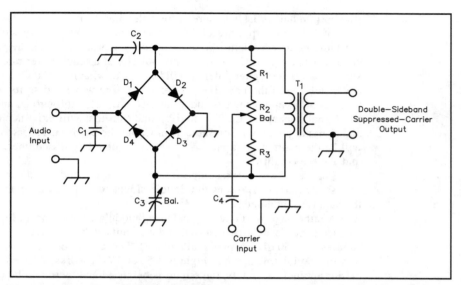

Figure 7-53 — Here is another common form of balanced modulator. C₃ and R₂ are adjusted for maximum carrier suppression.

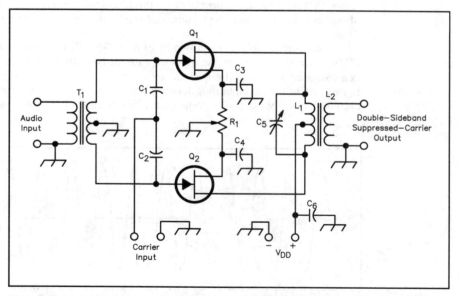

Figure 7-54 — This circuit is an active balanced modulator using two FET devices. R₁ is adjusted for maximum carrier suppression, and C₅ is adjusted for maximum double-sideband suppressed-carrier output.

ance, allowing no audio to appear at the output.

Consider the carrier input for a moment. Injection voltage is supplied to each gate in a parallel fashion. The input to each gate is of equal amplitude and of the same phase, and the output circuit is connected for push-pull operation. Currents through each half of the tank are equal and opposite. The signals will effectively cancel and the output will be zero.

Let's analyze the circuit with both the audio and carrier energy applied. Since we have a push-pull input arrangement for the audio information, the bias for the FETs varies at an audio rate. The audio signal applied to one gate is 180° out of phase with the other. While one of the devices is forward biased, the other is reverse biased. The input to each device is the audio signal plus the carrier signal. Sum and difference frequencies are developed at the output to produce a double-sideband, suppressed-carrier signal.

Removing the Unwanted Sideband

We now have a signal that contains both the upper and lower sidebands of what was an amplitude-modulated signal with carrier. The next step in generating our SSB signal is to remove one of the sidebands. Looking back at Figure 7-49, we see that the next stage after the balanced modulator is the filter. This circuit does just as its name suggests — it filters out one of the sidebands.

An example of a simple crystal filter is shown in **Figure 7-55**. The two crystals would be separated by approximately 2 kHz to provide a bandwidth suitable for passing one sideband and not the other. The curve to the right of the filter shows the shape of the filter response. More-elaborate filters using four and six crystals will give steeper slopes on the response curve, without affecting the bandwidth near the top of the curve. As shown at B, a filter with more crystals has a narrower response at the – 60 dB point. The filter at B has better "skirt selectivity" than the one at A. Two half-lattice filters of the type shown at A are connected back to back to form the filter at B. Crystal-lattice filters of this type are available commercially for frequencies up to 40 MHz or so.

Amplification for the Modulator

The last stage shown in the block diagram in Figure 7-49 is the linear amplifier. Since the modulation process occurs at low power levels in a conventional transmitter, it is necessary that any amplifiers following the balanced modulator be linear. In the low-power stages of a transmitter, high voltage gain and maximum linearity are quite a bit more important than efficiency.

Generating an FM Signal

Any type of modulation that changes the phase angle of the transmitted sine wave is called *angle modulation*. Frequency modulation and phase modulation are the two common types of angle modulation. Most methods of producing FM will fall into two general categories. They are direct FM and indirect FM. As you might expect, each has its advantages and disadvantages. Let's look at the direct-FM method first.

Direct FM

A **reactance modulator** is a simple and satisfactory device for producing FM

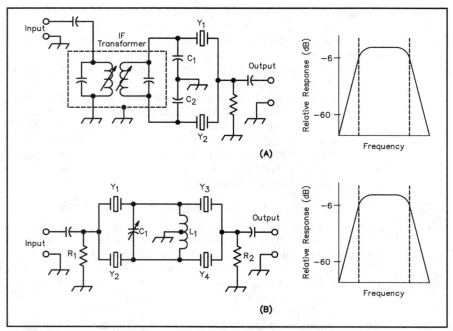

Figure 7-55—A half-lattice band-pass filter is shown at A. B shows two half-lattice filters in cascade. The filter response is shown to the right of each filter.

in an amateur transmitter. This is a vacuum tube or transistor connected to the RF tank circuit of an oscillator so that it acts as a variable inductance or capacitance. The only way to produce a true emission F3E signal is with a reactance modulator on the transmitter oscillator.

Figure 7-56 is a representative circuit. Gate 1 of the modulator MOSFET is connected across the oscillator tank circuit (C_1 and L_1) through resistor R_1 and blocking capacitor C_2. C_3 represents the input capacitance of the modulator transistor.

R_1 is made large compared to the reactance of C_3, so the RF current through R_1 and C_3 will be practically in phase with the RF voltage appearing at the terminals of the tank circuit. The voltage across C_3 will lag the current by 90°, however. The RF current in the drain circuit of the modulator will be in phase with the gate voltage, and consequently is 90° behind the current through C_3, or lagging the RF-tank-circuit voltage by 90°. This lagging current is drawn through the oscillator tank, giving the same effect as though an inductor were connected across the tank. The frequency increases in proportion to the amplitude of the lagging modulator current. The audio voltage, introduced through a radio-frequency choke, varies the transconductance of the transistor and thereby varies the RF drain current.

The modulator sensitivity (frequency change per unit change in modulating voltage) depends on the transconductance of the modulator transistor. Sensitivity

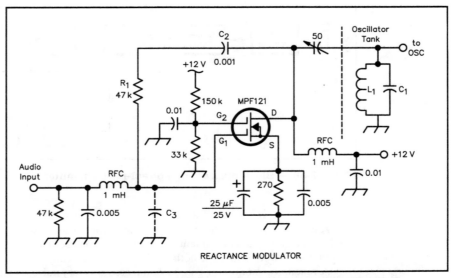

Figure 7-56 — A reactance modulator using a high-transconductance MOSFET.

increases when R_1 is made smaller in comparison with the reactance of C_3. It also increases with an increased L/C ratio in the oscillator tank circuit. For highest carrier stability, however, it is desirable to use the largest tank capacitance that will permit you to obtain the required deviation, while keeping within the limits of linear operation.

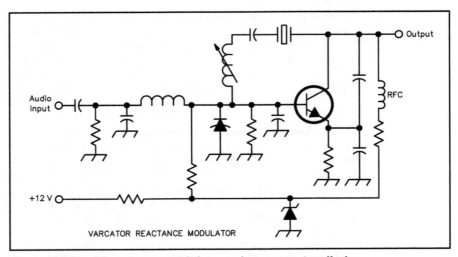

Figure 7-57 — A reactance modulator using a varactor diode.

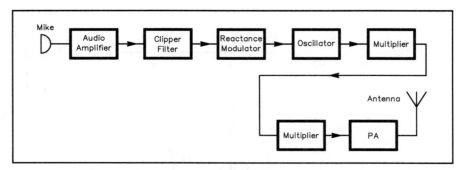

Figure 7-58 — Block diagram of a direct-FM transmitter.

A change in any of the modulator-transistor voltages will cause a change in RF drain current, and consequently, a frequency change. Therefore, you should use a regulated power supply for both modulator and oscillator.

A reactance modulator can be connected to a crystal oscillator, as shown in **Figure 7-57**. The resulting signal will be more phase modulated than frequency modulated, however, since varying the frequency of a crystal oscillator will only produce a small amount of frequency deviation. Notice that this particular circuit uses a varactor diode to change the circuit capacitance by means of a bias voltage and the modulating signal. So, you should recognize that a reactance modulator acts as either a variable inductor or a variable capacitor in the tank-circuit oscillator.

Figure 7-58 is a block diagram of a system that uses a reactance modulator to shift an oscillator frequency and generate an FM signal directly. Successive multiplier stages provide output on the desired frequency, which is then amplified by a power amplifier stage.

With any reactance modulator, the modulated oscillator is usually operated on a relatively low frequency, so that a high order of carrier stability can be secured. Frequency multipliers are used to provide the final desired output frequency. It is important to note that when the frequency is multiplied, so is the frequency deviation. So the amount of deviation produced by the modulator must be adjusted carefully to give the proper deviation at the final output frequency.

Indirect FM

The same type of reactance-modulator circuit that is used to vary the oscillator-tank tuning for an FM system can be used to vary the amplifier-tank tuning, and thus vary the *phase* of the tank current, to produce *phase modulation (PM)*. Hence, the modulator circuit of Figure 7-57 or Figure 7-58 can be used for PM (emission G3E) if the reactance transistor or tube works on an amplifier tank instead of directly on a self-controlled oscillator. A **phase modulator** varies the tuning of an amplifier tank circuit to produce a PM signal. From a practical view, FM and PM are the same.

The phase shift that occurs when a circuit is detuned from resonance depends on the amount of detuning and the circuit Q. The higher the Q, the smaller the amount of detuning needed to secure a given amount of phase shift. If the Q is at least 10, the relationship between phase shift and detuning (in kilohertz either side of the resonant frequency) will be substantially linear over a phase-shift range of about 25°. From the standpoint of modulator sensitivity, the Q of the tuned circuit on which the modulator operates should be as high as possible. On the other hand, the effective Q of the circuit will not be very high if the amplifier is delivering power to a load, since the load resistance reduces the Q. There must, therefore, be a compromise between modulator sensitivity and RF power output from the modulated amplifier. An optimum Q figure appears to be about 20; this allows reasonable loading of the modulated amplifier, and the necessary tuning variation can be secured from a reactance modulator without difficulty. It is advisable to modulate at a low power level.

Reactance modulation of an amplifier stage usually results in simultaneous amplitude modulation because the modulated stage is detuned from resonance as the phase is shifted. This must be eliminated by feeding the modulated signal through an amplitude limiter or one or more "saturating" stages; that is, amplifiers that are operated Class C and driven hard enough so that variations in the amplitude of the input excitation produce no appreciable variations in the output amplitude.

The actual frequency deviation increases with the modulating audio frequency in PM. (Higher audio frequencies produce greater frequency deviation.) Therefore, it is necessary to cut off the frequencies above about 3000 Hz before modulation takes place. If this is not done, unnecessary sidebands will be generated at frequencies considerably removed from the carrier frequency.

In an FM system, the frequency deviation does not increase with modulating audio frequency. In this case, an audio shaping network, called a *preemphasis network*, is added to an FM transmitter to attenuate the lower audio frequencies. This spreads the audio signal energy evenly across the audio band. Preemphasis applied to an FM transmitter gives the deviation characteristic of PM. The reverse process, called *deemphasis*, is used at the receiver to restore the audio to its original relative proportions. A PM transmitter does not need this preemphasis network.

The indirect method of generating FM shown in **Figure 7-59** is currently popu-

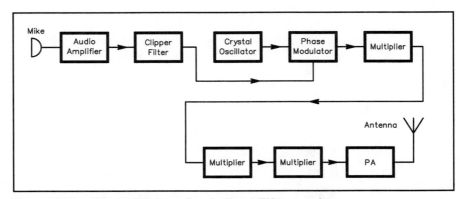

Figure 7-59 — Block diagram of an indirect-FM transmitter.

lar. Shaped audio is applied to a phase modulator to generate PM. Since the amount of deviation produced is very small, a large number of multiplier stages is necessary to achieve wideband deviation at the operating frequency.

[That completes your study of practical circuits for your Amateur Extra class license exam. Before you go on to the next chapter, though, study questions E7E01 through E7E06 in Chapter 10. Review this section as needed.]

Amplitude-compandored single sideband (ACSSB) — An SSB system that uses a logarithmic amplifier to compress voice signals at the transmitter and an inverse logarithmic amplifier to expand the voice signals in the receiver.

Amplitude modulation — A method of superimposing an information signal on an RF carrier wave in which the amplitude of the RF envelope (carrier and sidebands) is varied in relation to the information signal strength.

Bandwidth — As related to a transmitted signal, that frequency range that the signal occupies around a center frequency. Bandwidth increases with increasing information rate.

Baud — A unit of signaling speed, used to describe data transmission rates. One bit per second for single-channel binary-coded signals.

Circular polarization — Describes an electromagnetic wave in which the electric and magnetic fields are rotating. If the electric field vector is rotating in a clockwise sense, then it is called right-hand polarization and if the electric field vector is rotating in a counterclockwise sense, it is called left-hand polarization.

Compandoring — In an ACSSB system, the process of *com*pressing voice signals in a transmitter and *expand*ing them in a receiver.

Deviation — The peak difference between an instantaneous frequency of the modulated wave and the unmodulated-carrier frequency in an FM system.

Deviation ratio — The ratio of the maximum frequency deviation to the maximum modulating frequency in an FM system.

Direct sequence — A spread-spectrum communications system where a very fast binary bit stream is used to shift the phase of an RF carrier.

Electric field — A region through which an electric force will act on an electrically charged object.

Electric force — A push or pull exerted through space by on electrically charged object on another.

Electromagnetic waves — A disturbance moving through space or materials in the form of changing electric and magnetic fields.

Emission designators — A method of identifying the characteristics of a signal from a radio transmitter using a series of three characters following the ITU system.

Emission types — A method of identifying the signals from a radio transmitter using a "plain English" format that simplifies the ITU **emission designators**.

Facsimile — The process of scanning pictures or images and converting the information into signals that can be used to form a likeness of the copy in another location.

Frequency, f — The number of complete cycles of a wave occurring in a unit of time.

Frequency hopping — A spread-spectrum communications system where the center frequency of a conventional carrier is altered many times a second in accordance with a pseudorandom list of channels.

Frequency modulation — A method of superimposing an information signal on an RF carrier wave in which the instantaneous frequency of an RF carrier wave is varied in relation to the information signal strength.

Linear polarization — Describes the orientation of the electric-field component of an electromagnetic wave. The electric field can be vertical or horizontal with respect to the Earth's surface, resulting in either a vertically or a horizontally polarized wave. (Also called **plane polarization**.)

Magnetic field — A region through which a magnetic force will act on a magnetic object.

Magnetic force — A push or pull exerted through space by one magnetically charged object on another.

Modulation index — The ratio of the maximum frequency deviation of the modulated wave to the instantaneous frequency of the modulating signal.

Peak envelope power (PEP) — An expression used to indicate the maximum power level in a signal. It is found by squaring the RMS voltage at the envelope peak, and dividing by the load resistance.

Peak envelope voltage (PEV) — The maximum peak voltage occurring in a complex waveform.

Peak negative value — On a signal waveform, the maximum displacement from the zero line in the negative direction.

Peak positive value — On a signal waveform, the maximum displacement from the zero line in the positive direction.

Peak-to-peak (P-P) value — On a signal waveform, the maximum displacement between the peak positive value and the peak negative value.

Peak-to-peak (P-P) voltage — A measure of the voltage taken between the negative and positive peaks on a cycle.

Peak voltage — A measure of voltage on an ac waveform taken from the centerline (0 V) and the maximum positive or negative level.

Period, T — The time it takes to complete one cycle of an ac waveform.

Phase modulation — A method of superimposing an information signal on an RF carrier wave in which the phase of an RF carrier wave is varied in relation to the information signal strength.

Pilot tone — In an ACSSB system, a 3.1-kHz tone transmitted with the voice signal to allow a mobile receiver to lock onto the signal. The pilot tone is also used to control the inverse logarithmic amplifier gain.

Plane polarization — Describes the orientation of the electric-field component of an electromagnetic wave. The electric field can be vertical or horizontal with respect to the Earth's surface, resulting in either a vertically or a horizontally polarized wave. (Also called **linear polarization**.)

Polarization — A property of an electromagnetic wave that describes the orientation of the electric field of the wave.

Pseudonoise (PN) — A binary sequence designed to appear to be random (contain an approximately equal number of ones and zeros). Pseudonoise is generated by a digital circuit and mixed with digital information to produce a direct-sequence spread-spectrum signal.

Pulse-amplitude modulation (PAM) — A pulse-modulation system where the amplitude of a standard pulse is varied in relation to the information-signal amplitude at any instant.

Pulse modulation — Modulation of an RF carrier by a series of pulses. These pulses convey the information that has been sampled from an analog signal.

Pulse-position modulation (PPM) — A pulse-modulation system where the position (timing) of the pulses is varied from a standard value in relation to the information-signal amplitude at any instant.

Pulse-width modulation (PWM) — A pulse-modulation system where the width of a pulse is varied from a standard value in relation to the information-signal amplitude at any instant.

Root-mean-square (RMS) voltage — A measure of the effective value of an ac voltage.

Sawtooth wave — A waveform consisting of a linear ramp and then a return to the original value. It is made up of sine waves at a fundamental frequency and all harmonics.

Sine wave — A single-frequency waveform that can be expressed in terms of the mathematical sine function.

Single-sideband, suppressed-carrier signal — A radio signal in which only one of the two sidebands generated by amplitude modulation is transmitted. The other sideband and the RF carrier wave are removed before the signal is transmitted.

Slow-scan television — A TV system used by amateurs to transmit pictures within a signal bandwidth allowed on the HF bands by the FCC.

Spread-spectrum modulation — A signal-transmission technique where the transmitted carrier is spread out over a wide bandwidth.

Square wave — A periodic waveform that alternates between two values, and spends an equal time at each level. It is made up of sine waves at a fundamental frequency and all odd harmonics.

SIGNALS AND EMISSIONS

The syllabus for Element 4 specifies five groups of loosely related topics. There will be five questions from this material on your Amateur Extra class exam. There are five syllabus sections and five groups of questions in the E8 subelement of the Amateur Extra class question pool.

E8A   AC waveforms: sine wave, square wave, sawtooth wave; AC measurements: peak, peak-to-peak and root-mean-square (RMS) value, peak-envelope-power (PEP) relative to average

E8B   FCC emission designators versus emission types; modulation symbols and transmission characteristics; modulation methods; modulation index; deviation ratio; pulse modulation; width; position

E8C   Digital signals: CW; baudot; ASCII; packet; AMTOR; Clover; information rate vs bandwidth

E8D   Amplitude compandored single-sideband (ACSSB); spread-spectrum communications

E8E   Peak amplitude (positive and negative); peak-to-peak values: measurement; Electromagnetic radiation; wave polarization; signal-to-noise (S/N) ratio

When you have studied the information in each section of this chapter, use the examination questions listed in Chapter 10 as directed, to review your understanding of the material.

AC WAVEFORMS

Sine Waves

The basic ac waveform is the **sine wave**. A sine wave represents a single **frequency**. To visualize a sine wave, let's imagine a wheel, like a bicycle wheel. We will paint a dot on the edge of the wheel at one point, so we can watch that spot as the wheel spins. If you look at the wheel edge, the spot will just seem to move up and down, as illustrated in **Figure 8-1A**. B pictures the wheel from one side, with the spot shown stopped at several points around the circle. If you can imagine the pattern at A as the wheel moves sideways to the right, the spot will trace a sine wave, as shown at C. This sine wave also represents the output from an ac generator, or alternator as it is called. One full rotation corresponds to a complete cycle (360 degrees).

Figure 8-1 is not only a mechanical analogy of alternator operation; it is a mathematical model as well. In the mathematical model, a line drawn from the axle to our paint spot is a rotating vector. The graph describes the vector ordinate (value along the Y or vertical axis) as it varies with time. Twice during each cycle a sine wave passes through zero; once while going positive, and once while going negative.

The time required to complete one cycle is called the **period, T**. The frequency of the sine wave is the reciprocal of the period:

$$f = \frac{1}{T}$$

(Equation 8-1)

Sawtooth Waves

A **sawtooth wave**, as shown in **Figure 8-2A**, is so named because it closely resembles the teeth on a saw blade. It is characterized by a rise time significantly faster than the fall time (or vice versa). A sawtooth wave is made up of a sine wave

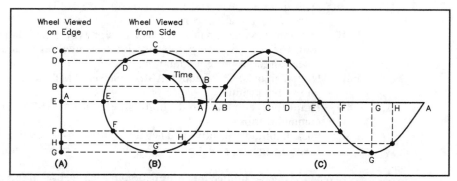

Figure 8-1 — This diagram illustrates the relationship between a sine wave and an object rotating in a circle. You can see how various points on the circle relate to sine-wave values.

at the fundamental frequency and sine waves at all the harmonic frequencies as well. When a sawtooth voltage is applied to the horizontal deflection plates of an oscilloscope, the electron beam sweeps slowly across the screen during the slowly changing portion of the waveform and then flies quickly back during the rapidly changing portion of the signal. This type of waveform is desired to obtain a linear sweep in an oscilloscope.

Square Waves

A **square wave** is one that abruptly changes back and forth between two voltage levels and remains an equal time at each level. See Figure 8-2B. (If the wave spends an unequal time at each level, it is known as a rectangular wave.) A square wave is made up of sine waves at the fundamental and all the odd harmonic frequencies.

[Study examination questions E8A01 through E8A07 in Chapter 10. Review this section as needed.]

AC MEASUREMENTS

The time dependence of alternating-current waveforms raises questions about defining and measuring values of voltage, current and power. Because these parameters change from one instant to the next, one might wonder, for example, which point on the cycle characterizes the voltage or current for the entire cycle. Since the wave is positive for exactly the same time it is negative in value, you might even wonder if the value shouldn't be zero. Actually, the average dc voltage and current are zero! A dc meter connected to an ac voltage would read zero, although you may be able to notice some slight flutter on a sensitive meter. To get some idea of how useful the ac voltage might be, we will have to connect a diode in series with the meter lead and use a specially calibrated meter scale.

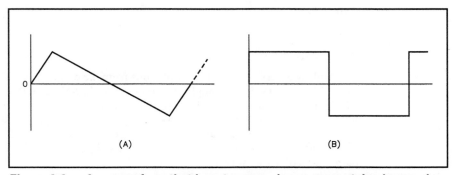

Figure 8-2 — Any waveform that is not a pure sine wave contains harmonics. Sawtooth waves (A) consist of both odd and even harmonics as well as the fundamental. Square waves (B) consist of only fundamental and odd-harmonic frequencies.

AC Voltage and Current

When viewing a sine wave on an oscilloscope, the easiest dimension to measure is the total vertical displacement, or **peak-to-peak (P-P) voltage**. The maximum positive or negative potential is called the **peak voltage**. In a symmetrical waveform it has half the value of the peak-to-peak amplitude.

When an ac voltage is applied to a resistor, the resistor will dissipate energy in the form of heat, just as if the voltage were dc. The dc voltage that would cause identical heating in the ac-excited resistor is called the **root-mean-square (RMS)** or effective value of the ac voltage. The phrase root-mean-square describes the mathematical process of actually calculating the effective value. The method involves squaring the peak values for a large number of points along the waveform (a calculus procedure), then finding the average of the squared values, and taking the square root of that number. The RMS voltage of any waveform can also be determined by measuring the heating effect in a resistor. For sine waves, the following relationships hold:

$$V_{peak} = V_{RMS} \times \sqrt{2} = V_{RMS} \times 1.414 \qquad \text{(Equation 8-2)}$$

and

$$V_{RMS} = \frac{V_{peak}}{\sqrt{2}} = V_{peak} \times 0.707 \qquad \text{(Equation 8-3)}$$

If we consider only the positive or negative half of a cycle, it is possible to calculate an average value for a sine-wave voltage. Meter movements respond to this average value rather than either the peak or RMS values of a voltage because of the inertia inherent in the needle and magnet. Often, the meter has a scale that is calibrated to read RMS values, even though the needle is actually responding to the average value. That is okay, as long as the waveform you are measuring is a pure sine wave. Other, more complex, waveforms will not give a true reading, however. The mathematical relationships between average, peak and RMS values for a sine wave are given by:

$$V_{avg} = V_{peak} \times 0.637 \qquad \text{(Equation 8-4)}$$

and

$$V_{avg} = V_{RMS} \times 0.900 \qquad \text{(Equation 8-5)}$$

Unless otherwise specified or obvious from the context, ac voltage is rendered as an RMS value. For example, the household 120-V ac outlet provides 120-V RMS, 169.7-V peak and 339.4-V P-P. The voltage at your household outlets varies with the amount of load that the power company must supply. It will vary around the nominal 120-V value, and is sometimes even specified as 110. Of course this means that the peak and P-P values also vary. **Figure 8-3** illustrates the voltage parameters of a sine wave.

Peak Amplitude and Peak-To-Peak Values

When an oscilloscope is used to view varying voltages, the easiest dimension

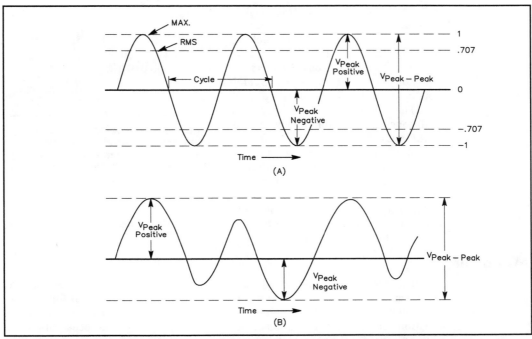

Figure 8-3 — It is easy to measure the peak negative, peak positive and peak-to-peak values of a waveform on an oscilloscope display. A shows a sine waveform and B shows a complex waveform.

to measure is the total vertical displacement, or **peak-to-peak (P-P)** voltage. The maximum positive or negative displacement from the center line is called peak voltage. Figure 8-3 shows that the **peak positive** and **peak negative** voltages need not have the same value. Of course, we can apply the same terminology to current or power values; we are not just talking about voltage values.

peak-to-peak = peak positive – peak negative   (Equation 8-6)

If the peak positive value plus the peak negative value equals zero for a waveform, that waveform is called a symmetrical waveform. In other words, the relationship between the peak-to-peak voltage and the peak voltage amplitude for a symmetrical waveform is 2:1. A sine wave is one example of a symmetrical waveform. For such waveforms:

peak-to-peak = 2 × peak positive = –2 × peak negative   (Equation 8-7)

Peak-to-peak voltage measurements are particularly useful for the characterization of linear devices. Generally, "linear" circuits are linear only over some finite range of input voltages. Inputs outside of this range will produce unwanted, spurious outputs. Because it is important to avoid such nonlinear operation, the maximum peak-to-peak input amplitude is an important criterion for evaluating linear (Class A) amplifiers.

For sine-wave signals, you can easily calculate the peak (and peak-to-peak) voltage if you know the RMS reading from an ac voltmeter. Just remember that the peak value is given by multiplying the RMS value by $\sqrt{2}$ (1.414) as shown by Equation 8-2. You can use the same equation to calculate the peak current, if you know the RMS current value.

The significant dimension of a multitone signal (a complex waveform) is the **peak envelope voltage (PEV)**, shown in **Figure 8-4**. PEV is important in calculating the power in a modulated signal, such as that from an amateur SSB transmitter.

All that has been said about volt-

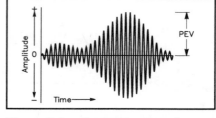

Figure 8-4 — A complex waveform, made up of several individual sine-wave signals. Peak envelope voltage (PEV) is an important parameter for a composite waveform.

age measurements applies also to current (provided the load is resistive) because the waveshapes are identical.

AC Power

The terms RMS, average and peak have different meanings when they refer to ac power. The reason is that while voltage and current are sinusoidal functions of time, power is the product of voltage and current, and this product is a sine squared function. The calculus operations that define RMS, average and peak values will naturally yield different results when applied to this new function. The relationships between ac voltage, current and power are as follows:

$$V_{RMS} \times I_{RMS} = P_{avg} \qquad \text{(Equation 8-8)}$$

Note that this calculation does not give a value for RMS power. The average power used to heat a resistor is equal to the dc power required to produce the same heat. RMS power has no physical significance!

For continuous sine wave signals:

$$V_{peak} \times I_{peak} = P_{peak} = 2 \times P_{avg} \qquad \text{(Equation 8-9)}$$

Unfortunately, the situation is more complicated in radio work. We seldom have a steady sine-wave signal being produced by a transmitter. The waveform varies with time, in order to carry some useful information for us. The peak power output of a radio transmitter, then, is the power averaged over the RF cycle having the greatest amplitude. Modulated signals are not purely sinusoidal because they are composites of two or more audio tones. The cycle-to-cycle variation is small enough, however, that sine-wave measurement techniques produce accurate results. In the context of radio signals, then, peak power means maximum average power. **Peak envelope power (PEP)** is the parameter most often used to express the maximum signal level. To compute the PEP of a waveform such as that sketched in Figure 8-4, multiply the PEV by 0.707 to obtain the RMS value, square the result and divide by the load resistance.

If you use a power amplifier in your amateur station, and want to ensure that you do not exceed the maximum allowable power, you might want to have a peak-reading power output meter in the feed line to your antenna. Such a meter will indicate the peak envelope power output (PEP) from your SSB transmitter and amplifier, as long as they are properly adjusted. Remember that you calculate power by multiplying the voltage and current values. Most wattmeters measure voltage, and are designed to be used in a system with a particular characteristic impedance, such as 50 Ω. You can also use a peak-reading voltmeter or an oscilloscope to monitor the output power from your single-sideband station. It is easy to calculate the PEP output using this voltage reading.

PEP is defined as the *average* power during one RF cycle at a modulation peak of the transmitted waveform. Calculate the average power by first dividing the peak voltage reading by $\sqrt{2}$ (1.414). (This is the same as multiplying the peak value by 0.707.) See Equation 8-3.

Calculate the peak envelope power by squaring the RMS voltage and dividing by the characteristic impedance of the system.

$$PEP = \frac{\left(V_{RMS}\right)^2}{Z}$$ (Equation 8-10)

For example, suppose you used an oscilloscope to monitor the output from your transmitter, and measured the peak-to-peak output voltage as 60. The transmitter is connected to a 50-Ω dummy load to ensure a good impedance match. That is equal to a peak output voltage of 30. Calculate the RMS value for this peak reading by using Equation 8-3.

$$V_{RMS} = 0.707 \times V_{PEAK} = 0.707 \times 30 \text{ V} = 21.2 \text{ V}$$

Then, from Equation 8-10, you can calculate the PEP output.

$$PEP = \frac{\left(V_{RMS}\right)^2}{Z} = \frac{\left(21.2 \text{ V}\right)^2}{50 \text{ }\Omega} = \frac{449 \text{ V}^2}{50 \text{ }\Omega} = 8.98 \text{ W} = 9 \text{ W}$$

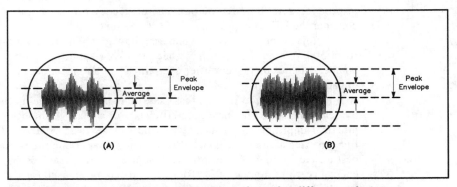

Figure 8-5 — Two envelope patterns that show the difference between average and peak levels. In each case, the RF amplitude (current or voltage) is plotted as a function of time. In B, the average level has been increased. That will raise the average output power compared to the peak value.

SSB Power

Envelope peaks occur only sporadically during voice transmission and have no relationship with meter readings. The meters respond to the amplitude (current or voltage) of the signal, averaged over several cycles of the modulation envelope.

The ratio of peak-to-average amplitude varies widely with voices of different characteristics. In the case shown in **Figure 8-5**, the average amplitude (found graphically) is such that the peak-to-average ratio of amplitudes is almost 3:1. Typical ratio values range from 2:1 to more than 10:1. So the PEP of an SSB signal may be about 2 or 3 times greater than the average power output. It may even be more than that, depending on the voice characteristics of the person speaking into the microphone.

[Turn to Chapter 10 and study questions E8A08 through E8A15. Also study questions E8E01 through E8E07. Review this section as needed.]

AMPLIFIER EFFICIENCY

We are always concerned about the efficiency of a circuit. Since a power amplifier is a large drain on a power supply, it is a good idea to pay a little extra attention to the efficiency of a high-power amplifier. The goal is to transfer as much power to the load (antenna) as possible. Some of the amplifier input power will always be dissipated in the amplifier circuitry, however. Some of the input power will be converted to heat, for example. We can express the efficiency of an amplifier as the output power divided by the input power, times 100%.

$$\text{Efficiency} = \frac{P_{OUT}}{P_{IN}} \times 100\%$$

(Equation 8-11)

The input power includes the dc input power (dc supply voltage times current) plus the drive power. For practical purposes, we can often ignore the drive power, because it will be a small percentage of the total input power. For example, suppose a certain amplifier requires 2000 W dc input power and 50 W RF drive power to produce a 1500-W RF output. The efficiency of that amplifier is 73% if you consider the RF drive power, or 75% if you ignore it.

At this point you may be wondering how much input power might be required for a 500-W amplifier you have been thinking about designing. The answer, of course, depends on the amplifier operating class and the modes for which you want to use the amplifier.

In Chapter 7 you learned that Class A amplifiers are the most linear, and also the least efficient. The maximum efficiency of a Class A amplifier is 50%, but for practical amplifiers the efficiency is usually closer to 25 or 30%. A Class AB amplifier can often attain more than 50% efficiency. Class B amplifiers have a theoretical best efficiency of 65%, and practical amplifiers often reach 60%. Class C amplifiers can reach as high as 80% efficient. They are often used for CW operation.

If the 500-W amplifier mentioned earlier will operate in Class AB, then we can pick an approximate efficiency of 50% to calculate the input power. Solving Equation 8-11 for input power, we have:

$$P_{IN} = \frac{P_{OUT}}{\text{Efficiency}} \times 100\% \qquad \text{(Equation 8-12)}$$

$$P_{IN} = \frac{500\,W}{50\%} \times 100\% = 1000\,W$$

[Turn to Chapter 10 now and study questions E8A16 and E8A17. Review this section if you have any trouble answering these questions.]

FCC EMISSION DESIGNATORS

The International Telecommunication Union (ITU) has developed a special system of identifiers to specify the types of signals (emissions) permitted to amateurs and other users of the radio spectrum. This system designates emissions according to their necessary bandwidth and their classification. While a complete **emission designator** might include up to five characters, generally only three of them are used.

The designators begin with a letter that tells what type of modulation is being used. The second character is a number that describes the signal used to modulate the carrier. The third character specifies the type of information being transmitted.

Table 8-1 summarizes the most common characters for each of the three symbols that make up an emission designator. Some of the more common combinations are:

- N0N — Unmodulated carrier
- A1A — Morse code telegraphy using amplitude modulation
- A3E — Double-sideband, full-carrier, amplitude-modulated telephony
- J3E — Amplitude-modulated, single-sideband, suppressed-carrier telephony
- J3F — Amplitude-modulated, single-sideband, suppressed-carrier television
- F3E — Frequency-modulated telephony
- G3E — Phase-modulated telephony
- F1B — Telegraphy using frequency-shift keying without a modulating audio tone (FSK RTTY). F1B is designed for automatic reception.
- F2B — Telegraphy produced by modulating an FM transmitter with audio tones (AFSK RTTY). F2B is also designed for automatic reception.
- F1D — FM data transmission, such as packet radio.

You can assemble an emission designator by selecting one character from each of the three sets, based on your knowledge of the transmission system. For example, suppose you know that a certain signal is produced by an AM (double-sideband, full carrier) transmitter. That is represented by the letter A for the first character. If the transmitter is modulated with a single-channel signal containing quantized or digital information without the use of a modulating subcarrier, you would select the number 1 as the second character. Finally, suppose the resulting signal is a telegraphy signal primarily intended for aural (by ear) reception rather than for machine or computer (automatic) copy. Then the third character of our emission symbol will be A. So we have just completely described an A1A signal, which is Morse code telegraphy!

Table 8-1
Partial List of Emissions Designators

(1) First Symbol — Modulation Type

Unmodulated carrier	N
Double sideband full carrier	A
Single sideband reduced carrier	R
Single sideband suppressed carrier	J
Vestigial sidebands	C
Frequency modulation	F
Phase modulation	G
Various forms of pulse modulation	P, K, L, M, Q, V, W, X

(2) Second Symbol — Nature of Modulating Signals

No modulating signal	0
A single channel containing quantized or digital information without the use of a modulating subcarrier	1
A single channel containing quantized or digital information with the use of a modulating subcarrier	2
A single channel containing analog information	3
Two or more channels containing quantized or digital information	7
Two or more channels containing analog information	8

(3) Third Symbol — Type of Transmitted Information

No information transmitted	N
Telegraphy — for aural reception	A
Telegraphy — for automatic reception	B
Facsimile	C
Data transmission, telemetry, telecommand	D
Telephony	E
Television	F

Emission Types

Part 97, the FCC Rules governing Amateur Radio, refers to **emission types** rather than **emission designators**. The emission types are CW, phone, RTTY, data, image, MCW (modulated continuous wave), SS (spread spectrum), pulse and test. Any signal may be described by both an emission designator or an emission type.

While emission types are fewer in number and easier to remember, they are a somewhat less descriptive means of identifying a signal. There is still a need for emission designators in amateur work, and Part 97 does reference the designators. In fact, the official FCC definitions for the emission types include references to the symbols that make up acceptable emission designators for each emission type.

The US Code of Federal Regulations, Title 47, consists of telecommunications rules numbered Parts 0 through 200. These Parts contain specific rules for many telecommunications services the FCC administers. Part 2, Section 2.201,

Emission, modulation and transmission characteristics, spells out the details of the ITU emission designators, as the FCC applies them in the US.

Facsimile and Television Emission Designators

Facsimile is the transmission of fixed images or pictures by electronic means, with the intent to reproduce the images in a permanent (printed) form. By contrast, television is the transmission of transient images of fixed or moving objects. For a discussion of facsimile and **slow-scan television** operation see Chapter 2.

There are several emission designators that may be used for facsimile and television signals, depending on how the signals are produced:
- A3C for full carrier, amplitude-modulated facsimile signals with a single information channel.
- F3C for frequency-modulated facsimile signals with a single information channel.
- A3F for full-carrier, amplitude-modulated television signals with a single information channel.
- C3F for vestigial-sideband TV signals.
- F3F for frequency-modulated television signals with a single information channel.
- J3F for single-sideband, suppressed-carrier TV signals.

An examination of §97.305 of the FCC Rules reveals that these image emission types go together. (See Chapter 1.) Where one is permitted, they are all permitted. Bandwidth restrictions for these modes are covered in §97.307 of the FCC Rules.

When it comes to emissions designators, things are not always what they seem. Slow-scan TV signals consist of a series of audio tones that correspond to sync and picture-brightness levels. When SSTV is sent using an SSB transmitter, the emission is actually frequency modulation, although with the designators, J3F is the appropriate symbol. See Chapter 2 for details. Fast-scan TV, by contrast, is normally amplitude modulation or vestigial sideband (like broadcast TV). At microwave frequencies, FM is also used for fast-scan TV in the amateur bands — as it is in the TV satellites.

[Review your understanding of ITU emission designators and types by turning to Chapter 10 and studying questions E8B01 through E8B08. Review this section as needed.]

MODULATION METHODS

To pass your Extra class exam, you will need to know what type of modulator circuit is used to produce the various types of radio signals. In Chapter 7, Practical Circuits, we described the circuit and operation of several modulator types. If you find that you don't remember the circuit details for these modulator types, you should go back to the appropriate section in that chapter. Block diagrams of various transmitter systems are included in this section to help you understand how the pieces fit together.

Amplitude Modulation

Double-sideband, full-carrier, **amplitude modulation** (emission A3E) can be

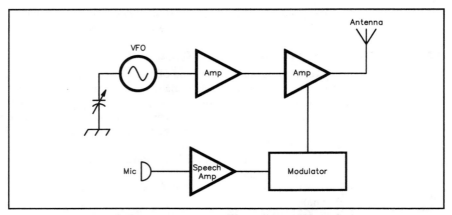

Figure 8-6 — Block diagram of an amplitude-modulated transmitter.

realized by simply modulating the supply voltage to an amplifier stage. See **Figure 8-6**. The two signals will be mixed, so the output from the modulated stage will include the input radio frequency, the modulating frequency, the sum of the two signals (upper sideband), the difference between them (lower sideband). One way to do this is to apply the modulating signal to the plate or collector supply voltage of a Class-C RF amplifier. Other methods require the modulating signal to be applied to the grid or base circuit of a Class-A or -AB amplifier. The main thing to remember is that the amplitude or strength of the output signal is changing in step with the amplitude of a modulating signal. If the frequency of that signal is also changing (as it would be for your voice), then the sideband frequencies are changing.

Single Sideband

A **single-sideband, suppressed-carrier signal** (emission J3E) is much like an AM signal, with one important exception. By transmitting only one of the sidebands, and eliminating the carrier, SSB occupies a much smaller bandwidth. It is the most-used method for transmitting voice signals on the HF amateur bands.

SSB signals can be generated in a two-step process. In the first step, a double-sideband, suppressed-carrier signal is generated in a balanced modulator. Remember that in a balanced modulator the input AF and RF signals do not appear at the output — only the sum (upper sideband) and difference (lower sideband) frequencies appear. In the second step, the unwanted sideband is filtered out, leaving only the desired one. **Figure 8-7** illustrates the essentials of such a sideband transmitter. Some transmitters use one crystal with the oscillator, but switch in a different filter to eliminate the other sideband.

Frequency or Phase Modulation

Frequency modulation (emission F3E) operates on an entirely different principle than amplitude modulation. With FM, the signal frequency is varied above

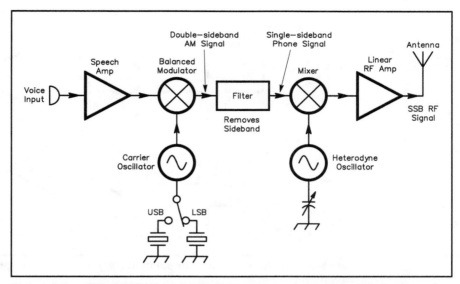

Figure 8-7 — This block diagram illustrates how a balanced modulator can be used to generate a double-sideband AM signal. The filter removes one sideband to produce a single-sideband signal.

and below the carrier frequency at a rate equal to the modulating-signal frequency. For example, if a 1000-Hz tone is used to modulate a transmitter, the carrier frequency will vary above and below the center frequency 1000 times per second. The amount of frequency change, however, depends on the instantaneous amplitude of the modulating signal. This frequency change is called **deviation**. A certain signal might produce a 5-kHz deviation. If another signal, with only half the amplitude of the first, were used to modulate the transmitter, it would produce a 2.5-kHz deviation. From this example, you can see that the deviation is proportional to the modulating signal amplitude.

Direct FM can be produced by a reactance modulator. The reactance modulator is connected to an oscillator in such a way as to act as a variable inductance or capacitance. When a modulating signal is applied, the oscillator frequency varies. **Figure 8-8A** shows how such a system is arranged.

Phase modulation (emission G3E, sometimes called indirect FM) can be realized by using a phase modulator. A phase modulator is similar to a reactance modulator in that it appears to be a variable inductance or capacitance when modulation is applied. The difference between the two is where they are found in the transmitter. Whereas the reactance modulator controls an oscillator, the phase modulator acts on a buffer or amplifier stage. This arrangement is shown at Figure 8-8B.

FM Terminology

You will need to understand two terms that refer to FM systems and operation: **deviation ratio** and **modulation index**. They may seem to be almost the same —

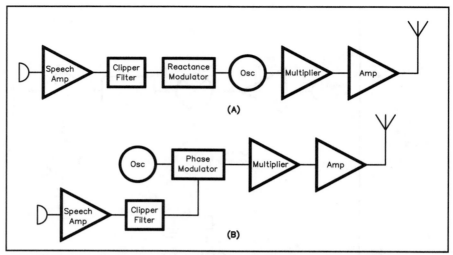

Figure 8-8 — Direct frequency modulation (FM) is shown at A; phase modulation (PM) at B.

indeed, they are closely related. Pay special attention to the equations given to calculate these quantities, and you should be able to distinguish between them.

Deviation Ratio

In an FM system, the ratio of the maximum carrier-frequency deviation to the highest modulating frequency is called the deviation ratio. It is a constant value for a given system, calculated by:

$$\text{deviation ratio} = \frac{D_{max}}{M} \qquad \text{(Equation 8-13)}$$

where:

D_{max} = peak deviation in hertz (half the difference between the maximum and minimum carrier-frequency values at 100% modulation).

M = maximum modulating frequency in hertz.

In the case of narrow-band FM, peak deviation at 100% modulation is 5 kHz. The maximum modulating frequency is 3 kHz. Therefore:

$$\text{deviation ratio} = \frac{D_{max}}{M} = \frac{5 \text{ kHz}}{3 \text{ kHz}} = 1.67$$

Notice that since both frequencies were given in kilohertz we did not have to change them to hertz before doing the calculation. The important thing is that they both be in the same units.

Modulation Index

The ratio of the maximum carrier-frequency deviation to the (instantaneous) modulating frequency is called the modulation index. That is:

$$\text{modulation index} = \frac{D_{max}}{m} \qquad \text{(Equation 8-14)}$$

where:

D_{max} = peak deviation in hertz.

m = modulating frequency in hertz at any given instant.

For example, suppose the peak frequency deviation of an FM transmitter is 3000 Hz either side of the carrier frequency. The modulation index when the carrier is modulated by a 1000-Hz sine wave is:

$$\text{modulation index} = \frac{D_{max}}{m} = \frac{3000 \text{ Hz}}{1000 \text{ Hz}} = 3$$

When modulated with a 3000-Hz sine wave with the same peak deviation (3000 Hz), the index would be 1; with a 100-Hz modulating wave and the same 3000-Hz peak deviation, the index would be 30, and so on.

In a phase modulator, the modulation index is constant regardless of the modulating frequency, as long as the amplitude is held constant. In other words, a 2-kHz tone will produce twice as much deviation as a 1-kHz tone if the amplitudes of the tones are equal. This may seem confusing at first; just think of the peak deviation as the variable in Equation 8-14.

By contrast, the modulation index varies inversely with the modulating frequency in a frequency modulator, as Equation 8-14 shows. A higher modulating frequency results in a lower modulation index, if the peak deviation remains the same. The actual deviation depends only on the amplitude of the modulating signal and is independent of frequency. Thus, a 2-kHz tone will produce the same deviation as a 1-kHz tone if the amplitudes of the tones are equal. The modulation index in the case of the 1-kHz tone is double that for the case of the 2-kHz tone.

Notice that with either an FM or a PM system, the deviation ratio and modulation index are independent of the frequency of the modulated RF carrier. It doesn't matter if the transmitter is a 10-meter FM rig or a 2-meter rig.

[At this point you should turn to Chapter 10 and study examination questions E8B09 through E8B13. Review this section as needed.]

PULSE MODULATION

There are several methods used for transmitting information as a series of brief RF pulses. These methods are collectively known as **pulse modulation**, and they vary in how the modulating-signal information is conveyed by the pulses. In **pulse-width modulation (PWM)**, also known as pulse-duration modulation (PDM), the duration (width) of the transmitted pulses varies with the applied modulation. In **pulse-position modulation (PPM)** the modulated signal is produced by varying the position of each pulse with respect to some fixed timebase. This means the modulating signal changes the timing of the transmitted pulses. Both systems start with a sample of the information signal. This sampling can be done by using a sawtooth waveform with a rapid rise time and a frequency that is several times higher than the information-signal frequency. **Figure 8-9** shows how this sawtooth

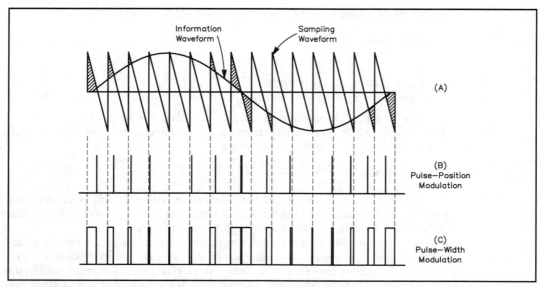

Figure 8-9 — An information signal and a sampling waveform are shown at A. B shows how the sampled information signal can produce pulse-position modulation, and C shows how it can produce pulse-width modulation.

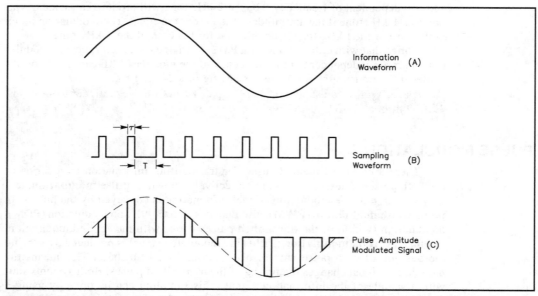

Figure 8-10 — A shows an information signal, B is a sampling waveform and C is a pulse-amplitude-modulated signal.

waveform can be used to trigger a sampling circuit and measure the difference between the information signal amplitude and the sampling wave.

While both PWM and PPM use pulses of uniform amplitude, the third common form of pulse modulation, **pulse-amplitude modulation (PAM)**, changes the amplitude of each pulse to reflect the modulating signal amplitude. **Figure 8-10** illustrates how PAM signals are generated. PAM is less immune to noise than are either of the other methods, and so it is seldom considered for communication systems, although it may be the easiest method to visualize and understand.

Notice that in all three cases, the duty cycle of the transmission is very low. A pulse of relatively short duration is transmitted, with a relatively long period of time separating each pulse. This causes the peak power of a pulse-modulated signal to be much greater than its average power.

Our examples have used a sine-wave-like information signal. If the information signal that is being sampled to produce pulse modulation is a voice, then the transmitted signal will be a representation of that speech pattern. When the signal is demodulated in the receiver, the voice audio will be reproduced. A voice signal can be made to vary the duration, position, amplitude or code of a standard pulse to produce a pulse-modulated signal. For example, in a pulse-width modulation system, the standard pulse is varied in duration by an amount that depends on the voice waveform at a particular instant.

Amateurs use pulse-position modulation and pulse-width modulation systems for radio control purposes. The transmitted signal contains coded information about the position of a control device and the receiver decodes the information to control the position of a servo motor.

Pulse-modulation techniques find applications in some other electronics circuits as well. Switch-mode power supplies and switching regulators commonly use a pulse-width modulator IC. In a switching regulator for a power supply the output voltage is usually higher than the desired regulated output voltage. The pulse-width modulator IC turns the switching transistor on and off at the proper time to ensure smooth regulation. The switching usually occurs rapidly, up to several hundred kilohertz.

[Turn to Chapter 10 and study examination questions E8B14 through E8B16. Review this section as needed.]

DIGITAL COMMUNICATIONS

By the time you reach the Amateur Extra license examination, you have been introduced to a number of methods for transmitting digital information. For the Amateur Extra class exam you will need to know a little bit about the digital codes available to amateurs. Morse code, the grandfather of all digital codes, consists of elements having unequal length (dots and dashes). The other codes you should be familiar with are Baudot, ASCII and AMTOR. Each of these codes use elements that have equal length.

Baudot Code

The Baudot code, also known as International Telegraph Alphabet Number 2 (or ITA2) was the only digital code (other than Morse) allowed on the amateur bands until

1980. The Baudot code uses five information bits, with additional bits (called "start" and "stop" bits) to indicate the beginning and end of each character. These five data bits can be arranged into 32 different combinations, so there are only 32 characters possible with the Baudot code. Some of these combinations can be used twice, and the letters (LTRS) and figures (FIGS) characters shift between those combinations. This is how all 26 letters of the alphabet, ten numbers and punctuation can all be included in the Baudot code. The Baudot code only allows upper-case letters, however.

AMTOR

Radioteletype (RTTY) communications are plagued with problems of fading and noise unless something is done to mitigate these effects. Many techniques have been tried to overcome fading and noise, but one of the most effective techniques is called *time diversity*. With time diversity, the same signal is sent at different times; the signals experience different fading and noise conditions, and the likelihood of the signal reaching its destination uncorrupted is greatly increased. Time diversity is the basis of AMTOR or *AMateur Teleprinting Over Radio*.

AMTOR was derived from a commercial system, SITOR, which was designed for use in the Maritime Mobile Service. SITOR is used for ship-to-ship, ship-to-shore and between a ship and a subscriber to the international telex network. In the Maritime Mobile Service, this type of system was devised as a means of improving communications between five-bit asynchronous teleprinters using the ITA2 (Baudot) code. The system converts the five-bit Baudot code to a seven-bit code for transmission. Ordinarily, a seven-bit code could have up to 2^7 or 128 possible combinations. The code used for AMTOR is limited, however, in order to make every character have a constant ratio of four mark bits to three space bits. This constant mark-to-space ratio provides AMTOR with a means of error detection; if a received character does not have this constant ratio, the receiving station knows it is erroneous.

Out of the 128 possible combinations of seven bits, the constant mark/space ratio limits the number of possible combinations to 35. Thirty-two of the possible characters are used for the 32 standard ITA2 (Baudot) characters, and there are three left over for use as *service information signals*. These three unique combinations are called Idle Signal Alpha, Idle Signal Beta and Repeat Request (or RQ).

The actual code used for AMTOR characters is not all that different from the Baudot code. What makes AMTOR different is the way the code is used; the error-detection features of AMTOR depend on the exact implementation of certain procedures. These procedures are spelled out in CCIR Recommendations 476-3, 476-4 and 625. To completely understand AMTOR emissions, you must understand these special operating procedures.

Mode A (ARQ)

The station that initiates a Mode A AMTOR QSO is known as the Master Station (MS). The other station is known as the Slave Station (SS). The Master Station retains that identity throughout the contact, even when it is the receiving station. The MS sends the selective call identifier of the called station in blocks of three characters, listening between blocks. Four-letter call identifiers are used in AMTOR; these calls are normally derived from the first character and the last three letters of the station call sign. For example, the AMTOR identifier for W9NGT would be

WNGT. When the Slave Station (SS) recognizes its identifier, it signals the Master Station that it is ready. The Master Station then becomes the *Information Sending Station* (ISS) and the Slave Station becomes the *Information Receiving Station* (IRS). Data transmission begins when the IRS signals that it is ready.

AMTOR Mode A uses an error-control system in which the receiving station automatically requests repeats when needed. Mode A is also known as ARQ, which stands for *Automatic Repeat reQuest*.

Mode A is a synchronous system; blocks of three characters are sent from the ISS to the IRS. After contact is established, the ISS sends its message in groups of three characters and pauses between groups for a reply from the IRS. Each character is sent at a rate of 100 bauds, amounting to 70 ms for one character or 210 ms for a three-character block. The block repetition cycle is 450 ms, so there are 240 ms during each cycle that the ISS is not sending. This 240-ms period is taken up by propagation time from the ISS to the IRS, 70 ms for the IRS to send its service information signal, and propagation time for the service signal back to the ISS. The IRS requests repeats of three-character information blocks that do not have the appropriate mark/space ratio.

When the ISS is done sending, it can enable the other station to become the ISS by sending the three-character sequence FIGS Z B. The ISS may end the contact by sending an "end of communication signal" consisting of three Idle Signal Alpha characters.

On the air, AMTOR Mode A signals have a characteristic "chirp-chirp" sound. Depending on propagation, you may only be able to hear one side of the contact.

Because of the 210/240-ms on/off cycle, Mode A can be used with some transmitters operating at full power levels. Baudot and ASCII RTTY require that the transmitter operate at a 100% duty cycle and in that case many transmitters must be reduced by 50% to 25% of full power.

Mode B (FEC)

When transmitting to no particular station (calling CQ, net or bulletin operation) there is no (one) station to act as IRS, so there is no station to request repeats of corrupted information. Even if there was one IRS, the ability (or inability) of this one station to receive information may not be representative of other stations desiring to copy the transmission.

Mode B uses a simple error-control technique of sending each character twice. This system is called *Forward Error Correction*, or FEC. If the repetition was sent immediately, a single noise burst or rapid fade could eradicate both characters. Burst errors can be virtually eliminated by delaying the repetition for a period thought to exceed the duration of most noise bursts. In Mode B AMTOR, the first transmission of a character (called the DX) is followed by four other characters, after which the retransmission (called the RX) of the original character occurs. At 70 ms per character, this leaves 280 ms between the end of the first transmission and the beginning of the second.

In Mode B, the receiving station tests for the constant 4/3 mark/space ratio and prints only unmutilated DX or RX characters; an error symbol or space is printed if both characters are mutilated.

The sending station transmitter must be capable of 100% duty cycle for Mode B. It may be necessary to reduce transmitter output power by 50% to 25% of full rating.

ASCII

ASCII stands for American National Standard Code for Information Interchange; ASCII is the commonly used code for computer systems. The ASCII code uses seven information bits, so more characters can be defined than with the five-bit Baudot code. In fact, there are 128 combinations possible with this seven-bit code. The seven data bits in each character make it possible for the ASCII character set to provide upper- and lower-case letters, numbers, punctuation and special characters. The ASCII code does not need a shift code, like the Baudot code does, to change between the letters and figures case.

An eighth bit is often included with the ASCII characters. This is called a *parity bit*, and is used to provide a form of error control. The eighth bit is a 1 or a 0 to maintain either an even number of 1s in the character string, or an odd number of 1s. The parity of the characters, then, can be set either to *even* or *odd*. If you don't need the error control provided by a parity check, you can also use the eighth bit to extend the character set to 256 characters instead of 128.

Packet Radio

Packet radio is communications for the computer age. As computers appeared in more and more shacks, computer programs were written to allow computers to send and receive CW and RTTY. That was great, but writing computer programs for CW and RTTY today is quite a waste of technology! Instead of writing computer programs to emulate keyers and teleprinters, some farsighted hams developed a new amateur mode of communications that unleashes the power of the computer. That mode is packet radio.

Being a child of the computer age, packet radio has the computer-age features that you would expect.

- It is data communications; high-speed, error-free packet-radio communications lends itself to the transfer of large amounts of data.
- It is fast; much faster than the highest speed CW or RTTY. This high speed makes packet ideal for meteor-scatter communications.
- It is error-free; no "hits" or "misses" caused by propagation variances or electrical interference.
- It is spectrum efficient; many stations can share one frequency at the same time.
- It is networking; packet stations can be linked together to send messages over long distances.
- It is message storage; packet-radio bulletin board systems (PBBS) provide storage of messages for later retrieval.

[Now turn to Chapter 10 and study examination questions E8C01 through E8C06. Review this section if you have difficulty with any of these questions.]

Information Rate and Bandwidth

Bandwidth of CW Signals

A Morse code (CW) signal produced by turning an AM transmitter on and off is described by the emission designator A1A. The **bandwidth** of a CW signal is determined by two factors: the speed of the CW being sent and the shape of the keying envelope. The usual equation for this calculation is:

$$Bw = B \times K \qquad \text{(Equation 8-15)}$$

where:

Bw is the necessary bandwidth of the signal.
B is the speed of the transmission in bauds.
K is a factor relating to the shape of the keying envelope.

To solve this equation, we must find values for B and K.

Morse code speed is usually expressed in words per minute (wpm), so we need to convert wpm to bauds. Since one **baud** equals one signal element per second, we must determine the number of signal elements in a Morse code word. (A signal element is the duration of one dot or one interelement space. Dashes count as three signal elements and the space between words is seven elements.) The standard word used for this calculation is "PARIS," which contains 50 signal elements (including the seven elements to the start of the next word). This results in the equation:

$$\frac{1 \text{ word}}{\text{minute}} = \frac{50 \text{ elements}}{\text{minute}} = \frac{50 \text{ elements}}{60 \text{ seconds}} = 0.83 \text{ bauds} \qquad \text{(Equation 8-16)}$$

From that we derive:

$$wpm \times 0.83 = bauds \qquad \text{(Equation 8-17)}$$

which can also be expressed as:

$$bauds = \frac{wpm}{1.2} \qquad \text{(Equation 8-18)}$$

Then, for CW signals Equation 8-15 becomes:

$$Bw = \left(\frac{wpm}{1.2}\right) \times K \qquad \text{(Equation 8-19)}$$

The second variable, K, is a measure of keying shape. As CW rise and fall times get shorter (more abrupt, harder keying), K gets larger. This is because signals with short rise and fall times contain more harmonics than longer, softer envelopes. (An ideal square wave contains an infinite number of odd harmonics; an ideal sine wave has no harmonic content.) The more harmonics in the keying envelope, the greater the bandwidth of the resulting CW signal. On paths with good strong signals, soft keying can be used, and K will be around 3. On fading paths, harder keying is necessary, resulting in a K of 5.

Since most Amateur Radio contacts are made over paths with at least some fading, amateur transmitters are usually adjusted to have keying rise and fall times on the order of 5 milliseconds. For this keying envelope, the K factor is about 4.8, and the resulting equation is:

$$Bw = \left(\frac{wpm}{1.2}\right) \times 4.8 = wpm \times 4 \qquad \text{(Equation 8-20)}$$

This equation is a good rule of thumb for calculating the necessary bandwidth of an amateur CW transmission.

As an example, suppose you are sending Morse code at a speed of 13 wpm. The bandwidth of the transmitted signal is:

Bw = wpm × 4 = 13 × 4 = 52 Hz.

Bandwidth of Binary Data Signals

Most amateur Baudot, AMTOR and ASCII transmissions employ frequency shift keying (FSK). In FSK systems, the transmitter uses one frequency to represent one state and another frequency to represent the other binary state. By shifting between these two frequencies (called the mark and space frequencies), the transmitter sends binary data. The difference between the mark frequency and the space frequency is called the shift. FSK signals can be generated either by shifting a transmitter oscillator or by injecting two audio tones, separated by the correct shift, into the microphone input of a single-sideband transmitter. If an FSK data emission is produced by these two methods, the emission designators are F1D and J2D, respectively. A J2D signal generated by a correctly adjusted SSB transmitter appears identical to an F1D signal. The necessary bandwidth of that signal is determined by the frequency shift used and the speed at which data is transmitted. The bandwidth is not affected by the code that is being transmitted. The equation relating necessary bandwidth to shift and data rate is:

$$Bw = (K \times shift) + B \qquad\qquad (Equation\ 8\text{-}21)$$

where:
   Bw is the necessary bandwidth in hertz.
   K is a constant that depends on the allowable signal distortion and transmission
      path. For most practical Amateur Radio communications, K = 1.2.
   Shift is the frequency shift in hertz.
   B is the data rate in bauds.

For example, the bandwidth of a 170-Hz shift, 300-baud ASCII signal transmitted as a J2D emission is:

Bw = (1.2 × 170 Hz) + 300 = 504 Hz

This is a necessary bandwidth of about 0.5 kHz.

[Before proceeding to the next section, turn to Chapter 10. Study examination questions E8C07 through E8C10. Review this section as needed.]

AMPLITUDE-COMPANDORED SINGLE SIDEBAND

Human speech has a dynamic range greater than 40 dB. That is, the booming voice of a lecturer may be 10,000 times louder than the voice of someone whispering in the audience. Even the peaks and nulls of normal conversation exhibit a dynamic range of 20 dB. Modulation systems must be able to accurately encode and later decode these amplitude variations to recover all of the original voice information. Failure to do so results in amplitude distortion and loss of intelligibility.

Communications systems must also contend with noise. Noise masks desired signals, reducing intelligibility. In any communications system, the S/N (signal-to-noise) ratio plays an important part in overall performance — the higher the S/N ratio, the better a system can accurately convey desired information.

In amplitude-modulation systems, S/N ratio can be improved by increasing the desired signal level relative to the noise. One obvious way to accomplish this is by increasing transmitter power. Another, more economical approach results from analysis of speech-waveform amplitude patterns. Since the average energy content of human speech is much less than its peak energy, a modulation method that decreases the peak-to-average ratio of transmitted speech would improve the S/N ratio.

Logarithmic amplifiers are often used to perform this task. **Figure 8-11** is a graphical representation of how a logarithmic amplifier works. Assume the input signal, X, and output, Y, are mathematically related by the function $Y = f(X)$. Further, assume this function is logarithmic. Such a function might take the form, $Y = \log(X)$. As X changes from amplitude level x_1 to x_2, Y changes from y_1 to y_2. The difference between output amplitudes, ΔY, is smaller than the difference between the input amplitudes, ΔX. Then we say Y is compressed. After transmission, the compressed signal is expanded back to its original shape. In the receiver, this is done with an expanding amplifier, which has an output function $B = g(A)$. It performs an inverse log function to get X back. The combined compression and expansion process is known as **compandoring**. When incorporated into a single-sideband transmission system, **amplitude-compandored single sideband (ACSSB)** offers significant S/N ratio improvement. ACSSB was originally

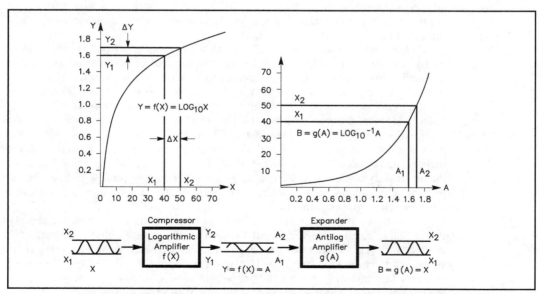

Figure 8-11 — These graphs illustrate the mathematical relationships in an amplitude-compandoring logarithmic amplifier and expander. See the text for a detailed discussion.

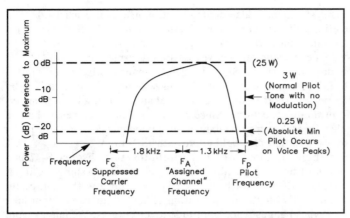

Figure 8-12 — This graph represents the frequency spectrum of an (upper-sideband) ACSSB signal.

developed to provide narrowband channels in the land-mobile VHF bands. It can be transmitted in a much narrower bandwidth than FM — in fact, a single FM channel can easily accommodate four ACSSB signals.

Baseband (audio) analog signal processing is at the heart of ACSSB. **Figure 8-12** represents a typical ACSSB signal. The passband is not flat. Preemphasis has been introduced (12 dB per octave) in the transmitter audio stages to accentuate higher frequencies. Conversely, the receiver uses a built-in deemphasis network to effectively flatten the passband. This process is similar to the preemphasis and deemphasis found in typical FM systems, and has the same aim — to improve the S/N ratio by reducing high-frequency hiss.

ACSSB systems use a **pilot tone** located 3.1 kHz away from the suppressed carrier. It is generated as an audio tone, and mixed with compressed voice information to form the base-band ACSSB signal. Steep-skirted low-pass filters keep higher-frequency audio components from interfering with the pilot; a cutoff frequency of approximately 2.8 kHz is used. On receive, audio and pilot information is processed separately. The strength of the recovered pilot tone is used to adjust the gain of audio expansion amplifiers.

Whenever the microphone PTT button is pressed, a full-power pilot tone is transmitted for $1/4$ second (250 ms). During this time, audio from the microphone is disabled. This allows an ACSSB receiver to lock quickly to the pilot tone, automatically tuning the receiver to the proper frequency. This is especially helpful in mobile systems. After $1/4$ second, the pilot tone amplitude drops approximately 9 dB, and voice transmission begins. The amplitude of the pilot varies with the information being sent — as the instantaneous voice signal becomes greater, the pilot is reduced in amplitude. Similarly, low-level voice energy causes the pilot to increase amplitude. The receiver uses this pilot tone to expand the signal, restoring the original dynamic range. For instance, a weak incoming pilot will drive the expansion amplifier in an ACSSB receiver audio stage to maximum gain, while a stronger pilot reduces gain, and thus intensity, of the audio output.

The effects of compandoring are most pronounced on weak or fading signals. Rapid fading causes received signal strength to drop too sharply for normal AGC circuits to respond. When this happens in an ACSSB receiver, the recovered audio signal and the pilot tone fade equally. Expansion amplifiers in the ACSSB receiver interpret this diminished pilot tone as a command to increase stage gain. Thus, momentarily weak signals are boosted, virtually eliminating flutter.

When extremely weak ACSSB signals are being received, however, both the recovered voice information and pilot are heavily shrouded in noise. Since the pilot requires less bandwidth than the audio signal, narrow pilot filters are used. These filters pass less noise than the wider audio filters, so the pilot tone S/N ratio will be higher. This, in turn, means the decoded pilot retains more information. Even though the audio channel may be very noisy, clean pilot tone information properly adjusts the gain of the logarithmic audio expansion amplifiers to accurately reconstruct the dynamic range of the original voice information. The result is gruff-sounding, because the missing voice information (tonal texture) has been replaced by noise. Still, this weak ACSSB signal is intelligible, while an equivalent FM signal would be lost in the noise.

[Turn to Chapter 10 now and study examination questions E8D01 through E8D05. Review this section as needed.]

SPREAD-SPECTRUM COMMUNICATIONS

The common rule of thumb for judging the efficiency of a modulation scheme is to examine how tightly it concentrates the signal for a given rate of information. While the compactness of the signal appeals to the conventional wisdom, **spread-spectrum modulation** techniques take the exact opposite approach — that of spreading the signal out over a very wide bandwidth.

Communications signals can be greatly increased in bandwidth (by factors of 10 to 10,000) by combining them with binary sequences. The exact techniques will be discussed later. This spreading has two beneficial effects. The first effect is dilution of the signal energy on a given frequency, so that while occupying a very large bandwidth, the power density present at any point within the spread signal is very slight. (See **Figure 8-13**.) The amount of signal dilution depends on several factors, such as transmitting power, distance from the transmitter and the width of the spread signal. The dilution may result in the signal being below the noise floor of a conventional receiver, and thus invisible to it, while the signal can still be received with a spread-spectrum receiver!

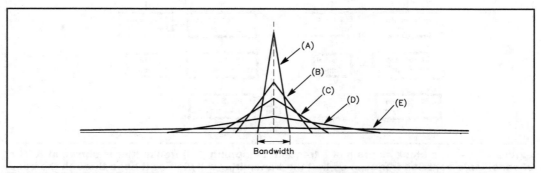

Figure 8-13 — This drawing is a graphic representation of the distribution of power as the signal bandwidth increases. The unspread signal (A) contains most of its energy around a center frequency. As the bandwidth increases (B), the power about the center frequency falls. At C and D, the energy is distributed in the spread signal's wider bandwidth. At E, the signal energy is spread over a very wide bandwidth, and there is little power at any one frequency.

The second beneficial effect of the signal-spreading process is that the spread-spectrum receiver can reject strong undesired signals — even those much stronger than the desired spread-spectrum-signal power density. This is because the receiver has a copy of the spreading sequence and uses it to "despread" the signal. Nonspread signals are then suppressed in the processing. The effectiveness of this interference-rejection property has made spread-spectrum a popular military antijamming technique.

Conventional signals such as narrowband FM, SSB and CW are rejected by a spread-spectrum receiver, as are other spread-spectrum signals not bearing the desired coding sequence. The result is a type of private channel, one where only the spread-spectrum signal using the correct **pseudonoise (PN)** coding signal will be accepted by the spread-spectrum receiver. A two-party conversation can take place, or if the code sequence is known to a number of people (as it is in amateur applications), net-type operations are possible.

The use of different binary sequences allows several spread-spectrum systems to operate independently within the same amateur band. This is a form of frequency sharing called code-division multiple access. If the system parameters are chosen judiciously and if the right conditions exist, conventional users in the same amateur band will experience very little interference from spread-spectrum users. This allows more signals to be packed into a band; however, each additional signal, conventional or spread spectrum, will add some interference for all users.

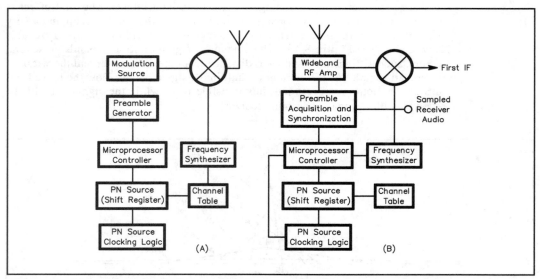

Figure 8-14 — The block diagram of a frequency-hopping (FH) transmitter is shown at A. The channel table contains frequencies that the transmitter will visit as it hops through the band. These channels are selected to avoid interference to fixed band users (such as repeaters). The preamble employs the falling edge of an audio tone to trigger hopping. Conventional modulation (such as SSB) is employed. The block diagram at B is for an FH receiver. Preamble acquisition and synchronization trigger the beginning of the hopping mode and keep the receiver channel changes in step with the transmitter.

Types of Spread Spectrum

There are many ways to cause a carrier to spread; however, all spread-spectrum systems can be viewed as a combination of two modulation processes. First, the information to be transmitted is applied to the carrier. A conventional form of modulation, either analog or digital, is commonly used for this step. Second, the carrier is modulated by the spreading code, causing it to spread out over a large bandwidth. Four spreading techniques are commonly used in military and space communications, but amateurs are currently only authorized to use two of the four techniques: frequency hopping and direct sequence.

Frequency Hopping

Frequency hopping (FH) is a form of spreading where the center frequency of a conventional carrier is altered many times per second in accordance with a

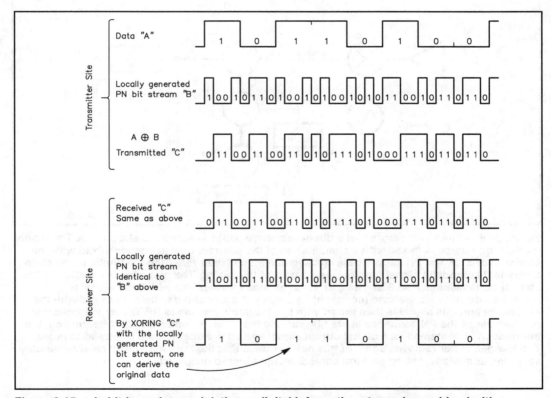

Figure 8-15 — In bit-inversion modulation, a digital information stream is combined with a pseudonoise bit stream, which is clocked at four times the information rate. The combination is the EXCLUSIVE-OR sum of the two. Notice that an information bit of one inverts the PN bits in the combination, while an information bit of zero causes the PN bits to be transmitted without inversion. The combination bit stream has the speed characteristics of the original PN stream, so it has a wider bandwidth than the information stream.

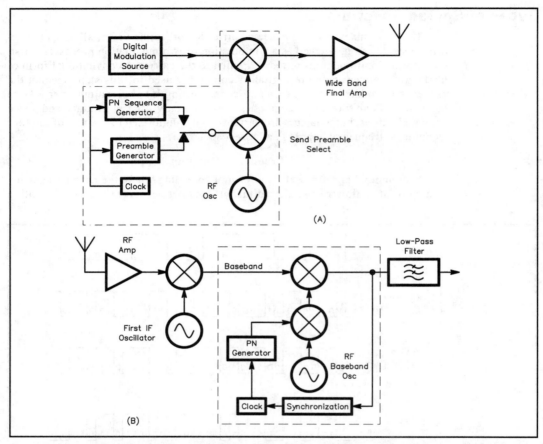

Figure 8-16 — The block diagram of a direct sequence (DS) transmitter is shown at A. The digital modulation source is mixed with a combination of the pseudonoise sequence mixed with the carrier oscillator. The PN sequence is clocked at a much faster rate than the digital modulation; a very fast composite signal emerges as a result of the mixing. The preamble is selected at the start of transmission. Part B shows a direct sequence receiver. The wideband signal is translated down to a baseband (common) frequency. A baseband oscillator is mixed with the PN source and this signal is then mixed with the incoming baseband RF. The synchronization process keeps the PN sequence in step by varying the clock for optimal lock. After mixing, the information is contained as a digital output signal and all the interference is spread to noise. The low-pass filter removes some of this noise. Notice that the transmitter and receiver employ very similar designs, one to perform spreading, the other to despread.

pseudorandom list of channels. See **Figure 8-14**. The amount of time the signal is present on any single channel is called the dwell time. To avoid interference both to and from conventional frequency users, the dwell time must be very short, typically less than 10 milliseconds.

Direct Sequence

In **direct sequence (DS)** spread spectrum, a very fast binary bit stream is used to shift the phase of an RF carrier. The binary sequence is designed to appear to be random (that is, a mix of approximately equal numbers of zeroes and ones), but the sequence is generated by a digital circuit. This binary sequence can be duplicated and synchronized at the transmitter and receiver. Such sequences are called **pseudonoise** or **PN**; noise, because the signal appears random, but *pseudo* noise because the signal only *appears* to be random — it can be duplicated at the receiver. Each PN code bit is called a *chip*. The phase shifting is commonly done in a balanced mixer that typically shifts the RF carrier between 0 and 180 degrees; this is called binary phase-shift keying (BPSK). Other types of phase-shift keying are also used. For example, quadrature phase-shift keying (QPSK) shifts between four different phases. The circuit that controls the spreading sequence of a spread-spectrum transmission is called a binary linear-feedback shift register.

DS spread spectrum is typically used to transmit digital information. See **Figures 8-15** and **8-16**. A common practice in DS systems is to mix the digital information stream with the PN code. The result of this mixing causes the PN code to be inverted for a number of chips to represent a one or left unchanged for an information bit of zero. This modulation process is called bit-inversion modulation.

[Now turn to Chapter 10 and study exam questions E8D06 through E8D11. Review this section if you have any problems.]

ELECTROMAGNETIC WAVES

All **electromagnetic waves** are moving fields of **electric** and **magnetic force**. Their lines of force are at right angles to each other, and are also both perpendicular to the direction of travel. See **Figure 8-17**. They can have any position with respect to the Earth. The plane containing the continuous lines of electric and magnetic force is called the wave front. Another way of visualizing this concept is to think of the wave front as being a fixed point on the moving wave.

Electromagnetic Radiation

Electricity requires a conductor to carry an electron current through a circuit. Electromagnetic waves move easily through the vacuum of free space. The **electric** and **magnetic fields** that constitute the wave do not require a conductor to carry them.

Radio waves travel through dielectric materials with ease. Waves cannot penetrate a good conductor, however. Instead of penetrating the conductor as they encounter it, the magnetic field generates current in the conductor surface. These induced currents are called eddy currents.

Wave Polarization

Polarization refers to the direction of the electric lines of force of a radio wave. See **Figure 8-18**. If the electric lines of force are parallel to the Earth, we call this a horizontally polarized radio wave. In a horizontally polarized wave, the electric lines of force are horizontal and the magnetic lines are vertical. A radio

Figure 8-17 —
Representation of electric and magnetic lines of force in a radio wave. Arrows indicate instantaneous directions of the fields for a wave traveling toward you, out of the page. Reversing the direction of one set of lines would reverse the direction of travel, but if you reversed the direction of both sets the wave would still be coming out of the page.

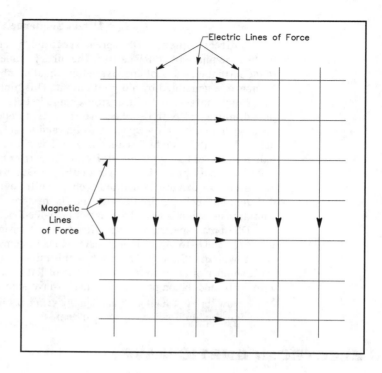

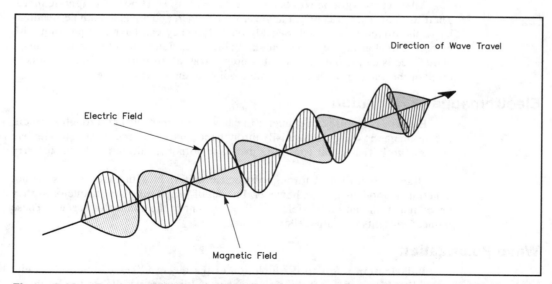

Figure 8-18 — Representation of the magnetic and electric fields of a vertically polarized radio wave. In this diagram, the electric field is in a vertical plane and the magnetic field is in a horizontal plane.

wave is vertically polarized if its electric lines of force are perpendicular to the Earth (vertical). In this case the magnetic lines are horizontal.

For the most part, polarization is determined by the type of transmitting antenna used, and its orientation. On one hand, for example, a Yagi antenna with its elements parallel to the Earth's surface transmits a horizontally polarized radio wave. On the other hand, an amateur mobile whip antenna, mounted vertically on an automobile, radiates a vertically polarized wave.

It is possible to generate waves with rotating field lines. This condition, where the electric field lines are continuously rotating through horizontal and vertical orientations, is called **circular polarization**. It is particularly helpful to use circular polarization in satellite communication, where polarization tends to shift.

Polarization that does not rotate is called **linear polarization** or **plane polarization**. Horizontal and vertical polarization are examples of linear polarization. (In space, of course, horizontal and vertical have no convenient reference.) Circular polarization is usable with linearly polarized antennas at the other end of the circuit. There will be some small loss in this case, however. If you use a vertically polarized antenna to receive a horizontally polarized radio wave (or vice versa) over a line-of-sight ground-wave path, you can expect the received signal strength to be reduced by more than 20 dB as compared to using an antenna with the same polarization as the wave. With propagation paths that use sky waves, this effect may disappear completely.

[Now study question numbers E8E08 through E8E13 in Chapter 10. Review this section as needed.]

SIGNAL-TO-NOISE RATIO

Receiver noise performance is established primarily in the RF amplifier and/or mixer stages. Low-noise, active devices should be used in the receiver front end to obtain good performance. The unwanted noise, in effect, masks the weaker signals and makes them difficult or impossible to copy. Noise generated in the receiver front end is amplified in the succeeding stages along with the signal energy. Therefore, in the interest of sensitivity, internal noise should be kept as low as possible.

Don't confuse external noise (man-made and atmospheric noise, which comes in on the antenna lead) with receiver noise during discussions of noise performance. The ratio of external noise to the incoming signal level does have a lot to do with reception. It is because external noise levels are quite high on the 160 through 20-meter bands that emphasis is seldom placed on low-internal-noise receivers for those bands. The primary source of noise heard on an HF receiver with an antenna connected is atmospheric noise. As the operating frequency is increased from 15 meters up through the microwave spectrum, however, the matter of receiver noise becomes a primary consideration. At these higher frequencies the receiver noise almost always exceeds that from external sources, especially at 2 meters and above.

Receiver noise is produced by the movement of electrons in any substance (such as wires, resistors and transistors) in the receiver circuitry. Electrons move in a random fashion colliding with relatively immobile ions that make up the bulk of the material. The final result of this effect is that in most substances there is no net current in any particular direction on a long-term average, but rather a series of random pulses. These pulses produce what is called thermal-agitation noise, or simply *thermal noise*.

Thermal-noise power is directly proportional to bandwidth and absolute temperature (in kelvins). For that reason, narrow-band systems exhibit better noise performance than do wide-band systems.

[Congratulations. You have now studied all of the material about signals and emissions for your Extra class exam. Before you go on to Chapter 9, study examination questions E8E14 and E8E15 in Chapter 10. Review this section as needed.]

CHAPTER 9
KEYWORDS
KEYWORDS
KEYWORDS

Antenna — An electric circuit designed specifically to radiate the energy applied to it in the form of electromagnetic waves. An antenna is reciprocal; a wave moving past it will induce a current in the circuit also. Antennas are used to transmit and receive radio waves.

Antenna bandwidth — A range of frequencies over which the antenna SWR will be below some specified value.

Antenna efficiency — The ratio of the radiation resistance to the total resistance of an antenna system, including losses.

Base loading — The technique of inserting a coil at the bottom of an electrically short vertical antenna in order to cancel the capacitive reactance of the antenna, producing a resonant antenna system.

Beamwidth — As related to directive antennas, the width (measured in degrees) of the major lobe between the two directions at which the relative power is one half (–3 dB) of the value at the peak of the lobe.

Center loading — A technique for adding a series inductor at or near the center of an antenna element in order to cancel the capacitive reactance of the antenna. This technique is usually used with elements that are less than $^1/_4$ wavelength.

Circular polarization — Describes an electromagnetic wave in which the electric and magnetic fields are rotating. If the electric field vector is rotating in a clockwise sense, then it is called right-hand circular polarization and if the electric field is rotating in a counterclockwise sense, it is called left-hand circular polarization. Note that the polarization sense is determined by standing behind the antenna for a signal being transmitted, or in front of it for a signal being received.

Delta match — A method for impedance matching between an open-wire transmission line and a half-wave radiator that is not split at the center. The feed-line wires are fanned out to attach to the antenna wire symmetrically around the center point. The resulting connection looks somewhat like a capital Greek delta.

Dielectric — An insulating material. A dielectric is a medium in which it is possible to maintain an electric field with little or no additional direct-current energy supplied after the field has been established.

Dielectric constant (ε) — Relative figure of merit for an insulating material. This is the property that determines how much electric energy can be stored in a unit volume of the material per volt of applied potential.

Dipole antenna — An antenna with two elements in a straight line that are fed in the center; literally, two poles. For amateur work, dipoles are usually operated at half-wave resonance.

Effective isotropic radiated power (EIRP) — A measure of the power radiated from an antenna system. EIRP takes into account transmitter output power, feed-line losses and other system losses, and antenna gain as compared to an isotropic radiator.

Folded dipole — An antenna consisting of two (or more) parallel, closely spaced halfwave wires connected at their ends. One of the wires is fed at its center.

Gain — An increase in the effective power radiated by an antenna in a certain desired direction. This is at the expense of power radiated in other directions.

Gamma match — A method for matching the impedance of a feed line to a half-wave radiator that is split in the center (such as a dipole). It consists of an adjustable arm that is mounted close to the driven element and in parallel with it near the feed point. The connection looks somewhat like a capital Greek gamma.

Horizontal polarization — Describes an electromagnetic wave in which the electric field is horizontal, or parallel to the Earth's surface.

Isotropic radiator — An imaginary antenna in free space that radiates equally in all directions (a spherical radiation pattern). It is used as a reference to compare the gain of various real antennas.

Loading coil — An inductor that is inserted in an antenna element or transmission line for the purpose of producing a resonant system at a specific frequency.

Major lobe of radiation — A three-dimensional area that contains the maximum radiation peak in the space around an antenna. The field strength decreases from the peak level, until a point is reached where it starts to increase again. The area described by the radiation maximum is known as the major lobe.

Matching stub — A section of transmission line used to tune an antenna element to resonance or to aid in obtaining an impedance match between the feed point and the feed line.

Minor lobe of radiation — Those areas of an antenna pattern where there is some increase in radiation, but not as much as in the major lobe. Minor lobes normally appear at the back and sides of the antenna.

Nonresonant rhombic antenna — A diamond-shaped antenna consisting of sides that are each at least one wavelength long. The feed line is connected to one end of the diamond, and there is a terminating resistance of approximately 800 Ω at the opposite end. The antenna has a unidirectional radiation pattern.

Parabolic (dish) antenna — An antenna reflector that is a portion of a parabolic curve. Used mainly at UHF and higher frequencies to obtain high gain and narrow beamwidth when excited by one of a variety of driven elements placed at the dish focus to illuminate the reflector.

Radiation resistance — The equivalent resistance that would dissipate the same amount of power as is radiated from an antenna. It is calculated by dividing the radiated power by the square of the RMS antenna current.

Reflection coefficient (ρ) — The ratio of the reflected voltage at a given point on a transmission line to the incident voltage at the same point. The reflection coefficient is also equal to the ratio of reflected and incident currents.

Resonant rhombic antenna — A diamond-shaped antenna consisting of sides that are each at least one wavelength long. The feed line is connected to one end of the diamond, and the opposite end is left open. The antenna has a bidirectional radiation pattern.

Top loading — The addition of inductive reactance (a coil) or capacitive reactance (a capacitance hat) at the end of a driven element opposite the feed point. It is intended to increase the electrical length of the radiator.

Traps — Parallel LC networks inserted in an antenna element to provide multiband operation.

Velocity factor — An expression of how fast a radio wave will travel through a material. It is usually stated as a fraction of the speed the wave would have in free space (where the wave would have its maximum velocity). Velocity factor is also sometimes specified as a percentage of the speed of a radio wave in free space.

Vertical polarization — Describes an electromagnetic wave in which the electric field is vertical, or perpendicular to the Earth's surface.

ANTENNAS AND FEED LINES

There are several sections to this chapter, covering material you need to know in order to pass your Amateur Extra exam. But as you learn this material you will be able to do far more than simply pass an exam. The more you learn about antennas and feed lines, the better able you will be to experiment with new antenna and impedance-matching ideas. Experimentation with various types of antennas will enhance your enjoyment of our hobby. There are many more types of antennas than can be described in this book. Of the antenna types mentioned here, little detail is given about some. There is enough information to enable you to pass the Element 4 exam, but not so much as to be confusing. When you need to learn more about a particular antenna type, you will want to refer to *The ARRL Antenna Book* or one of the other good antenna reference books that are available.

After you have studied the information in each section of this chapter, use the examination questions in Chapter 10 to check your understanding of the material. If you are unable to answer a question correctly, go back and review the appropriate part of this chapter.

Your exam will include five questions from the material in this chapter. There are five groups of questions in the Antennas and Feed Lines subelement of the Element 4 Question Pool. These questions cover the range of topics listed in the syllabus for this subelement:

THE ISOTROPIC RADIATOR

An **isotropic radiator** is a theoretical antenna that is assumed to radiate equally in all directions. This hypothetical antenna has a radiation pattern that is omnidirectional, because the signal is equal in all directions. An isotropic radiator is imagined as a single point in space, far from the effects of Earth ground or other influencing factors. Of course no such antenna actually exists, but it is useful to assume one does for comparison with real antennas. The radiation from a practical antenna never has the same intensity in all directions. The intensity may even be near zero in some directions from the antenna; in others it will probably be greater than you would expect from an antenna that did radiate equally well in all directions.

As you might expect, the solid (three-dimensional) radiation pattern of an isotropic radiator is a sphere, since the field strength is the same in all directions. In any plane containing the isotropic antenna, the pattern is a circle with the antenna at its center. See **Figure 9-1**. (This pattern is reduced by 2.14 dB for comparison with the dipole pattern shown in **Figure 9-2**. Radiation-pattern plots can be scaled to just about any value desired for comparisons.) From a theoretical point of view, the isotropic radiator is useful as a reference antenna for comparison with the radiation patterns from real antennas. It is especially useful when comparing the gains of various directional antennas.

An isotropic radiator has no directivity at all, because the radiated signal strength is the same in all directions. We can also say the isotropic radiator has no gain in any direction for this same reason.

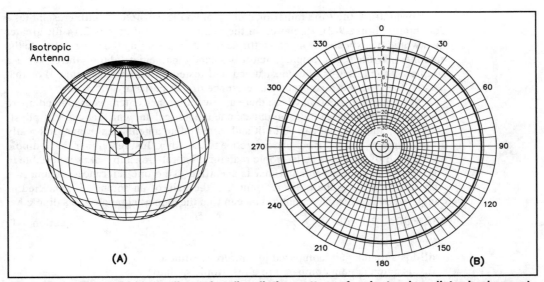

Figure 9-1 — The solid (three-dimensional) radiation pattern of an isotropic radiator is shown at A. B shows a plot of the radiation pattern of the isotropic radiator in any plane. This pattern has been reduced by 2.14 dB for comparison with the dipole pattern shown in Figure 9-2B.

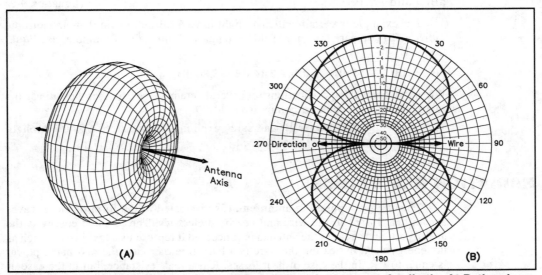

Figure 9-2 — At A, the solid (three dimensional) directive pattern of a dipole. At B, the plane directive diagram of a dipole. The solid line shows the direction of the wire.

By contrast, the solid radiation pattern of an ideal **dipole antenna** resembles a doughnut (**Figure 9-2**). Radiation in the main lobe of a dipole is 2.14 dB greater than would be expected from an isotropic radiator (assuming the same power to the antennas). While the isotropic radiator is a handy mathematical tool, the dipole is a simple antenna that can be constructed and tested on an antenna range. For that reason, the dipole is also used as a reference of comparison.

So we have a situation where there are two reference antennas used to compare the radiation patterns of other antennas: one theoretical antenna that can't be physically built, and one that can be built and tested on an antenna test range. It's really quite simple to convert from one reference to the other. Just remember that a dipole has 2.14 dB of gain over an isotropic radiator. That relationship makes it possible to convert the gain of an antenna that is specified using one reference antenna to a gain specified using the other reference. If the gain of an antenna is specified as compared to an isotropic radiator, you can find the gain as compared to a dipole by:

$$dBd = dBi - 2.14 \text{ dB} \qquad \text{(Equation 9-1)}$$

where:

dBd is antenna gain compared to a reference dipole.
dBi is antenna gain compared to an isotropic radiator.

If an antenna gain is specified as compared to a dipole, you can find the gain as compared to an isotropic radiator by:

$$dBi = dBd + 2.14 \qquad \text{(Equation 9-2)}$$

For example, a certain antenna might have 6 dB more gain than an isotropic radiator. To compare the gain of this antenna with the gain of a dipole, use Equation 9-1.

$$dBd = dBi - 2.14 \text{ dB} = 6 \text{ dBi} - 2.14 \text{ dB} = 3.86 \text{ dB}$$

When you compare specifications for several antennas, be sure that they all use the same reference antenna for comparison.

[Now turn to Chapter 10 and study examination questions E9A01 through E9A05. Review this section as needed.]

RADIATION RESISTANCE

The energy supplied to an **antenna** is dissipated in the form of radio waves and in heat losses in the wire and nearby dielectrics. The radiated energy is the useful part, and as far as the antenna is concerned it represents a loss just as much as the energy used in heating the wire is a loss. In either case the dissipated power is equal to I^2R. In the case of heat losses, R is a real resistance, but in the case of radiation, R is an assumed resistance, which, if present, would dissipate the power actually radiated from the antenna. This assumed resistance is called the **radiation**

resistance. The total power loss in the antenna is therefore equal to $I^2 (R_R + R)$, where R_R is the radiation resistance and R is the real, or ohmic, resistance.

In the ordinary half-wave dipole antenna operated at amateur frequencies, the power lost as heat in the conductor does not exceed a few percent of the total power supplied to the antenna. This is because the RF resistance of copper wire even as small as number 14 is very low compared with the radiation resistance of an antenna that is reasonably clear of surrounding objects and is not too close to the ground. Therefore, we can assume that the ohmic loss in a reasonably well located antenna is negligible, and that all of the resistance shown by the antenna is radiation resistance. As a radiator of electromagnetic waves, such an antenna is a highly efficient device.

The value of radiation resistance, as measured at the center of a half-wave antenna, depends on a number of factors. One is the location of the antenna with respect to other objects, particularly the earth. Another is the length/diameter ratio of the conductor used. In free space — with the antenna remote from everything else — the radiation resistance of a resonant $1/2$-wavelength dipole antenna made of an infinitely thin conductor is approximately 73 Ω. The concept of a free-space antenna forms a convenient basis for calculation because the modifying effect of the ground can be taken into account separately. If the antenna is at least several wavelengths away from ground and other objects, it can be considered to be in free space insofar as its own electrical properties are concerned. This condition can be met easily with antennas in the VHF and UHF range. At these frequencies, antennas are small and a wavelength may be only a few feet (or less) so it is easy to mount the antenna several wavelengths above ground.

As the antenna is made thicker, the radiation resistance decreases. For most wire antennas it is close to 65 Ω. The radiation resistance will usually lie between 55 and 60 Ω for antennas constructed of rod or tubing.

The actual value of the radiation resistance has no appreciable effect on the radiation efficiency of a practical antenna. This is because the ohmic resistance is only on the order of 1 Ω with the conductors used for thick antennas. The ohmic resistance does not become important until the radiation resistance drops to very low values — say less than 10 Ω — as may be the case when several antenna elements are coupled to form an array.

The radiation resistance of a resonant antenna is the "load" for the transmitter or for the RF transmission line connecting the transmitter and antenna. Its value is important, therefore, in determining the way in which the antenna and transmitter or line are coupled. Most modern transmitters require a 50-Ω load. To transfer the maximum amount of power possible, the transmitter output impedance, the transmission-line characteristic impedance and the radiation resistance should all be equal, or matched by means of an appropriate impedance-matching network.

ANTENNA EFFICIENCY

As you try to optimize your Amateur Radio station, you may want to begin with the antenna system. Improvements to the antenna system will probably have a greater impact on your station than any other changes you can make, considering cost and time invested. **Antenna efficiency** may be one of your first considerations. The efficiency of an antenna is given by:

$$\text{Efficiency} = \frac{R_R}{R_T} \times 100\% \qquad\qquad \text{(Equation 9-3)}$$

where:

R_R = radiation resistance.
R_T = total resistance.

The total resistance includes radiation resistance, resistance in conductors and dielectrics (including the resistance of loading coils, if used), and the resistance of the grounding system, usually referred to as "ground resistance."

A half-wave antenna operates at very high efficiency because the conductor resistance is negligible compared with the radiation resistance. In the case of a $^1/_4$-wavelength vertical antenna, with one side of the feed line connected to ground, the ground resistance usually is not negligible. If the antenna is short (compared with a quarter wavelength) the resistance of the necessary loading coil may become appreciable. To attain an efficiency comparable with that of a half-wave antenna in a grounded one having a height of $^1/_4$ wavelength or less, great care must be used to reduce both ground resistance and the resistance of any required loading inductors. Without a fairly elaborate grounding system, the efficiency is not likely to exceed 50% and may be much less, particularly at heights below $^1/_4$ wavelength. A $^1/_4$-wavelength ground-mounted vertical antenna normally includes many radial wires, laid out as spokes on a wheel, with the antenna element in the center.

If a half-wave dipole antenna has a radiation resistance of 70 Ω and a total resistance of 75 Ω, Equation 9-3 tells us the efficiency of the antenna:

$$\text{Efficiency} = \frac{R_R}{R_T} \times 100\% = \frac{70\ \Omega}{75\ \Omega} \times 100\% = 93\%$$

[Before proceeding, turn to Chapter 10 and study questions E9A06 through E9A09 and questions E9A14 and E9A15. Review this section as needed.]

FOLDED DIPOLE ANTENNAS

A **folded dipole** antenna is a wire antenna consisting of two (or more) parallel wires that are closely spaced and connected at their ends. One of the wires is fed at its center.

In **Figure 9-3**, suppose for the moment that the upper conductor between points B and C is disconnected and removed. The system is then a simple center-fed dipole, and the current direction along the antenna and line at a given instant is as shown by the arrows. Next, restore the upper conductor between B and C. The current in the top section will flow away from B and toward C at the instant shown for the antenna current.

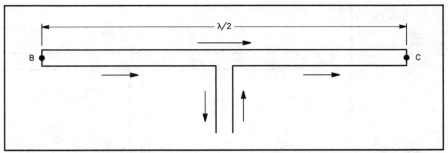

Figure 9-3 — This diagram illustrates the direction of current in a folded dipole.

This may seem confusing and be opposite to the direction you would expect the current to flow on that portion of the line. Just remember that for a sine wave, the current direction is reversed in alternate half-wave sections along a wire. Because of the way the second wire is "folded," however, the currents in the two conductors of the antenna are actually flowing in the same direction. Although the antenna physically resembles a transmission line, it is not actually a line. The antenna element merely consists of two parallel conductors carrying current in the same direction. If it were acting like a transmission line, the currents would be flowing in opposite directions. The connections at the ends of the two conductors are assumed to be of negligible length.

A half-wave dipole formed in this way will have the same directional properties and total radiation resistance as an ordinary dipole. The transmission line is connected to only one of the conductors, however. You should expect that the antenna will "look" different, with respect to its input impedance, as viewed by the line.

The effect on the impedance at the antenna input terminals can be visualized quite readily. The center impedance of the dipole as a whole is the same as the impedance of a single-conductor dipole — that is, approximately 73 Ω. A given amount of power will therefore cause a definite value of current, I. In the ordinary half-wave dipole this current flows at the junction of the line and antenna. In the folded dipole the same total current also flows, but is equally divided between two conductors in parallel. The current in each conductor is therefore I/2. Consequently, the line "sees" a higher impedance because it is delivering the same power at only half the current.

Ohm's Law reveals that the new value of impedance is equal to four times the impedance of a simple dipole. The input-terminal impedance at the center of a two-conductor folded dipole is about 300 Ω. If more wires are added in parallel the current continues to divide between them and the terminal impedance is raised still

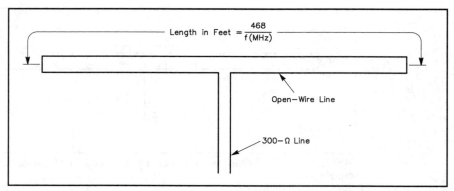

Figure 9-4 — This diagram shows the construction information for a half-wave folded dipole.

more. This explanation is a simplified one based on the assumption that the conductors are close together and have the same diameter.

Another advantage of the folded dipole is that it has a low SWR over a wider frequency range than a normal dipole. The term **antenna bandwidth** refers generally to the range of frequencies over which an antenna can be used to obtain good performance. The bandwidth is often referenced to some SWR value, such as, "The 2:1 SWR bandwidth is 3.5 to 3.8 MHz." This means that the SWR between 3.5 and 3.8 MHz will be 2:1 or lower.

The two-wire system shown in **Figure 9-4** is an especially useful one because the input impedance is so close to 300 Ω that it can be fed directly with 300-Ω twin-lead or open-wire line without any other matching arrangement.

[Study questions E9A10 and E9A12 in Chapter 10. Review this section as needed.]

TRAP ANTENNAS

By using tuned circuits of appropriate design strategically placed in a dipole, the antenna can be made to show what is essentially fundamental resonance at a number of different frequencies. The general principle is illustrated by **Figure 9-5**. The two inner lengths of wire, X, together form a simple dipole resonant at the highest band desired, say 14 MHz. The tuned circuit L_1-C_1 (called a **trap**) is also resonant at this frequency, and when connected as shown offers a very high impedance to RF current of that frequency. Effectively, therefore, these two tuned circuits act as insulators for the inner dipole, and the outer sections beyond L_1-C_1 are inactive.

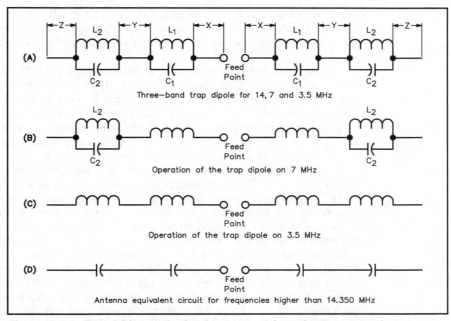

Figure 9-5 — Part A shows the basic construction of an antenna that uses two sets of traps, for operation on three frequency bands. Parts B and C show the inductive loading of the antenna on successively lower-frequency bands. D shows the capacitive loading that would result if the antenna were operated on a higher frequency, although the antenna will not normally be used on frequencies higher than the inner-dipole-section resonance.

On the next lower frequency band of interest, say 7 MHz, L_1-C_1 shows an inductive reactance and is the electrical equivalent of a coil. The two sections marked Y are now added electrically to X-X so that, together with the loading coils represented by the inductive reactance of L_1-C_1, the system is resonant at 7 MHz out to the ends of the Y sections. This part of the antenna is equivalent to a loaded dipole on 7 MHz and will exhibit about the same impedance at the feed point as a simple dipole for that band. The tuned circuit L_2-C_2 is resonant at 7 MHz and acts as a high impedance for this frequency, so the 7-MHz dipole is in turn insulated, for all practical purposes, from the remaining outer parts of the antenna.

Carrying the same reasoning one step further, L_2-C_2 shows inductive reactance on the next lower frequency band, 3.5 MHz, and is equivalent to a coil on that band. The length of the added sections, Z-Z, together with the two sets of equivalent loading coils indicated in part C, makes the whole system resonant as a loaded dipole on 3.5 MHz. A single transmission line having a characteristic impedance of the same order as the feed-point impedance of a simple dipole can be connected at the center of the antenna. This line will be satisfactorily matched on all three bands, and so will operate at a low SWR on all three. A line of 50-Ω impedance will work just fine.

Since the tuned circuits have some inherent losses, the efficiency of this system depends on the Q of the tuned circuits. Low-loss (high-Q) coils should be used, and the capacitor losses likewise should be kept as low as possible. With tuned circuits that are good in this respect — comparable with the low-loss components used in transmitter tank circuits, for example — the reduction in efficiency as compared with the efficiency of a simple dipole is small, but tuned circuits of low Q can lose an appreciable portion of the power supplied to the antenna.

The lengths of the added antenna sections Y and Z must, in general, be determined experimentally. The length required for resonance in a given band depends on the length/diameter ratio of the antenna conductor and on the L/C ratio of the trap acting as a loading coil. The effective reactance of an LC circuit at half the frequency to which it is resonant is equal to $^2/_3$ the reactance of the inductance at the resonant frequency. For example, if L_1-C_1 resonates at 14 MHz and L_1 has an inductive reactance of 300 Ω at that frequency, the inductive reactance of the circuit at 7 MHz will be equal to $^2/_3 \times 300 = 200$ Ω. The added antenna section, Y, would have to be cut to the proper length to resonate at 7 MHz with this amount of loading. Since any reasonable L/C ratio can be used in the trap without affecting its performance materially at the resonant frequency, the L/C ratio can be varied to control the added antenna length required. The added section will be shorter with high-inductance trap circuits and longer with high-capacitance traps.

Trap dipoles have two major disadvantages. Because the trap dipole is a multiband antenna, it can do a good job of radiating harmonics. Further, during operation on the lower frequency bands, the series inductance (loading) from the traps raises the antenna Q. That means less bandwidth for a given SWR limit.

Of course quarter-wavelength vertical antennas can also use trap construction. Only one trap is needed for each band in that case, since only the vertical portion of the antenna uses traps. The ground system under the antenna does not use traps.

MOBILE ANTENNAS FOR HF

Mobile antennas are usually vertically mounted whip antennas 8 feet or less in length. As the operating frequency is lowered, the feed-point impedance of a fixed-length antenna appears to be a decreasing resistance in series with an increasing capacitive reactance. This capacitive reactance must be tuned out, which indicates the use of a series inductive reactance, or **loading coil**. The amount of inductance required is determined by the desired operating frequency and where the coil is placed in the antenna.

Base loading requires the lowest value of inductance for a given antenna length, and as the coil is moved farther up the whip, the necessary value increases. This is

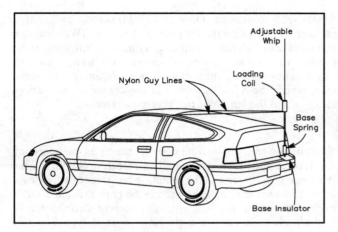

Figure 9-6 — This drawing shows a typical bumper-mounted HF-mobile antenna. Note the nylon guy lines.

because the capacitance between the portion of the whip above the coil and the car body decreases (higher capacitive reactance), requiring more inductance to tune the antenna to resonance. One advantage of placing the coil at least part way up the whip, however, is that the current distribution is improved, and that increases the radiation resistance. The major disadvantage is that the requirement for a larger loading coil means that the coil losses will be greater, although this is offset somewhat by lower current through the larger coil. **Center loading** has been generally accepted as a good compromise in mobile-antenna design.

Figure 9-6 shows a typical bumper-mounted, center-loaded whip antenna suitable for operation in the HF range. The antenna could also be mounted on the car body proper (such as a fender). The base spring acts as a shock absorber for the base of the whip, since the continual flexing while the car is in motion would otherwise weaken the antenna. A short, heavy, mast section is mounted between the base spring and loading coil. Some models have a mechanism that allows the antenna to be tipped over for adjustment or for fastening to the roof of the car when not in use.

It is also advisable to extend a couple of guy lines from the base of the loading coil to clips or hooks fastened to the rain trough on the roof of the car. Nylon fishing line of about 40-pound test strength is suitable for this purpose. The guy lines act as safety cords and also reduce the swaying motion of the antenna considerably. The feed line to the transmitter is connected to the bumper and base of the antenna. Good low-resistance connections are important here.

Tune-up of the antenna is usually accomplished by changing the height of the adjustable whip section above the precut loading coil. A noise bridge or "SWR analyzer" can aid the process of adjusting the antenna for operation on a particular frequency. Without one of those devices, you can tune the receiver and try to determine where the signals seem to peak up. Once you find this frequency, check the SWR with the transmitter on, and find the frequency of lowest SWR. Shorten the adjustable section to increase the resonant frequency or make it longer to lower the frequency. It is important that the antenna be 10 feet or more away from surrounding objects such as overhead wires, since considerable detuning can occur. Once you find the setting where the SWR is lowest at the center of the desired operating frequency range, record the length of the adjustable section.

Loading Coils

The difficulty in constructing suitable loading coils increases as the operating frequency is lowered for typical antenna lengths used in mobile work. Since the required resonating inductance gets larger and the radiation resistance decreases at lower frequencies, most of the power may be dissipated in the coil resistance and in other ohmic losses. This is one reason why it is advisable to buy a commercially made loading coil with the highest power rating possible, even though you may only be considering low-power operation. Percentage-wise, the coil losses in the higher power loading coils are usually less, with subsequent improvement in radiating efficiency, regardless of the power level used. Of course, this same philosophy also applies to homemade loading coils.

The primary goal here is to provide a coil with the highest possible Q. This means the coil should have a high ratio of reactance to resistance, so that heating losses will be minimized. High-Q coils require a large conductor, "air-wound" construction, large spacing between turns, the best insulating material available, a diameter not less than half the length of the coil (this is not always mechanically feasible) and a minimum of metal in the field.

Once the antenna is tuned to resonance, the input impedance at the antenna terminals will look like a pure resistance. Neglecting losses, this value drops from nearly 15 Ω on 15 meters to 0.1 Ω on 160 meters for an 8-foot whip. When coil and other losses are included, the input resistance increases to approximately 20 Ω on 160 meters and 16 Ω on 15 meters. These values are for relatively high-efficiency systems. From this, you can see that the radiating efficiency is much poorer on 160 meters than on 15 meters under typical conditions.

Since most modern gear is designed to operate into a 50-Ω impedance, a matching network may be necessary with some mobile antennas. This can take the form of either a broad-band transformer, a tapped coil or an LC matching network. With homemade or modified designs, the tapped-coil arrangement is perhaps the easiest to build, while the broad-band transformer requires no adjustment. As the losses go up, so does the input resistance, and in less efficient systems the matching network may be eliminated.

The Equivalent Circuit of a Typical Mobile Antenna

Antenna resonance is defined as the frequency at which the input impedance at the antenna terminals is a pure resistance. The shortest length at which this

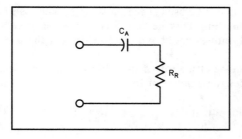

Figure 9-7 — At frequencies below the resonant frequency, the whip antenna will show capacitive reactance as well as resistance. R_R is the radiation resistance, and C_A represents the capacitive reactance.

occurs for a vertical antenna over a ground plane is a quarter wavelength at the operating frequency; the impedance value for this length (neglecting losses) is about 36 Ω. The idea of resonance can be extended to antennas shorter (or longer) than a quarter wavelength, and means only that the input impedance is purely resistive. When the frequency is reduced, the antenna looks like a series RC circuit, as shown in **Figure 9-7**.

The capacitive reactance can be canceled out by connecting an equivalent inductive reactance, L_L, in series as shown in **Figure 9-8**. This arrangement tunes the system to resonance at a particular frequency. The price you pay for the shortened antenna is decreased bandwidth.

Mobile Antenna Efficiency

Antenna efficiency was discussed earlier in this chapter. In that section the importance of minimizing ohmic losses was discussed. If you have trouble understanding the material here, review that section.

For lowest loss, an amateur would use a self-resonant antenna, such as a dipole or a quarter-wavelength vertical. That is not possible, of course, for HF mobile operation, except at the upper end of the range.

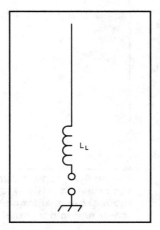

Mobile antenna losses at HF, for the most part, are caused by two factors: ground return resistance and loading coil losses. To minimize ground losses, the transmission line should be connected to the metal automobile body through a low resistance and low reactance connection. A good way to do this is with a short length of ground strap or coaxial-cable braid.

Figure 9-8 — At frequencies lower than the resonant frequency, the capacitive reactance of a whip antenna can be canceled by adding an equivalent inductive reactance, in the form of a loading coil, in series with the antenna.

Another method that can be used successfully to reduce loading-coil losses is a technique called **top loading**. This method calls for a "capacitive hat" to be added above the loading coil, either just above the coil or near the top of the antenna whip.

The added capacitance at the top of the whip allows a smaller value of load inductance. This reduces the amount of loss in the system, and improves the antenna radiation efficiency.

[Now turn to Chapter 10 and study questions E9D08 through E9D10, E9D13 and E9D14. Review this section as needed.]

ANTENNA GAIN, BEAMWIDTH AND RADIATION PATTERNS

In a perfect directional antenna, the radio energy would be concentrated in the *forward* direction only. This is known as the **major lobe of radiation**. (Most beams also have **minor lobes** in the back and side directions.)

Figure 9-9 is an example of a radiation pattern for a beam antenna, illustrating major and minor pattern lobes. The greater the number of elements and the longer the distance between elements (up to an optimum spacing), the narrower the radiated beam. By reducing radiation in the side and back directions and concentrating it instead into a narrow beam in the forward direction, a beam antenna can have more effective radiated power than a dipole. The ratio (expressed in decibels) between the signal radiated from a beam in the direction of the main lobe and the signal radiated from a reference antenna (usually a dipole) in the same direction and at the same transmitting location is called the **gain** of the beam. A typical beam might have 6 dB of gain compared to a dipole, which means that it makes your signal sound four times (6 dB) louder than if you were using a dipole with the same transmitter .

The gain of directional antennas is the result of concentrating the radio wave in one direction at the expense of radiation in other directions. Since practical antennas are not perfect, there is always some radiation in undesired directions as well. A plot of relative field strength in all horizontal directions is called the horizontal-radiation pattern. The vertical-radiation pattern is a similar plot of the field strength in the vertical plane.

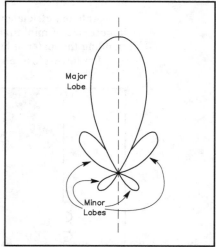

Figure 9-9 — This diagram represents the radiation pattern for a hypothetical beam antenna, illustrating major and minor radiation lobes.

Measuring an antenna horizontal radiation pattern accurately requires that the antenna be placed well away from any obstructions for a distance of at least several wavelengths. On a typical antenna test range, a low-power transmitter feeds a signal to a standard antenna. The antenna to be tested is connected to a receiver, and the received signal strengths are recorded as the test antenna is rotated through 360°. The received signal strengths at various points are plotted, and a radiation pattern results. Vertical radiation patterns are more difficult to measure.

Modern computers make it possible to analyze antenna designs by mathematically modeling the antenna. Computer analysis allows us to study how performance changes as the height of the antenna changes or what effects different ground conditions will have. There are a number of programs in common use for this type of analysis. Most of them are derived from a program developed at US government laboratories, called *NEC*, short for "Numerical Electromagnetics Code." This complex program, originally written for mainframe computers, uses a modeling technique called the "method of moments." The antenna wire (or tubing elements) is modeled as a series of segments, and a distinct value of current through each segment is computed. The field resulting from the RF current in each segment is evaluated, along with the effects from other mutually coupled segments.

Most of the programs available to amateurs for modeling antenna performance are based on the full *NEC* program. Programs that use the method of moments analysis can predict antenna performance, provided the antenna is modeled properly for the program input. Most of the programs can even produce a set of radiation patterns for the antenna.

If the radiation pattern is calculated for the plane of the antenna elements, it is called an E-plane radiation pattern. (It is called the E-plane because the electric field of the signal radiated from the antenna is oriented in this plane.) If the radiation pattern is calculated for the plane perpendicular to the elements, it is called the H-plane radiation pattern. (The symbol H is used to represent the magnetic field, which is oriented perpendicular — 90° — to the electric field.) The free-space polarization of an antenna is determined by the orientation of the electric field of the signal radiated from an antenna.

When we consider real antennas near the Earth, we usually refer to the antenna polarization relative to the ground. For example, if the electric field of the radiation from an antenna is oriented parallel to the surface of the Earth, we say the antenna is *horizontally polarized*. If the electric field is oriented perpendicular to the surface of the Earth, the antenna is *vertically polarized*. The simplest way to determine the polarization of an antenna is by looking at the orientation of the elements. The electric field will be oriented in the same direction as the antenna elements. So if a Yagi antenna has elements oriented parallel to the ground, it will produce horizontally polarized radiation, and if the antenna elements are oriented perpendicular to the ground, it will produce vertically polarized radiation.

Analyzing Radiation Patterns

An antenna radiation pattern contains a wealth of information about the antenna and its expected performance. By learning to recognize certain types of patterns and what they represent, you will be able to identify many important antenna characteristics, and to compare design variations.

Always look at the signal-strength scale on a pattern. The scales are usually expressed in decibels, but the actual values used can vary quite a bit. The scale is usually selected to show desired pattern details. Different scales change the shape of the pattern, however, so it is difficult or misleading to compare radiation patterns of several antennas unless the scales are the same.

The radiation pattern is usually drawn so it just touches the outer circle at its maximum strength. The scale can then be used to measure the relative strength of signals radiated in any direction. The signal-strength scale represents a relative comparison then. To compare one antenna with another, you should also look for a note indicating the basis for the outer ring, or 0 dB circle on the pattern. A dipole pattern plotted on a scale that uses 0 dB = +2.15 dBi can't be compared directly with a Yagi pattern plotted on a scale that uses 0 dB = +6 dBi. (The units dBi tell us the signal powers are compared to an *isotropic radiator*, which is a theoretical ideal point-source antenna located in free space. An isotropic radiator transmits signals equally in all directions, so the radiation pattern for this antenna is a perfect sphere.)

The **beamwidth** is the angular distance between the points on either side of the main lobe, at which the gain is 3 dB below the maximum. See **Figure 9-10**. A three-element beam, for example, might be found to have a beamwidth of 50°. This means if you turn your beam plus or minus 25° from the optimum heading, the signal you receive (and the signal received from your transmitter) will drop by 3 dB.

Take a look at **Figure 9-11**. This is a free-space pattern, meaning it is taken as being completely isolated from any Earth effects. If the same pattern were considered relative to Earth, however, we would call it an *azimuthal pattern*, or a *horizontal pattern*. (Azimuth refers to compass directions.) In that case we would probably label the angles from 0° to 360° instead of 0° to + and – 180°. Keep in mind that the pattern doesn't tell us the polarization of the radiation from the antenna. Calling it an azimuthal

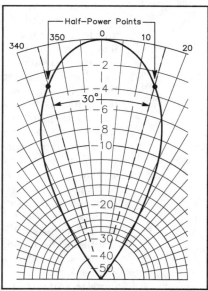

Figure 9-10 — The beamwidth of an antenna is the angular distance between the directions at which the received or transmitted power is one half (–3 dB) the maximum power.

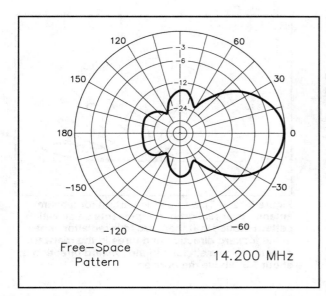

Figure 9-11 — This free-space radiation pattern is the H-plane pattern of a beam antenna for operation on the 20-meter band. Viewed in relation to the Earth, this pattern represents an azimuthal, or horizontal radiation pattern.
The text describes how to read front-to-back and front-to-side ratios from the graph.

pattern helps avoid any confusion about the antenna polarization.

The major lobe of radiation from the antenna of Figure 9-11 points to the right, and is centered along the 0° axis. You can make a pretty good estimate of the beamwidth of this antenna by carefully reading the graph. Notice that angles are marked off every 15°, and the –3 dB circle is the first one inside the outer circle. It looks like the pattern crosses the –3 dB circle at points about 25° either side of 0°. So we can estimate the beamwidth of this antenna as 50°.

We are usually interested in the *front-to-back ratio* of our beam antennas. This is a comparison of the signal strength in the forward direction compared to the strength in a direction 180° from that. This number gives us a good idea of how well an antenna will reject an interfering signal coming in from a direction opposite the desired direction. In the case of the example shown in Figure 9-11, we can read the front-to-back ratio by finding the maximum value of the minor lobe at 180°. This maximum appears to be about half way between the –12 dB and –24 dB circles, so we would estimate it to be about 18 dB below the main lobe.

Front-to-side ratio is another antenna parameter that is often of interest. You can probably guess that this refers to the strength of signals coming in from a direction 90° away from the main lobe. We can estimate the front-to-side ratio of the pattern shown in Figure 9-11 by reading the value off the graph. You can see there is a minor radiation lobe off each side of the antenna, and the maximum strength of each lobe is a bit more than 12 dB below the main-lobe maximum. A front-to-side ratio of 14 dB looks like a pretty good estimate for this pattern.

When we look at the radiation pattern in a single plane we are not getting the whole picture. Antennas radiate signals in all directions, so we would have to look at the radiation pattern in three dimensions to get the complete picture. Showing such patterns on a two-dimensional page is rather complicated, however. Because there is usually some symmetry to the pattern, we usually take a second

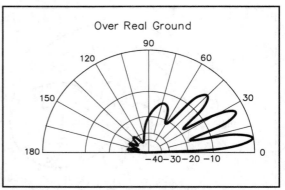

Figure 9-12 — This radiation pattern for a beam antenna over real ground represents an elevation pattern. Notice that there are four radiation lobes in the forward direction and three in the rearward direction. The main lobe in the forward direction is about 7.5° above the horizon.

view of the radiation in a plane perpendicular to the first plane. In the case of an antenna in free space, we might call this second view an H-plane pattern, because it is taken in the plane of the radio wave magnetic field. If we consider the antenna in relation to Earth, the second view is called a *vertical radiation pattern*, or an *elevation pattern*.

An elevation pattern over real ground is usually shown for half the circle. Any radiation that would have gone down into the ground is reflected back into space above the Earth, and that energy is added to the pattern above the antenna. **Figure 9-12** shows an elevation pattern over real ground.

One of the first things you should notice about the pattern shown in Figure 9-12 is that there are four lobes in the forward direction, at various angles above the horizon. The main radiation lobe is at an angle of about 7.5° above the horizon. The other forward-direction lobes appear at about 25°, 45° and 75°.

There are three minor radiation lobes off the back of the antenna shown in Figure 9-12. The graph may be a bit difficult to read accurately here, but the largest of these lobes appears to cross the –30 dB circle of the pattern, so we can estimate the front-to-back ratio of the antenna at about 28 dB for this antenna.

[We've covered quite a bit of material since a review in the question pool. This is a good time to go to Chapter 10 and study questions E9A11, E9A13, E9B01 through E9B04, E9B09 through E9B11, E9C11, E9C14 and E9C15 before proceeding. If you are uncertain about any of the answers to these questions, be sure to review this section.]

How to Calculate Beamwidth

The solid radiation pattern of an antenna can't be adequately shown by plotting field-strength data on a sheet of graph paper. Two cross-sectional (plane) diagrams are often used to picture the radiation pattern, then. These two diagrams are called the E-plane and the H-plane patterns. The E-plane pattern lies in the plane of the electric lines of force, which are designated using the E symbol. The H-plane pattern lies in the plane of the magnetic lines of force, which are designated using the H symbol. The E lines are taken to represent the polarization of the radio waves from the antenna, and are parallel to the antenna elements. The H plane is perpendicular to the antenna elements. For a horizontal dipole, the E plane will be parallel to the Earth and represents the azimuthal pattern of the antenna. The H plane, then, is perpendicular to the Earth, and represents the vertical pattern.

If the radiation beam is well defined (not much power in minor lobes), then there is an approximate formula relating the antenna gain to the measured half-power beamwidth of the E- and H-plane radiation patterns. This formula uses a lossless isotropic radiator as the gain reference. The formula is:

$$\text{Gain Ratio} = \frac{41,253}{\theta_E \times \theta_H} \qquad \text{(Equation 9-4)}$$

where θ_E and θ_H are the half-power beamwidths, in degrees, of the E- and H-plane patterns, respectively. Gain refers to a gain ratio, and is *not* expressed in decibels. The gain reference can be changed to that of an actual dipole, rather than for a lossless isotropic radiator, by changing the constant 41,253 to 25,000. Thus, the formula becomes

$$\text{Gain Ratio} = \frac{25,000}{\theta_E \times \theta_H} \qquad \text{(Equation 9-5)}$$

It isn't hard to solve either of these equations for beamwidth, as long as we make some simplifying assumptions. If the pattern is symmetrical around the E and H planes, we can set the two angles equal. This gives a θ^2 term in the denominator, so it is basic algebra to solve for beamwidth. For the isotropic reference antenna:

$$\text{Gain Ratio} = \frac{41,253}{\theta^2} \qquad \text{(Equation 9-6)}$$

$$\theta = \text{Beamwidth} = \sqrt{\frac{41,253}{\text{Gain Ratio}}} = \frac{203}{\sqrt{\text{Gain Ratio}}} \qquad \text{(Equation 9-7)}$$

where the half-power beamwidth is in degrees and the gain is a ratio (*not expressed in decibels*). Again, this equation is an approximation for an antenna with symmetrical horizontal and vertical-plane patterns and no significant minor lobes. From this equation, it is easy to see that as antenna gain *increases*, the beamwidth becomes *narrower*, or *decreases*.

If we are given gain in dB, we must first convert to a gain ratio. The gain ratio here refers to the ratio of power radiated from the antenna in the direction of the main lobe divided by the power in that same direction from the reference antenna. (The reference antenna is usually either an isotropic radiator or a dipole antenna.) A decibel is ten times the logarithm of a power ratio:

$$dB = 10 \log \left(\frac{P_2}{P_1} \right) \qquad \text{(Equation 9-8)}$$

So if we divide both sides by 10 we have:

$$\frac{dB}{10} = \log (\text{Gain Ratio}) \qquad \text{(Equation 9-9)}$$

To get rid of the logarithm we take the antilog of both sides. The antilog is the "inverse logarithm." The log of 20 is 1.3, so the antilog of 1.3 is 20. We can also think of the antilog of a number as 10 raised to the number ($10^{1.3} = 20$). So if we take the antilog of both sides of our equation, we have

$$10^{\frac{dB}{10}} = \text{Gain Ratio} \qquad \text{(Equation 9-10)}$$

This means that if we know the gain of an antenna is 20 dBi, we can solve for the ratio:

$$\text{Gain Ratio} = 10^{\frac{20}{10}} = 10^2 = 100$$

Then use Equation 9-7 to calculate the beamwidth:

$$\text{Beamwidth} = \frac{203}{\sqrt{\text{Gain Ratio}}} = \frac{203}{\sqrt{100}} = \frac{203}{10}$$

Beamwidth = 20.3°

Keep in mind that Equation 9-7 is used when the antenna gain is compared with an isotropic radiator (dBi). If you have an antenna gain specified in dBd (compared to a dipole), you can use Equation 9-5 or you can convert the dBd figure to one specified in dBi.

[Practice these beamwidth calculations by going to Chapter 10 now and answering examination questions E9D03 and E9D06. Review this section if you have difficulty with either of those questions.]

Antenna Optimization

Any antenna design represents some compromises. You may be able to modify the design of a particular antenna to improve some desired characteristic, if you are aware of the trade-offs. As we mentioned earlier, the method of moments computer modeling techniques that have become popular can be a great help in deciding which design modifications will produce the "best" antenna for your situation.

When you evaluate the gain of an antenna, you (or the computer modeling program) will have to take into account a number of parameters. You will have to include the antenna radiation resistance, any loss resistance in the elements and impedance-matching components as well as the E-field and H-field radiation patterns.

You should also evaluate the antenna across the entire frequency band for which it is designed. You may discover that while the gain seems high, it may decrease rapidly as you move away from the single design frequency, and the performance may be poor over the remainder of the band. (You may be willing to make that trade-off if all your operating on that band is within a narrow frequency range.) You may also discover that the feed-point impedance varies widely as you change frequency across the desired band. This will make it difficult to design a single impedance-matching system for the antenna. You may also discover that the front-to-back ratio varies excessively across the band, resulting in too much variation in the rearward pattern lobes.

The forward gain of a Yagi antenna can be increased by using a longer boom, spreading the elements farther apart or adding more elements. Of course there are practical limitations on how long you can make the boom for any antenna! The element lengths will have to be adjusted to retune them when the boom length changes. You can design a Yagi antenna for maximum forward gain, but in that case the feed-point impedance usually becomes too low, and the SWR bandwidth will decrease.

Effects of Ground on Radiation Patterns

The radiation pattern of an antenna over real ground is always affected by the electrical conductivity and dielectric constant of the soil, and most importantly by the height of a horizontally polarized antenna over ground. Signals reflected from the ground combine with the signals radiated directly from the antenna. If the signals are in phase when they combine, the signal strength will be increased, but if they are out of phase, the strength will be decreased. These ground reflections affect the radiation pattern for many wavelengths out from the antenna.

This is especially true of the far-field pattern of a vertically polarized antenna, which in theory will produce a low takeoff angle for the radiation. (The *far-field pattern* refers to signal strengths measured several wavelengths away from the antenna. Signals reflected from the ground as far as 100 wavelengths from the antenna will effect the far-field pattern.) The low-angle radiation from a vertically polarized antenna mounted over seawater will be much stronger than for a similar antenna mounted over rocky soil, for example. (The far-field, low-angle radiation pattern for a horizontally polarized antenna will not be significantly affected, however.)

There are several things you can do to reduce the near-field ground losses of a vertically polarized antenna system. Adding more radials is one common technique. If you can only manage a modest on-ground radial system under an eighth-wavelength, inductively loaded vertical antenna, a wire-mesh screen about an eighth-wavelength square is a good compromise.

Many amateurs prefer to raise the vertical antenna above ground and use an elevated-radial *counterpoise* system. The counterpoise takes the place of a direct connection to the ground. An elevated counterpoise under a vertically polarized antenna can reduce the near-field ground losses, as compared to on-ground radial systems that are much more extensive.

There is little you can do to improve the far-field, low-angle radiation pattern of a vertically polarized antenna if the ground under the antenna has poor conductivity. Most measures that people might try will only be applied close to the antenna, and will only affect the near-field pattern. For example, you can water the ground under the antenna, but unless you watered the ground for 100 or more wavelengths, it won't improve the ground conductivity for these distant ground reflections. (Even then, it would only be a temporary fix, or would have to be repeated often enough to keep the ground wet.) Adding more radials or extending the radials more than a quarter wavelength won't help either, unless you are able to build an extensive ground screen under the antenna out to 10 or more wavelengths!

[Turn to Chapter 10 now and study questions E9B05 through E9B08 and questions E9C12 and E9C13. Review this section if you have difficulty with any of these questions.]

PHASED VERTICAL ANTENNAS

For your Amateur Extra exam, you will need to know the various pattern shapes that can be obtained using an antenna system that consists of two vertical antennas fed with various phase relationships. **Figure 9-13** shows patterns for a number of common spacings. The patterns assume that the antennas are identical and are fed with equal current. You could probably even predict patterns at other spacings after studying the general trends illustrated in Figure 9-13. The two antennas are arranged along the vertical axis for each of those figures. The antenna toward the top of the pattern has the lagging excitation.

By studying the patterns shown in Figure 9-13 you can see that when the two antennas are fed in phase, a pattern that is broadside to the elements always results. If the antennas are $^1/_4$ wavelength apart and fed in phase the pattern is elliptical, like a slightly flattened circle. This system is substantially omnidirectional. If the antennas are $^1/_2$ wavelength apart and fed in phase the pattern is a figure 8 that is broadside to the antennas.

At spacings of less than $^5/_8$ wavelength, with the elements fed 180° out of phase, the maximum radiation lobe is in line with the antennas. For example if the antennas are $^1/_8$ wavelength apart, $^1/_4$ wavelength apart or $^1/_2$ wavelength and fed 180° out of phase, the pattern is a figure 8 that is in line with the antennas.

With intermediate amounts of phase difference, the results cannot be stated so simply. Patterns evolve that are not symmetrical in all four quadrants. If the antennas are $^1/_4$ wavelength apart and fed 90° out of phase an interesting pattern results.

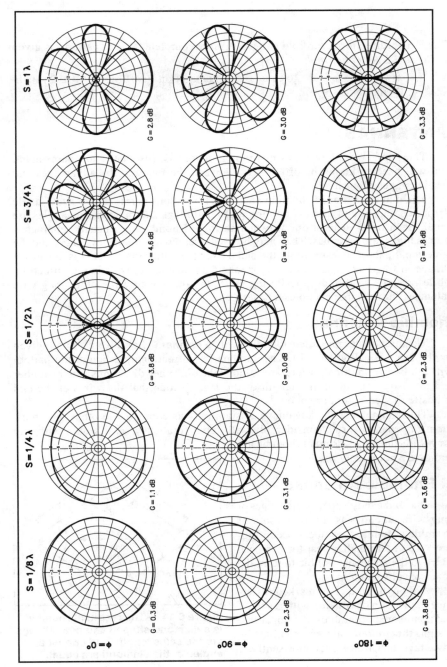

Figure 9-13 — Horizontal directive patterns of two phased verticals, spaced and phased as indicated. In these plots, the elements are aligned with the vertical axis, and the uppermost element is the one of lagging phase at angles other than zero degrees. The two elements are assumed to be the same length, with exactly equal currents. The gain values associated with each pattern indicate the gain of the two vertical antennas over a single vertical.

This is a unidirectional cardioid pattern. A more complete table of patterns is given in *The ARRL Antenna Book*.

[Now turn to Chapter 10 and study examination questions E9C01 through E9C06. Review this section as needed.]

RHOMBIC ANTENNAS

There are two major types of rhombic antennas: resonant and nonresonant. You will need to know the differences between the two and the advantages and disadvantages of each.

The diamond-shaped rhombic antenna can be considered as two acute-angle Vs placed end-to-end; there are four equal-length legs and the opposite angles are equal. Each leg is at least one wavelength long. Rhombics are usually characterized as high-gain antennas. They may be used over a wide frequency range, and the directional pattern is essentially the same over the entire range. This is because a change in frequency causes the major lobe from one leg to shift in one direction, while the lobe from the opposite leg shifts the other way. Like other long-wire antennas, the rhombic occupies a large area.

Resonant Rhombic Antennas

The direction of maximum radiation with a **resonant rhombic antenna** is given by the arrows in **Figure 9-14**. The antenna is bidirectional, although the pattern will not be symmetrical. There are minor lobes in other directions; their number and intensity depend on the length of the legs. Notice that the wires at the end opposite the feed point are open.

This antenna has the advantage of simplicity as compared to other rhombics. Input impedance varies considerably with input frequency, as it would with any long-wire antenna. Resonant rhombic antennas are not widely used.

Nonresonant Rhombic Antennas

One of the most sophisticated types of long-wire antennas is the **nonresonant rhombic antenna**. As you can see in **Figure 9-15**, a terminating resistor has been added to the resonant rhombic to form the nonresonant version.

Although there is no marked difference in the gain obtainable with resonant and nonresonant rhombics of comparable design, the nonresonant antenna has the advantage that it presents an essentially

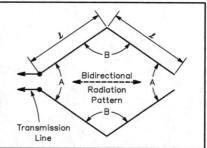

Figure 9-14 — The resonant rhombic is a diamond-shaped antenna. All legs are the same length, and opposite angles of the diamond are equal.

resistive and constant load over a wide frequency range. In addition, nonresonant operation makes the antenna substantially unidirectional, with good gain and a good front-to-back ratio. The unterminated or resonant rhombic is always bidirectional, so the main effect of adding a terminating resistance is that it changes the pattern from essentially bidirectional to essentially unidirectional. In a sense, you can consider the power to be dissipated in the terminating resistor as simply power that would have been radiated in the other direction had the resistor not been there, so the

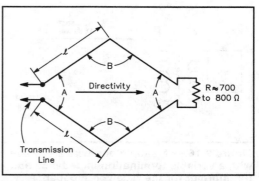

Figure 9-15 — The nonresonant rhombic antenna has a terminating resistor added at the end opposite the feed line. The main effect of this resistor is to change the pattern from one that is primarily bidirectional to one that is primarily unidirectional.

fact that some of the power (about one-third) is used up in heating the resistor does not mean an actual loss in the desired direction.

Looking into the input end of an ordinary nonresonant rhombic antenna, the characteristic impedance is on the order of 700 to 800 ohms when properly terminated in a resistance at the far end. The required terminating resistance is approximately equal to the antenna characteristic impedance, or slightly larger in value. The correct value normally will be about 800 ohms; for most work this value will ensure that the operation will not be far from optimum.

A nonreactive resistor should be used to terminate a rhombic antenna. Wirewound resistors are not suitable because they have far too much inductance and distributed capacitance. To allow a safety factor, the total rated power dissipation of the resistor or resistors should be equal to at least half the transmitter power output.

There probably aren't many amateurs using rhombic antennas. The primary disadvantage of these antennas is that they require a very large area, since each leg should be a minimum of one wavelength. Four tall, sturdy supports are required for a single antenna, and the direction of radiation from it can't be changed readily.

Beverage Antennas

Nearly every antenna installed by radio amateurs is used for both receiving and transmitting. One common exception is a type of "travelling wave antenna" invented by H.H. Beverage. The Beverage antenna acts like a long transmission line, with one lossy conductor (the Earth) and one good conductor (the wire). Like a nonresonant rhombic antenna, a Beverage antenna has a terminating resistor at

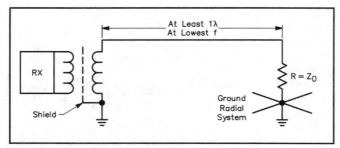

Figure 9-16 — A simple one-wire Beverage antenna with a variable termination impedance and a matching transformer for the receiver impedance.

the end farthest from the radio. The terminating resistor has a value equal to the characteristic impedance of the Beverage antenna, and connects to ground, as shown in **Figure 9-16**.

Beverage antennas are effective directional antennas for 160 meters, and also are used on 80 meters. They are less effective at higher frequencies, however, and are seldom used on 40 meters and shorter wavelength bands. Beverage antennas should be at least 1 wavelength long at the lowest operating frequency. Longer antennas provide increased gain and directivity. Beverage antennas are installed at relatively low heights, normally 8 to 10 feet above ground. They should form a relatively straight line extending from the radio shack toward the desired direction of maximum received signal strength.

The characteristic impedance of a Beverage antenna depends on the height above ground and the wire diameter, but not the antenna length. Typical values are between 400 and 600 ohms. It is also important to have a good RF ground for the resistor connection. One way to accomplish this is to lay radial waves on (or just under) the ground at the termination end. *The ARRL Antenna Book* and *ON4UN's Low Band DXing* are two books that contain additional information about Beverage antennas. Both are published by ARRL.

[Turn to Chapter 10 and study examination questions E9C07 through E9C10. Also study question E9C16. Review this section as needed.]

ANTENNAS FOR SPACE AND SATELLITE COMMUNICATIONS

Chapter 2 covers the use of Amateur Radio satellites. Here we consider those special requirements for antennas used in space radio communications. But first, let's quickly review some antenna fundamentals.

Gain and Antenna Size

"The larger the antenna (in wavelengths) the greater the gain" is a good guideline. For a properly designed Yagi, that means the longer the boom, the greater the gain.

The gain of a **parabolic**, or "**dish**," **antenna** is directly proportional to the square of the dish diameter, and directly proportional to the square of the frequency. That means the gain will increase by 6 dB if either the reflector diameter or the operating frequency is doubled.

How Much Gain is Required?

You might think more gain is better. That may be true when you are transmitting in a fixed direction. If you are trying to communicate through a rapidly moving satellite, however, it is not true. The sharper pattern of a higher-gain antenna can cause aiming problems, so an antenna with less gain — and the wider beamwidth that goes with it — may be more to your advantage. You'll want to find the "right" compromise between gain and beamwidth.

Effective isotropic radiated power (EIRP) is the standard measure of power radiation for space communications. It takes into account the antenna gain and the amount of transmitter power available at the antenna. EIRP is measured in dBW (decibels referenced to 1 watt), and an isotropic radiator is used for comparison. For example, a station has a 13.2-dBi-gain antenna, 2.2-dB feed-line loss and 10-W transmitter output power. You should remember that a decibel is ten times the logarithm of a power ratio:

$$dB = 10 \log \left(\frac{P_2}{P_1} \right) \qquad \text{(Equation 9-8)}$$

For our example, the reference power is 1 W, so $P_1 = 1$ W.

$$dB = 10 \log \left(\frac{10 \, W}{1 \, W} \right) = 10 \times 1 = 10 \, dBW$$

The EIRP of our station is the sum of all of these factors.

EIRP = +13.2 dBi + (–2.2 dB) + (+10 dBW) = +21 dBW

The same EIRP could be obtained at a station with a 10.3-dBi gain antenna, 2.3-dB feed-line loss and 20-W transmitter output. Using Equation 9-8 to express the transmitter power in dBW, we get +13 dBW. The advantage of the station in the second example is that the antenna beamwidth is not as sharp, so aiming is easier. There are many combinations of antenna gain and transmitter power that will work for space communications, as long as you meet the minimum EIRP requirements.

Several factors determine the minimum EIRP you'll need for your Earth station. To start with, there is the satellite. The spacecraft altitude (height) and the types of antennas on board are important factors. The frequency in use determines the path loss; the higher the frequency the greater the loss. You must also consider the satellite receiver sensitivity, because a less-sensitive receiver will require higher transmitter uplink power. The same factors apply for telecommand (remote control) except that higher signal-to-noise ratios are usually desirable. Once these factors are known, minimum uplink antenna gain can be calculated.

You will also have to consider several factors when you design your receiving antenna system. The required gain will depend on the satellite altitude as well as the satellite transmitter output power and the satellite transmitting antennas.

[Now turn to Chapter 10 and study examination questions E9D01, E9D02 and E9D04. Review this section as needed.]

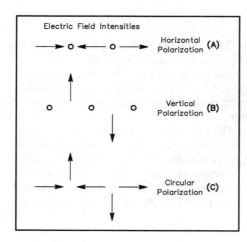

Figure 9-17 — These drawings represent the varying electric field intensity of an electromagnetic wave. The arrows represent the maximum intensity and the circles represent points when the intensity is zero. The actual intensity of the wave varies continuously between these maximum and zero points. When horizontally polarized (A) and vertically polarized (B) fields are combined with a phase difference of 90°, the result is circular polarization (C).

What About Polarization?

Best results in space radio communication are obtained not by using **horizontal** or **vertical polarization**, but by using a combination of the two called **circular polarization**. In horizontal and vertical polarization, the electric field builds to a maximum in one direction, subsides to zero, builds to a maximum in the opposite direction, subsides to zero and continues as shown in **Figure 9-17A** and 9-17B.

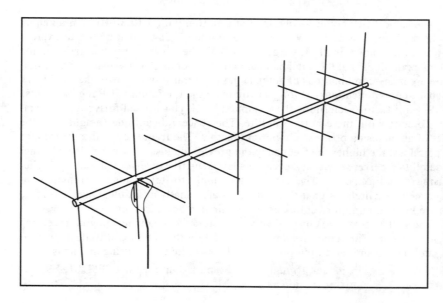

Figure 9-18 — Two Yagi antennas built on the same boom, with elements placed perpendicular to each other form the basis of a circularly polarized antenna. The driven elements are located at the same position along the boom, so they lie on the same plane, which is perpendicular to the boom. The driven elements are then fed 90° out of phase.

Vertical polarization means the electric field is oriented perpendicular to the Earth and **horizontal polarization** means the electric field is oriented parallel to the Earth. When two equal signals, one horizontally polarized and one vertically polarized, are combined with a phase difference of 90°, the result is a circularly polarized wave. See Figure 9-17C.

You can construct a circularly polarized antenna from two dipoles or Yagis mounted at 90° with respect to each other and fed 90° out of phase. **Figure 9-18** shows an example of a circularly polarized antenna made from two Yagi antennas. The two driven elements must be at the same position along the boom for this antenna. The driven elements are on the same plane, which is perpendicular to the boom, and to the direction of maximum signal.

There is one other antenna factor that you should be aware of if you plan to operate through a satellite. For terrestrial communications, beam antennas are mounted with the boom parallel to the horizon, and a rotator turns the antenna to any desired compass heading, or *azimuth*. As a satellite passes your location, however, it may be high above the horizon for part of the time. To point your antenna at the satellite, then, you will also have to be able to elevate it above horizontal. You'll need a second rotator to change the antenna *elevation* angle to track the satellite across the sky.

[Before proceeding to the next section, turn to Chapter 10 and study examination questions E9D05 and E9D07. Review this section as needed.]

MATCHING ANTENNAS TO FEED LINES

There are many techniques for matching transmission lines to antennas. This section covers the basic principles of the delta, gamma, hairpin and stub-matching systems. *The ARRL Antenna Book* contains additional information about these and other impedance-matching techniques.

The Delta Match

If you try to feed a half-wave dipole with an open-wire feed line, you will face a problem. The center impedance of the dipole is too low to be matched directly by any practical type of air-insulated parallel-conductor line. It is possible to find a value of impedance between two points on the antenna that can be matched to an open-wire line when a "fanned" section or **delta match** is used to couple the line and antenna. The antenna is not broken in the center, so there is no center insulator. Also, the delta-match connection is symmetrical about the center of the antenna. This principle is illustrated in **Figure 9-19**.

The delta match gives us a way to match a high-impedance transmission line to a lower impedance antenna. The line connects to the driven element in two places, spaced a fraction of a wavelength on each side of the element center. When the proper dimensions are unknown, the delta match is awkward to adjust because both the length and width of the delta must be varied. An additional disadvantage is that there is always some radiation from the delta. This is because the conductors are not close enough together to meet the requirement (for negligible radiation) that the spacing should be very small in comparison with the wavelength.

The Gamma Match

A commonly used method for matching a coaxial feed line to the driven element of a parasitic array is the **gamma match**. Shown in **Figure 9-20**, the gamma match has considerable flexibility in impedance matching ratio. Because this match is inherently unbalanced, no balun is needed. The gamma match gives us a way to match an unbalanced feed line to an antenna. The feed line attaches at the center of the driven element and at a fraction of a wavelength to one side of center.

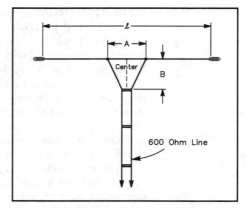

Figure 9-19 — The delta matching system is used to match a high-impedance transmission line to a lower-impedance antenna. The feed line attaches to the driven element in two places, spaced a fraction of a wavelength on each side of the element center.

Electrically speaking, the gamma conductor and the associated antenna conductor can be considered as a section of transmission line shorted at the end. Since it is shorter than $1/4$ wavelength, the gamma match has inductive reactance; this means that if the antenna itself is exactly resonant at the operating frequency, the input impedance of the gamma will show inductive reactance as well as resistance. Any reactance must be tuned out if a good match to the transmission line is to be realized. This can be done in two ways. The antenna can be shortened to obtain a value of capacitive reactance that will reflect through the matching system to cancel the inductive reactance at the input terminals, or a capacitance of the proper value can be inserted in series at the input terminals as shown in Figure 9-20.

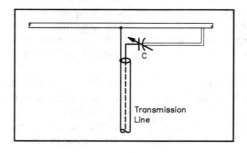

Figure 9-20 — The gamma matching system is used to match an unbalanced feed line to an antenna. The feed line attaches to the center of the driven element and to a point that is a fraction of a wavelength to one side of center. As a rule, the gamma-matching capacitor value should be approximately 7 pF per meter of wavelength at the operating frequency (about 140 pF for a 20-meter antenna and 70 pF for a 10-meter antenna).

Gamma matches have been widely used for matching coaxial cable to all-metal parasitic beams for a number of years. Because this technique is well suited to "plumber's delight" construction, in which all the metal parts are electrically and mechanically connected, the gamma match has become quite popular for amateur arrays.

Because of the many variable factors — driven-element length, gamma rod length, rod diameter, spacing between rod and driven element, and value of series capacitance — a number of combinations will provide the desired match. The maximum capacitance value should be approximately 7 pF per meter of wavelength. For example, 140 pF would be about the right value for 20-meter operation. On 10 meters, you would want to use a 70-pF capacitor. A more detailed discussion of the gamma match can be found in *The ARRL Antenna Book*.

The Hairpin Match

The *hairpin* matching system is a popular method of matching a feed line to a Yagi antenna. **Figure 9-21A** illustrates this technique. To use a hairpin match, the driven element must be tuned so it has a capacitive reactance at the desired operating frequency. (This means the element is a little too short for resonance.) The hairpin adds some inductive reactance to match the feed-point impedance.

Figure 9-21B shows the equivalent lumped-constant network for a typical hairpin matching system on a 3-element Yagi. R_A and C_A represent the antenna feed-point

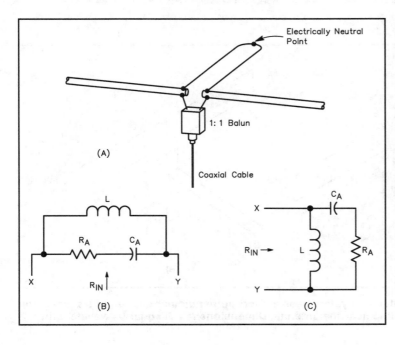

Figure 9-21 — The driven element of a Yagi antenna can be fed with a hairpin matching system, as shown in Part A. B shows the lumped-constant equivalent circuit, where R_A and C_A represent the antenna feed-point impedance, and L represents the parallel inductance of the hairpin. Points X and Y represent the feed-line connection. When the equivalent circuit is redrawn as shown in Part C, we can see that L and C_A form a simple L network to match the feed-line characteristic impedance to the antenna resistance, R_A.

impedance. L is the parallel inductance of the hairpin. When the network is redrawn, as shown in Figure 9-21C, you can see that the circuit is the equivalent of an L network.

The Stub Match

In some cases, it is possible to match a transmission line and antenna by connecting an appropriate reactance in parallel with them. Reactances formed from sections of transmission line are called **matching stubs**. Those stubs are designated either as open or closed, depending on whether the free end is an open or short circuit. An impedance match can be obtained by connecting the feed line at an appropriate point along the matching stub, as shown in **Figure 9-22**. The system illustrated here is sometimes referred to as the universal stub system.

Matching stubs have the advantage that they can be used even when the load is considerably reactive. That is a particularly useful characteristic when the antenna is not a multiple of a quarter-wavelength long, as in the case of a $^5/_8$-wavelength radiator.

A stub match can be used with coaxial cable, as illustrated in **Figure 9-23**. In a practical installation, the junction of the transmission line and stub would be a T connector, to keep the stub and feed line perpendicular to each other.

[Turn to Chapter 10 at this time, and study examination questions E9D11 and E9D12. Also study questions E9E01 through E9E04. Review this section as needed.]

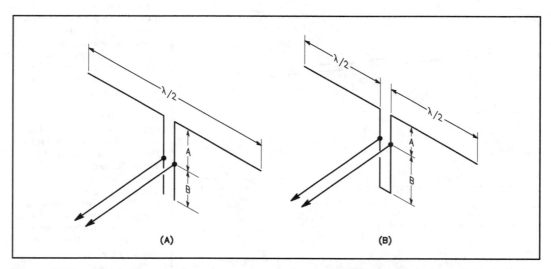

(A)

(B)

Figure 9-22 — The stub matching system uses a short perpendicular section of transmission line connected to the feed line near the antenna. Dimensions A + B equal $^1/_4$ wavelength.

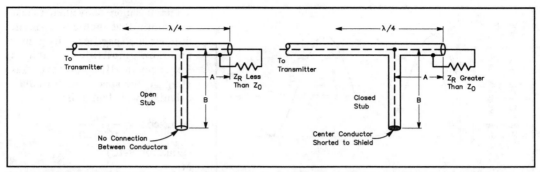

Figure 9-23 — Open and closed stubs can be used for matching to coaxial lines. Dimensions A + B equal ¹/₄ wavelength.

VELOCITY FACTOR AND ELECTRICAL LENGTH OF A TRANSMISSION LINE

Radio waves travel through space at a speed of about 300,000,000 meters per second (approximately 186,000 miles per second). If a wave travels through anything but a vacuum, its speed is always less than that.

Waves used in radio communication may have frequencies from about 10,000 to several billion hertz (Hz). Suppose the frequency of the wave shown in **Figure 9-24** is 30,000,000 Hz, or 30 megahertz (MHz). One cycle is completed in 1/30,000,000 second (33 nanoseconds). This time is called the *period* of the wave. The wave is traveling at 300,000,000 meters per second, so it will move only 10 meters during the time the current is going through one complete cycle. The electromagnetic field 10 meters away from the source is caused by the current that was flowing one period earlier in time. The field 20 meters away is caused by the current that was flowing two periods earlier, and so on.

Wavelength is the distance between two points of the same phase (for example, peaks) in consecutive cycles. This distance must be measured along the direction of wave travel. In the example found in the previous paragraph, the wavelength is 10 meters. The formula for wavelength is:

$$\lambda = \frac{v}{f}$$

(Equation 9-11)

where:
  λ = wavelength.
  v = velocity of wave.
  f = frequency of wave.

Equations normally specify basic units, such as velocity in m/s or ft/s and frequency in Hz. When we are working with very large or very small numbers it is

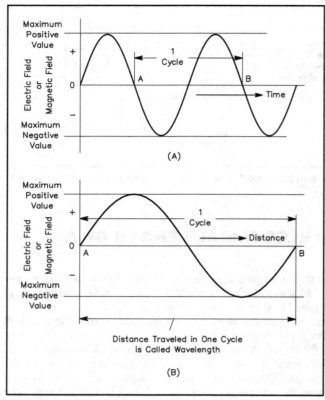

Maximum Positive Value

+

Electric Field or Magnetic Field

0

Time

Maximum Negative Value

1 Cycle

A

B

(A)

Maximum Positive Value

+

Electric Field or Magnetic Field

0

Distance

Maximum Negative Value

1 Cycle

A

B

Distance Traveled in One Cycle is Called Wavelength

(B)

Figure 9-24 — The instantaneous amplitude of both fields (electric and magnetic) varies sinusoidally with time, as shown in these graphs. Since the fields travel at constant velocity, the graphs also represents the instantaneous distribution of field intensity along the wave path. The distance between two points of equal phase, such as A-B, is the length of the wave.

sometimes more convenient to learn a formula that includes a smaller number and specifies units in more common multiples, such as frequency in MHz. For waves traveling in free space, the formula to calculate wavelength is:

$$\lambda(\text{meters}) = \frac{300}{f(\text{MHz})}$$

(Equation 9-12)
or

$$\lambda(\text{feet}) = \frac{984}{f(\text{MHz})}$$

(Equation 9-13)

Wavelength in a Wire

An alternating voltage applied to a feed line would give rise to the sort of current shown in **Figure 9-25**. (Feed lines are also known as *transmission lines*, because they transfer the radio energy from the transmitter to the antenna — and from the antenna to the receiver.) If the frequency of the ac voltage is 10 MHz, each cycle will take 0.1 microsecond. Therefore, a complete current cycle will be present along each 30 meters of line (assuming free-space velocity). This distance is one wavelength. Current observed at B occurs just one cycle later in time than the current at A. To put it another way, the current initiated at A does not appear at B, one wavelength away, until the applied voltage has had time to go through a complete cycle.

In Figure 9-25, the series of drawings shows how the instantaneous current might appear if we could take snapshots of it at quarter-cycle intervals. The current travels out from the input end of the line in waves.

At any selected point on the line, the current goes through its complete range of ac values in the time of one cycle just as it does at the input end.

Velocity of Propagation

In the previous example, we assumed that energy traveled along the line at the velocity of light. The actual velocity is very close to that of light if the insulation between the conductors of the line is solely air. The presence of **dielectric** materials other than air reduces the velocity, since electromagnetic waves travel more slowly in materials other than a vacuum. Because of this, the length of line in one wavelength will depend on the velocity of the wave as it moves along the line.

The ratio of the actual velocity at which a signal travels along a line to the speed of light in a vacuum is called the **velocity factor**.

$$V = \frac{\text{speed of wave (in line)}}{\text{speed of light (in vacuum)}}$$

(Equation 9-14)

where V = velocity factor.

The velocity factor is also related to the **dielectric constant**, ε, by:

$$V = \frac{1}{\sqrt{\varepsilon}}$$ (Equation 9-15)

where:

V = velocity factor.
ε = dielectric constant.

For example, several popular types of coaxial cable have a polyethylene dielectric, which has a dielectric constant of 2.3. For those types of coaxial cable, we can use Equation 9-15 to calculate the velocity factor of the line.

$$V = \frac{1}{\sqrt{\varepsilon}} = \frac{1}{\sqrt{2.3}} = \frac{1}{1.5} = 0.66$$

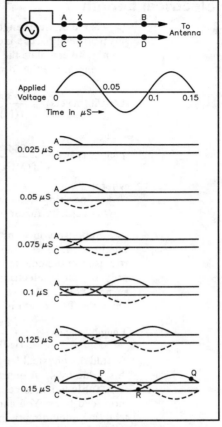

Figure 9-25 — Instantaneous current along a transmission line at successive time intervals. The period of the wave (the time for one cycle) is 0.1 microsecond.

Electrical Length

The electrical length of a transmission line (or antenna) is not the same as its physical length. The electrical length is measured in wavelengths at a given frequency. To calculate the physical length of a transmission line that is electrically one wavelength, use the formulas:

$$\text{Length (meters)} = \frac{300}{f(\text{MHz})} \times V \qquad \text{(Equation 9-16)}$$

or

$$\text{Length (feet)} = \frac{984}{f(\text{MHz})} \times V \qquad \text{(Equation 9-17)}$$

where:
   f = operating frequency (in MHz).
   V = velocity factor.

Suppose you want a section of RG-8 coaxial cable that is one-quarter wavelength at 14.1 MHz. What is its physical length? The answer depends on the dielectric used in the coaxial cable. RG-8 is manufactured with polyethylene or foamed polyethylene dielectric; velocity factors for the two versions are 0.66 and 0.80, respectively. We'll use the polyethylene line with a velocity factor of 0.66 for our example. The length in meters of one wavelength is given by Equation 9-16:

$$\text{Length (meters)} = \frac{300}{f(\text{MHz})} \times V = \frac{300}{14.1} \times 0.66 = 21.3 \times 0.66 = 14.1 \text{ m}$$

To find the physical length for a quarter wavelength section of line, we must divide this value by 4. A quarter wavelength section of this coax is 3.52 meters.

Table 9-1 lists velocity factors and other characteristics for some other common feed lines. You can calculate the physical length of a section of any type of feed line, including twin lead and ladder line, at some specific frequency as long as you know the velocity factor.

In review, the lower the velocity factor, the slower a radio-frequency wave moves through the line. The lower the velocity factor, the shorter a line is for the same electrical length at a given frequency. One wavelength in a practical line is always shorter than a wavelength in free space.

Feed Line Loss

When you select a feed line for a particular antenna, you must consider some conflicting factors and make a few trade-offs. For example, most amateurs want to use a relatively inexpensive feed line. We also want a feed line that does not lose an appreciable amount of signal energy, however. For many applications, coaxial cable seems to be a good choice, but parallel-conductor feed lines generally have lower loss and may provide some other advantages, such as operation with high SWR conditions on the line.

Table 9-1

Characteristics of Commonly Used Transmission Lines

Type of line	Z_0 Ohms	Vel %	pF per foot	OD (inches)	Diel. Material	Max Operating volts (RMS)	Loss in dB per 100 ft at 30 MHz
RG-8	52.0	66	29.5	0.405	PE	3700	1.4
RG-8 Foam	50.0	78	26.0	0.405	Foam PE	600	0.9
RG-8X	52.0	78	26.0	0.242	Foam PE	300	—
RG-58	53.5	66	28.5	0.195	PE	1400	2.4
RG-58 Foam	50.0	78	26.0	0.195	Foam PE	300	2.1
RG-174	50.0	66	30.8	0.100	PE	1100	5.5
Aluminum Jacket Foam Dielectric							
1/2 inch	50.0	81	25.0	0.500		2500	0.4
3/4 inch	50.0	81	25.0	0.750		4000	0.32
7/8 inch	50.0	81	25.0	0.875		4500	0.26
Parallel Conductor							
Open wire, #12	600	97	—	—		—	0.1
300-Ω twin lead	300.0	80	5.8	—		—	—
Open wire, "Window" Type Ladder Line							
1/2 inch	300.0	95	—	—		—	0.16
1 inch	450.0	95	—	—		—	0.16

Dielectric Designation	Name	Temperature Limits
PE	Polyethylene	−65° to +80°C
Foam PE	Foamed Polyethylene	−65° to +80°C
PTFE	Polyetrafluoroethylene (Teflon)	−65° to +250°C

From *The ARRL Antenna Book*, 18th Ed, p 24-17.

Line loss increases as the operating frequency increases, so on the lower-frequency HF bands, you may decide to use a less-expensive coaxial cable with a higher loss than you would on 10 meters. Open-wire or ladder-line feed lines generally have lower loss than coaxial cables at any particular frequency. Table 9-1 includes approximate loss values for 100 feet of the various feed lines at 30 MHz. Again, these values vary significantly as the frequency changes.

The line loss values given in Table 9-1 are for a feed line with load and input impedances that are matched to the feed line characteristic impedance. (A perfectly matched line condition means the SWR will be 1:1.) If the impedances are not matched, the loss will be somewhat larger than the values listed. (In the case of a mismatched line, the SWR will be greater than 1:1.)

The voltage **reflection coefficient** is the ratio of the reflected voltage at some point on a feed line to the incident voltage at the same point. It is also equal to the ratio of reflected current to incident current at some point on the line. The reflection coefficient is determined by the relationship between the feed line characteristic impedance and the actual load impedance. The reflection coefficient is a complex quantity, having both amplitude and phase. It is generally designated by the lower case Greek letter ρ (rho), although some professional literature uses the capital Greek letter Γ (gamma). The reflection coefficient is a good parameter to describe the interactions at the load end of a mismatched transmission line.

[Before proceeding, study examination questions E9E07 through E9E15 in Chapter 10. Review this section as needed.]

PROPERTIES OF OPEN AND SHORTED FEED-LINE SECTIONS

For this section you will need to study **Figures 9-26** and **9-27**. In those diagrams you will see the relationships between impedance, voltage and current illustrated for open and shorted sections of various lengths of feed line. The impedance "seen" looking into various lengths of feed line are indicated directly above the chart. Curves above the axis marked with R, X_L and X_C indicate the relative value of the impedance presented at the input. Circuit symbols indicate the equivalent circuits for the lines at that particular length. Standing waves of voltage (E) and current (I) are shown above each line. Remember that Z = E / I, by Ohm's Law, so you can use the curves above each piece of line to estimate the input impedance of a given line length.

Impedance is minimum and current is maximum at all odd quarter-wavelength points ($^1/_4$, $^3/_4$ and so on) as measured from the input end along an open line. A $^1/_4$-wavelength feed line that is open at one end presents a very low impedance to a signal generator.

At all even quarter-wavelength points ($^1/_2$, 1, $^3/_2$ and so on), the impedance is maximum and *voltage* is maximum. A $^1/_2$-wavelength feed line that is open at one end presents a very high impedance to a signal generator. You should also notice that the impedance is resistive at multiples of a quarter wavelength.

At points in between quarter-wavelength marks, the impedance is either capacitive or inductive, as shown near the top of Figures 9-26 and 9-27. For example, between 0 and $^1/_4$ wavelength the impedance is capacitive, so a $^1/_8$-wavelength feed line that is open at one end presents an impedance that is capacitive. Between $^1/_4$ and $^1/_2$ wavelength the impedance is inductive, so a $^3/_8$-wavelength line that is open at one end presents an impedance that is inductive. As you can see on the graph, the impedance alternates between capacitive and inductive values.

Shorted lines can be considered in a similar manner. See Figure 9-27. For example, a $^1/_8$-wavelength line that is shorted at one end presents an inductive reactance to a signal generator. A $^1/_4$-wavelength feed line that is shorted at one end presents a very high impedance to a signal generator. A $^1/_2$-wavelength feed line that is shorted at one end presents a very low impedance to a signal generator. Also notice that the impedance is resistive at multiples of a quarter wavelength.

You can find more information on transmission lines in *The ARRL Antenna Book*.

[Complete your study of this chapter by turning to Chapter 10 and studying examination questions E9E05 and E9E06. Review this section as needed. If you have gone through this book chapter by chapter, you have now completed your study of the material for your Amateur Extra class examination. Congratulations, and good luck with your exam!]

(See Figures 9-26 and 9-27 on the next pages.)

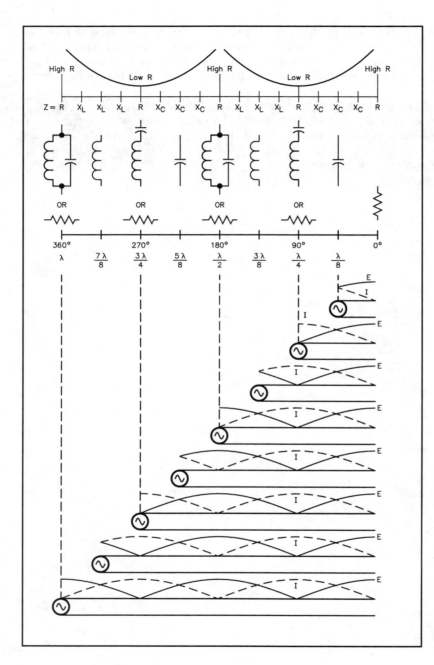

Figure 9-26 — This diagram summarizes the characteristics of open-ended transmission lines. Voltage standing waves are shown as solid lines above each length of cable, and current standing waves are shown as dashed lines.

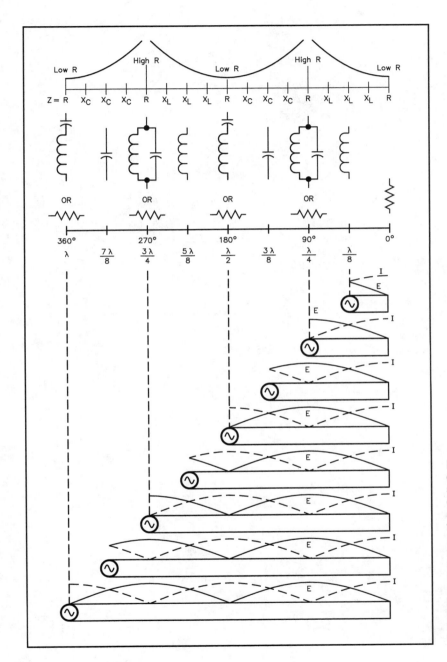

Figure 9-27 — This diagram summarizes the characteristics of short-circuited transmission lines. Voltage standing waves are shown as solid lines above each length of cable, and current standing waves are shown as dashed lines.

ELEMENT 4 (AMATEUR EXTRA CLASS) QUESTION POOL—WITH ANSWERS

This chapter contains the complete question pool for the Amateur Extra (Element 4) exam. Before you read the questions and answers printed in this chapter, *be sure to read the text in Chapters 1 through 9*. Use the questions in this chapter as review exercises, when the text tells you to study them. Don't try to memorize all the questions and answers.

This question pool, released by the Volunteer Examiner Coordinators' Question Pool Committee, is for use on exams beginning April 15, 2000. The pool is scheduled to be used until June 30, 2002. Changes to FCC Rules and other factors may result in earlier revisions. Such changes will be announced in *QST* and other Amateur Radio publications. Normally, the Question Pool Committee will simply withdraw outdated questions from the question pool when such Rules changes occur.

How Many Questions?

The FCC specifies that an Element 4 exam must include 50 questions. This question pool is divided into nine sections, called subelements. (A subelement is a portion of the exam element, in this case Element 4.) Although it's not an FCC requirement, the Question Pool Committee specifies the number of questions from each subelement that should appear on your test. For example, there should be seven questions from the Commission's Rules section, Subelement E1. **Table 10-1** summarizes the number of questions from each subelement that make up an Element 4 exam. The number of questions to be used from each subelement also appears at the beginning of that subelement in the question pool.

The Volunteer Examiner Coordinators' Question Pool Committee has broken the subelements into smaller groups. There are the same number of groups as there are questions from each subelement, and the committee intends for one question to come from each of the smaller groups. This is not an FCC requirement, however.

There is a list of topics printed in **bold** type at the beginning of each small group. This list represents the *syllabus*, or study guide, topics for that section. The entire Amateur Extra syllabus is printed at the end of the Introduction chapter.

The small groups are listed alphabetically within each subelement. For example, since there are seven exam questions from Subelement E1, these sections are labeled E1A through E1G. Five exam questions come from the Circuit Components subelement, so that subelement has sections labeled E6A through E6E.

The question numbers used in the question pool relate to the syllabus or study guide printed at the end of the Introduction chapter. The syllabus is an outline of topics covered by the exam. Each question number begins with an E. This indicates the question is from the Amateur Extra question pool. Next is a number to indicate which subelement the question is from. These numbers will range from 1 to 9. Following this number is a letter to indicate which group the question is from in that subelement. Each question number ends with a two-digit number to specify its position in the set. So question number E2A01 is the first question in the A group of the second subelement. Question E9B08 is the eighth question in the B group of the ninth subelement.

Who Picks the Questions?

The FCC allows Volunteer Examiner Teams to select the questions that will be used on amateur exams. If your test is coordinated by the ARRL/VEC, your test will be prepared by the VEC, or using a computer program supplied by the VEC. All VECs and examiners must

---

Table 10-1
Amateur Extra Exam Content

Subelement	Topic	Number of Questions
E1	Commission's Rules	7
E2	Operating Procedures	4
E3	Radio-Wave Propagation	3
E4	Amateur Radio Practices	5
E5	Electrical Principles	9
E6	Circuit Components	5
E7	Practical Circuits	7
E8	Signals and Emissions	5
E9	Antennas and Feed Lines	5

use the questions, answers and distracters (incorrect answers) printed here.

This question pool contains more than ten times the number of questions necessary to make up an exam. This ensures that the examiners have sufficient questions to choose from when they make up an exam.

Who Gives the Test?

All Amateur Radio license exams are given by teams of three or more Volunteer Examiners (VEs). Each of the examiners is accredited by a Volunteer Examiner Coordinator (VEC) to give exams under their program. A VEC is an organization that has entered into an agreement with the FCC to coordinate the efforts of VEs. The VEC reviews the paperwork for each exam session, and then forwards the information to the FCC. The FCC then issues new or upgraded Amateur Radio licenses to those who qualify for them.

Question Pool Format

The rest of this chapter contains the entire Element 4 question pool. We have printed the answer key to these questions along the edge of the page. There is a line to indicate where you should fold the page to hide the answer key while you study. After making your best effort to answer the questions, you can look at the answers to check your understanding. We also have included page references along with the answers. These page numbers indicate where you will find the text discussion related to each question. If you have any problems with a question, refer to the page listed for that question. You may have to study beyond the listed page number to review all the related material. We also have included references to sections of Part 97 for the Commission's Rules Subelement, E1. This is to help you identify the particular rule citations for these questions. The complete text of the FCC Rules in Part 97 is included in *The ARRL's FCC Rule Book*.

The Question Pool Committee included all the drawings for the question pool on two pages. Some of these drawings were picked up from the previous Advanced Question Pool. That is why some drawing figure numbers begin with A. The drawing numbers picked up from the previous Amateur Extra Question Pool begin with E. We have placed a copy of these pages at the beginning of the question pool. In addition, the individual figures appear with the applicable questions in the Question Pool.

Good luck with your studies! With a bit of time devoted to reviewing this book, you'll soon be ready for your Amateur Extra exam!

Element 4 (Extra)

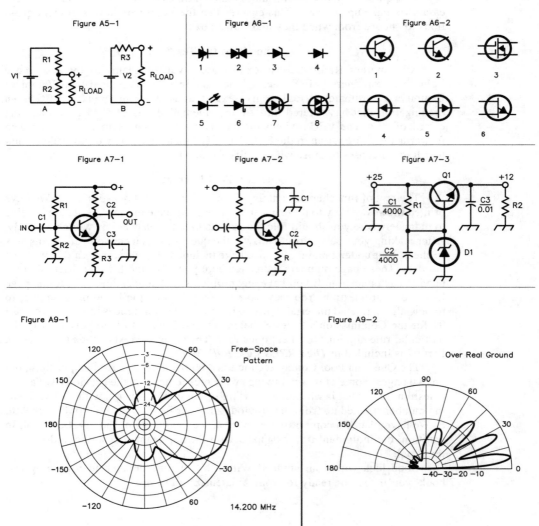

Figure A5-1

Figure A6-1

Figure A6-2

Figure A7-1

Figure A7-2

Figure A7-3

Figure A9-1

Free-Space Pattern

14.200 MHz

Figure A9-2

Over Real Ground

As prepared by Question Pool Committee
for all examinations administered
April 15, 2000 through June 30, 2002

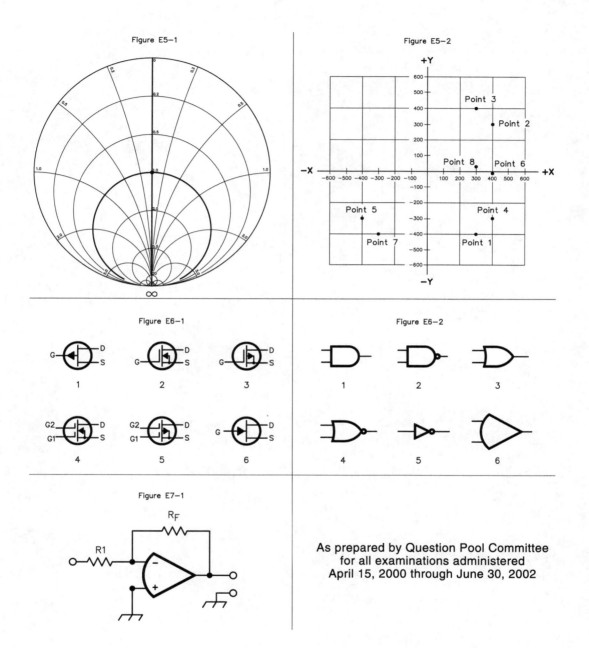

Figure E5—1

Figure E5—2

Figure E6—1

Figure E6—2

Figure E7—1

As prepared by Question Pool Committee
for all examinations administered
April 15, 2000 through June 30, 2002

ELEMENT 4 (AMATEUR EXTRA CLASS) QUESTION POOL—WITH ANSWERS

As released by the Question Pool Committee,
National Conference of Volunteer Examiner Coordinators

February 17, 2000

Subelement E1 — Commission's Rules
[7 Exam Questions — 7 Groups]

E1A Operating standards: frequency privileges for Extra class amateurs; emission standards; message forwarding; frequency sharing between ITU Regions; FCC modification of station license; 30-meter band sharing; stations aboard ships or aircraft; telemetry; telecommand of an amateur station; authorized telecommand transmissions; definitions of image, pulse and test

E1A01
What exclusive frequency privileges in the 80-meter band are authorized to Extra class control operators?
A. 3525-3775 kHz
B. 3500-3525 kHz
C. 3700-3750 kHz
D. 3500-3550 kHz

E1A02
What exclusive frequency privileges in the 75-meter band are authorized to Extra class control operators?
A. 3775-3800 kHz
B. 3800-3850 kHz
C. 3750-3775 kHz
D. 3800-3825 kHz

E1A03
What exclusive frequency privileges in the 40-meter band are authorized to Extra class control operators?
A. 7000-7025 kHz
B. 7000-7050 kHz
C. 7025-7050 kHz
D. 7100-7150 kHz

Answer Key

Page numbers tell you where to look in this book for more information.

Subelement E1

Numbers in [square brackets] indicate sections in Part 97, the Amateur Radio Rules.

E1A01
(B)
[97.301b]
Page 1-4

E1A02
(C)
[97.301b]
Page 1-4

E1A03
(A)
[97.301b]
Page 1-4

E1A04 (D) [97.301b] Page 1-4	**E1A04** What exclusive frequency privileges in the 20-meter band are authorized to Extra class control operators? A. 14.100-14.175 MHz and 14.150-14.175 MHz B. 14.000-14.125 MHz and 14.250-14.300 MHz C. 14.025-14.050 MHz and 14.100-14.150 MHz D. 14.000-14.025 MHz and 14.150-14.175 MHz
E1A05 (C) [97.301b] Page 1-4	**E1A05** What exclusive frequency privileges in the 15-meter band are authorized to Extra class control operators? A. 21.000-21.200 MHz and 21.250-21.270 MHz B. 21.050-21.100 MHz and 21.150-21.175 MHz C. 21.000-21.025 MHz and 21.200-21.225 MHz D. 21.000-21.025 MHz and 21.250-21.275 MHz
E1A06 (D) [97.307c] Page 1-5	**E1A06** What must an amateur licensee do if a spurious emission from his or her station causes harmful interference to the reception of another radio station? A. Pay a fine each time it happens B. Submit a written explanation to the FCC C. Forfeit the station license if it happens more than once D. Eliminate or reduce the interference
E1A07 (A) [97.307d] Page 1-5	**E1A07** What is the maximum mean power permitted for any spurious emission from a transmitter or external RF power amplifier transmitting on a frequency below 30 MHz? A. 50 mW B. 100 mW C. 10 mW D. 10 W
E1A08 (B) [97.307d] Page 1-5	**E1A08** How much below the mean power of the fundamental emission must any spurious emissions from a station transmitter or external RF power amplifier transmitting on a frequency below 30 MHz be attenuated? A. At least 10 dB B. At least 40 dB C. At least 50 dB D. At least 100 dB
E1A09 (C) [97.307e] Page 1-5	**E1A09** How much below the mean power of the fundamental emission must any spurious emissions from a transmitter or external RF power amplifier transmitting on a frequency between 30 and 225 MHz be attenuated? A. At least 10 dB B. At least 40 dB C. At least 60 dB D. At least 100 dB

E1A10
What is the maximum mean power permitted for any spurious emission from a transmitter having a mean power of 25 W or less on frequencies between 30 and 225 MHz?
A. 5 microwatts
B. 10 microwatts
C. 20 microwatts
D. 25 microwatts

E1A10
(D)
[97.307e]
Page 1-5

E1A11
If a packet bulletin board station in a message forwarding system inadvertently forwards a message that is in violation of FCC rules, who is accountable for the rules violation?
A. The control operator of the packet bulletin board station
B. The control operator of the originating station and conditionally the first forwarding station
C. The control operators of all the stations in the system
D. The control operators of all the stations in the system not authenticating the source from which they accept communications

E1A11
(B)
[97.219b&d]
Page 1-4

E1A12
If your packet bulletin board station inadvertently forwards a communication that violates FCC rules, what is the first action you should take?
A. Discontinue forwarding the communication as soon as you become aware of it
B. Notify the originating station that the communication does not comply with FCC rules
C. Notify the nearest FCC Field Engineer's office
D. Discontinue forwarding all messages

E1A12
(A)
[97.219c]
Page 1-4

E1A13
Why might the FCC modify an amateur station license?
A. To relieve crowding in certain bands
B. To better prepare for a time of national emergency
C. To enforce a radio quiet zone within one mile of an airport
D. To promote the public interest, convenience and necessity

E1A13
(D)
[97.27a]
Page 1-7

E1A14
If an amateur station is installed on board a ship or aircraft and is separate from the main radio installation, what condition must be met before the station is operated?
A. Its operation must be approved by the master of the ship or the pilot in command of the aircraft
B. Its antenna must be separate from the main ship or aircraft antennas, transmitting only when the main radios are not in use
C. It must have a power supply that is completely independent of the main ship or aircraft power supply
D. Its operator must have an FCC Marine or Aircraft endorsement on his or her amateur license

E1A14
(A)
[97.11a]
Page 1-7

E1A15
What type of FCC-issued license or permit is required to transmit amateur communications from a vessel registered in the US while in international waters?
A. Any amateur license with an FCC Marine or Aircraft endorsement
B. Any amateur license or reciprocal permit for alien amateur licensee
C. Any General class or higher license
D. An Extra class license

E1A15
(B)
[97.11]
Page 1-7

E1A16
(D)
[97.211b]
Page 1-17

E1A16
When may a station use special codes intended to obscure the meaning of messages?
A. Never under any circumstances
B. Only when a Special Temporary Authority has been obtained from the FCC
C. Only when an Extra class operator is controlling the station
D. When sending telecommand messages to a station in space operation

E1B Station restrictions: restrictions on station locations; restricted operation; teacher as control operator; station antenna structures; definition and operation of remote control and automatic control; control link

E1B01
(A)
[97.13a]
Page 1-8

E1B01
Which of the following factors might restrict the physical location of an amateur operator's station equipment or antenna structure?
A. The land may have environmental importance; or it is significant in American history, architecture or culture
B. The location's political or societal importance
C. The location's geographical or horticultural importance
D. The location's international importance, requiring consultation with one or more foreign governments before installation

E1B02
(A)
[97.13b]
Page 1-8

E1B02
Outside of what distance from an FCC monitoring facility may an amateur station be located without concern for protecting the facility from harmful interference?
A. 1 mile
B. 3 miles
C. 10 miles
D. 30 miles

E1B03
(C)
[97.13a]
Page 1-8

E1B03
What must be done before an amateur station is placed within an officially designated wilderness area or wildlife preserve, or an area listed in the National Register of Historical Places?
A. A proposal must be submitted to the National Park Service
B. A letter of intent must be filed with the National Audubon Society
C. An Environmental Assessment must be submitted to the FCC
D. A form FSD-15 must be submitted to the Department of the Interior

E1B04
(A)
[97.121a]
Page 1-7

E1B04
If an amateur station interferes with the reception of broadcast stations on a well-engineered receiver, during what hours shall the amateur station NOT be operated on the interfering frequencies?
A. Daily from 8 PM to 10:30 PM local time and additionally from 10:30 AM to 1 PM on Sunday
B. Daily from 6 PM to 12 AM local time and additionally from 8 AM to 5 PM on Sunday
C. Daily for any continuous span of at least 2.5 hours and for at least 5 continuous hours on Sunday
D. Daily for any continuous span of at least 6 hours and for at least 9 continuous hours on Sunday

E1B05
If an amateur station causes interference to the reception of a domestic broadcast station with a receiver of good engineering design, on what frequencies may the operation of the amateur station be restricted?
A. On the frequency used by the domestic broadcast station
B. On all frequencies below 30 MHz
C. On all frequencies above 30 MHz
D. On the interfering amateur frequency or frequencies

E1B05
(D)
[97.121a]
Page 1-7

E1B06
When may a paid professional teacher be the control operator of an amateur station used in the teacher's classroom?
A. Only when the teacher is not paid during periods of time when an amateur station is used
B. Only when the classroom is in a correctional institution
C. Only when the station is used by that teacher as a part of classroom instruction at an educational institution
D. Only when the station is restricted to making contacts with similar stations at other educational institutions

E1B06
(C)
[97.113c]
Page 1-9

E1B07
Who may accept compensation when acting as a control operator in a classroom?
A. Any licensed amateur
B. Only teachers at educational institutions
C. Only teachers at correctional institutions
D. Only students at educational or correctional institutions

E1B07
(B)
[97.113c]
Page 1-9

E1B08
What limits must state and local authorities observe when legislating height and dimension restrictions for amateur antenna structures?
A. FAA regulations specify a minimum height for amateur antenna structures located near airports
B. FCC regulations specify a 200 foot minimum height for amateur antenna structures
C. State and local restrictions of amateur antenna structures are not allowed
D. PRB-1 specifies that authorities must reasonably accommodate the installation of amateur antenna structures

E1B08
(D)
[97.15b]
Page 1-9

E1B09
If an amateur antenna structure is located in a valley or canyon, what height restrictions apply?
A. The structure must not extend more than 200 feet above average height of the terrain
B. The structure must be no higher than 200 feet above ground level at its site
C. There are no height restrictions since the structure would not be a hazard to aircraft in a valley or canyon
D. The structure must not extend more than 200 feet above the top of the valley or canyon

E1B09
(B)
[97.15]
Page 1-8

E1B10
(D)
[97.15]
Page 1-8

E1B10
What kind of approval is required before erecting an amateur antenna located near an airport as defined in the FCC rules?
A. The FAA and FCC both must approve any type of antenna structure located near an airport
B. Approval must be obtained from the airport manager
C. Approval must be obtained from the local zoning authorities
D. The FCC must approve an antenna structure that is higher than 20 feet above any natural or existing man made structure

E1B11
(C)
[97.15]
Page 1-9

E1B11
What special restrictions does the FCC impose on amateur antennas mounted on motor vehicles?
A. Such antennas may not extend more than 15 feet above the roof of the vehicle
B. Complex antennas, such as a Yagi or quad beam, may not be installed on motor vehicles
C. None
D. Such antennas must comply with the recommendations of the vehicle manufacturer

E1B12
(C)
[97.15a]
Page 1-8

E1B12
What must an amateur obtain before installing an antenna structure more than 200 feet high?
A. An environmental assessment
B. A Special Temporary Authorization
C. Prior FCC approval
D. An effective radiated power statement

E1B13
(D)
[97.3a38]
Page 1-10

E1B13
What is meant by a remotely controlled station?
A. A station operated away from its regular home location
B. Control of a station from a point located other than at the station transmitter
C. A station operating under automatic control
D. A station controlled indirectly through a control link

E1B14
(A)
[97.201d,
97.203d,
97.205d]
Page 1-10

E1B14
Which kind of station operation may not be automatically controlled?
A. Control of a model craft
B. Beacon operation
C. Auxiliary operation
D. Repeater operation

E1B15
(A)
[97.3a6]
Page 1-10

E1B15
What is meant by automatic control of a station?
A. The use of devices and procedures for control so that a control operator does not have to be present at a control point
B. A station operating with its output power controlled automatically
C. Remotely controlling a station such that a control operator does not have to be present at the control point at all times
D. The use of a control link between a control point and a locally controlled station

E1B16
How do the control operator responsibilities of a station under automatic control differ from one under local control?
A. Under local control there is no control operator
B. Under automatic control a control operator is not required to be present at a control point
C. Under automatic control there is no control operator
D. Under local control a control operator is not required to be present at a control point

E1B17
What is a control link?
A. A device that automatically controls an unattended station
B. An automatically operated link between two stations
C. The means of control between a control point and a remotely controlled station
D. A device that limits the time of a station's transmission

E1B18
What is the term for apparatus to effect remote control between a control point and a remotely controlled station?
A. A tone link
B. A wire control
C. A remote control
D. A control link

E1C Reciprocal operating: definition of reciprocal operating permit; purpose of reciprocal agreement rules; alien control operator privileges; identification; application for reciprocal permit; reciprocal permit license term (Note: This includes CEPT and IARP.)

E1C01
What is an FCC authorization for alien reciprocal operation?
A. An FCC authorization to a holder of an amateur license issued by certain foreign governments to operate an amateur station in the US
B. An FCC permit to allow a US licensed amateur to operate in a foreign nation, except Canada
C. An FCC permit allowing a foreign licensed amateur to handle third-party traffic between the US and the amateur's own nation
D. An FCC agreement with another country allowing the passing of third-party traffic between amateurs of the two nations

E1C02
Who is eligible for an FCC authorization for alien reciprocal operation?
A. Anyone holding a valid amateur license issued by a foreign government
B. Any non-US citizen holding an amateur license issued by a foreign government with which the US has a reciprocal operating arrangement
C. Anyone holding a valid amateur license issued by a foreign government with which the US has a reciprocal operating arrangement
D. Any non-US citizen holding a valid amateur or shortwave listener's license issued by a foreign government

E1B16
(B)
[97.3a6]
Page 1-10

E1B17
(C)
[97.3a38]
Page 1-10

E1B18
(D)
[97.3a38]
Page 1-10

E1C01
(A)
[97.5c,d,e, 97.107]
Page 1-10

E1C02
(B)
[97.107]
Page 1-10

E1C03
(C)
[97.107]
Page 1-11

E1C03
What operator frequency privileges are authorized by an FCC authorization for alien reciprocal operation?
A. Those authorized to a holder of the equivalent US amateur license, unless the FCC specifies otherwise by endorsement on the authorization
B. Those that the holder of the permit would have in their own country
C. Those authorized to US amateurs that the holder of the permit would have in their own country, unless the FCC specifies otherwise
D. Only those frequencies approved by the International Amateur Radio Union, unless the FCC specifies otherwise

E1C04
(D)
[97.119g]
Page 1-11

E1C04
What additional station identification, in addition to his or her own call sign, does an alien operator supply when operating in the US under an FCC authorization for alien reciprocal operation?
A. No additional identification is required
B. The grid-square locator closest to his or her present location is included before the call
C. The serial number of the permit and the call-letter district number of the station location is included before the call
D. The letter-numeral indicating the station location in the US is included before their own call and closest city and state

E1C05
(A)
[97.107]
Page 1-10

E1C05
When may a US citizen holding a foreign amateur license obtain an FCC authorization for alien reciprocal operation?
A. Never; US citizens are not eligible
B. When the citizen has imported his or her equipment from the foreign country
C. When the citizen has never held a US amateur license
D. When the citizen has no current US amateur license

E1C06
(A)
[97.107]
Page 1-10

E1C06
Which of the following would disqualify a foreign amateur from being eligible for a US authorization for alien reciprocal operation?
A. Holding only an amateur license issued by a country but not being a citizen of that country
B. Citizenship in their own country but not US citizenship
C. Holding only an amateur license issued by their own country but holding no US amateur license
D. Holding an amateur license issued by their own country granting them frequency privileges beyond US Extra class privileges

E1C07
(B)
[97.107a]
Page 1-11

E1C07
What special document is required before a Canadian citizen holding a Canadian amateur license may operate in the US?
A. All aliens, including Canadians, must obtain an FCC authorization for alien reciprocal operation
B. No special document is required
C. The citizen must have an FCC-issued validation of their Canadian license
D. The citizen must have an FCC-issued Certificate of US License Grant without Examination to operate for a period longer than 10 days

E1C08
What operating privileges does a properly licensed alien amateur have in the US, if the US and the alien amateur's home country have a multilateral or bilateral reciprocal operating agreement?
A. All privileges of their home license
B. All privileges of a US Amateur Extra license
C. Those granted by their home license that match US privileges, not to exceed the operating privileges of an Amateur Extra license
D. Those granted by their home license that match US privileges authorized to amateurs operating in ITU Region 1

E1C09
From which locations may a licensed alien amateur be a control operator?
A. Only locations within the boundaries of the 50 United States
B. Only locations listed as the primary station location on a US amateur license
C. Only locations on ground within the US and its territories; no shipboard or aeronautical mobile operation is permitted
D. Any location where the amateur service is regulated by the FCC

E1C10
Which of the following multilateral or bilateral operating arrangements allow US amateurs to operate in many European countries and alien amateurs from many European countries to operate in the US?
A. CEPT agreement
B. IARP agreement
C. ITU agreement
D. All these choices are correct

E1C11
Which of the following multilateral or bilateral operating arrangements allow US amateurs and many Central and South American amateurs to operate in each others' countries?
A. CEPT agreement
B. IARP agreement
C. ITU agreement
D. All of these choices are correct

E1D Radio Amateur Civil Emergency Service (RACES): definition; purpose; station registration; station license required; control operator requirements; control operator privileges; frequencies available; limitations on use of RACES frequencies; points of communication for RACES operation; permissible communications

E1D01
What is RACES?
A. An amateur network for providing emergency communications during athletic races
B. The Radio Amateur Civil Emergency Service
C. The Radio Amateur Corps for Engineering Services
D. An amateur network for providing emergency communications during boat or aircraft races

E1C08	
(C)	
[97.107b]	
Page 1-11	

E1C09
(D)
[97.5c]
Page 1-11

E1C10
(A)
[97.5d]
Page 1-11

E1C11
(B)
[97.5e]
Page 1-11

E1D01
(B)
[97.3a37]
Page 1-13

E1D02
(A)
[97.3a37]
Page 1-13

E1D02
What is the purpose of RACES?
A. To provide civil-defense communications during emergencies
B. To provide emergency communications for boat or aircraft races
C. To provide routine and emergency communications for athletic races
D. To provide routine and emergency military communications

E1D03
(C)
[97.407a]
Page 1-13

E1D03
With what other organization must an amateur station be registered before RACES operation is permitted?
A. The Amateur Radio Emergency Service
B. The US Department of Defense
C. A civil defense organization
D. The FCC Field Operations Bureau

E1D04
(C)
[97.407a]
Page 1-14

E1D04
Which amateur stations may be operated in RACES?
A. Only Extra class amateur stations
B. Any licensed amateur station (except a station licensed to a Technician)
C. Any licensed amateur station certified by the responsible civil defense organization
D. Any licensed amateur station (except a station licensed to a Technician) certified by the responsible civil defense organization

E1D05
(D)
[97.21a1]
Page 1-14

E1D05
Application for modification of a RACES license must be made on what FCC form, and sent to what FCC office?
A. Form 605, sent to Washington, DC
B. Form 605, sent to Gettysburg, PA
C. Form 610-A, sent to Washington, DC
D. A Club Call Sign Administrator must submit the information to the FCC in an electronic batch file

E1D06
(D)
[97.407a]
Page 1-14

E1D06
Who may be the control operator of a RACES station?
A. Anyone who holds an FCC-issued amateur license other than Novice
B. Only an Extra class licensee
C. Anyone who holds an FCC-issued amateur license other than Novice and is certified by a civil defense organization
D. Anyone who holds an FCC-issued amateur license and is certified by a civil defense organization

E1D07
(A)
[97.407b]
Page 1-14

E1D07
What additional operator privileges are granted to an Extra class operator registered with RACES?
A. None
B. CW operations on 5167.5 kHz
C. Unattended HF packet-radio station operations
D. 237-MHz civil defense band operations

E1D08
What frequencies are normally available for RACES operation?
A. Only those frequencies authorized to civil defense organizations
B. Only those frequencies authorized to emergency military communications
C. Only the top 25 kHz of each amateur frequency band
D. All frequencies available to the amateur service

E1D09
What type of emergency can cause limits to be placed on the frequencies available for RACES operation?
A. An emergency in which the President invokes the War Emergency Powers under the provisions of the Communications Act of 1934
B. An emergency in only one state in the US would limit RACES operations to a single HF frequency band
C. An emergency confined to a 25-mile area would limit RACES operations to a single VHF band
D. An emergency involving no immediate danger of loss of life

E1D10
With what stations may amateur RACES stations communicate?
A. Any RACES stations and any amateur stations except stations licensed to Technician class operators
B. Any RACES stations and certain other stations authorized by the responsible civil defense official
C. Any amateur station or a station in the Disaster Communications Service
D. Any amateur station and any military emergency station

E1D11
What are permissible communications in RACES?
A. Any type of communications when there is no emergency
B. Any Amateur Radio Emergency Service communications
C. National defense or immediate safety of people and property and communications authorized by the area civil defense organization
D. National defense and security or immediate safety of people and property communications authorized by the President

E1D08
(D)
[97.407b]
Page 1-14

E1D09
(A)
[97.407b]
Page 1-14

E1D10
(B)
[97.407c,d]
Page 1-14

E1D11
(C)
[97.407e]
Page 1-14

E1E Amateur Satellite Service: definition; purpose; station license required for space station; frequencies available; telecommand operation: definition; eligibility; telecommand station (definition); space telecommand station; special provisions; telemetry: definition; special provisions; space station: definition; eligibility; special provisions; authorized frequencies (space station); notification requirements; earth operation: definition; eligibility {97.209(a)}; authorized frequencies (Earth station)

E1E01
(C)
[97.3a3]
Page 1-16

E1E01
What is the Amateur Satellite Service?
A. A radio navigation service using stations on earth satellites for the same purposes as those of the amateur service
B. A radio communication service using stations on earth satellites for weather information gathering
C. A radio communication service using stations on earth satellites for the same purpose as those of the amateur service
D. A radio location service using stations on earth satellites for amateur radar experimentation

E1E02
(A)
[97.207]
Page 1-18

E1E02
Which HF amateur bands have frequencies available for space operation?
A. Only 40 m, 20 m, 17 m, 15 m, 12 m and 10 m
B. Only 40 m, 30 m, 20 m, 15 m and 10 m
C. Only 40 m, 30 m, 20 m, 15 m, 12 m and 10 m
D. All HF bands, but only in the Extra class segments

E1E03
(D)
[97.207]
Page 1-18

E1E03
Which of the following types of communications may space stations transmit?
A. Automatic retransmission of signals from Earth stations and other space stations
B. One way communications
C. Telemetry consisting of specially coded messages
D. All of these choices are correct

E1E04
(B)
[97.3a44]
Page 1-16

E1E04
What type of amateur station operation transmits communications used to initiate, modify or terminate the functions of a space station?
A. Space operation
B. Telecommand operation
C. Earth operation
D. Control operation

E1E05
(D)
[97.211a]
Page 1-16

E1E05
Which amateur stations are eligible to be telecommand stations?
A. Any except those of Technician licensees
B. Only those of Extra class licensees
C. Only a station operated by the space station licensee
D. Any station designated by the space station licensee

E1E06
What term does the FCC use for space-to-earth transmissions used to communicate the results of measurements made by a space station?
A. Data transmission
B. Frame check sequence
C. Telemetry
D. Telecommand

E1E07
What is the term used to describe the operation of an amateur station that is more than 50 km above the earth's surface?
A. EME station operation
B. Space station operation
C. Downlink station operation
D. Ionospheric station operation

E1E08
Which amateur stations are eligible for space operation?
A. Any except those of Technician licensees
B. Only those of General, Advanced or Extra class licensees
C. Only those of Extra class licensees
D. Any amateur station

E1E09
Before initiating space station transmissions, by when must the licensee of the station give the FCC prior written pre-space notification?
A. Before 3 months and before 72 hours
B. Before 6 months and before 3 months
C. Before 12 months and before 3 months
D. Before 27 months and before 5 months

E1E10
After space station transmissions are initiated, by when must the licensee of the station give the FCC written in-space notification?
A. Within 24 hours
B. Within 72 hours
C. Within 7 days
D. Within 30 days

E1E11
After space station transmissions are terminated, by when must the licensee of the station normally give the FCC written post-space notification?
A. No later than 48 hours
B. No later than 72 hours
C. No later than 7 days
D. No later than 3 months

E1E06
(C)
[97.207f]
Page 1-17

E1E07
(B)
[97.3a40]
Page 1-16

E1E08
(D)
[97.207a]
Page 1-17

E1E09
(D)
[97.207g]
Page 1-18

E1E10
(C)
[97.207h]
Page 1-18

E1E11
(D)
[97.207i]
Page 1-18

E1E12
(B)
[97.3a16]
Page 1-16

E1E12
What term describes an amateur station located on or within 50 km of earth's surface intended for communications with space stations?
A. Telecommand station
B. Earth station
C. Telemetry station
D. Auxiliary station

E1F Volunteer Examiner Coordinators (VECs): definition; VEC qualifications; VEC agreement; scheduling examinations; coordinating VEs; reimbursement for expenses {97.527}; accrediting VEs; question pools; Volunteer Examiners (VEs): definition; requirements; accreditation; reimbursement for expenses; VE conduct; preparing an examination; examination elements; definition of code and written elements; preparation responsibility; examination requirements; examination credit; examination procedure; examination administration; temporary operating authority

E1F01
(C)
[97.521]
Page 1-19

E1F01
What is a Volunteer Examiner Coordinator?
A. A person who has volunteered to administer amateur license examinations
B. A person who has volunteered to prepare amateur license examinations
C. An organization that has entered into an agreement with the FCC to coordinate amateur license examinations given by Volunteer Examiners
D. An organization that has entered into an agreement with the FCC to coordinate the preparation of amateur license examinations

E1F02
(A)
[97.519,
97.521,
97.523]
Page 1-19

E1F02
Which of the following is NOT among the functions of a VEC?
A. Prepare and administer amateur operator license examinations, grade examinee's answers and inform examinees of their pass/fail results
B. Collect FCC Forms 605 documents and test results from the administering VEs
C. Assure that all desiring an amateur operator license examination are registered without regard to race, sex, religion or national origin
D. Cooperate in maintaining a pool of questions for each written amateur examination element

E1F03
(B)
[97.521]
Page 1-19

E1F03
Which of the following is NOT among the qualifying requirements to be a VEC?
A. Be an organization that exists for the purpose of furthering the amateur service
B. Be engaged in the manufacture and/or sale of amateur station equipment or amateur license preparation materials
C. Agree to coordinate examinations for all classes of amateur operator licenses
D. Agree to administer amateur operator license examinations in accordance with FCC Rules throughout at least one call-letter district

E1F04
(B)
[97.519a]
Page 1-19

E1F04
What organization coordinates the preparation and administration of amateur license examinations?
A. The FCC
B. A VEC
C. A group of three or more volunteers
D. A local radio club

E1F05
Under what circumstances may a VEC refuse to accredit a person as a Volunteer
Examiner?
A. If the VEC determines that questions of the person's integrity or honesty could
   compromise amateur license examinations
B. If the VEC determines that the person is a Volunteer Examiner for another VEC
C. If the prospective VE is not a member of a club actively engaged in the preparation
   and administration of amateur license examinations
D. If the prospective VE is a citizen of a foreign country

E1F06
Where are the questions listed that must be used in all written US amateur license
examinations?
A. In the instructions each VEC gives to their VEs
B. In an FCC-maintained question pool
C. In the VEC-maintained question pool
D. In the appropriate FCC Report and Order

E1F07
What is an accredited VE?
A. An amateur operator who is approved by three or more fellow VEs to administer
   amateur license examinations
B. An amateur operator who is approved by a VEC to administer amateur operator
   license examinations
C. An amateur operator who administers amateur license examinations for a fee
D. An amateur operator who is approved by an FCC staff member to administer amateur
   license examinations

E1F08
What is the VE accreditation process?
A. General and higher class licensees are automatically allowed to conduct amateur
   license examinations once their license is granted
B. The FCC tests volunteers who wish to conduct amateur license examinations
C. A prospective VE requests permission from three or more already accredited VEs to
   administer amateur license examinations
D. Each VEC ensures its Volunteer Examiner applicants meet FCC requirements to serve
   as VEs

E1F09
Which persons seeking to be VEs cannot be accredited?
A. Persons holding less than an Advanced class license
B. Persons less than 21 years of age
C. Persons who have ever had their amateur licenses suspended or revoked
D. Persons who are employees of the federal government

E1F10
For what type of services may a VE be reimbursed for out-of-pocket expenses?
A. Preparing, processing or administering amateur license examinations
B. Teaching and administering amateur license study courses
C. None; a VE cannot be reimbursed for out-of-pocket expenses
D. Purchasing and distributing amateur license preparation materials

E1F05
(A)
[97.525a4]
Page 1-21

E1F06
(C)
[97.523]
Page 1-20

E1F07
(B)
[97.525]
Page 1-21

E1F08
(D)
[97.509b1,
97.525]
Page 1-21

E1F09
(C)
[97.509b4]
Page 1-21

E1F10
(A)
[97.527a]
Page 1-23

E1F11
(A)
[97.509e,
97.527b]
Page 1-23

E1F11
How much money beyond reimbursement for out-of-pocket expenses may a person accept for serving as a VE?
A. None
B. Up to the national minimum hourly wage times the number of hours spent serving as a VE
C. Up to the maximum fee per applicant set by the FCC each year
D. As much as applicants are willing to donate

E1F12
(B)
[97.507a,b,c]
Page 1-24

E1F12
Who may prepare an Element 2 amateur operator license examination?
A. A VEC that selects questions from the appropriate FCC bulletin
B. A Technician, General, Advanced, or Extra class VE or a qualified supplier that selects questions from the appropriate VEC question pool
C. An Extra class VE who selects questions from the appropriate FCC bulletin
D. The FCC, which selects questions from the appropriate VEC question pool

E1F13
(C)
[97.507a,b,c]
Page 1-24

E1F13
Who may prepare an Element 3 amateur operator license examination?
A. Only an Extra class VE who selects questions from the appropriate FCC bulletin
B. A VEC that selects questions from the appropriate FCC bulletin
C. An Advanced or Extra class VE or a qualified supplier that selects questions from the appropriate VEC question pool
D. The FCC, which selects questions from the appropriate VEC question pool

E1F14
(D)
[97.507a,b,c]
Page 1-24

E1F14
Who may prepare an Element 4 amateur operator license examination?
A. The FCC, which selects questions from the appropriate VEC question pool
B. A VEC that selects questions from the appropriate FCC bulletin
C. An Extra class VE that selects questions from the appropriate FCC bulletin
D. An Extra class VE or a qualified supplier who selects questions from the appropriate VEC question pool

E1F15
(C)
[97.505a6]
Page 1-25

E1F15
What amateur operator license examination credit must be given for a valid Certificate of Successful Completion of Examination (CSCE)?
A. Only the written elements the CSCE indicates the examinee passed
B. Only the telegraphy elements the CSCE indicates the examinee passed
C. Each element the CSCE indicates the examinee passed
D. No credit

E1F16
(A)
[97.509c]
Page 1-25

E1F16
Where must Volunteer Examiners be while they are conducting an amateur license examination?
A. They must all be present and observing the candidate(s) throughout the entire examination
B. They must all leave the room after handing out the exams to allow the candidate(s) to concentrate on the exam material
C. They may be anywhere as long as at least one VE is present and is observing the candidate(s) throughout the entire examination
D. They may be anywhere as long as they are listed as having participated in the examination

E1F17
Who is responsible for the proper conduct and necessary supervision during an amateur operator license examination session?
A. The VEC coordinating the session
B. The FCC
C. The administering Volunteer Examiners
D. The Volunteer Examiner in charge of the session

E1F18
What should a VE do if a candidate fails to comply with the examiner's instructions during an amateur operator license examination?
A. Warn the candidate that continued failure to comply will result in termination of the examination
B. Immediately terminate the candidate's examination
C. Allow the candidate to complete the examination, but invalidate the results
D. Immediately terminate everyone's examination and close the session

E1F19
What must be done with the test papers of each element completed by the candidates(s) at an amateur operator license examination?
A. They must be collected and graded by the administering VEs within 10 days of the examination
B. They must be collected and sent to the coordinating VEC for grading within 10 days of the examination
C. They must be collected and graded immediately by the administering VEs
D. They must be collected and sent to the FCC for grading within 10 days of the examination

E1F20
What must the VEs do if an examinee for an amateur operator license does not score a passing grade on all examination elements needed for an upgrade?
A. Return the application document to the examinee and inform the examinee of the grade(s)
B. Return the application document to the examinee and inform the examinee which questions were incorrectly answered
C. Simply inform the examinee of the failure(s)
D. Inform the examinee which questions were incorrectly answered and show how the questions should have been answered

E1G   Type acceptance of external RF power amplifiers and external RF power amplifier kits; Line A; National Radio Quiet Zone; business communications; definition and operation of spread spectrum; auxiliary station operation

E1G01
How many external RF amplifiers of a particular design capable of operation below 144 MHz may an unlicensed, non-amateur build or modify in one calendar year without obtaining a grant of Certification?
A. 1
B. 5
C. 10
D. None

E1F17
(C)
[97.509c]
Page 1-25

E1F18
(B)
[97.509c]
Page 1-25

E1F19
(C)
[97.509h]
Page 1-25

E1F20
(A)
[97.509j]
Page 1-25

E1G01
(D)
[97.315a]
Page 1-28

E1G02
(B)
[97.315c]
Page 1-30

E1G02
If an RF amplifier manufacturer was granted Certification for one of its amplifier models for amateur use, what would this allow the manufacturer to market?
A. All current models of their equipment
B. Only that particular amplifier model
C. Any future amplifier models
D. Both the current and any future amplifier models

E1G03
(A)
[97.315b5]
Page 1-30

E1G03
Under what condition may an equipment dealer sell an external RF power amplifier capable of operation below 144 MHz if it has not been granted FCC Certification?
A. If it was purchased in used condition from an amateur operator and is sold to another amateur operator for use at that operator's station
B. If it was assembled from a kit by the equipment dealer
C. If it was imported from a manufacturer in a country that does not require type acceptance of RF power amplifiers
D. If it was imported from a manufacturer in another country, and it was type accepted by that country's government

E1G04
(D)
[97.317a1]
Page 1-30

E1G04
Which of the following is one of the standards that must be met by an external RF power amplifier if it is to qualify for a grant of FCC Certification?
A. It must produce full legal output when driven by not more than 5 watts of mean RF input power
B. It must be capable of external RF switching between its input and output networks
C. It must exhibit a gain of 0 dB or less over its full output range
D. It must satisfy the spurious emission standards when operated at its full output power

E1G05
(D)
[97.317a2]
Page 1-30

E1G05
Which of the following is one of the standards that must be met by an external RF power amplifier if it is to qualify for a grant of Certification?
A. It must produce full legal output when driven by not more than 5 watts of mean RF input power
B. It must be capable of external RF switching between its input and output networks
C. It must exhibit a gain of 0 dB or less over its full output range
D. It must satisfy the spurious emission standards when placed in the "standby" or "off" position, but is still connected to the transmitter

E1G06
(C)
[97.317b]
Page 1-30

E1G06
Which of the following is one of the standards that must be met by an external RF power amplifier if it is to qualify for a grant of Certification?
A. It must produce full legal output when driven by not more than 5 watts of mean RF input power
B. It must exhibit a gain of at least 20 dB for any input signal
C. It must not be capable of operation on any frequency between 24 MHz and 35 MHz
D. Any spurious emissions from the amplifier must be no more than 40 dB stronger than the desired output signal

E1G07
Which of the following is one of the standards that must be met by an external RF power amplifier if it is to qualify for a grant of Certification?
A. It must have a time-delay circuit to prevent it from operating continuously for more than ten minutes
B. It must satisfy the spurious emission standards when driven with at least 50 W mean RF power (unless a higher drive level is specified)
C. It must not be capable of modification by an amateur operator without voiding the warranty
D. It must exhibit no more than 6 dB of gain over its entire operating range

E1G08
Which of the following would disqualify an external RF power amplifier from being granted Certification?
A. Any accessible wiring which, when altered, would permit operation of the amplifier in a manner contrary to FCC Rules
B. Failure to include a schematic diagram and theory of operation manual that would permit an amateur to modify the amplifier
C. The capability of being switched by the operator to any amateur frequency below 24 MHz
D. Failure to produce 1500 watts of output power when driven by at least 50 watts of mean input power

E1G09
Which of the following would disqualify an external RF power amplifier from being granted Certification?
A. Failure to include controls or adjustments that would permit the amplifier to operate on any frequency below 24 MHz
B. Failure to produce 1500 watts of output power when driven by at least 50 watts of mean input power
C. Any features designed to facilitate operation in a telecommunication service other than the Amateur Service
D. The omission of a schematic diagram and theory of operation manual that would permit an amateur to modify the amplifier

E1G10
Which of the following would disqualify an external RF power amplifier from being granted Certification?
A. The omission of a safety switch in the high-voltage power supply to turn off the power if the cabinet is opened
B. Failure of the amplifier to exhibit more than 15 dB of gain over its entire operating range
C. The omission of a time-delay circuit to prevent the amplifier from operating continuously for more than ten minutes
D. The inclusion of instructions for operation or modification of the amplifier in a manner contrary to the FCC Rules

E1G07
(B)
[97.317a3]
Page 1-28

E1G08
(A)
[97.317c1]
Page 1-30

E1G09
(C)
[97.317c8]
Page 1-30

E1G10
(D)
[97.317c3]
Page 1-30

E1G11
Which of the following would disqualify an external RF power amplifier from being granted Certification?
A. Failure to include a safety switch in the high-voltage power supply to turn off the power if the cabinet is opened
B. The amplifier produces 3 dB of gain for input signals between 26 MHz and 28 MHz
C. The inclusion of a schematic diagram and theory of operation manual that would permit an amateur to modify the amplifier
D. The amplifier produces 1500 watts of output power when driven by at least 50 watts of mean input power

Subelement E2

Subelement E2 — Operating Procedures

[4 Exam Questions — 4 Groups]

E2A Amateur Satellites: Orbital mechanics; Frequencies available for satellite operation; Satellite hardware; Operating through amateur satellites

E2A01
What is the direction of an ascending pass for an amateur satellite?
A. From west to east
B. From east to west
C. From south to north
D. From north to south

E2A02
What is the direction of a descending pass for an amateur satellite?
A. From north to south
B. From west to east
C. From east to west
D. From south to north

E2A03
What is the period of an amateur satellite?
A. The point of maximum height of a satellite's orbit
B. The point of minimum height of a satellite's orbit
C. The amount of time it takes for a satellite to complete one orbit
D. The time it takes a satellite to travel from perigee to apogee

E2A04
What are the receiving and retransmitting frequency bands used for Mode A in amateur satellite operations?
A. Satellite receiving on 10 meters and retransmitting on 2 meters
B. Satellite receiving on 70 centimeters and retransmitting on 2 meters
C. Satellite receiving on 70 centimeters and retransmitting on 10 meters
D. Satellite receiving on 2 meters and retransmitting on 10 meters

E2A05
What are the receiving and retransmitting frequency bands used for Mode B in amateur satellite operations?
A. Satellite receiving on 10 meters and retransmitting on 2 meters
B. Satellite receiving on 70 centimeters and retransmitting on 2 meters
C. Satellite receiving on 70 centimeters and retransmitting on 10 meters
D. Satellite receiving on 2 meters and retransmitting on 10 meters

E2A06
What are the receiving and retransmitting frequency bands used for Mode J in amateur satellite operations?
A. Satellite receiving on 70 centimeters and retransmitting on 2 meters
B. Satellite receiving on 2 meters and retransmitting on 10 meters
C. Satellite receiving on 2 meters and retransmitting on 70 centimeters
D. Satellite receiving on 70 centimeters and transmitting on 10 meters

E2A07
What are the receiving and retransmitting frequency bands used for Mode L in amateur satellite operations?
A. Satellite receiving on 70 centimeters and retransmitting on 10 meters
B. Satellite receiving on 10 meters and retransmitting on 70 centimeters
C. Satellite receiving on 70 centimeters and retransmitting on 23 centimeters
D. Satellite receiving on 23 centimeters and retransmitting on 70 centimeters

E2A08
What is a linear transponder?
A. A repeater that passes only linear or CW signals
B. A device that receives and retransmits signals of any mode in a certain passband
C. An amplifier that varies its output linearly in response to input signals
D. A device which responds to satellite telecommands and is used to activate a linear sequence of events

E2A09
What is the name of the effect which causes the downlink frequency of a satellite to vary by several kHz during a low-earth orbit because the distance between the satellite and ground station is changing?
A. The Kepler effect
B. The Bernoulli effect
C. The Einstein effect
D. The Doppler effect

E2A10
Why does the received signal from a Phase 3 amateur satellite exhibit a fairly rapid pulsed fading effect?
A. Because the satellite is rotating
B. Because of ionospheric absorption
C. Because of the satellite's low orbital altitude
D. Because of the Doppler effect

E2A05
(B)
Page 2-7

E2A06
(C)
Page 2-7

E2A07
(D)
Page 2-7

E2A08
(B)
Page 2-5

E2A09
(D)
Page 2-7

E2A10
(A)
Page 2-8

E2A11
(B)
Page 2-8

E2A11
What type of antenna can be used to minimize the effects of spin modulation and Faraday rotation?
A. A nonpolarized antenna
B. A circularly polarized antenna
C. An isotropic antenna
D. A log-periodic dipole array

E2B Television: fast scan television (FSTV) standards; slow scan television (SSTV) standards; facsimile (fax) communications

E2B01
(A)
Page 2-9

E2B01
How many times per second is a new frame transmitted in a fast-scan television system?
A. 30
B. 60
C. 90
D. 120

E2B02
(C)
Page 2-9

E2B02
How many horizontal lines make up a fast-scan television frame?
A. 30
B. 60
C. 525
D. 1050

E2B03
(D)
Page 2-9

E2B03
How is the interlace scanning pattern generated in a fast-scan television system?
A. By scanning the field from top to bottom
B. By scanning the field from bottom to top
C. By scanning from left to right in one field and right to left in the next
D. By scanning odd numbered lines in one field and even numbered ones in the next

E2B04
(B)
Page 2-9

E2B04
What is blanking in a video signal?
A. Synchronization of the horizontal and vertical sync pulses
B. Turning off the scanning beam while it is traveling from right to left and from bottom to top
C. Turning off the scanning beam at the conclusion of a transmission
D. Transmitting a black and white test pattern

E2B05
(D)
Page 2-12

E2B05
What is the bandwidth of a vestigial sideband AM fast-scan television transmission?
A. 3 kHz
B. 10 kHz
C. 25 kHz
D. 6 MHz

E2B06
What is the standard video level, in percent PEV, for black?
A. 0%
B. 12.5%
C. 70%
D. 100%

E2B06
(C)
Page 2-11

E2B07
What is the standard video level, in percent PEV, for blanking?
A. 0%
B. 12.5%
C. 75%
D. 100%

E2B07
(C)
Page 2-11

E2B08
Which of the following is NOT a common method of transmitting accompanying audio with amateur fast-scan television?
A. Amplitude modulation of the video carrier
B. Frequency-modulated sub-carrier
C. A separate VHF or UHF audio link
D. Frequency modulation of the video carrier

E2B08
(A)
Page 2-8

E2B09
What is facsimile?
A. The transmission of characters by radioteletype that form a picture when printed
B. The transmission of still pictures by slow-scan television
C. The transmission of video by amateur television
D. The transmission of printed pictures for permanent display on paper

E2B09
(D)
Page 2-13

E2B10
What is the modern standard scan rate for a facsimile picture transmitted by an amateur station?
A. 240 lines per minute
B. 50 lines per minute
C. 150 lines per second
D. 60 lines per second

E2B10
(A)
Page 2-14

E2B11
What is the approximate transmission time per frame for a facsimile picture transmitted by an amateur station at 240 lpm?
A. 6 minutes
B. 3.3 minutes
C. 6 seconds
D. 1/60 second

E2B11
(B)
Page 2-14

E2B12
In facsimile, what device converts variations in picture brightness and darkness into voltage variations?
A. An LED
B. A Hall-effect transistor
C. A photodetector
D. An optoisolator

E2B12
(C)
Page 2-13

E2C Contest and DX operating; spread-spectrum transmissions; automatic HF forwarding.

E2C01
(A)
Page 2-16

E2C01
What would be the ideal operating strategy for a worldwide DX contest during a solar minimum instead of a solar maximum?
A. 160-40 meters would be emphasized during the evening; 20 meters during daylight hours
B. There would be little to no strategic difference
C. 80 meters would support worldwide communication during mid-day hours
D. 10 and 15 meters should be tried one hour before sunset

E2C02
(A)
Page 2-16

E2C02
When operating during a contest, which of these standards should you generally follow?
A. Always listen before transmitting, be courteous and do not cause harmful interference to other communications
B. Always reply to other stations calling CQ at least as many times as you call CQ
C. When initiating a contact, always reply with the call sign of the station you are calling followed by your own call sign
D. Always include your signal report, name and transmitter power output in any exchange with another station

E2C03
(B)
Page 2-16

E2C03
What is one of the main purposes for holding on-the-air operating contests?
A. To test the dollar-to-feature value of station equipment during difficult operating circumstances
B. To enhance the communicating and operating skills of amateurs in readiness for an emergency
C. To measure the ionospheric capacity for refracting RF signals under varying conditions
D. To demonstrate to the FCC that amateur station operation is possible during difficult operating circumstances

E2C04
(C)
Page 2-16

E2C04
Which of the following is typical of operations during an international amateur DX contest?
A. Calling CQ is always done on an odd minute and listening is always done on an even minute
B. Contacting a DX station is best accomplished when the WWV K index is above a reading of 8
C. Some DX operators use split frequency operations (transmitting on a frequency different from the receiving frequency)
D. DX contacts during the day are never possible because of known band attenuation from the sun

E2C05
(D)
Page 2-18

E2C05
If a DX station asks for your grid square locator, what should be your reply?
A. The square of the power fed to the grid of your final amplifier and your current city, state and country
B. The DX station's call sign followed by your call sign and your RST signal report
C. The subsection of the IARU region in which you are located based upon dividing the entire region into a grid of squares 10 km wide
D. Your geographic "Maidenhead" grid location (e.g., FN31AA) based on your current latitude and longitude

E2C06
What does a "Maidenhead" grid square refer to?
A. A two-degree longitude by one degree latitude square, as part of a world wide numbering system
B. A one-degree longitude by one degree latitude square, beginning at the South Pole
C. An antenna made of wire grid used to amplify low-angle incoming signals while reducing high-angle incoming signals
D. An antenna consisting of a screen or grid positioned directly beneath the radiating element

E2C06
(A)
Page 2-18

E2C07
During a VHF/UHF contest, in which band section would you expect to find the highest level of contest activity?
A. At the top of each band, usually in a segment reserved for contests
B. In the middle of each band, usually on the national calling frequency
C. At the bottom of each band, usually in the weak signal segment
D. In the middle of the band, usually 25 kHz above the national calling frequency

E2C07
(C)
Page 2-16

E2C08
Which of the following frequency ranges is reserved by "gentlemen's agreement" for DX contacts during international 6-meter contests?
A. 50.000 to 50.025 MHz
B. 50.050 to 50.075 MHz
C. 50.075 to 50.100 MHz
D. 50.100 to 50.125 MHz

E2C08
(D)
Page 2-16

E2C09
If you are in the US calling a station in Texas on a frequency of 1832 kHz and a station replies that you are "in the window," what does this mean?
A. You are operating out of the band privileges of your license
B. You are calling at the wrong time of day to be within the window of frequencies that can be received in Texas at that time
C. You are transmitting in a frequency segment that is reserved for international DX contacts by "gentlemen's agreement"
D. Your modulation has reached an undesirable level and you are interfering with another contact

E2C09
(C)
Page 2-16

E2C10
Why are received spread-spectrum signals so resistant to interference?
A. Signals not using the spectrum-spreading algorithm are suppressed in the receiver
B. The high power used by a spread-spectrum transmitter keeps its signal from being easily overpowered
C. The receiver is always equipped with a special digital signal processor (DSP) interference filter
D. If interference is detected by the receiver it will signal the transmitter to change frequencies

E2C10
(A)
Page 2-20

E2C11
(D)
Page 2-19

E2C11
How does the spread-spectrum technique of frequency hopping (FH) work?
A. If interference is detected by the receiver it will signal the transmitter to change frequencies
B. If interference is detected by the receiver it will signal the transmitter to wait until the frequency is clear
C. A pseudo-random binary bit stream is used to shift the phase of an RF carrier very rapidly in a particular sequence
D. The frequency of an RF carrier is changed very rapidly according to a particular pseudo-random sequence

E2C12
(C)
Page 2-21

E2C12
What is the most common data rate used for HF packet communications?
A. 48 bauds
B. 110 bauds
C. 300 bauds
D. 1200 bauds

E2D   Digital Operating: HF digital communications (ie, PacTOR, CLOVER, AMTOR, PSK31, HF packet); packet clusters; HF digital bulletin boards

E2D01
(B)
Page 2-21

E2D01
What is the most common method of transmitting data emissions below 30 MHz?
A. DTMF tones modulating an FM signal
B. FSK (frequency-shift keying) of an RF carrier
C. AFSK (audio frequency-shift keying) of an FM signal
D. Key-operated on/off switching of an RF carrier

E2D02
(A)
Page 2-21

E2D02
What do the letters "FEC" mean as they relate to AMTOR operation?
A. Forward Error Correction
B. First Error Correction
C. Fatal Error Correction
D. Final Error Correction

E2D03
(C)
Page 2-21

E2D03
How is Forward Error Correction implemented?
A. By transmitting blocks of 3 data characters from the sending station to the receiving station which the receiving station acknowledges
B. By transmitting a special FEC algorithm which the receiving station uses for data validation
C. By transmitting each data character twice, since there is no specific acknowledgment of reception
D. By varying the frequency shift of the transmitted signal according to a predefined algorithm

E2D04
(B)
Page 2-22

E2D04
What does "CMD:" mean when it is displayed on the video monitor of a packet station?
A. The TNC is ready to exit the packet terminal program
B. The TNC is in command mode, ready to receive instructions from the keyboard
C. The TNC will exit to the command mode on the next keystroke
D. The TNC is in KISS mode running TCP/IP, ready for the next command

E2D05
What is the Baudot code?
A. A code used to transmit data only in modern computer-based data systems using seven data bits
B. A binary code consisting of eight data bits
C. An alternate name for Morse code
D. The "International Telegraph Alphabet Number 2" (ITA2) which uses five data bits

E2D06
If an oscilloscope is connected to a TNC or terminal unit and is displaying two crossed ellipses, one of which suddenly disappears, what would this indicate about the observed signal?
A. The phenomenon known as "selective fading" has occurred
B. One of the signal filters has saturated
C. The receiver should be retuned, as it has probably moved at least 5 kHz from the desired receive frequency
D. The mark and space signal have been inverted and the receiving equipment has not yet responded to the change

E2D07
Which of the following systems is used to transmit high-quality still images by radio?
A. AMTOR
B. Baudot RTTY
C. AMTEX
D. Facsimile

E2D08
What special restrictions are imposed on facsimile (fax) transmissions?
A. None; they are allowed on all amateur frequencies
B. They are restricted to 7.245 MHz, 14.245 MHz, 21.345 MHz, and 28.945 MHz
C. They are allowed in phone band segments if their bandwidth is no greater than that of a voice signal of the same modulation type
D. They are not permitted above 54 MHz

E2D09
What is the name for a bulletin transmission system that includes a special header to allow receiving stations to determine if the bulletin has been previously received?
A. ARQ mode A
B. FEC mode B
C. AMTOR
D. AMTEX

E2D10
What is a Packet Cluster Bulletin Board?
A. A packet bulletin board devoted primarily to serving a special interest group
B. A group of general-purpose packet bulletin boards linked together in a "cluster"
C. A special interest cluster of packet bulletin boards devoted entirely to packet radio computer communications
D. A special interest telephone/modem bulletin board devoted to amateur DX operations

E2D05
(D)
Page 2-20

E2D06
(A)
Page 2-21

E2D07
(D)
Page 2-13

E2D08
(C)
Page 2-13

E2D09
(D)
Page 2-21

E2D10
(A)
Page 2-22

E2D11
(C)
Page 2-21

E2D11
Which of the following statements comparing HF and 2-meter packet operations
is NOT true?
A. HF packet typically uses an FSK signal with a data rate of 300 bauds; 2-meter packet
   uses an AFSK signal with a data rate of 1200 bauds
B. HF packet and 2-meter packet operations use the same code for information
   exchange
C. HF packet is limited to Extra class amateur licensees; 2 meter packet is open to all
   but Novice class amateur licensees
D. HF packet operations are limited to "CW/Data"-only band segments; 2-meter packet
   is allowed wherever FM operations are allowed

Subelement
E3

Subelement E3 — Radio Wave Propagation

[3 Exam Questions — 3 Groups]

E3A Earth-Moon-Earth (EME or moonbounce) communications; meteor scatter

E3A01
(D)
Page 3-4

E3A01
What is the maximum separation between two stations communicating by moonbounce?
A. 500 miles maximum, if the moon is at perigee
B. 2000 miles maximum, if the moon is at apogee
C. 5000 miles maximum, if the moon is at perigee
D. Any distance as long as the stations have a mutual lunar window

E3A02
(B)
Page 3-6

E3A02
What characterizes libration fading of an earth-moon-earth signal?
A. A slow change in the pitch of the CW signal
B. A fluttery, rapid irregular fading
C. A gradual loss of signal as the sun rises
D. The returning echo is several hertz lower in frequency than the transmitted signal

E3A03
(A)
Page 3-5

E3A03
What are the best days to schedule EME contacts?
A. When the moon is at perigee
B. When the moon is full
C. When the moon is at apogee
D. When the weather at both stations is clear

E3A04
(D)
Page 3-5

E3A04
What type of receiving system is required for EME communications?
A. Equipment with very low power output
B. Equipment with very low dynamic range
C. Equipment with very low gain
D. Equipment with very low noise figures

E3A05
What transmit and receive time sequencing is normally used on 144 MHz when attempting an earth-moon-earth contact?
A. Two-minute sequences, where one station transmits for a full two minutes and then receives for the following two minutes
B. One-minute sequences, where one station transmits for one minute and then receives for the following one minute
C. Two-and-one-half minute sequences, where one station transmits for a full 2.5 minutes and then receives for the following 2.5 minutes
D. Five-minute sequences, where one station transmits for five minutes and then receives for the following five minutes

E3A05
(A)
Page 3-9

E3A06
What transmit and receive time sequencing is normally used on 432 MHz when attempting an EME contact?
A. Two-minute sequences, where one station transmits for a full two minutes and then receives for the following two minutes
B. One-minute sequences, where one station transmits for one minute and then receives for the following one minute
C. Two and one half minute sequences, where one station transmits for a full 2.5 minutes and then receives for the following 2.5 minutes
D. Five minute sequences, where one station transmits for five minutes and then receives for the following five minutes

E3A06
(C)
Page 3-9

E3A07
What frequency range would you normally tune to find EME stations in the 2-meter band?
A. 144.000 - 144.001 MHz
B. 144.000 - 144.100 MHz
C. 144.100 - 144.300 MHz
D. 145.000 - 145.100 MHz

E3A07
(B)
Page 3-9

E3A08
What frequency range would you normally tune to find EME stations in the 70-cm band?
A. 430.000 - 430.150 MHz
B. 430.100 - 431.100 MHz
C. 431.100 - 431.200 MHz
D. 432.000 - 432.100 MHz

E3A08
(D)
Page 3-9

E3A09
When the earth's atmosphere is struck by a meteor, a cylindrical region of free electrons is formed at what layer of the ionosphere?
A. The E layer
B. The F1 layer
C. The F2 layer
D. The D layer

E3A09
(A)
Page 3-10

E3A10
(C)
Page 3-10

E3A10
Which range of frequencies is well suited for meteor-scatter communications?
A. 1.8 - 1.9 MHz
B. 10 - 14 MHz
C. 28 - 148 MHz
D. 220 - 450 MHz

E3A11
(C)
Page 3-12

E3A11
What transmit and receive time sequencing is normally used on 144 MHz when attempting a meteor-scatter contact?
A. Two-minute sequences, where one station transmits for a full two minutes and then receives for the following two minutes
B. One-minute sequences, where one station transmits for one minute and then receives for the following one minute
C. 15-second sequences, where one station transmits for 15 seconds and then receives for the following 15 seconds
D. 30-second sequences, where one station transmits for 30 seconds and then receives for the following 30 seconds

E3B Transequatorial; long path; gray line

E3B01
(A)
Page 3-12

E3B01
What is transequatorial propagation?
A. Propagation between two points at approximately the same distance north and south of the magnetic equator
B. Propagation between two points at approximately the same latitude on the magnetic equator
C. Propagation between two continents by way of ducts along the magnetic equator
D. Propagation between two stations at the same latitude

E3B02
(C)
Page 3-13

E3B02
What is the approximate maximum range for signals using transequatorial propagation?
A. 1000 miles
B. 2500 miles
C. 5000 miles
D. 7500 miles

E3B03
(C)
Page 3-13

E3B03
What is the best time of day for transequatorial propagation?
A. Morning
B. Noon
C. Afternoon or early evening
D. Late at night

E3B04
(A)
Page 3-14

E3B04
What type of propagation is probably occurring if a beam antenna must be pointed in a direction 180 degrees away from a station to receive the strongest signals?
A. Long-path
B. Sporadic-E
C. Transequatorial
D. Auroral

E3B05
On what amateur bands can long-path propagation provide signal enhancement?
A. 160 to 40 meters
B. 30 to 10 meters
C. 160 to 10 meters
D. 160 to 6 meters

E3B05
(D)
Page 3-15

E3B06
What amateur band consistently yields long-path enhancement using a modest antenna of relatively high gain?
A. 80 meters
B. 20 meters
C. 10 meters
D. 6 meters

E3B06
(B)
Page 3-15

E3B07
What is the typical reason for hearing an echo on the received signal of a station in Europe while directing your HF antenna toward the station?
A. The station's transmitter has poor frequency stability
B. The station's transmitter is producing spurious emissions
C. Auroral conditions are causing a direct and a long-path reflected signal to be received
D. There are two signals being received, one from the most direct path and one from long-path propagation

E3B07
(D)
Page 3-14

E3B08
What type of propagation is probably occurring if radio signals travel along the earth's terminator?
A. Transequatorial
B. Sporadic-E
C. Long-path
D. Gray-line

E3B08
(D)
Page 3-15

E3B09
At what time of day is gray-line propagation most prevalent?
A. Twilight, at sunrise and sunset
B. When the sun is directly above the location of the transmitting station
C. When the sun is directly overhead at the middle of the communications path between the two stations
D. When the sun is directly above the location of the receiving station

E3B09
(A)
Page 3-15

E3B10
What is the cause of gray-line propagation?
A. At midday the sun, being directly overhead, superheats the ionosphere causing increased refraction of radio waves
B. At twilight solar absorption drops greatly while atmospheric ionization is not weakened enough to reduce the MUF
C. At darkness solar absorption drops greatly while atmospheric ionization remains steady
D. At midafternoon the sun heats the ionosphere, increasing radio wave refraction and the MUF

E3B10
(B)
Page 3-15

E3B11
(C)
Page 3-15

E3B11
What communications are possible during gray-line propagation?
A. Contacts up to 2,000 miles only on the 10-meter band
B. Contacts up to 750 miles on the 6- and 2-meter bands
C. Contacts up to 8,000 to 10,000 miles on three or four HF bands
D. Contacts up to 12,000 to 15,000 miles on the 10- and 15-meter bands

E3C Auroral propagation; selective fading; radio-path horizon; take-off angle over flat or sloping terrain; earth effects on propagation

E3C01
(D)
Page 3-18

E3C01
What effect does auroral activity have upon radio communications?
A. The readability of SSB signals increases
B. FM communications are clearer
C. CW signals have a clearer tone
D. CW signals have a fluttery tone

E3C02
(C)
Page 3-17

E3C02
What is the cause of auroral activity?
A. A high sunspot level
B. A low sunspot level
C. The emission of charged particles from the sun
D. Meteor showers concentrated in the northern latitudes

E3C03
(D)
Page 3-17

E3C03
Where in the ionosphere does auroral activity occur?
A. At F-region height
B. In the equatorial band
C. At D-region height
D. At E-region height

E3C04
(A)
Page 3-18

E3C04
Which emission modes are best for auroral propagation?
A. CW and SSB
B. SSB and FM
C. FM and CW
D. RTTY and AM

E3C05
(B)
Page 3-20

E3C05
What causes selective fading?
A. Small changes in beam heading at the receiving station
B. Phase differences between radio-wave components of the same transmission, as experienced at the receiving station
C. Large changes in the height of the ionosphere at the receiving station ordinarily occurring shortly after either sunrise or sunset
D. Time differences between the receiving and transmitting stations

E3C06
Which emission modes suffer the most from selective fading?
A. CW and SSB
B. FM and double sideband AM
C. SSB and AMTOR
D. SSTV and CW

E3C06
(B)
Page 3-22

E3C07
How does the bandwidth of a transmitted signal affect selective fading?
A. It is more pronounced at wide bandwidths
B. It is more pronounced at narrow bandwidths
C. It is the same for both narrow and wide bandwidths
D. The receiver bandwidth determines the selective fading effect

E3C07
(A)
Page 3-22

E3C08
How much farther does the VHF/UHF radio-path horizon distance exceed the geometric horizon?
A. By approximately 15% of the distance
B. By approximately twice the distance
C. By approximately one-half the distance
D. By approximately four times the distance

E3C08
(A)
Page 3-2

E3C09
For a 3-element Yagi antenna with horizontally mounted elements, how does the main lobe takeoff angle vary with height above flat ground?
A. It increases with increasing height
B. It decreases with increasing height
C. It does not vary with height
D. It depends on E-region height, not antenna height

E3C09
(B)
Page 3-22

E3C10
What is the name of the high-angle wave in HF propagation that travels for some distance within the F2 region?
A. Oblique-angle ray
B. Pedersen ray
C. Ordinary ray
D. Heaviside ray

E3C10
(B)
Page 3-23

E3C11
What effect is usually responsible for propagating a VHF signal over 500 miles?
A. D-region absorption
B. Faraday rotation
C. Tropospheric ducting
D. Moonbounce

E3C11
(C)
Page 3-22

E3C12
What happens to an electromagnetic wave as it encounters air molecules and other particles?
A. The wave loses kinetic energy
B. The wave gains kinetic energy
C. An aurora is created
D. Nothing happens because the waves have no physical substance

E3C12
(A)
Page 3-22

Subelement E4 — Amateur Radio Practices

[5 Exam Questions — 5 Groups]

E4A Test equipment: spectrum analyzers (interpreting spectrum analyzer displays; transmitter output spectrum); logic probes (indications of high and low states in digital circuits; indications of pulse conditions in digital circuits)

E4A01
(C)
Page 4-10

E4A01
How does a spectrum analyzer differ from a conventional time-domain oscilloscope?
A. A spectrum analyzer measures ionospheric reflection; an oscilloscope displays electrical signals
B. A spectrum analyzer displays signals in the time domain; an oscilloscope displays signals in the frequency domain
C. A spectrum analyzer displays signals in the frequency domain; an oscilloscope displays signals in the time domain
D. A spectrum analyzer displays radio frequencies; an oscilloscope displays audio frequencies

E4A02
(D)
Page 4-10

E4A02
What does the horizontal axis of a spectrum analyzer display?
A. Amplitude
B. Voltage
C. Resonance
D. Frequency

E4A03
(A)
Page 4-10

E4A03
What does the vertical axis of a spectrum analyzer display?
A. Amplitude
B. Duration
C. Frequency
D. Time

E4A04
(A)
Page 4-12

E4A04
Which test instrument is used to display spurious signals from a radio transmitter?
A. A spectrum analyzer
B. A wattmeter
C. A logic analyzer
D. A time-domain reflectometer

E4A05
(B)
Page 4-13

E4A05
Which test instrument is used to display intermodulation distortion products from an SSB transmitter?
A. A wattmeter
B. A spectrum analyzer
C. A logic analyzer
D. A time-domain reflectometer

E4A06
Which of the following is NOT something you would determine with a spectrum analyzer?
A. The degree of isolation between the input and output ports of a 2-meter duplexer
B. Whether a crystal is operating on its fundamental or overtone frequency
C. The speed at which a transceiver switches from transmit to receive when being used for packet radio
D. The spectral output of a transmitter

E4A06
(C)
Page 4-14

E4A07
What is an advantage of using a spectrum analyzer to observe the output from a VHF transmitter?
A. There are no advantages; an inexpensive oscilloscope can display the same information
B. It displays all frequency components of the transmitted signal
C. It displays a time-varying representation of the modulation envelope
D. It costs much less than any other instrumentation useful for such measurements

E4A07
(B)
Page 4-11

E4A08
What advantage does a logic probe have over a voltmeter for monitoring the status of a logic circuit?
A. It has many more leads to connect to the circuit than a voltmeter
B. It can be used to test analog and digital circuits
C. It can read logic circuit voltage more accurately than a voltmeter
D. It is smaller and shows a simplified readout

E4A08
(D)
Page 4-15

E4A09
Which test instrument is used to directly indicate high and low digital states?
A. An ohmmeter
B. An electroscope
C. A logic probe
D. A Wheatstone bridge

E4A09
(C)
Page 4-15

E4A10
What can a logic probe indicate about a digital logic circuit?
A. A short-circuit fault
B. An open-circuit fault
C. The resistance between logic modules
D. The high and low logic states

E4A10
(D)
Page 4-15

E4A11
Which test instrument besides an oscilloscope is used to indicate pulse conditions in a digital logic circuit?
A. A logic probe
B. An ohmmeter
C. An electroscope
D. A Wheatstone bridge

E4A11
(A)
Page 4-15

E4B Frequency measurement devices (i.e., frequency counter, oscilloscope Lissajous figures, dip meter); meter performance limitations; oscilloscope performance limitations; frequency counter performance limitations

E4B01
(B)
Page 4-2

E4B01
What is a frequency standard?
A. A frequency chosen by a net control operator for net operations
B. A device used to produce a highly accurate reference frequency
C. A device for accurately measuring frequency to within 1 Hz
D. A device used to generate wide-band random frequencies

E4B02
(A)
Page 4-2

E4B02
What does a frequency counter do?
A. It makes frequency measurements
B. It produces a reference frequency
C. It measures FM transmitter deviation
D. It generates broad-band white noise

E4B03
(B)
Page 4-2

E4B03
What factors limit the accuracy, frequency response and stability of a frequency counter?
A. Number of digits in the readout, speed of the logic and time base stability
B. Time base accuracy, speed of the logic and time base stability
C. Time base accuracy, temperature coefficient of the logic and time base stability
D. Number of digits in the readout, external frequency reference and temperature coefficient of the logic

E4B04
(C)
Page 4-2

E4B04
How can the accuracy of a frequency counter be improved?
A. By using slower digital logic
B. By improving the accuracy of the frequency response
C. By increasing the accuracy of the time base
D. By using faster digital logic

E4B05
(C)
Page 4-3

E4B05
If a frequency counter with a time base accuracy of +/– 1.0 ppm reads 146,520,000 Hz, what is the most the actual frequency being measured could differ from the reading?
A. 165.2 Hz
B. 14.652 kHz
C. 146.52 Hz
D. 1.4652 MHz

E4B06
(A)
Page 4-3

E4B06
If a frequency counter with a time base accuracy of +/– 0.1 ppm reads 146,520,000 Hz, what is the most the actual frequency being measured could differ from the reading?
A. 14.652 Hz
B. 0.1 MHz
C. 1.4652 Hz
D. 1.4652 kHz

E4B07
If a frequency counter with a time base accuracy of +/– 10 ppm reads 146,520,000 Hz, what is the most the actual frequency being measured could differ from the reading?
A. 146.52 Hz
B. 10 Hz
C. 146.52 kHz
D. 1465.20 Hz

E4B08
If a frequency counter with a time base accuracy of +/– 1.0 ppm reads 432,100,000 Hz, what is the most the actual frequency being measured could differ from the reading?
A. 43.21 MHz
B. 10 Hz
C. 1.0 MHz
D. 432.1 Hz

E4B09
If a frequency counter with a time base accuracy of +/– 0.1 ppm reads 432,100,000 Hz, what is the most the actual frequency being measured could differ from the reading?
A. 43.21 Hz
B. 0.1 MHz
C. 432.1 Hz
D. 0.2 MHz

E4B10
If a frequency counter with a time base accuracy of +/– 10 ppm reads 432,100,000 Hz, what is the most the actual frequency being measured could differ from the reading?
A. 10 MHz
B. 10 Hz
C. 4321 Hz
D. 432.1 Hz

E4B11
If a 100 Hz signal is fed to the horizontal input of an oscilloscope and a 150 Hz signal is fed to the vertical input, what type of Lissajous figure should be displayed on the screen?
A. A looping pattern with 100 loops horizontally and 150 loops vertically
B. A rectangular pattern 100 mm wide and 150 mm high
C. A looping pattern with 3 loops horizontally and 2 loops vertically
D. An oval pattern 100 mm wide and 150 mm high

E4B12
What is a dip-meter?
A. A field-strength meter
B. An SWR meter
C. A variable LC oscillator with metered feedback current
D. A marker generator

E4B07
(D)
Page 4-3

E4B08
(D)
Page 4-3

E4B09
(A)
Page 4-3

E4B10
(C)
Page 4-3

E4B11
(C)
Page 4-7

E4B12
(C)
Page 4-4

E4B13
(D)
Page 4-4

E4B13
What does a dip-meter do?
A. It accurately indicates signal strength
B. It measures frequency accurately
C. It measures transmitter output power accurately
D. It gives an indication of the resonant frequency of a circuit

E4B14
(B)
Page 4-4

E4B14
How does a dip-meter function?
A. Reflected waves at a specific frequency desensitize a detector coil
B. Power coupled from an oscillator causes a decrease in metered current
C. Power from a transmitter cancels feedback current
D. Harmonics from an oscillator cause an increase in resonant circuit Q

E4B15
(D)
Page 4-4

E4B15
What two ways could a dip-meter be used in an amateur station?
A. To measure resonant frequency of antenna traps and to measure percentage of modulation
B. To measure antenna resonance and to measure percentage of modulation
C. To measure antenna resonance and to measure antenna impedance
D. To measure resonant frequency of antenna traps and to measure a tuned circuit resonant frequency

E4B16
(B)
Page 4-5

E4B16
What types of coupling occur between a dip-meter and a tuned circuit being checked?
A. Resistive and inductive
B. Inductive and capacitive
C. Resistive and capacitive
D. Strong field

E4B17
(A)
Page 4-4

E4B17
For best accuracy, how tightly should a dip-meter be coupled with a tuned circuit being checked?
A. As loosely as possible
B. As tightly as possible
C. First loosely, then tightly
D. With a jumper wire between the meter and the circuit to be checked

E4B18
(A)
Page 4-9

E4B18
What factors limit the accuracy, frequency response and stability of an oscilloscope?
A. Accuracy and linearity of the time base and the linearity and bandwidth of the deflection amplifiers
B. Tube face voltage increments and deflection amplifier voltage
C. Accuracy and linearity of the time base and tube face voltage increments
D. Deflection amplifier output impedance and tube face frequency increments

E4C Receiver performance characteristics (i.e., phase noise, desensitization, capture effect, intercept point, noise floor, dynamic range {blocking and IMD}, image rejection, MDS, signal-to-noise-ratio); intermodulation and cross-modulation interference

E4C01
What is the effect of excessive phase noise in a receiver local oscillator?
A. It limits the receiver ability to receive strong signals
B. It reduces the receiver sensitivity
C. It decreases the receiver third-order intermodulation distortion dynamic range
D. It allows strong signals on nearby frequencies to interfere with reception of weak signals

E4C01
(D)
Page 4-20

E4C02
What is the term for the reduction in receiver sensitivity caused by a strong signal near the received frequency?
A. Desensitization
B. Quieting
C. Cross-modulation interference
D. Squelch gain rollback

E4C02
(A)
Page 4-22

E4C03
What causes receiver desensitization?
A. Audio gain adjusted too low
B. Strong adjacent-channel signals
C. Squelch gain adjusted too high
D. Squelch gain adjusted too low

E4C03
(B)
Page 4-22

E4C04
What is one way receiver desensitization can be reduced?
A. Shield the receiver from the transmitter causing the problem
B. Increase the transmitter audio gain
C. Decrease the receiver squelch gain
D. Increase the receiver bandwidth

E4C04
(A)
Page 4-22

E4C05
What is the capture effect?
A. All signals on a frequency are demodulated by an FM receiver
B. All signals on a frequency are demodulated by an AM receiver
C. The strongest signal received is the only demodulated signal
D. The weakest signal received is the only demodulated signal

E4C05
(C)
Page 4-23

E4C06
What is the term for the blocking of one FM-phone signal by another stronger FM-phone signal?
A. Desensitization
B. Cross-modulation interference
C. Capture effect
D. Frequency discrimination

E4C06
(C)
Page 4-23

E4C07
(A)
Page 4-23

E4C07
With which emission type is capture effect most pronounced?
A. FM
B. SSB
C. AM
D. CW

E4C08
(D)
Page 4-16

E4C08
What is meant by the noise floor of a receiver?
A. The weakest signal that can be detected under noisy atmospheric conditions
B. The amount of phase noise generated by the receiver local oscillator
C. The minimum level of noise that will overload the receiver RF amplifier stage
D. The weakest signal that can be detected above the receiver internal noise

E4C09
(B)
Page 4-17

E4C09
What is the blocking dynamic range of a receiver that has an 8-dB noise figure and an IF bandwidth of 500 Hz if the blocking level (1-dB compression point) is −20 dBm?
A. −119 dBm
B. 119 dB
C. 146 dB
D. −146 dBm

E4C10
(C)
Page 4-17

E4C10
What is meant by the dynamic range of a communications receiver?
A. The number of kHz between the lowest and the highest frequency to which the receiver can be tuned
B. The maximum possible undistorted audio output of the receiver, referenced to one milliwatt
C. The ratio between the minimum discernible signal and the largest tolerable signal without causing audible distortion products
D. The difference between the lowest-frequency signal and the highest-frequency signal detectable without moving the tuning knob

E4C11
(A)
Page 4-19

E4C11
What type of problems are caused by poor dynamic range in a communications receiver?
A. Cross modulation of the desired signal and desensitization from strong adjacent signals
B. Oscillator instability requiring frequent retuning, and loss of ability to recover the opposite sideband, should it be transmitted
C. Cross modulation of the desired signal and insufficient audio power to operate the speaker
D. Oscillator instability and severe audio distortion of all but the strongest received signals

E4C12
(B)
Page 4-21

E4C12
What part of a superheterodyne receiver determines the image rejection ratio of the receiver?
A. Product detector
B. RF amplifier
C. AGC loop
D. IF filter

E4C13
If you measured the MDS of a receiver, what would you be measuring?
A. The meter display sensitivity (MDS), or the responsiveness of the receiver S-meter to all signals
B. The minimum discernible signal (MDS), or the weakest signal that the receiver can detect
C. The minimum distorting signal (MDS), or the strongest signal the receiver can detect without overloading
D. The maximum detectable spectrum (MDS), or the lowest to highest frequency range of the receiver

E4C13
(B)
Page 4-16

E4C14
How does intermodulation interference between two repeater transmitters usually occur?
A. When the signals from the transmitters are reflected out of phase from airplanes passing overhead
B. When they are in close proximity and the signals mix in one or both of their final amplifiers
C. When they are in close proximity and the signals cause feedback in one or both of their final amplifiers
D. When the signals from the transmitters are reflected in phase from airplanes passing overhead

E4C14
(B)
Page 4-23

E4C15
How can intermodulation interference between two repeater transmitters in close proximity often be reduced or eliminated?
A. By using a Class C final amplifier with high driving power
B. By installing a terminated circulator or ferrite isolator in the feed line to the transmitter and duplexer
C. By installing a band-pass filter in the antenna feed line
D. By installing a low-pass filter in the antenna feed line

E4C15
(B)
Page 4-24

E4C16
If a receiver tuned to 146.70 MHz receives an intermodulation-product signal whenever a nearby transmitter transmits on 146.52 MHz, what are the two most likely frequencies for the other interfering signal?
A. 146.34 MHz and 146.61 MHz
B. 146.88 MHz and 146.34 MHz
C. 146.10 MHz and 147.30 MHz
D. 73.35 MHz and 239.40 MHz

E4C16
(A)
Page 4-25

E4D Noise suppression: ignition noise; alternator noise (whine); electronic motor noise; static; line noise

E4D01
What is one of the most significant problems associated with mobile transceivers?
A. Ignition noise
B. Doppler shift
C. Radar interference
D. Mechanical vibrations

E4D01
(A)
Page 4-27

E4D02
(A)
Page 4-27

E4D02
What is the proper procedure for suppressing electrical noise in a mobile transceiver?
A. Apply shielding and filtering where necessary
B. Insulate all plane sheet metal surfaces from each other
C. Apply antistatic spray liberally to all non-metallic surfaces
D. Install filter capacitors in series with all DC wiring

E4D03
(C)
Page 4-28

E4D03
Where can ferrite beads be installed to suppress ignition noise in a mobile transceiver?
A. In the resistive high-voltage cable
B. Between the starter solenoid and the starter motor
C. In the primary and secondary ignition leads
D. In the antenna lead to the transceiver

E4D04
(C)
Page 4-28

E4D04
How can ensuring good electrical contact between connecting metal surfaces in a vehicle reduce ignition noise?
A. It reduces the frequency of the ignition spark
B. It helps radiate the ignition noise away from the vehicle
C. It encourages lower frequency electrical resonances in the vehicle
D. It reduces static buildup on the vehicle body

E4D05
(B)
Page 4-28

E4D05
How can alternator whine be minimized?
A. By connecting the radio's power leads to the battery by the longest possible path
B. By connecting the radio's power leads to the battery by the shortest possible path
C. By installing a high-pass filter in series with the radio's DC power lead to the vehicle's electrical system
D. By installing filter capacitors in series with the DC power lead

E4D06
(D)
Page 4-28

E4D06
How can conducted and radiated noise caused by an automobile alternator be suppressed?
A. By installing filter capacitors in series with the DC power lead and by installing a blocking capacitor in the field lead
B. By connecting the radio to the battery by the longest possible path and installing a blocking capacitor in both leads
C. By installing a high-pass filter in series with the radio's power lead and a low-pass filter in parallel with the field lead
D. By connecting the radio's power leads directly to the battery and by installing coaxial capacitors in the alternator leads

E4D07
(B)
Page 4-29

E4D07
How can you reduce noise from an electric motor?
A. Install a ferrite bead on the AC line used to power the motor
B. Install a brute-force, AC-line filter in series with the motor leads
C. Install a bypass capacitor in series with the motor leads
D. Use a ground-fault current interrupter in the circuit used to power the motor

E4D08
What is a major cause of atmospheric static?
A. Sunspots
B. Thunderstorms
C. Airplanes
D. Meteor showers

E4D08
(B)
Page 4-29

E4D09
How can you determine if a line-noise interference problem is being generated within your home?
A. Check the power-line voltage with a time-domain reflectometer
B. Observe the AC waveform on an oscilloscope
C. Turn off the main circuit breaker and listen on a battery-operated radio
D. Observe the power-line voltage on a spectrum analyzer

E4D09
(C)
Page 4-30

E4D10
What type of signal is picked up by electrical wiring near a radio transmitter?
A. A common-mode signal at the frequency of the radio transmitter
B. An electrical-sparking signal
C. A differential-mode signal at the AC-line frequency
D. Harmonics of the AC-line frequency

E4D10
(A)
Page 4-31

E4D11
What type of equipment cannot be used to locate power line noise?
A. An AM receiver with a directional antenna
B. An FM receiver with a directional antenna
C. A hand-held RF sniffer
D. An ultrasonic transducer, amplifier and parabolic reflector

E4D11
(B)
Page 4-31

E4E Component mounting techniques (i.e., surface, dead bug {raised}, circuit board); direction finding: techniques and equipment; fox hunting

E4E01
What circuit construction technique uses leadless components mounted between circuit board pads?
A. Raised mounting
B. Integrated circuit mounting
C. Hybrid device mounting
D. Surface mounting

E4E01
(D)
Page 4-32

E4E02
What is the main drawback of a wire-loop antenna for direction finding?
A. It has a bidirectional pattern broadside to the loop
B. It is non-rotatable
C. It receives equally well in all directions
D. It is practical for use only on VHF bands

E4E02
(A)
Page 4-33

E4E03
(B)
Page 4-33

E4E03
What pattern is desirable for a direction-finding antenna?
A. One which is non-cardioid
B. One with good front-to-back and front-to-side ratios
C. One with good top-to-bottom and side-to-side ratios
D. One with shallow nulls

E4E04
(C)
Page 4-37

E4E04
What is the triangulation method of direction finding?
A. The geometric angle of ground waves and sky waves from the signal source are used to locate the source
B. A fixed receiving station plots three beam headings from the signal source on a map
C. Beam headings from several receiving stations are used to plot the signal source on a map
D. A fixed receiving station uses three different antennas to plot the location of the signal source

E4E05
(D)
Page 4-33

E4E05
Why is an RF attenuator desirable in a receiver used for direction finding?
A. It narrows the bandwidth of the received signal
B. It eliminates the effects of isotropic radiation
C. It reduces loss of received signals caused by antenna pattern nulls
D. It prevents receiver overload from extremely strong signals

E4E06
(A)
Page 4-34

E4E06
What is a sense antenna?
A. A vertical antenna added to a loop antenna to produce a cardioid reception pattern
B. A horizontal antenna added to a loop antenna to produce a cardioid reception pattern
C. A vertical antenna added to an Adcock antenna to produce an omnidirectional reception pattern
D. A horizontal antenna added to an Adcock antenna to produce an omnidirectional reception pattern

E4E07
(D)
Page 4-36

E4E07
What type of antenna is most useful for sky-wave reception in radio direction finding?
A. A log-periodic dipole array
B. An isotropic antenna
C. A circularly-polarized antenna
D. An Adcock antenna

E4E08
(C)
Page 4-33

E4E08
What is a loop antenna?
A. A large circularly-polarized antenna
B. A small coil of wire tightly wound around a toroidal ferrite core
C. Several turns of wire wound in the shape of a large open coil
D. Any antenna coupled to a feed line through an inductive loop of wire

E4E09
How can the output voltage of a loop antenna be increased?
A. By reducing the permeability of the loop shield
B. By increasing the number of wire turns in the loop and reducing the area of the loop structure
C. By reducing either the number of wire turns in the loop or the area of the loop structure
D. By increasing either the number of wire turns in the loop or the area of the loop structure

E4E09
(D)
Page 4-34

E4E10
Why is an antenna system with a cardioid pattern desirable for a direction-finding system?
A. The broad-side responses of the cardioid pattern can be aimed at the desired station
B. The deep null of the cardioid pattern can pinpoint the direction of the desired station
C. The sharp peak response of the cardioid pattern can pinpoint the direction of the desired station
D. The high-radiation angle of the cardioid pattern is useful for short-distance direction finding

E4E10
(B)
Page 4-34

E4E11
What type of terrain can cause errors in direction finding?
A. Homogeneous terrain
B. Smooth grassy terrain
C. Varied terrain
D. Terrain with no buildings or mountains

E4E11
(C)
Page 4-38

E4E12
What is the activity known as fox hunting?
A. Amateurs using receivers and direction-finding techniques attempt to locate a hidden transmitter
B. Amateurs using transmitting equipment and direction-finding techniques attempt to locate a hidden receiver
C. Amateurs helping the government track radio-transmitter collars attached to animals
D. Amateurs assemble stations using generators and portable antennas to test their emergency communications skills

E4E12
(A)
Page 4-33

Subelement E5 — Electrical Principles

[9 Exam Questions — 9 Groups]

E5A Characteristics of resonant circuits: Series resonance (capacitor and inductor to resonate at a specific frequency); Parallel resonance (capacitor and inductor to resonate at a specific frequency); half-power bandwidth

E5A01
(A)
Page 5-40

E5A01
What can cause the voltage across reactances in series to be larger than the voltage applied to them?
A. Resonance
B. Capacitance
C. Conductance
D. Resistance

E5A02
(C)
Page 5-40

E5A02
What is resonance in an electrical circuit?
A. The highest frequency that will pass current
B. The lowest frequency that will pass current
C. The frequency at which capacitive reactance equals inductive reactance
D. The frequency at which power factor is at a minimum

E5A03
(B)
Page 5-40

E5A03
What are the conditions for resonance to occur in an electrical circuit?
A. The power factor is at a minimum
B. Inductive and capacitive reactances are equal
C. The square root of the sum of the capacitive and inductive reactance is equal to the resonant frequency
D. The square root of the product of the capacitive and inductive reactance is equal to the resonant frequency

E5A04
(D)
Page 5-40

E5A04
When the inductive reactance of an electrical circuit equals its capacitive reactance, what is this condition called?
A. Reactive quiescence
B. High Q
C. Reactive equilibrium
D. Resonance

E5A05
(D)
Page 5-42

E5A05
What is the magnitude of the impedance of a series R-L-C circuit at resonance?
A. High, as compared to the circuit resistance
B. Approximately equal to capacitive reactance
C. Approximately equal to inductive reactance
D. Approximately equal to circuit resistance

E5A06
What is the magnitude of the impedance of a circuit with a resistor, an inductor and a capacitor all in parallel, at resonance?
A. Approximately equal to circuit resistance
B. Approximately equal to inductive reactance
C. Low, as compared to the circuit resistance
D. Approximately equal to capacitive reactance

E5A06
(A)
Page 5-43

E5A07
What is the magnitude of the current at the input of a series R-L-C circuit at resonance?
A. It is at a minimum
B. It is at a maximum
C. It is DC
D. It is zero

E5A07
(B)
Page 5-42

E5A08
What is the magnitude of the circulating current within the components of a parallel L-C circuit at resonance?
A. It is at a minimum
B. It is at a maximum
C. It is DC
D. It is zero

E5A08
(B)
Page 5-43

E5A09
What is the magnitude of the current at the input of a parallel R-L-C circuit at resonance?
A. It is at a minimum
B. It is at a maximum
C. It is DC
D. It is zero

E5A09
(A)
Page 5-43

E5A10
What is the relationship between the current through a resonant circuit and the voltage across the circuit?
A. The voltage leads the current by 90 degrees
B. The current leads the voltage by 90 degrees
C. The voltage and current are in phase
D. The voltage and current are 180 degrees out of phase

E5A10
(C)
Page 5-45

E5A11
What is the relationship between the current into (or out of) a parallel resonant circuit and the voltage across the circuit?
A. The voltage leads the current by 90 degrees
B. The current leads the voltage by 90 degrees
C. The voltage and current are in phase
D. The voltage and current are 180 degrees out of phase

E5A11
(C)
Page 5-45

E5A12
(A)
Page 5-49

E5A12
What is the half-power bandwidth of a parallel resonant circuit that has a resonant frequency of 1.8 MHz and a Q of 95?
A. 18.9 kHz
B. 1.89 kHz
C. 189 Hz
D. 58.7 kHz

E5A13
(C)
Page 5-49

E5A13
What is the half-power bandwidth of a parallel resonant circuit that has a resonant frequency of 7.1 MHz and a Q of 150?
A. 211 kHz
B. 16.5 kHz
C. 47.3 kHz
D. 21.1 kHz

E5A14
(A)
Page 5-49

E5A14
What is the half-power bandwidth of a parallel resonant circuit that has a resonant frequency of 14.25 MHz and a Q of 150?
A. 95 kHz
B. 10.5 kHz
C. 10.5 MHz
D. 17 kHz

E5A15
(D)
Page 5-49

E5A15
What is the half-power bandwidth of a parallel resonant circuit that has a resonant frequency of 21.15 MHz and a Q of 95?
A. 4.49 kHz
B. 44.9 kHz
C. 22.3 kHz
D. 222.6 kHz

E5A16
(C)
Page 5-49

E5A16
What is the half-power bandwidth of a parallel resonant circuit that has a resonant frequency of 3.7 MHz and a Q of 118?
A. 22.3 kHz
B. 76.2 kHz
C. 31.4 kHz
D. 10.8 kHz

E5A17
(C)
Page 5-49

E5A17
What is the half-power bandwidth of a parallel resonant circuit that has a resonant frequency of 14.25 MHz and a Q of 187?
A. 22.3 kHz
B. 10.8 kHz
C. 76.2 kHz
D. 13.1 kHz

E5B Exponential charge/discharge curves (time constants): definition; time constants in RL and RC circuits

E5B01
What is the term for the time required for the capacitor in an RC circuit to be charged to 63.2% of the supply voltage?
A. An exponential rate of one
B. One time constant
C. One exponential period
D. A time factor of one

E5B01
(B)
Page 5-14

E5B02
What is the term for the time required for the current in an RL circuit to build up to 63.2% of the maximum value?
A. One time constant
B. An exponential period of one
C. A time factor of one
D. One exponential rate

E5B02
(A)
Page 5-17

E5B03
What is the term for the time it takes for a charged capacitor in an RC circuit to discharge to 36.8% of its initial value of stored charge?
A. One discharge period
B. An exponential discharge rate of one
C. A discharge factor of one
D. One time constant

E5B03
(D)
Page 5-14

E5B04
The capacitor in an RC circuit is charged to what percentage of the supply voltage after two time constants?
A. 36.8%
B. 63.2%
C. 86.5%
D. 95%

E5B04
(C)
Page 5-14

E5B05
The capacitor in an RC circuit is discharged to what percentage of the starting voltage after two time constants?
A. 86.5%
B. 63.2%
C. 36.8%
D. 13.5%

E5B05
(D)
Page 5-14

E5B06
What is the time constant of a circuit having two 100-microfarad capacitors and two 470-kilohm resistors all in series?
A. 47 seconds
B. 101.1 seconds
C. 103 seconds
D. 220 seconds

E5B06
(A)
Page 5-11

E5B07	E5B07
(D)	What is the time constant of a circuit having two 220-microfarad capacitors and two
Page 5-11	1-megohm resistors all in parallel?
	A. 47 seconds
	B. 101.1 seconds
	C. 103 seconds
	D. 220 seconds

E5B08	E5B08
(C)	What is the time constant of a circuit having a 220-microfarad capacitor in series with a
Page 5-11	470-kilohm resistor?
	A. 47 seconds
	B. 80 seconds
	C. 103 seconds
	D. 220 seconds

E5B09	E5B09
(A)	How long does it take for an initial charge of 20 V DC to decrease to 7.36 V DC in a
Page 5-15	0.01-microfarad capacitor when a 2-megohm resistor is connected across it?
	A. 0.02 seconds
	B. 0.08 seconds
	C. 450 seconds
	D. 1350 seconds

E5B10	E5B10
(B)	How long does it take for an initial charge of 20 V DC to decrease to 0.37 V DC in a
Page 5-15	0.01-microfarad capacitor when a 2-megohm resistor is connected across it?
	A. 0.02 seconds
	B. 0.08 seconds
	C. 450 seconds
	D. 1350 seconds

E5B11	E5B11
(C)	How long does it take for an initial charge of 800 V DC to decrease to 294 V DC in a
Page 5-15	450-microfarad capacitor when a 1-megohm resistor is connected across it?
	A. 0.02 seconds
	B. 0.08 seconds
	C. 450 seconds
	D. 1350 seconds

E5C Impedance diagrams: Basic principles of Smith charts; impedance of RLC networks at specified frequencies

E5C01	E5C01
(A)	What type of graph can be used to calculate impedance along transmission lines?
Page 5-36	A. A Smith chart
	B. A logarithmic chart
	C. A Jones chart
	D. A radiation pattern chart

E5C02
What type of coordinate system is used in a Smith chart?
A. Voltage and current circles
B. Resistance and reactance circles
C. Voltage and current lines
D. Resistance and reactance lines

E5C02
(B)
Page 5-36

E5C03
What type of calculations can be performed using a Smith chart?
A. Beam headings and radiation patterns
B. Satellite azimuth and elevation bearings
C. Impedance and SWR values in transmission lines
D. Circuit gain calculations

E5C03
(C)
Page 5-36

E5C04
What are the two families of circles that make up a Smith chart?
A. Resistance and voltage
B. Reactance and voltage
C. Resistance and reactance
D. Voltage and impedance

E5C04
(C)
Page 5-36

E5C05
What type of chart is shown in Figure E5-1?
A. Smith chart
B. Free-space radiation directivity chart
C. Vertical-space radiation pattern chart
D. Horizontal-space radiation pattern chart

E5C05
(A)
Page 5-37

E5C06
On the Smith chart shown in Figure E5-1, what is the name for the large outer circle bounding the coordinate portion of the chart?
A. Prime axis
B. Reactance axis
C. Impedance axis
D. Polar axis

E5C06
(B)
Page 5-37

E5C07
On the Smith chart shown in Figure E5-1, what is the only straight line shown?
A. The reactance axis
B. The current axis
C. The voltage axis
D. The resistance axis

E5C07
(D)
Page 5-37

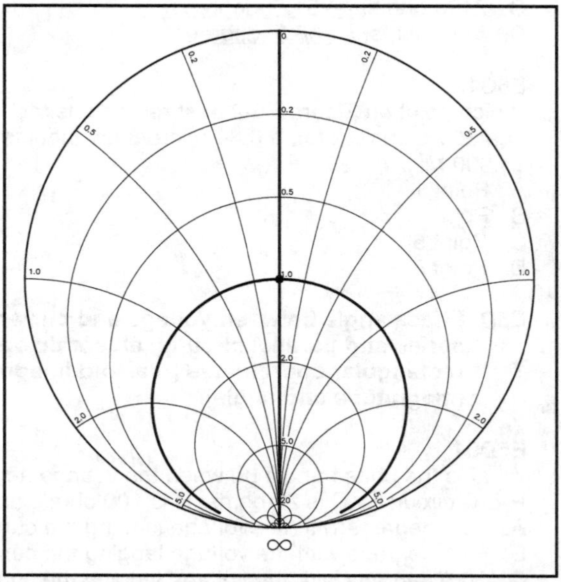

Figure E5-1 — Refer to questions E5C05, E5C06 and E5C07.

E5C08
What is the process of normalizing with regard to a Smith chart?
A. Reassigning resistance values with regard to the reactance axis
B. Reassigning reactance values with regard to the resistance axis
C. Reassigning resistance values with regard to the prime center
D. Reassigning prime center with regard to the reactance axis

E5C09
What is the third family of circles, which are added to a Smith chart during the process of solving problems?
A. Standing-wave ratio circles
B. Antenna-length circles
C. Coaxial-length circles
D. Radiation-pattern circles

E5C10
In rectangular coordinates, what is the impedance of a network comprised of a 10-microhenry inductor in series with a 40-ohm resistor at 500 MHz?
A. 40 + j31,400
B. 40 − j31,400
C. 31,400 + j40
D. 31,400 − j40

E5C11
In polar coordinates, what is the impedance of a network comprised of a 100-picofarad capacitor in parallel with a 4,000-ohm resistor at 500 kHz?
A. 2490 ohms, / 51.5 degrees
B. 4000 ohms, / 38.5 degrees
C. 2490 ohms, / −51.5 degrees
D. 5112 ohms, / −38.5 degrees

E5C12
Which point on Figure E5-2 best represents the impedance of a series circuit consisting of a 300-ohm resistor, a 0.64-microhenry inductor and an 85-picofarad capacitor at 24.900 MHz?
A. Point 1
B. Point 3
C. Point 5
D. Point 8

E5D Phase angle between voltage and current; impedances and phase angles of series and parallel circuits; algebraic operations using complex numbers: rectangular coordinates (real and imaginary parts); polar coordinates (magnitude and angle)

E5D01
What is the phase angle between the voltage across and the current through a series R-L-C circuit if XC is 25 ohms, R is 100 ohms, and XL is 100 ohms?
A. 36.9 degrees with the voltage leading the current
B. 53.1 degrees with the voltage lagging the current
C. 36.9 degrees with the voltage lagging the current
D. 53.1 degrees with the voltage leading the current

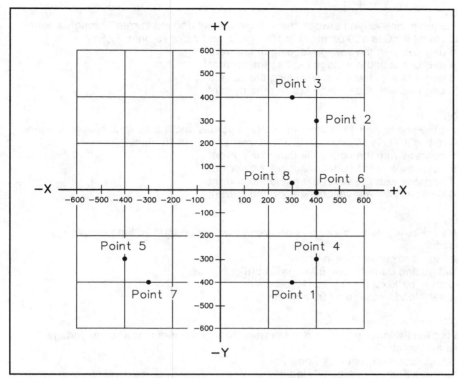

Figure E5-2 — Refer to question E5C12.

E5D02
What is the phase angle between the voltage across and the current through a series R-L-C circuit if XC is 500 ohms, R is 1 kilohm, and XL is 250 ohms?
A. 68.2 degrees with the voltage leading the current
B. 14.1 degrees with the voltage leading the current
C. 14.1 degrees with the voltage lagging the current
D. 68.2 degrees with the voltage lagging the current

E5D03
What is the phase angle between the voltage across and the current through a series R-L-C circuit if XC is 50 ohms, R is 100 ohms, and XL is 25 ohms?
A. 76 degrees with the voltage lagging the current
B. 14 degrees with the voltage leading the current
C. 76 degrees with the voltage leading the current
D. 14 degrees with the voltage lagging the current

E5D02
(C)
Page 5-25

E5D03
(D)
Page 5-25

E5D04
(A)
Page 5-25

E5D04
What is the phase angle between the voltage across and the current through a series R-L-C circuit if XC is 100 ohms, R is 100 ohms, and XL is 75 ohms?
A. 14 degrees with the voltage lagging the current
B. 14 degrees with the voltage leading the current
C. 76 degrees with the voltage leading the current
D. 76 degrees with the voltage lagging the current

E5D05
(D)
Page 5-25

E5D05
What is the phase angle between the voltage across and the current through a series R-L-C circuit if XC is 50 ohms, R is 100 ohms, and XL is 75 ohms?
A. 76 degrees with the voltage leading the current
B. 76 degrees with the voltage lagging the current
C. 14 degrees with the voltage lagging the current
D. 14 degrees with the voltage leading the current

E5D06
(D)
Page 5-18

E5D06
What is the relationship between the current through and the voltage across a capacitor?
A. Voltage and current are in phase
B. Voltage and current are 180 degrees out of phase
C. Voltage leads current by 90 degrees
D. Current leads voltage by 90 degrees

E5D07
(A)
Page 5-19

E5D07
What is the relationship between the current through an inductor and the voltage across an inductor?
A. Voltage leads current by 90 degrees
B. Current leads voltage by 90 degrees
C. Voltage and current are 180 degrees out of phase
D. Voltage and current are in phase

E5D08
(B)
Page 5-31

E5D08
In polar coordinates, what is the impedance of a network comprised of a 100-ohm-reactance inductor in series with a 100-ohm resistor?
A. 121 ohms, / 35 degrees
B. 141 ohms, / 45 degrees
C. 161 ohms, / 55 degrees
D. 181 ohms, / 65 degrees

E5D09
(D)
Page 5-31

E5D09
In polar coordinates, what is the impedance of a network comprised of a 100-ohm-reactance inductor, a 100-ohm-reactance capacitor, and a 100-ohm resistor all connected in series?
A. 100 ohms, / 90 degrees
B. 10 ohms, / 0 degrees
C. 10 ohms, / 100 degrees
D. 100 ohms, / 0 degrees

E5D10

In polar coordinates, what is the impedance of a network comprised of a 300-ohm-reactance capacitor, a 600-ohm-reactance inductor, and a 400-ohm resistor, all connected in series?

A. 500 ohms, / 37 degrees
B. 400 ohms, / 27 degrees
C. 300 ohms, / 17 degrees
D. 200 ohms, / 10 degrees

E5D11

In rectangular coordinates, what is the impedance of a network comprised of a 1.0-millihenry inductor in series with a 200-ohm resistor at 30 kHz?

A. 200 − j188
B. 200 + j188
C. 188 − j200
D. 188 + j200

E5D12

In rectangular coordinates, what is the impedance of a network comprised of a 10-millihenry inductor in series with a 600-ohm resistor at 10 kHz?

A. 628 + j600
B. 628 − j600
C. 600 + j628
D. 600 − j628

E5D13

In rectangular coordinates, what is the impedance of a network comprised of a 0.1-microfarad capacitor in series with a 40-ohm resistor at 50 kHz?

A. 40 + j32
B. 40 − j32
C. 32 − j40
D. 32 + j40

E5E Skin effect; electrostatic and electromagnetic fields

E5E01

What is the result of skin effect?

A. As frequency increases, RF current flows in a thinner layer of the conductor, closer to the surface
B. As frequency decreases, RF current flows in a thinner layer of the conductor, closer to the surface
C. Thermal effects on the surface of the conductor increase the impedance
D. Thermal effects on the surface of the conductor decrease the impedance

E5E02

What effect causes most of an RF current to flow along the surface of a conductor?

A. Layer effect
B. Seeburg effect
C. Skin effect
D. Resonance effect

E5D10
(A)
Page 5-31

E5D11
(B)
Page 5-29

E5D12
(C)
Page 5-29

E5D13
(B)
Page 5-30

E5E01
(A)
Page 5-48

E5E02
(C)
Page 5-48

E5E03
(A)
Page 5-48

E5E03
Where does almost all RF current flow in a conductor?
A. Along the surface of the conductor
B. In the center of the conductor
C. In a magnetic field around the conductor
D. In a magnetic field in the center of the conductor

E5E04
(D)
Page 5-48

E5E04
Why does most of an RF current flow within a few thousandths of an inch of its conductor's surface?
A. Because a conductor has AC resistance due to self-inductance
B. Because the RF resistance of a conductor is much less than the DC resistance
C. Because of the heating of the conductor's interior
D. Because of skin effect

E5E05
(C)
Page 5-48

E5E05
Why is the resistance of a conductor different for RF currents than for direct currents?
A. Because the insulation conducts current at high frequencies
B. Because of the Heisenburg Effect
C. Because of skin effect
D. Because conductors are non-linear devices

E5E06
(C)
Page 5-8

E5E06
What device is used to store electrical energy in an electrostatic field?
A. A battery
B. A transformer
C. A capacitor
D. An inductor

E5E07
(B)
Page 5-9

E5E07
What unit measures electrical energy stored in an electrostatic field?
A. Coulomb
B. Joule
C. Watt
D. Volt

E5E08
(B)
Page 5-9

E5E08
What is a magnetic field?
A. Current through the space around a permanent magnet
B. The space around a conductor, through which a magnetic force acts
C. The space between the plates of a charged capacitor, through which a magnetic force acts
D. The force that drives current through a resistor

E5E09
(D)
Page 5-9

E5E09
In what direction is the magnetic field oriented about a conductor in relation to the direction of electron flow?
A. In the same direction as the current
B. In a direction opposite to the current
C. In all directions; omnidirectional
D. In a direction determined by the left-hand rule

E5E10
What determines the strength of a magnetic field around a conductor?
A. The resistance divided by the current
B. The ratio of the current to the resistance
C. The diameter of the conductor
D. The amount of current

E5E11
What is the term for energy that is stored in an electromagnetic or electrostatic field?
A. Amperes-joules
B. Potential energy
C. Joules-coulombs
D. Kinetic energy

E5F Circuit Q; reactive power; power factor

E5F01
What is the Q of a parallel R-L-C circuit if the resonant frequency is 14.128 MHz, L is 2.7 microhenrys and R is 18 kilohms?
A. 75.1
B. 7.51
C. 71.5
D. 0.013

E5F02
What is the Q of a parallel R-L-C circuit if the resonant frequency is 4.468 MHz, L is 47 microhenrys and R is 180 ohms?
A. 0.00735
B. 7.35
C. 0.136
D. 13.3

E5F03
What is the Q of a parallel R-L-C circuit if the resonant frequency is 7.125 MHz, L is 8.2 microhenrys and R is 1 kilohm?
A. 36.8
B. 0.273
C. 0.368
D. 2.73

E5F04
What is the Q of a parallel R-L-C circuit if the resonant frequency is 7.125 MHz, L is 12.6 microhenrys and R is 22 kilohms?
A. 22.1
B. 39
C. 25.6
D. 0.0256

E5E10
(D)
Page 5-10

E5E11
(B)
Page 5-9

E5F01
(A)
Page 5-48

E5F02
(C)
Page 5-48

E5F03
(D)
Page 5-48

E5F04
(B)
Page 5-48

E5F05
(D)
Page 5-48

E5F05
What is the Q of a parallel R-L-C circuit if the resonant frequency is 3.625 MHz, L is 42 microhenrys and R is 220 ohms?
A. 23
B. 0.00435
C. 4.35
D. 0.23

E5F06
(C)
Page 5-49

E5F06
Why is a resistor often included in a parallel resonant circuit?
A. To increase the Q and decrease the skin effect
B. To decrease the Q and increase the resonant frequency
C. To decrease the Q and increase the bandwidth
D. To increase the Q and decrease the bandwidth

E5F07
(D)
Page 5-51

E5F07
What is the term for an out-of-phase, nonproductive power associated with inductors and capacitors?
A. Effective power
B. True power
C. Peak envelope power
D. Reactive power

E5F08
(B)
Page 5-50

E5F08
In a circuit that has both inductors and capacitors, what happens to reactive power?
A. It is dissipated as heat in the circuit
B. It goes back and forth between magnetic and electric fields, but is not dissipated
C. It is dissipated as kinetic energy in the circuit
D. It is dissipated in the formation of inductive and capacitive fields

E5F09
(A)
Page 5-51

E5F09
In a circuit where the AC voltage and current are out of phase, how can the true power be determined?
A. By multiplying the apparent power times the power factor
B. By subtracting the apparent power from the power factor
C. By dividing the apparent power by the power factor
D. By multiplying the RMS voltage times the RMS current

E5F10
(C)
Page 5-53

E5F10
What is the power factor of an R-L circuit having a 60 degree phase angle between the voltage and the current?
A. 1.414
B. 0.866
C. 0.5
D. 1.73

E5F11
How many watts are consumed in a circuit having a power factor of 0.2 if the input is
100-V AC at 4 amperes?
A. 400 watts
B. 80 watts
C. 2000 watts
D. 50 watts

E5F11
(B)
Page 5-53

E5F12
Why would the power used in a circuit be less than the product of the magnitudes of
the AC voltage and current?
A. Because there is a phase angle greater than zero between the current and voltage
B. Because there are only resistances in the circuit
C. Because there are no reactances in the circuit
D. Because there is a phase angle equal to zero between the current and voltage

E5F12
(A)
Page 5-50

E5G Effective radiated power; system gains and losses

E5G01
What is the effective radiated power of a repeater station with 50 watts transmitter
power output, 4-dB feed line loss, 2-dB duplexer loss, 1-dB circulator loss and 6-dBd
antenna gain?
A. 199 watts
B. 39.7 watts
C. 45 watts
D. 62.9 watts

E5G01
(B)
Page 5-54

E5G02
What is the effective radiated power of a repeater station with 50 watts transmitter
power output, 5-dB feed line loss, 3-dB duplexer loss, 1-dB circulator loss and 7-dBd
antenna gain?
A. 79.2 watts
B. 315 watts
C. 31.5 watts
D. 40.5 watts

E5G02
(C)
Page 5-54

E5G03
What is the effective radiated power of a station with 75 watts transmitter power output,
4-dB feed line loss and 10-dBd antenna gain?
A. 600 watts
B. 75 watts
C. 150 watts
D. 299 watts

E5G03
(D)
Page 5-54

E5G04
What is the effective radiated power of a repeater station with 75 watts transmitter
power output, 5-dB feed line loss, 3-dB duplexer loss, 1-dB circulator loss and 6-dBd
antenna gain?
A. 37.6 watts
B. 237 watts
C. 150 watts
D. 23.7 watts

E5G04
(A)
Page 5-54

E5G05
(D)
Page 5-54

E5G05
What is the effective radiated power of a station with 100 watts transmitter power output, 1-dB feed line loss and 6-dBd antenna gain?
A. 350 watts
B. 500 watts
C. 20 watts
D. 316 watts

E5G06
(B)
Page 5-54

E5G06
What is the effective radiated power of a repeater station with 100 watts transmitter power output, 5-dB feed line loss, 3-dB duplexer loss, 1-dB circulator loss and 10-dBd antenna gain?
A. 794 watts
B. 126 watts
C. 79.4 watts
D. 1260 watts

E5G07
(C)
Page 5-54

E5G07
What is the effective radiated power of a repeater station with 120 watts transmitter power output, 5-dB feed line loss, 3-dB duplexer loss, 1-dB circulator loss and 6-dBd antenna gain?
A. 601 watts
B. 240 watts
C. 60 watts
D. 79 watts

E5G08
(D)
Page 5-54

E5G08
What is the effective radiated power of a repeater station with 150 watts transmitter power output, 2-dB feed line loss, 2.2-dB duplexer loss and 7-dBd antenna gain?
A. 1977 watts
B. 78.7 watts
C. 420 watts
D. 286 watts

E5G09
(A)
Page 5-54

E5G09
What is the effective radiated power of a repeater station with 200 watts transmitter power output, 4-dB feed line loss, 3.2-dB duplexer loss, 0.8-dB circulator loss and 10-dBd antenna gain?
A. 317 watts
B. 2000 watts
C. 126 watts
D. 300 watts

E5G10
(B)
Page 5-54

E5G10
What is the effective radiated power of a repeater station with 200 watts transmitter power output, 2-dB feed line loss, 2.8-dB duplexer loss, 1.2-dB circulator loss and 7-dBd antenna gain?
A. 159 watts
B. 252 watts
C. 632 watts
D. 63.2 watts

E5G11
What term describes station output (including the transmitter, antenna and everything in between), when considering transmitter power and system gains and losses?
A. Power factor
B. Half-power bandwidth
C. Effective radiated power
D. Apparent power

E5G11
(C)
Page 5-54

E5H Replacement of voltage source and resistive voltage divider with equivalent voltage source and one resistor (Thevenin's Theorem).

E5H01
In Figure A5-1, what values of V2 and R3 result in the same voltage and current as when V1 is 8 volts, R1 is 8 kilohms, and R2 is 8 kilohms?
A. R3 = 4 kilohms and V2 = 8 volts
B. R3 = 4 kilohms and V2 = 4 volts
C. R3 = 16 kilohms and V2 = 8 volts
D. R3 = 16 kilohms and V2 = 4 volts

E5H01
(B)
Page 5-56

E5H02
In Figure A5-1, what values of V2 and R3 result in the same voltage and current as when V1 is 8 volts, R1 is 16 kilohms, and R2 is 8 kilohms?
A. R3 = 24 kilohms and V2 = 5.33 volts
B. R3 = 5.33 kilohms and V2 = 8 volts
C. R3 = 5.33 kilohms and V2 = 2.67 volts
D. R3 = 24 kilohms and V2 = 8 volts

E5H02
(C)
Page 5-56

E5H03
In Figure A5-1, what values of V2 and R3 result in the same voltage and current as when V1 is 8 volts, R1 is 8 kilohms, and R2 is 16 kilohms?
A. R3 = 5.33 kilohms and V2 = 5.33 volts
B. R3 = 8 kilohms and V2 = 4 volts
C. R3 = 24 kilohms and V2 = 8 volts
D. R3 = 5.33 kilohms and V2 = 8 volts

E5H03
(A)
Page 5-56

E5H04
In Figure A5-1, what values of V2 and R3 result in the same voltage and current as when V1 is 10 volts, R1 is 10 kilohms, and R2 is 10 kilohms?
A. R3 = 10 kilohms and V2 = 5 volts
B. R3 = 20 kilohms and V2 = 5 volts
C. R3 = 20 kilohms and V2 = 10 volts
D. R3 = 5 kilohms and V2 = 5 volts

E5H04
(D)
Page 5-56

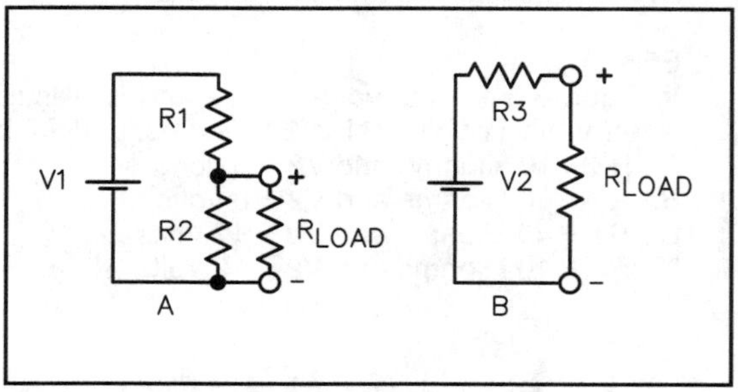

Figure A5-1 — Refer to questions E5H01 through E5H10.

Element 4 (Amateur Extra Class) Question Pool—With Answers   **10-67**

E5H05
(C)
Page 5-56

E5H05
In Figure A5-1, what values of V2 and R3 result in the same voltage and current as when V1 is 10 volts, R1 is 20 kilohms, and R2 is 10 kilohms?
A. R3 = 30 kilohms and V2 = 10 volts
B. R3 = 6.67 kilohms and V2 = 10 volts
C. R3 = 6.67 kilohms and V2 = 3.33 volts
D. R3 = 30 kilohms and V2 = 3.33 volts

E5H06
(A)
Page 5-56

E5H06
In Figure A5-1, what values of V2 and R3 result in the same voltage and current as when V1 is 10 volts, R1 is 10 kilohms, and R2 is 20 kilohms?
A. R3 = 6.67 kilohms and V2 = 6.67 volts
B. R3 = 6.67 kilohms and V2 = 10 volts
C. R3 = 30 kilohms and V2 = 6.67 volts
D. R3 = 30 kilohms and V2 = 10 volts

E5H07
(B)
Page 5-56

E5H07
In Figure A5-1, what values of V2 and R3 result in the same voltage and current as when V1 is 12 volts, R1 is 10 kilohms, and R2 is 10 kilohms?
A. R3 = 20 kilohms and V2 = 12 volts
B. R3 = 5 kilohms and V2 = 6 volts
C. R3 = 5 kilohms and V2 = 12 volts
D. R3 = 30 kilohms and V2 = 6 volts

E5H08
(B)
Page 5-56

E5H08
In Figure A5-1, what values of V2 and R3 result in the same voltage and current as when V1 is 12 volts, R1 is 20 kilohms, and R2 is 10 kilohms?
A. R3 = 30 kilohms and V2 = 4 volts
B. R3 = 6.67 kilohms and V2 = 4 volts
C. R3 = 30 kilohms and V2 = 12 volts
D. R3 = 6.67 kilohms and V2 = 12 volts

E5H09
(C)
Page 5-56

E5H09
In Figure A5-1, what values of V2 and R3 result in the same voltage and current as when V1 is 12 volts, R1 is 10 kilohms, and R2 is 20 kilohms?
A. R3 = 6.67 kilohms and V2 = 12 volts
B. R3 = 30 kilohms and V2 = 12 volts
C. R3 = 6.67 kilohms and V2 = 8 volts
D. R3 = 30 kilohms and V2 = 8 volts

E5H10
(A)
Page 5-56

E5H10
In Figure A5-1, what values of V2 and R3 result in the same voltage and current as when V1 is 12 volts, R1 is 20 kilohms, and R2 is 20 kilohms?
A. R3 = 10 kilohms and V2 = 6 volts
B. R3 = 40 kilohms and V2 = 6 volts
C. R3 = 40 kilohms and V2 = 12 volts
D. R3 = 10 kilohms and V2 = 12 volts

E5H11
What circuit principle describes the replacement of any complex two-terminal network of voltage sources and resistances with a single voltage source and a single resistor?
A. Ohm's Law
B. Kirchhoff's Law
C. Laplace's Theorem
D. Thevenin's Theorem

E5H11
(D)
Page 5-56

E5I Photoconductive principles and effects

E5I01
What is photoconductivity?
A. The conversion of photon energy to electromotive energy
B. The increased conductivity of an illuminated semiconductor junction
C. The conversion of electromotive energy to photon energy
D. The decreased conductivity of an illuminated semiconductor junction

E5I01
(B)
Page 5-4

E5I02
What happens to the conductivity of a photoconductive material when light shines on it?
A. It increases
B. It decreases
C. It stays the same
D. It becomes temperature dependent

E5I02
(A)
Page 5-4

E5I03
What happens to the resistance of a photoconductive material when light shines on it?
A. It increases
B. It becomes temperature dependent
C. It stays the same
D. It decreases

E5I03
(D)
Page 5-4

E5I04
What happens to the conductivity of a semiconductor junction when light shines on it?
A. It stays the same
B. It becomes temperature dependent
C. It increases
D. It decreases

E5I04
(C)
Page 5-4

E5I05
What is an optocoupler?
A. A resistor and a capacitor
B. A frequency modulated helium-neon laser
C. An amplitude modulated helium-neon laser
D. An LED and a phototransistor

E5I05
(D)
Page 5-5

E5I06
What is an optoisolator?
A. An LED and a phototransistor
B. A P-N junction that develops an excess positive charge when exposed to light
C. An LED and a capacitor
D. An LED and a solar cell

E5I06
(A)
Page 5-5

E5I07
(B)
Page 5-6

E5I07
What is an optical shaft encoder?
A. An array of neon or LED indicators whose light transmission path is controlled by a rotating wheel
B. An array of optocouplers whose light transmission path is controlled by a rotating wheel
C. An array of neon or LED indicators mounted on a rotating wheel in a coded pattern
D. An array of optocouplers mounted on a rotating wheel in a coded pattern

E5I08
(D)
Page 5-4

E5I08
What characteristic of a crystalline solid will photoconductivity change?
A. The capacitance
B. The inductance
C. The specific gravity
D. The resistance

E5I09
(C)
Page 5-4

E5I09
Which material will exhibit the greatest photoconductive effect when visible light shines on it?
A. Potassium nitrate
B. Lead sulfide
C. Cadmium sulfide
D. Sodium chloride

E5I10
(B)
Page 5-4

E5I10
Which material will exhibit the greatest photoconductive effect when infrared light shines on it?
A. Potassium nitrate
B. Lead sulfide
C. Cadmium sulfide
D. Sodium chloride

E5I11
(A)
Page 5-4

E5I11
Which material is affected the most by photoconductivity?
A. A crystalline semiconductor
B. An ordinary metal
C. A heavy metal
D. A liquid semiconductor

E5I12
(B)
Page 5-5

E5I12
What characteristic of optoisolators is often used in power supplies?
A. They have a low impedance between the light source and the phototransistor
B. They have a very high impedance between the light source and the phototransistor
C. They have a low impedance between the light source and the LED
D. They have a very high impedance between the light source and the LED

E5I13
What characteristic of optoisolators makes them suitable for use with a triac to form the solid-state equivalent of a mechanical relay for a 120 V AC household circuit?
A. Optoisolators provide a low impedance link between a control circuit and a power circuit
B. Optoisolators provide impedance matching between the control circuit and power circuit
C. Optoisolators provide a very high degree of electrical isolation between a control circuit and a power circuit
D. Optoisolators eliminate (isolate) the effects of reflected light in the control circuit

E5I13
(C)
Page 5-7

Subelement E6 — Circuit Components

Subelement E6

[5 Exam Questions — 5 Groups]

E6A **Semiconductor material: Germanium, Silicon, P-type, N-type; Transistor types: NPN, PNP, junction, unijunction, power; field-effect transistors (FETs): enhancement mode; depletion mode; MOS; CMOS; N-channel; P-channel**

E6A01
In what application is gallium arsenide used as a semiconductor material in preference to germanium or silicon?
A. In bipolar transistors
B. In high-power circuits
C. At microwave frequencies
D. At very low frequencies

E6A01
(C)
Page 6-5

E6A02
What type of semiconductor material contains more free electrons than pure germanium or silicon crystals?
A. N-type
B. P-type
C. Bipolar
D. Insulated gate

E6A02
(A)
Page 6-2

E6A03
What type of semiconductor material might be produced by adding some indium atoms to germanium crystals?
A. J-type
B. MOS-type
C. N-type
D. P-type

E6A03
(D)
Page 6-2

E6A04
What are the majority charge carriers in P-type semiconductor material?
A. Free neutrons
B. Free protons
C. Holes
D. Free electrons

E6A04
(C)
Page 6-3

E6A05
(C)
Page 6-3

E6A06
(C)
Page 6-15

E6A07
(A)
Page 6-14

E6A08
(D)
Page 6-15

E6A09
(D)
Page 6-16

E6A10
(C)
Page 6-16

E6A05
What is the name given to an impurity atom that adds holes to a semiconductor crystal structure?
A. Insulator impurity
B. N-type impurity
C. Acceptor impurity
D. Donor impurity

E6A06
What is the alpha of a bipolar transistor?
A. The change of collector current with respect to base current
B. The change of base current with respect to collector current
C. The change of collector current with respect to emitter current
D. The change of collector current with respect to gate current

E6A07
In Figure A6-2, what is the schematic symbol for a PNP transistor?
A. 1
B. 2
C. 4
D. 5

E6A08
What term indicates the frequency at which a transistor grounded base current gain has decreased to 0.7 of the gain obtainable at 1 kHz?
A. Corner frequency
B. Alpha rejection frequency
C. Beta cutoff frequency
D. Alpha cutoff frequency

E6A09
In Figure A6-2, what is the schematic symbol for a unijunction transistor?
A. 3
B. 4
C. 5
D. 6

E6A10
What are the elements of a unijunction transistor?
A. Gate, base 1 and base 2
B. Gate, cathode and anode
C. Base 1, base 2 and emitter
D. Gate, source and sink

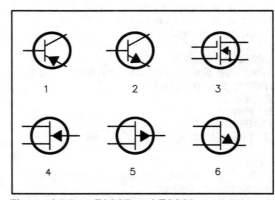

Figure A6-2 — E6A07 and E6A09.

E6A11
What is an enhancement-mode FET?
A. An FET with a channel that blocks voltage through the gate
B. An FET with a channel that allows a current when the gate voltage is zero
C. An FET without a channel to hinder current through the gate
D. An FET without a channel; no current occurs with zero gate voltage

E6A11
(D)
Page 6-19

E6A12
What is a depletion-mode FET?
A. An FET that has a channel with no gate voltage applied; a current flows with zero gate voltage
B. An FET that has a channel that blocks current when the gate voltage is zero
C. An FET without a channel; no current flows with zero gate voltage
D. An FET without a channel to hinder current through the gate

E6A12
(A)
Page 6-19

E6A13
In Figure E6-1, what is the schematic symbol for an N-channel dual-gate MOSFET?
A. 2
B. 4
C. 5
D. 6

E6A13
(B)
Page 6-18

E6A14
In Figure E6-1, what is the schematic symbol for a P-channel junction FET?
A. 1
B. 2
C. 3
D. 6

E6A14
(A)
Page 6-17

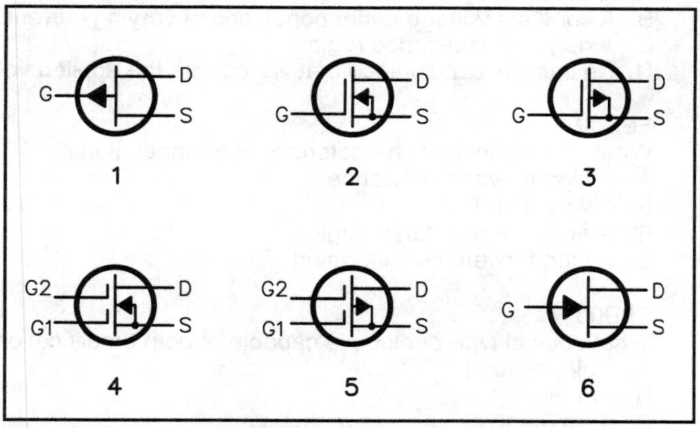

Figure E6-1 — Refer to questions E6A13 and E6A14.

E6A15
Why do many MOSFET devices have built-in gate-protective Zener diodes?
A. To provide a voltage reference for the correct amount of reverse-bias gate voltage
B. To protect the substrate from excessive voltages
C. To keep the gate voltage within specifications and prevent the device from overheating
D. To prevent the gate insulation from being punctured by small static charges or excessive voltages

E6A15
(D)
Page 6-18

E6A16
What do the initials CMOS stand for?
A. Common mode oscillating system
B. Complementary mica-oxide silicon
C. Complementary metal-oxide semiconductor
D. Complementary metal-oxide substrate

E6A16
(C)
Page 6-18

E6A17
(C)
Page 6-17

E6A17
How does the input impedance of a field-effect transistor compare with that of a bipolar transistor?
A. They cannot be compared without first knowing the supply voltage
B. An FET has low input impedance; a bipolar transistor has high input impedance
C. An FET has high input impedance; a bipolar transistor has low input impedance
D. The input impedance of FETs and bipolar transistors is the same

E6B Diodes: Zener, tunnel, varactor, hot-carrier, junction, point contact, PIN and light emitting; operational amplifiers (inverting amplifiers, noninverting amplifiers, voltage gain, frequency response, FET amplifier circuits, single-stage amplifier applications); phase-locked loops

E6B01
(B)
Page 6-10

E6B01
What is the principal characteristic of a Zener diode?
A. A constant current under conditions of varying voltage
B. A constant voltage under conditions of varying current
C. A negative resistance region
D. An internal capacitance that varies with the applied voltage

E6B02
(C)
Page 6-11

E6B02
What is the principal characteristic of a tunnel diode?
A. A high forward resistance
B. A very high PIV
C. A negative resistance region
D. A high forward current rating

E6B03
(C)
Page 6-11

E6B03
What special type of diode is capable of both amplification and oscillation?
A. Point contact
B. Zener
C. Tunnel
D. Junction

E6B04
(A)
Page 6-8

E6B04
What type of semiconductor diode varies its internal capacitance as the voltage applied to its terminals varies?
A. Varactor
B. Tunnel
C. Silicon-controlled rectifier
D. Zener

E6B05
(D)
Page 6-8

E6B05
In Figure A6-1, what is the schematic symbol for a varactor diode?
A. 8
B. 6
C. 2
D. 1

E6B06
What is a common use of a hot-carrier diode?
A. As balanced mixers in FM generation
B. As a variable capacitance in an automatic frequency control circuit
C. As a constant voltage reference in a power supply
D. As VHF and UHF mixers and detectors

E6B07
What limits the maximum forward current in a junction diode?
A. Peak inverse voltage
B. Junction temperature
C. Forward voltage
D. Back EMF

E6B08
Structurally, what are the two main categories of semiconductor diodes?
A. Junction and point contact
B. Electrolytic and junction
C. Electrolytic and point contact
D. Vacuum and point contact

E6B09
What is a common use for point contact diodes?
A. As a constant current source
B. As a constant voltage source
C. As an RF detector
D. As a high voltage rectifier

E6B10
In Figure A6-1, what is the schematic symbol for a light-emitting diode?
A. 1
B. 5
C. 6
D. 7

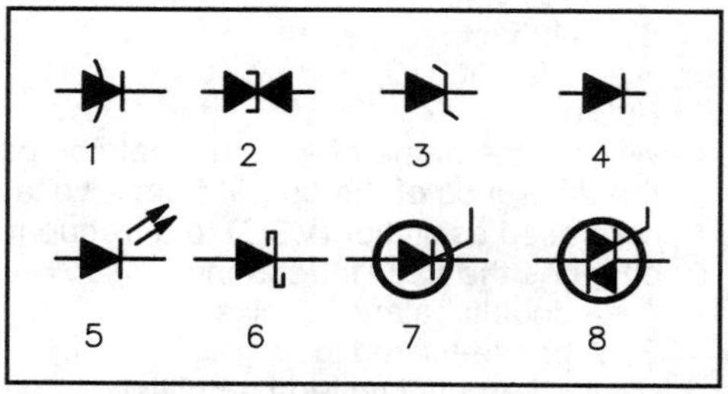

Figure A6-1 — Refer to questions E6B05 and E6B10.

E6B11
What is the phase relationship between the input and output signals of an inverting op-amp circuit?
A. 180 degrees out of phase
B. In phase
C. 90 degrees out of phase
D. 60 degrees out of phase

E6B12
What voltage gain can be expected from the circuit in Figure E7-1 when R1 is 10 ohms and RF is 47 kilohms?
A. 0.00021
B. 9400
C. 4700
D. 2350

E6B06	(D) Page 6-13
E6B07	(B) Page 6-7
E6B08	(A) Page 6-6
E6B09	(C) Page 6-13
E6B10	(B) Page 6-12
E6B11	(A) Page 6-21
E6B12	(C) Page 6-24

E6B13
(D)
Page 6-21

E6B13
How does the gain of a theoretically ideal operational amplifier vary with frequency?
A. It increases linearly with increasing frequency
B. It decreases linearly with increasing frequency
C. It decreases logarithmically with increasing frequency
D. It does not vary with frequency

E6B14
(A)
Page 6-19

E6B14
What essentially determines the output impedance of a FET common-source amplifier?
A. The drain resistor
B. The input impedance of the FET
C. The drain supply voltage
D. The gate supply voltage

E6B15
(D)
Page 6-24

E6B15
What will be the voltage of the circuit shown in Figure E7-1 if R1 is 1000 ohms and RF is 10,000 ohms and 2.3 volts is applied to the input?
A. 2.3 volts
B. 23 volts
C. −2.3 volts
D. −23 volts

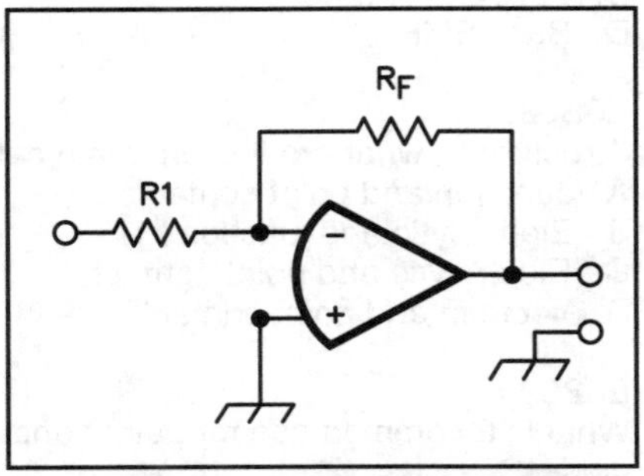

Figure E7-1 — Refer to questions E6B12 and E6B15.

E6B16
(B)
Page 6-24

E6B16
What is the name of a circuit that compares the difference of the output from a voltage-controlled oscillator (VCO) to a frequency standard and produces an error voltage that changes the VCO's frequency?
A. A doubly balanced mixer
B. A phase-locked loop
C. A differential voltage amplifier
D. A variable frequency oscillator

E6B17
(A)
Page 6-26

E6B17
What is the capture range of a phase-locked loop circuit?
A. The frequency range over which the circuit can lock
B. The voltage range over which the circuit can lock
C. The input impedance range over which the circuit can lock
D. The range of time it takes the circuit to lock

E6C TTL digital integrated circuits; CMOS digital integrated circuits; gates

E6C01
(C)
Page 6-28

E6C01
What is the recommended power supply voltage for TTL series integrated circuits?
A. 12 volts
B. 1.5 volts
C. 5 volts
D. 13.6 volts

E6C02
What logic state do the inputs of a TTL device assume if they are left open?
A. A high-logic state
B. A low-logic state
C. The device becomes randomized and will not provide consistent high or low-logic states
D. Open inputs on a TTL device are ignored

E6C03
What level of input voltage is high in a TTL device operating with a 5-volt power supply?
A. 2.0 to 5.5 volts
B. 1.5 to 3.0 volts
C. 1.0 to 1.5 volts
D. −5.0 to −2.0 volts

E6C04
What level of input voltage is low in a TTL device operating with a 5-volt power-supply?
A. −2.0 to −5.5 volts
B. 2.0 to 5.5 volts
C. 0.0 to 0.8 volts
D. −0.8 to 0.4 volts

E6C05
What is one major advantage of CMOS over other devices?
A. Small size
B. Low power consumption
C. Low cost
D. Ease of circuit design

E6C06
Why do CMOS digital integrated circuits have high immunity to noise on the input signal or power supply?
A. Larger bypass capacitors are used in CMOS circuit design
B. The input switching threshold is about two times the power supply voltage
C. The input switching threshold is about one-half the power supply voltage
D. Input signals are stronger

E6C07
In Figure E6-2, what is the schematic symbol for an AND gate?
A. 1
B. 2
C. 3
D. 4

E6C08
In Figure E6-2, what is the schematic symbol for a NAND gate?
A. 1
B. 2
C. 3
D. 4

E6C02
(A)
Page 6-28

E6C03
(A)
Page 6-28

E6C04
(C)
Page 6-28

E6C05
(B)
Page 6-29

E6C06
(C)
Page 6-29

E6C07
(A)
Page 6-28

E6C08
(B)
Page 6-28

E6C09
(B)
Page 6-28

E6C09
In Figure E6-2, what is the schematic symbol for an OR gate?
A. 2
B. 3
C. 4
D. 6

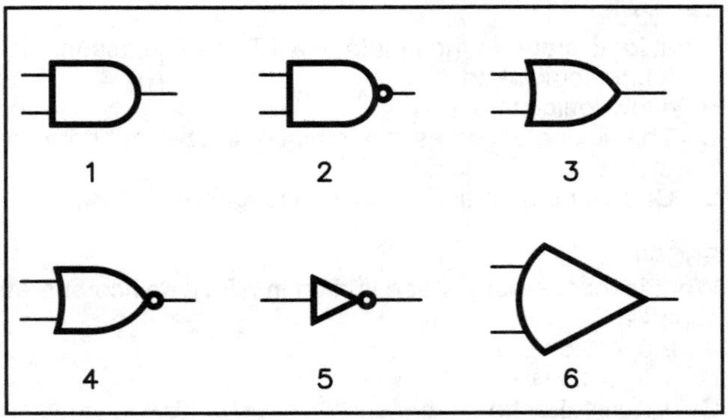

Figure E6-2 — Refer to questions E6C07 through E6C11.

E6C10
(D)
Page 6-28

E6C10
In Figure E6-2, what is the schematic symbol for a NOR gate?
A. 1
B. 2
C. 3
D. 4

E6C11
(C)
Page 6-28

E6C11
In Figure E6-2, what is the schematic symbol for a NOT gate?
A. 2
B. 4
C. 5
D. 6

E6D Vidicon and cathode-ray tube devices; charge-coupled devices (CCDs); liquid crystal displays (LCDs); toroids: permeability, core material, selecting, winding

E6D01
(D)
Page 6-30

E6D01
How is the electron beam deflected in a vidicon?
A. By varying the beam voltage
B. By varying the bias voltage on the beam forming grids inside the tube
C. By varying the beam current
D. By varying electromagnetic fields

E6D02
(D)
Page 6-31

E6D02
What is cathode ray tube (CRT) persistence?
A. The time it takes for an image to appear after the electron beam is turned on
B. The relative brightness of the display under varying conditions of ambient light
C. The ability of the display to remain in focus under varying conditions
D. The length of time the image remains on the screen after the beam is turned off

E6D03
(A)
Page 6-31

E6D03
If a cathode ray tube (CRT) is designed to operate with an anode voltage of 25,000 volts, what will happen if the anode voltage is increased to 35,000 volts?
A. The image size will decrease and the tube will produce X-rays
B. The image size will increase and the tube will produce X-rays
C. The image will become larger and brighter
D. There will be no apparent change

E6D04

Exceeding what design rating can cause a cathode ray tube (CRT) to generate X-rays?

A. The heater voltage
B. The anode voltage
C. The operating temperature
D. The operating frequency

E6D05

Which of the following is true of a charge-coupled device (CCD)?

A. Its phase shift changes rapidly with frequency
B. It is a CMOS analog-to-digital converter
C. It samples an analog signal and passes it in stages from the input to the output
D. It is used in a battery charger circuit

E6D06

What function does a charge-coupled device (CCD) serve in a modern video camera?

A. It stores photogenerated charges as signals corresponding to pixels
B. It generates the horizontal pulses needed for electron beam scanning
C. It focuses the light used to produce a pattern of electrical charges corresponding to the image
D. It combines audio and video information to produce a composite RF signal

E6D07

What is a liquid-crystal display (LCD)?

A. A modern replacement for a quartz crystal oscillator which displays its fundamental frequency
B. A display that uses a crystalline liquid to change the way light is refracted
C. A frequency-determining unit for a transmitter or receiver
D. A display that uses a glowing liquid to remain brightly lit in dim light

E6D08

What material property determines the inductance of a toroidal inductor with a 10-turn winding?

A. Core load current
B. Core resistance
C. Core reactivity
D. Core permeability

E6D09

By careful selection of core material, over what frequency range can toroidal cores produce useful inductors?

A. From a few kHz to no more than several MHz
B. From DC to at least 1000 MHz
C. From DC to no more than 3000 kHz
D. From a few hundred MHz to at least 1000 GHz

E6D10

What materials are used to make ferromagnetic inductors and transformers?

A. Ferrite and powdered-iron toroids
B. Silicon-ferrite toroids and shellac
C. Powdered-ferrite and silicon toroids
D. Ferrite and silicon-epoxy toroids

E6D04
(B)
Page 6-31

E6D05
(C)
Page 6-32

E6D06
(A)
Page 6-32

E6D07
(B)
Page 6-32

E6D08
(D)
Page 6-33

E6D09
(B)
Page 6-33

E6D10
(A)
Page 6-33

E6D11
(B)
Page 6-33

E6D11
What is one important reason for using powdered-iron toroids rather than ferrite toroids in an inductor?
A. Powdered-iron toroids generally have greater initial permeabilities
B. Powdered-iron toroids generally have better temperature stability
C. Powdered-iron toroids generally require fewer turns to produce a given inductance value
D. Powdered-iron toroids are easier to use with surface-mount technology

E6D12
(B)
Page 6-36

E6D12
What would be a good choice of toroid core material to make a common-mode choke (such as winding telephone wires or stereo speaker leads on a core) to cure an HF RFI problem?
A. Type 61 mix ferrite (initial permeability of 125)
B. Type 43 mix ferrite (initial permeability of 850)
C. Type 6 mix powdered iron (initial permeability of 8)
D. Type 12 mix powdered iron (initial permeability of 3)

E6D13
(C)
Page 6-36

E6D13
What devices are commonly used as parasitic suppressors at the input and output terminals of VHF and UHF amplifiers?
A. Electrolytic capacitors
B. Butterworth filters
C. Ferrite beads
D. Steel-core toroids

E6D14
(A)
Page 6-33

E6D14
What is a primary advantage of using a toroidal core instead of a linear core in an inductor?
A. Toroidal cores contain most of the magnetic field within the core material
B. Toroidal cores make it easier to couple the magnetic energy into other components
C. Toroidal cores exhibit greater hysteresis
D. Toroidal cores have lower Q characteristics

E6D15
(C)
Page 6-35

E6D15
How many turns will be required to produce a 1-mH inductor using a ferrite toroidal core that has an inductance index (A_L) value of 523?
A. 2 turns
B. 4 turns
C. 43 turns
D. 229 turns

E6D16
(A)
Page 6-35

E6D16
How many turns will be required to produce a 5-microhenry inductor using a powdered-iron toroidal core that has an inductance index (A_L) value of 40?
A. 35 turns
B. 13 turns
C. 79 turns
D. 141 turns

E6E Quartz crystal (frequency determining properties as used in oscillators and filters); monolithic amplifiers (MMICs)

E6E01
For single-sideband phone emissions, what would be the bandwidth of a good crystal lattice band-pass filter?
A. 6 kHz at –6 dB
B. 2.1 kHz at –6 dB
C. 500 Hz at –6 dB
D. 15 kHz at –6 dB

E6E02
For double-sideband phone emissions, what would be the bandwidth of a good crystal lattice band-pass filter?
A. 1 kHz at –6 dB
B. 500 Hz at –6 dB
C. 6 kHz at –6 dB
D. 15 kHz at –6 dB

E6E03
What is a crystal lattice filter?
A. A power supply filter made with interlaced quartz crystals
B. An audio filter made with four quartz crystals that resonate at 1-kHz intervals
C. A filter with wide bandwidth and shallow skirts made using quartz crystals
D. A filter with narrow bandwidth and steep skirts made using quartz crystals

E6E04
What technique is used to construct low-cost, high-performance crystal filters?
A. Choose a center frequency that matches the available crystals
B. Choose a crystal with the desired bandwidth and operating frequency to match a desired center frequency
C. Measure crystal bandwidth to ensure at least 20% coupling
D. Measure crystal frequencies and carefully select units with less than 10% frequency difference

E6E05
Which factor helps determine the bandwidth and response shape of a crystal filter?
A. The relative frequencies of the individual crystals
B. The center frequency chosen for the filter
C. The gain of the RF stage preceding the filter
D. The amplitude of the signals passing through the filter

E6E06
What is the piezoelectric effect?
A. Physical deformation of a crystal by the application of a voltage
B. Mechanical deformation of a crystal by the application of a magnetic field
C. The generation of electrical energy by the application of light
D. Reversed conduction states when a P-N junction is exposed to light

E6E01
(B)
Page 6-37

E6E02
(C)
Page 6-37

E6E03
(D)
Page 6-37

E6E04
(D)
Page 6-38

E6E05
(A)
Page 6-37

E6E06
(A)
Page 6-37

E6E07
(C)
Page 6-39

E6E07
Which of the following devices would be most suitable for constructing a receive preamplifier for 1296 MHz?
A. A 2N2222 bipolar transistor
B. An MRF901 bipolar transistor
C. An MSA-0135 monolithic microwave integrated circuit (MMIC)
D. An MPF102 N-junction field-effect transistor (JFET)

E6E08
(A)
Page 6-39

E6E08
Which device might be used to simplify the design and construction of a 3456-MHz receiver?
A. An MSA-0735 monolithic microwave integrated circuit (MMIC).
B. An MRF901 bipolar transistor
C. An MGF1402 gallium arsenide field-effect transistor (GaAsFET)
D. An MPF102 N-junction field-effect transistor (JFET)

E6E09
(D)
Page 6-38

E6E09
What type of amplifier device consists of a small "pill sized" package with an input lead, an output lead and 2 ground leads?
A. A gallium arsenide field-effect transistor (GaAsFET)
B. An operational amplifier integrated circuit (OAIC)
C. An indium arsenide integrated circuit (IAIC)
D. A monolithic microwave integrated circuit (MMIC)

E6E10
(B)
Page 6-39

E6E10
What typical construction technique do amateurs use when building an amplifier containing a monolithic microwave integrated circuit (MMIC)?
A. Ground-plane "ugly" construction
B. Microstrip construction
C. Point-to-point construction
D. Wave-soldering construction

E6E11
(A)
Page 6-39

E6E11
How is the operating bias voltage supplied to a monolithic microwave integrated circuit (MMIC)?
A. Through a resistor and RF choke connected to the amplifier output lead
B. MMICs require no operating bias
C. Through a capacitor and RF choke connected to the amplifier input lead
D. Directly to the bias-voltage (VCC IN) lead

Subelement E7 — Practical Circuits

[7 Exam Questions — 7 Groups]

E7A Digital logic circuits: Flip flops; Astable and monostable multivibrators; Gates (AND, NAND, OR, NOR); Positive and negative logic

E7A01
What is a bistable multivibrator circuit?
A. An "AND" gate
B. An "OR" gate
C. A flip-flop
D. A clock

E7A02
How many output level changes are obtained for every two trigger pulses applied to the input of a "T" flip-flop circuit?
A. None
B. One
C. Two
D. Four

E7A03
The frequency of an AC signal can be divided electronically by what type of digital circuit?
A. A free-running multivibrator
B. A bistable multivibrator
C. An OR gate
D. An astable multivibrator

E7A04
How many flip-flops are required to divide a signal frequency by 4?
A. 1
B. 2
C. 4
D. 8

E7A05
What is the characteristic function of an astable multivibrator?
A. It alternates between two stable states
B. It alternates between a stable state and an unstable state
C. It blocks either a 0 pulse or a 1 pulse and passes the other
D. It alternates between two unstable states

E7A06
What is the characteristic function of a monostable multivibrator?
A. It switches momentarily to the opposite binary state and then returns after a set time to its original state
B. It is a "clock" that produces a continuous square wave oscillating between 1 and 0
C. It stores one bit of data in either a 0 or 1 state
D. It maintains a constant output voltage, regardless of variations in the input voltage

E7A07
(B)
Page 7-3

E7A07
What logical operation does an AND gate perform?
A. It produces a logic "0" at its output only if all inputs are logic "1"
B. It produces a logic "1" at its output only if all inputs are logic "1"
C. It produces a logic "1" at its output if only one input is a logic "1"
D. It produces a logic "1" at its output if all inputs are logic "0"

E7A08
(D)
Page 7-5

E7A08
What logical operation does a NAND gate perform?
A. It produces a logic "0" at its output only when all inputs are logic "0"
B. It produces a logic "1" at its output only when all inputs are logic "1"
C. It produces a logic "0" at its output if some but not all of its inputs are logic "1"
D. It produces a logic "0" at its output only when all inputs are logic "1"

E7A09
(A)
Page 7-4

E7A09
What logical operation does an OR gate perform?
A. It produces a logic "1" at its output if any input is or all inputs are logic "1"
B. It produces a logic "0" at its output if all inputs are logic "1"
C. It only produces a logic "0" at its output when all inputs are logic "1"
D. It produces a logic "1" at its output if all inputs are logic "0"

E7A10
(C)
Page 7-5

E7A10
What logical operation does a NOR gate perform?
A. It produces a logic "0" at its output only if all inputs are logic "0"
B. It produces a logic "1" at its output only if all inputs are logic "1"
C. It produces a logic "0" at its output if any input is or all inputs are logic "1"
D. It produces a logic "1" at its output only when none of its inputs are logic "0"

E7A11
(C)
Page 7-2

E7A11
What is a truth table?
A. A table of logic symbols that indicate the high logic states of an op-amp
B. A diagram showing logic states when the digital device's output is true
C. A list of input combinations and their corresponding outputs that characterize the function of a digital device
D. A table of logic symbols that indicates the low logic states of an op-amp

E7A12
(D)
Page 7-5

E7A12
In a positive-logic circuit, what level is used to represent a logic 1?
A. A low level
B. A positive-transition level
C. A negative-transition level
D. A high level

E7A13
(A)
Page 7-5

E7A13
In a negative-logic circuit, what level is used to represent a logic 1?
A. A low level
B. A positive-transition level
C. A negative-transition level
D. A high level

E7B Amplifier circuits: Class A, Class AB, Class B, Class C, amplifier operating efficiency (ie, DC input versus PEP), transmitter final amplifiers; amplifier circuits: tube, bipolar transistor, FET

E7B01
For what portion of a signal cycle does a Class AB amplifier operate?
A. More than 180 degrees but less than 360 degrees
B. Exactly 180 degrees
C. The entire cycle
D. Less than 180 degrees

E7B02
Which class of amplifier provides the highest efficiency?
A. Class A
B. Class B
C. Class C
D. Class AB

E7B03
Where on the load line should a solid-state power amplifier be operated for best efficiency and stability?
A. Just below the saturation point
B. Just above the saturation point
C. At the saturation point
D. At 1.414 times the saturation point

E7B04
How can parasitic oscillations be eliminated from a power amplifier?
A. By tuning for maximum SWR
B. By tuning for maximum power output
C. By neutralization
D. By tuning the output

E7B05
How can even-order harmonics be reduced or prevented in transmitter amplifiers?
A. By using a push-push amplifier
B. By using a push-pull amplifier
C. By operating Class C
D. By operating Class AB

E7B06
What can occur when a nonlinear amplifier is used with a single-sideband phone transmitter?
A. Reduced amplifier efficiency
B. Increased intelligibility
C. Sideband inversion
D. Distortion

E7B01
(A)
Page 7-18

E7B02
(C)
Page 7-18

E7B03
(A)
Page 7-24

E7B04
(C)
Page 7-25

E7B05
(B)
Page 7-20

E7B06
(D)
Page 7-17

E7B07
(C)
Page 7-26

E7B07
How can a vacuum-tube power amplifier be neutralized?
A. By increasing the grid drive
B. By feeding back an in-phase component of the output to the input
C. By feeding back an out-of-phase component of the output to the input
D. By feeding back an out-of-phase component of the input to the output

E7B08
(B)
Page 7-19

E7B08
What tank-circuit Q is required to reduce harmonics to an acceptable level?
A. Approximately 120
B. Approximately 12
C. Approximately 1200
D. Approximately 1.2

E7B09
(B)
Page 7-23

E7B09
In Figure A7-1, what is the purpose of R1 and R2?
A. Load resistors
B. Fixed bias
C. Self bias
D. Feedback

E7B10
(D)
Page 7-23

E7B10
In Figure A7-1, what is the purpose of C3?
A. AC feedback
B. Input coupling
C. Power supply decoupling
D. Emitter bypass

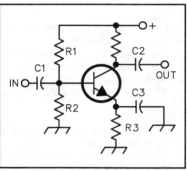

E7B11
(D)
Page 7-23

E7B11
In Figure A7-1, what is the purpose of R3?
A. Fixed bias
B. Emitter bypass
C. Output load resistor
D. Self bias

Figure A7-1 — Refer to questions E7B09, E7B10 and E7B11.

E7B12
(A)
Page 7-25

E7B12
In Figure A7-2, what is the purpose of R?
A. Emitter load
B. Fixed bias
C. Collector load
D. Voltage regulation

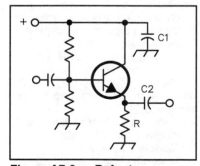

E7B13
(A)
Page 7-25

E7B13
In Figure A7-2, what is the purpose of C2?
A. Output coupling
B. Emitter bypass
C. Input coupling
D. Hum filtering

Figure A7-2 — Refer to questions E7B12 and E7B13.

E7B14
What is the purpose of D1 in the
circuit shown in Figure A7-3?
A. Line voltage stabilization
B. Voltage reference
C. Peak clipping
D. Hum filtering

E7B15
What is the purpose of Q1 in the
circuit shown in Figure A7-3?
A. It increases the output ripple
B. It provides a constant load for the
   voltage source
C. It increases the current-handling
   capability
D. It provides D1 with current

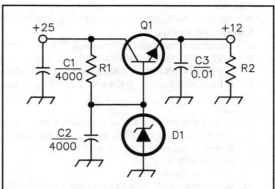

Figure A7-3 — Refer to questions E7B14,
E7B15 and E7B16.

E7B16
What is the purpose of C2 in the circuit shown in Figure A7-3?
A. It bypasses hum around D1
B. It is a brute force filter for the output
C. To self resonate at the hum frequency
D. To provide fixed DC bias for Q1

E7B14
(B)
Page 7-36

E7B15
(C)
Page 7-36

E7B16
(A)
Page 7-36

**E7C Impedance-matching networks: Pi, L, Pi-L; filter circuits: constant K, M-
derived, band-stop, notch, crystal lattice, pi-section, T-section, L-section,
Butterworth, Chebyshev, elliptical; filter applications (audio, IF, digital signal
processing {DSP})**

E7C01
How are the capacitors and inductors of a low-pass filter pi-network arranged between
the network's input and output?
A. Two inductors are in series between the input and output and a capacitor is
   connected between the two inductors and ground
B. Two capacitors are in series between the input and output and an inductor is
   connected between the two capacitors and ground
C. An inductor is in parallel with the input, another inductor is in parallel with the
   output, and a capacitor is in series between the two
D. A capacitor is in parallel with the input, another capacitor is in parallel with the
   output, and an inductor is in series between the two

E7C02
What is an L-network?
A. A network consisting entirely of four inductors
B. A network consisting of an inductor and a capacitor
C. A network used to generate a leading phase angle
D. A network used to generate a lagging phase angle

E7C01
(D)
Page 7-27

E7C02
(B)
Page 7-28

E7C03
(C)
Page 7-30

E7C03
A T-network with series capacitors and a parallel (shunt) inductor has which of the following properties?
A. It transforms impedances and is a low-pass filter
B. It transforms reactances and is a low-pass filter
C. It transforms impedances and is a high-pass filter
D. It transforms reactances and is a high-pass filter

E7C04
(A)
Page 7-30

E7C04
What advantage does a pi-L-network have over a pi-network for impedance matching between the final amplifier of a vacuum-tube type transmitter and a multiband antenna?
A. Greater harmonic suppression
B. Higher efficiency
C. Lower losses
D. Greater transformation range

E7C05
(C)
Page 7-28

E7C05
How does a network transform one impedance to another?
A. It introduces negative resistance to cancel the resistive part of an impedance
B. It introduces transconductance to cancel the reactive part of an impedance
C. It cancels the reactive part of an impedance and changes the resistive part
D. Network resistances substitute for load resistances

E7C06
(B)
Page 5-46

E7C06
What value capacitor would be required to tune a 20-microhenry inductor to resonate in the 80-meter band?
A. 150 picofarads
B. 100 picofarads
C. 200 picofarads
D. 100 microfarads

E7C07
(D)
Page 7-31

E7C07
Which filter type is described as having ripple in the passband and a sharp cutoff?
A. A Butterworth filter
B. An active LC filter
C. A passive op-amp filter
D. A Chebyshev filter

E7C08
(C)
Page 7-31

E7C08
What are the distinguishing features of an elliptical filter?
A. Gradual passband rolloff with minimal stop-band ripple
B. Extremely flat response over its passband, with gradually rounded stop-band corners
C. Extremely sharp cutoff, with one or more infinitely deep notches in the stop band
D. Gradual passband rolloff with extreme stop-band ripple

E7C09
What kind of audio filter would you use to attenuate an interfering carrier signal while receiving an SSB transmission?
A. A band-pass filter
B. A notch filter
C. A pi-network filter
D. An all-pass filter

E7C09
(B)
Page 7-32

E7C10
What characteristic do typical SSB receiver IF filters lack that is important to digital communications?
A. Steep amplitude-response skirts
B. Passband ripple
C. High input impedance
D. Linear phase response

E7C10
(D)
Page 7-34

E7C11
What kind of digital signal processing audio filter might be used to remove unwanted noise from a received SSB signal?
A. An adaptive filter
B. A notch filter
C. A Hilbert-transform filter
D. A phase-inverting filter

E7C11
(A)
Page 7-34

E7C12
What kind of digital signal processing filter might be used in generating an SSB signal?
A. An adaptive filter
B. A notch filter
C. A Hilbert-transform filter
D. An elliptical filter

E7C12
(C)
Page 7-34

E7C13
Which type of filter would be the best to use in a 2-meter repeater duplexer?
A. A crystal filter
B. A cavity filter
C. A DSP filter
D. An L-C filter

E7C13
(B)
Page 7-30

E7D Oscillators: types, applications, stability; voltage-regulator circuits: discrete, integrated and switched mode

E7D01
What are three major oscillator circuits often used in Amateur Radio equipment?
A. Taft, Pierce and negative feedback
B. Colpitts, Hartley and Taft
C. Taft, Hartley and Pierce
D. Colpitts, Hartley and Pierce

E7D01
(D)
Page 7-42

E7D02
(C)
Page 7-42

E7D02
What condition must exist for a circuit to oscillate?
A. It must have a gain of less than 1
B. It must be neutralized
C. It must have positive feedback sufficient to overcome losses
D. It must have negative feedback sufficient to cancel the input

E7D03
(A)
Page 7-42

E7D03
How is the positive feedback coupled to the input in a Hartley oscillator?
A. Through a tapped coil
B. Through a capacitive divider
C. Through link coupling
D. Through a neutralizing capacitor

E7D04
(C)
Page 7-42

E7D04
How is the positive feedback coupled to the input in a Colpitts oscillator?
A. Through a tapped coil
B. Through link coupling
C. Through a capacitive divider
D. Through a neutralizing capacitor

E7D05
(D)
Page 7-44

E7D05
How is the positive feedback coupled to the input in a Pierce oscillator?
A. Through a tapped coil
B. Through link coupling
C. Through a neutralizing capacitor
D. Through capacitive coupling

E7D06
(B)
Page 7-45

E7D06
Which type of oscillator circuits are commonly used in a VFO?
A. Pierce and Zener
B. Colpitts and Hartley
C. Armstrong and deForest
D. Negative feedback and Balanced feedback

E7D07
(B)
Page 7-47

E7D07
Why must a very stable reference oscillator be used as part of a phase-locked loop (PLL) frequency synthesizer?
A. Any amplitude variations in the reference oscillator signal will prevent the loop from locking to the desired signal
B. Any phase variations in the reference oscillator signal will produce phase noise in the synthesizer output
C. Any phase variations in the reference oscillator signal will produce harmonic distortion in the modulating signal
D. Any amplitude variations in the reference oscillator signal will prevent the loop from changing frequency

E7D08
What is one characteristic of a linear electronic voltage regulator?
A. It has a ramp voltage as its output
B. The pass transistor switches from the "off" state to the "on" state
C. The control device is switched on or off, with the duty cycle proportional to the line or load conditions
D. The conduction of a control element is varied in direct proportion to the line voltage or load current

E7D08
(D)
Page 7-34

E7D09
What is one characteristic of a switching electronic voltage regulator?
A. The conduction of a control element is varied in direct proportion to the line voltage or load current
B. It provides more than one output voltage
C. The control device is switched on or off, with the duty cycle proportional to the line or load conditions
D. It gives a ramp voltage at its output

E7D09
(C)
Page 7-35

E7D10
What device is typically used as a stable reference voltage in a linear voltage regulator?
A. A Zener diode
B. A tunnel diode
C. An SCR
D. A varactor diode

E7D10
(A)
Page 7-35

E7D11
What type of linear regulator is used in applications requiring efficient utilization of the primary power source?
A. A constant current source
B. A series regulator
C. A shunt regulator
D. A shunt current source

E7D11
(B)
Page 7-38

E7D12
What type of linear voltage regulator is used in applications requiring a constant load on the unregulated voltage source?
A. A constant current source
B. A series regulator
C. A shunt current source
D. A shunt regulator

E7D12
(D)
Page 7-38

E7D13
To obtain the best temperature stability, approximately what operating voltage should be used for the reference diode in a linear voltage regulator?
A. 2 volts
B. 3 volts
C. 6 volts
D. 10 volts

E7D13
(C)
Page 7-36

E7D14 (B) Page 7-41	E7D14 What are the important characteristics of a three-terminal regulator? A. Maximum and minimum input voltage, minimum output current and voltage B. Maximum and minimum input voltage, maximum output current and voltage C. Maximum and minimum input voltage, minimum output current and maximum output voltage D. Maximum and minimum input voltage, minimum output voltage and maximum output current
E7D15 (A) Page 7-35	E7D15 What type of voltage regulator limits the voltage drop across its junction when a specified current passes through it in the reverse-breakdown direction? A. A Zener diode B. A three-terminal regulator C. A bipolar regulator D. A pass-transistor regulator
E7D16 (C) Page 7-40	E7D16 What type of voltage regulator contains a voltage reference, error amplifier, sensing resistors and transistors, and a pass element in one package? A. A switching regulator B. A Zener regulator C. A three-terminal regulator D. An op-amp regulator

E7E Modulators: reactance, phase, balanced; detectors; mixer stages; frequency synthesizers

E7E01 (B) Page 7-61	E7E01 How is an F3E FM-phone emission produced? A. With a balanced modulator on the audio amplifier B. With a reactance modulator on the oscillator C. With a reactance modulator on the final amplifier D. With a balanced modulator on the oscillator
E7E02 (C) Page 7-61	E7E02 How does a reactance modulator work? A. It acts as a variable resistance or capacitance to produce FM signals B. It acts as a variable resistance or capacitance to produce AM signals C. It acts as a variable inductance or capacitance to produce FM signals D. It acts as a variable inductance or capacitance to produce AM signals
E7E03 (C) Page 7-64	E7E03 How does a phase modulator work? A. It varies the tuning of a microphone preamplifier to produce FM signals B. It varies the tuning of an amplifier tank circuit to produce AM signals C. It varies the tuning of an amplifier tank circuit to produce FM signals D. It varies the tuning of a microphone preamplifier to produce AM signals

E7E04
How can a single-sideband phone signal be generated?
A. By using a balanced modulator followed by a filter
B. By using a reactance modulator followed by a mixer
C. By using a loop modulator followed by a mixer
D. By driving a product detector with a DSB signal

E7E05
What audio shaping network is added at a transmitter to proportionally attenuate the lower audio frequencies, giving an even spread to the energy in the audio band?
A. A de-emphasis network
B. A heterodyne suppressor
C. An audio prescaler
D. A pre-emphasis network

E7E06
What audio shaping network is added at a receiver to restore proportionally attenuated lower audio frequencies?
A. A de-emphasis network
B. A heterodyne suppressor
C. An audio prescaler
D. A pre-emphasis network

E7E07
What is the mixing process?
A. The elimination of noise in a wideband receiver by phase comparison
B. The elimination of noise in a wideband receiver by phase differentiation
C. The recovery of the intelligence from a modulated RF signal
D. The combination of two signals to produce sum and difference frequencies

E7E08
What are the principal frequencies that appear at the output of a mixer circuit?
A. Two and four times the original frequency
B. The sum, difference and square root of the input frequencies
C. The original frequencies and the sum and difference frequencies
D. 1.414 and 0.707 times the input frequency

E7E09
What occurs in a receiver when an excessive amount of signal energy reaches the mixer circuit?
A. Spurious mixer products are generated
B. Mixer blanking occurs
C. Automatic limiting occurs
D. A beat frequency is generated

E7E10
What type of frequency synthesizer circuit uses a stable voltage-controlled oscillator, programmable divider, phase detector, loop filter and a reference frequency source?
A. A direct digital synthesizer
B. A hybrid synthesizer
C. A phase-locked loop synthesizer
D. A diode-switching matrix synthesizer

E7E04	(A) Page 7-56
E7E05	(D) Page 7-65
E7E06	(A) Page 7-65
E7E07	(D) Page 7-50
E7E08	(C) Page 7-50
E7E09	(A) Page 7-50
E7E10	(C) Page 7-47

E7E11
(A)
Page 7-48

E7E11
What type of frequency synthesizer circuit uses a phase accumulator, lookup table, digital to analog converter and a low-pass antialias filter?
A. A direct digital synthesizer
B. A hybrid synthesizer
C. A phase-locked loop synthesizer
D. A diode-switching matrix synthesizer

E7E12
(D)
Page 7-48

E7E12
What are the main blocks of a direct digital frequency synthesizer?
A. A variable-frequency crystal oscillator, phase accumulator, digital to analog converter and a loop filter
B. A stable voltage-controlled oscillator, programmable divider, phase detector, loop filter and a digital to analog converter
C. A variable-frequency oscillator, programmable divider, phase detector and a low-pass antialias filter
D. A phase accumulator, lookup table, digital to analog converter and a low-pass antialias filter

E7E13
(B)
Page 7-48

E7E13
What information is contained in the lookup table of a direct digital frequency synthesizer?
A. The phase relationship between a reference oscillator and the output waveform
B. The amplitude values that represent a sine-wave output
C. The phase relationship between a voltage-controlled oscillator and the output waveform
D. The synthesizer frequency limits and frequency values stored in the radio memories

E7E14
(C)
Page 7-49

E7E14
What are the major spectral impurity components of direct digital synthesizers?
A. Broadband noise
B. Digital conversion noise
C. Spurs at discrete frequencies
D. Nyquist limit noise

E7E15
(A)
Page 7-47

E7E15
What are the major spectral impurity components of phase-locked loop synthesizers?
A. Broadband noise
B. Digital conversion noise
C. Spurs at discrete frequencies
D. Nyquist limit noise

E7F Digital frequency divider circuits; frequency marker generators; frequency counters

E7F01
What is the purpose of a prescaler circuit?
A. It converts the output of a JK flip-flop to that of an RS flip-flop
B. It multiplies an HF signal so a low-frequency counter can display the operating frequency
C. It prevents oscillation in a low-frequency counter circuit
D. It divides an HF signal so a low-frequency counter can display the operating frequency

E7F02
How many states does a decade counter digital IC have?
A. 2
B. 10
C. 20
D. 100

E7F03
What is the function of a decade counter digital IC?
A. It produces one output pulse for every ten input pulses
B. It decodes a decimal number for display on a seven-segment LED display
C. It produces ten output pulses for every input pulse
D. It adds two decimal numbers

E7F04
What additional circuitry is required in a 100-kHz crystal-controlled marker generator to provide markers at 50 and 25 kHz?
A. An emitter-follower
B. Two frequency multipliers
C. Two flip-flops
D. A voltage divider

E7F05
If a 1-MHz oscillator is used with a divide-by-ten circuit to make a marker generator, what will the output be?
A. A 1-MHz sinusoidal signal with harmonics every 100 kHz
B. A 100-kHz signal with harmonics every 100 kHz
C. A 1-MHz square wave with harmonics every 1 MHz
D. A 100-kHz signal modulated by a 10-kHz signal

E7F06
What is a crystal-controlled marker generator?
A. A low-stability oscillator that "sweeps" through a ban of frequencies
B. An oscillator often used in aircraft to determine the craft's location relative to the inner and outer markers at airports
C. A high-stability oscillator whose output frequency and amplitude can be varied over a wide range
D. A high-stability oscillator that generates a series of reference signals at known frequency intervals

E7F01	(D) Page 7-13
E7F02	(B) Page 7-11
E7F03	(A) Page 7-11
E7F04	(C) Page 7-12
E7F05	(B) Page 7-13
E7F06	(D) Page 7-11

E7F07
(A)
Page 7-13

E7F07
What type of circuit does NOT make a good marker generator?
A. A sinusoidal crystal oscillator
B. A crystal oscillator followed by a class C amplifier
C. A TTL device wired as a crystal oscillator
D. A crystal oscillator and a frequency divider

E7F08
(C)
Page 7-11

E7F08
What is the purpose of a marker generator?
A. To add audio markers to an oscilloscope
B. To provide a frequency reference for a phase locked loop
C. To provide a means of calibrating a receiver's frequency settings
D. To add time signals to a transmitted signal

E7F09
(A)
Page 7-13

E7F09
What does the accuracy of a frequency counter depend on?
A. The internal crystal reference
B. A voltage-regulated power supply with an unvarying output
C. Accuracy of the AC input frequency to the power supply
D. Proper balancing of the power-supply diodes

E7F10
(C)
Page 7-13

E7F10
How does a frequency counter determine the frequency of a signal?
A. It counts the total number of pulses in a circuit
B. It monitors a WWV reference signal for comparison with the measured signal
C. It counts the number of input pulses in a specific period of time
D. It converts the phase of the measured signal to a voltage which is proportional to the frequency

E7F11
(A)
Page 7-13

E7F11
What is the purpose of a frequency counter?
A. To indicate the frequency of the strongest input signal which is within the counter's frequency range
B. To generate a series of reference signals at known frequency intervals
C. To display all frequency components of a transmitted signal
D. To compare the difference between the input and a voltage-controlled oscillator and produce an error voltage

E7G Active audio filters: characteristics; basic circuit design; preselector applications

E7G01
(B)
Page 7-14

E7G01
What determines the gain and frequency characteristics of an op-amp RC active filter?
A. The values of capacitances and resistances built into the op-amp
B. The values of capacitances and resistances external to the op-amp
C. The input voltage and frequency of the op-amp's DC power supply
D. The output voltage and smoothness of the op-amp's DC power supply

E7G02
What causes ringing in a filter?
A. The slew rate of the filter
B. The bandwidth of the filter
C. The filter shape, as measured in the frequency domain
D. The gain of the filter

E7G03
What are the advantages of using an op-amp instead of LC elements in an audio filter?
A. Op-amps are more rugged and can withstand more abuse than can LC elements
B. Op-amps are fixed at one frequency
C. Op-amps are available in more varieties than are LC elements
D. Op-amps exhibit gain rather than insertion loss

E7G04
What type of capacitors should be used in an op-amp RC active filter circuit?
A. Electrolytic
B. Disc ceramic
C. Polystyrene
D. Paper dielectric

E7G05
How can unwanted ringing and audio instability be prevented in a multisection op-amp RC audio filter circuit?
A. Restrict both gain and Q
B. Restrict gain, but increase Q
C. Restrict Q, but increase gain
D. Increase both gain and Q

E7G06
What parameter must be selected when designing an audio filter using an op-amp?
A. Bandpass characteristic
B. Desired current gain
C. Temperature coefficient
D. Output-offset overshoot

E7G07
The design of a preselector involves a trade-off between bandwidth and what other factor?
A. The amount of ringing
B. Insertion loss
C. The number of parts
D. The choice of capacitors or inductors

E7G02
(C)
Page 7-15

E7G03
(D)
Page 7-14

E7G04
(C)
Page 7-14

E7G05
(A)
Page 7-15

E7G06
(A)
Page 7-14

E7G07
(B)
Page 7-14

E7G08
When designing an op-amp RC active filter for a given frequency range and Q, what steps are typically followed when selecting the external components?
A. Standard capacitor values are chosen first, the resistances are calculated, then resistors of the nearest standard value are used
B. Standard resistor values are chosen first, the capacitances are calculated, then capacitors of the nearest standard value are used
C. Standard resistor and capacitor values are used, the circuit is tested, then additional resistors are added to make any adjustments
D. Standard resistor and capacitor values are used, the circuit is tested, then additional capacitors are added to make any adjustments

E7G09
When designing an op-amp RC active filter for a given frequency range and Q, why are the external capacitance values usually chosen first, then the external resistance values calculated?
A. An op-amp will perform as an active filter using only standard external capacitance values
B. The calculations are easier to make with known capacitance values rather than with known resistance values
C. Capacitors with unusual capacitance values are not widely available, so standard values are used to begin the calculations
D. The equations for the calculations can only be used with known capacitance values

E7G10
What are the principal uses of an op-amp RC active filter in amateur circuitry?
A. High-pass filters used to block RFI at the input to receivers
B. Low-pass filters used between transmitters and transmission lines
C. Filters used for smoothing power-supply output
D. Audio filters used for receivers

E7G11
Where should an op-amp RC active audio filter be placed in an amateur receiver?
A. In the IF strip, immediately before the detector
B. In the audio circuitry immediately before the speaker or phone jack
C. Between the balanced modulator and frequency multiplier
D. In the low-level audio stages

Subelement E8 — Signals And Emissions

[5 Exam Questions — 5 Groups]

E8A AC waveforms: sine wave, square wave, sawtooth wave; AC measurements: peak, peak-to-peak and root-mean-square (RMS) value, peak-envelope-power (PEP) relative to average

E8A01
Starting at a positive peak, how many times does a sine wave cross the zero axis in one complete cycle?
A. 180 times
B. 4 times
C. 2 times
D. 360 times

E8A01
(C)
Page 8-2

E8A02
What is a wave called that abruptly changes back and forth between two voltage levels and remains an equal time at each level?
A. A sine wave
B. A cosine wave
C. A square wave
D. A sawtooth wave

E8A02
(C)
Page 8-3

E8A03
What sine waves added to a fundamental frequency make up a square wave?
A. A sine wave 0.707 times the fundamental frequency
B. All odd and even harmonics
C. All even harmonics
D. All odd harmonics

E8A03
(D)
Page 8-3

E8A04
What type of wave is made up of a sine wave of a fundamental frequency and all its odd harmonics?
A. A square wave
B. A sine wave
C. A cosine wave
D. A tangent wave

E8A04
(A)
Page 8-3

E8A05
What is a sawtooth wave?
A. A wave that alternates between two values and spends an equal time at each level
B. A wave with a straight line rise time faster than the fall time (or vice versa)
C. A wave that produces a phase angle tangent to the unit circle
D. A wave whose amplitude at any given instant can be represented by a point on a wheel rotating at a uniform speed

E8A05
(B)
Page 8-2

E8A06
What type of wave has a rise time significantly faster than the fall time (or vice versa)?
A. A cosine wave
B. A square wave
C. A sawtooth wave
D. A sine wave

E8A06
(C)
Page 8-2

E8A07
(A)
Page 8-2

E8A07
What type of wave is made up of sine waves of a fundamental frequency and all harmonics?
A. A sawtooth wave
B. A square wave
C. A sine wave
D. A cosine wave

E8A08
(B)
Page 8-4

E8A08
What is the peak voltage at a common household electrical outlet?
A. 240 volts
B. 170 volts
C. 120 volts
D. 340 volts

E8A09
(C)
Page 8-4

E8A09
What is the peak-to-peak voltage at a common household electrical outlet?
A. 240 volts
B. 120 volts
C. 340 volts
D. 170 volts

E8A10
(A)
Page 8-4

E8A10
What is the RMS voltage at a common household electrical power outlet?
A. 120-V AC
B. 340-V AC
C. 85-V AC
D. 170-V AC

E8A11
(A)
Page 8-4

E8A11
What is the RMS value of a 340-volt peak-to-peak pure sine wave?
A. 120-V AC
B. 170-V AC
C. 240-V AC
D. 300-V AC

E8A12
(C)
Page 8-4

E8A12
What is the equivalent to the root-mean-square value of an AC voltage?
A. The AC voltage found by taking the square of the average value of the peak AC voltage
B. The DC voltage causing the same heating of a given resistor as the peak AC voltage
C. The AC voltage causing the same heating of a given resistor as a DC voltage of the same value
D. The AC voltage found by taking the square root of the average AC value

E8A13
What would be the most accurate way of determining the RMS voltage of a complex waveform?
A. By using a grid dip meter
B. By measuring the voltage with a D'Arsonval meter
C. By using an absorption wavemeter
D. By measuring the heating effect in a known resistor

E8A14
For many types of voices, what is the approximate ratio of PEP to average power during a modulation peak in a single-sideband phone signal?
A. 2.5 to 1
B. 25 to 1
C. 1 to 1
D. 100 to 1

E8A15
In a single-sideband phone signal, what determines the PEP-to-average power ratio?
A. The frequency of the modulating signal
B. The speech characteristics
C. The degree of carrier suppression
D. The amplifier power

E8A16
What is the approximate DC input power to a Class B RF power amplifier stage in an FM-phone transmitter when the PEP output power is 1500 watts?
A. 900 watts
B. 1765 watts
C. 2500 watts
D. 3000 watts

E8A17
What is the approximate DC input power to a Class AB RF power amplifier stage in an unmodulated carrier transmitter when the PEP output power is 500 watts?
A. 250 watts
B. 600 watts
C. 800 watts
D. 1000 watts

E8B FCC emission designators versus emission types; modulation symbols and transmission characteristics; modulation methods; modulation index; deviation ratio; pulse modulation: width; position

E8B01
What is emission A3C?
A. Facsimile
B. RTTY
C. ATV
D. Slow Scan TV

E8A13	(D) Page 8-4
E8A14	(A) Page 8-8
E8A15	(B) Page 8-8
E8A16	(C) Page 8-8
E8A17	(D) Page 8-8
E8B01	(A) Page 8-11

E8B02
(B)
Page 8-11

E8B02
What type of emission is produced when an AM transmitter is modulated by a facsimile signal?
A. A3F
B. A3C
C. F3F
D. F3C

E8B03
(C)
Page 8-11

E8B03
What does a facsimile transmission produce?
A. Tone-modulated telegraphy
B. A pattern of printed characters designed to form a picture
C. Printed pictures by electrical means
D. Moving pictures by electrical means

E8B04
(D)
Page 8-11

E8B04
What is emission F3F?
A. Modulated CW
B. Facsimile
C. RTTY
D. Television

E8B05
(D)
Page 8-11

E8B05
What type of emission is produced when an SSB transmitter is modulated by a slow-scan television signal?
A. J3A
B. F3F
C. A3F
D. J3F

E8B06
(B)
Page 8-10

E8B06
If the first symbol of an ITU emission designator is J, representing a single-sideband, suppressed-carrier signal, what information about the emission is described?
A. The nature of any signal multiplexing
B. The type of modulation of the main carrier
C. The maximum permissible bandwidth
D. The maximum signal level, in decibels

E8B07
(C)
Page 8-10

E8B07
If the second symbol of an ITU emission designator is 1, representing a single channel containing quantized, or digital information, what information about the emission is described?
A. The maximum transmission rate, in bauds
B. The maximum permissible deviation
C. The nature of signals modulating the main carrier
D. The type of information to be transmitted

E8B08
If the third symbol of an ITU emission designator is D, representing data transmission, telemetry or telecommand, what information about the emission is described?
A. The maximum transmission rate, in bauds
B. The maximum permissible deviation
C. The nature of signals modulating the main carrier
D. The type of information to be transmitted

E8B09
How can the unwanted sideband be removed from a double-sideband signal generated by a balanced modulator to produce a single-sideband phone signal?
A. By filtering
B. By heterodyning
C. By mixing
D. By neutralization

E8B10
How does the modulation index of a phase-modulated emission vary with RF carrier frequency (the modulated frequency)?
A. It increases as the RF carrier frequency increases
B. It decreases as the RF carrier frequency increases
C. It varies with the square root of the RF carrier frequency
D. It does not depend on the RF carrier frequency

E8B11
In an FM-phone signal having a maximum frequency deviation of 3000 Hz either side of the carrier frequency, what is the modulation index when the modulating frequency is 1000 Hz?
A. 3
B. 0.3
C. 3000
D. 1000

E8B12
What is the modulation index of an FM-phone transmitter producing an instantaneous carrier deviation of 6 kHz when modulated with a 2-kHz modulating frequency?
A. 6000
B. 3
C. 2000
D. 1/3

E8B13
What is the deviation ratio of an FM-phone signal having a maximum frequency swing of plus or minus 5 kHz and accepting a maximum modulation rate of 3 kHz?
A. 60
B. 0.16
C. 0.6
D. 1.66

E8B08
(D)
Page 8-10

E8B09
(A)
Page 8-12

E8B10
(D)
Page 8-14

E8B11
(A)
Page 8-15

E8B12
(B)
Page 8-15

E8B13
(D)
Page 8-14

E8B14
(A)
Page 8-17

E8B14
In a pulse-modulation system, why is the transmitter's peak power much greater than its average power?
A. The signal duty cycle is less than 100%
B. The signal reaches peak amplitude only when voice modulated
C. The signal reaches peak amplitude only when voltage spikes are generated within the modulator
D. The signal reaches peak amplitude only when the pulses are also amplitude modulated

E8B15
(C)
Page 8-15

E8B15
What is one way that voice is transmitted in a pulse-width modulation system?
A. A standard pulse is varied in amplitude by an amount depending on the voice waveform at that instant
B. The position of a standard pulse is varied by an amount depending on the voice waveform at that instant
C. A standard pulse is varied in duration by an amount depending on the voice waveform at that instant
D. The number of standard pulses per second varies depending on the voice waveform at that instant

E8B16
(A)
Page 8-17

E8B16
What function does a pulse-width modulator perform in a switching regulator power supply?
A. It turns the switch transistor on and off at the proper time to ensure smooth regulation
B. It increases and decreases the load current at the proper time to ensure smooth regulation
C. It increases or decreases the frequency of the input voltage to ensure that AC pulses are sent at regular intervals to the rectifier
D. It turns the rectifier on and off at regular intervals to avoid overheating the power supply

E8C Digital signals: CW; baudot; ASCII; packet; AMTOR; Clover; information rate vs bandwidth

E8C01
(D)
Page 8-17

E8C01
What digital code consists of elements having unequal length?
A. ASCII
B. AX.25
C. Baudot
D. Morse code

E8C02
(B)
Page 8-17

E8C02
What are some of the differences between the Baudot digital code and ASCII?
A. Baudot uses four data bits per character, ASCII uses eight; Baudot uses one character as a shift code, ASCII has no shift code
B. Baudot uses five data bits per character, ASCII uses eight; Baudot uses one character as a shift code, ASCII has no shift code
C. Baudot uses six data bits per character, ASCII uses eight; Baudot has no shift code, ASCII uses one character as a shift code
D. Baudot uses seven data bits per character, ASCII uses eight; Baudot has no shift code, ASCII uses one character as a shift code

E8C03
What is one advantage of using the ASCII code for data communications?
A. It includes built-in error-correction features
B. It contains fewer information bits per character than any other code
C. It is possible to transmit both upper and lower case text
D. It uses one character as a "shift" code to send numeric and special characters

E8C03
(C)
Page 8-20

E8C04
What digital communications system is well suited for meteor-scatter communications?
A. ACSSB
B. Packet radio
C. AMTOR
D. Spread spectrum

E8C04
(B)
Page 8-20

E8C05
What type of error control system does Mode A AMTOR use?
A. Each character is sent twice
B. The receiving station checks the calculated frame check sequence (FCS) against the transmitted FCS
C. The receiving station checks the calculated frame parity against the transmitted parity
D. The receiving station automatically requests repeats when needed

E8C05
(D)
Page 8-18

E8C06
What type of error control system does Mode B AMTOR use?
A. Each character is sent twice
B. The receiving station checks the calculated frame check sequence (FCS) against the transmitted FCS
C. The receiving station checks the calculated frame parity against the transmitted parity
D. The receiving station automatically requests repeats when needed

E8C06
(A)
Page 8-19

E8C07
What is the necessary bandwidth of a 13-WPM international Morse code emission A1A transmission?
A. Approximately 13 Hz
B. Approximately 26 Hz
C. Approximately 52 Hz
D. Approximately 104 Hz

E8C07
(C)
Page 8-21

E8C08
What is the necessary bandwidth for a 170-hertz shift, 300-baud ASCII emission J2D transmission?
A. 0 Hz
B. 0.3 kHz
C. 0.5 kHz
D. 1.0 kHz

E8C08
(C)
Page 8-22

E8C09
(D)
Page 8-22

E8C09
What is the necessary bandwidth of a 1000-Hz shift, 1200-baud ASCII emission F1D transmission?
A. 1000 Hz
B. 1200 Hz
C. 440 Hz
D. 2400 Hz

E8C10
(A)
Page 8-22

E8C10
What is the necessary bandwidth of a 4800-Hz frequency shift, 9600-baud ASCII emission F1D transmission?
A. 15.36 kHz
B. 9.6 kHz
C. 4.8 kHz
D. 5.76 kHz

E8D Amplitude compandored single-sideband (ACSSB); spread-spectrum communications

E8D01
(C)
Page 8-23

E8D01
What is amplitude compandored single-sideband?
A. Reception of single-sideband signal with a conventional CW receiver
B. Reception of single-sideband signal with a conventional FM receiver
C. Single-sideband signal incorporating speech compression at the transmitter and speech expansion at the receiver
D. Single-sideband signal incorporating speech expansion at the transmitter and speech compression at the receiver

E8D02
(A)
Page 8-23

E8D02
What is meant by compandoring?
A. Compressing speech at the transmitter and expanding it at the receiver
B. Using an audio-frequency signal to produce pulse-length modulation
C. Combining amplitude and frequency modulation to produce a single-sideband signal
D. Detecting and demodulating a single-sideband signal by converting it to a pulse-modulated signal

E8D03
(A)
Page 8-24

E8D03
What is the purpose of a pilot tone in an amplitude-compandored single-sideband system?
A. It permits rapid tuning of a mobile receiver
B. It replaces the suppressed carrier at the receiver
C. It permits rapid change of frequency to escape high-powered interference
D. It acts as a beacon to indicate the present propagation characteristic of the band

E8D04
(D)
Page 8-24

E8D04
What is the approximate frequency of the pilot tone in an amplitude-compandored single-sideband system?
A. 1 kHz
B. 5 MHz
C. 455 kHz
D. 3 kHz

E8D05
How many more voice transmissions can be packed into a given frequency band for amplitude-compandored single-sideband systems over conventional FM-phone systems?
A. 2
B. 4
C. 8
D. 16

E8D05
(B)
Page 8-24

E8D06
What term describes a wide-bandwidth communications system in which the RF carrier varies according to some predetermined sequence?
A. Amplitude compandored single sideband
B. AMTOR
C. Time-domain frequency modulation
D. Spread-spectrum communication

E8D06
(D)
Page 8-25

E8D07
What spread-spectrum communications technique alters the center frequency of a conventional carrier many times per second in accordance with a pseudo-random list of channels?
A. Frequency hopping
B. Direct sequence
C. Time-domain frequency modulation
D. Frequency compandored spread-spectrum

E8D07
(A)
Page 8-27

E8D08
What spread-spectrum communications technique uses a very fast binary bit stream to shift the phase of an RF carrier?
A. Frequency hopping
B. Direct sequence
C. Binary phase-shift keying
D. Phase compandored spread-spectrum

E8D08
(B)
Page 8-29

E8D09
What controls the spreading sequence of an amateur spread-spectrum transmission?
A. A frequency-agile linear amplifier
B. A crystal-controlled filter linked to a high-speed crystal switching mechanism
C. A binary linear-feedback shift register
D. A binary code which varies if propagation changes

E8D09
(C)
Page 8-29

E8D10
Why are spread-spectrum communications so resistant to interference?
A. Interfering signals are removed by a frequency-agile crystal filter
B. Spread-spectrum transmitters use much higher power than conventional carrier-frequency transmitters
C. Spread-spectrum transmitters can "hunt" for the best carrier frequency to use within a given RF spectrum
D. Only signals using the correct spreading sequence are received

E8D10
(D)
Page 8-26

E8D11
Why do spread-spectrum communications interfere so little with conventional channelized communications in the same band?
A. A spread-spectrum transmitter avoids channels within the band which are in use by conventional transmitters
B. Spread-spectrum signals appear only as low-level noise in conventional receivers
C. Spread-spectrum signals change too rapidly to be detected by conventional receivers
D. Special crystal filters are needed in conventional receivers to detect spread-spectrum signals

E8E Peak amplitude (positive and negative); peak-to-peak values: measurements; Electromagnetic radiation; wave polarization; signal-to-noise (S/N) ratio

E8E01
What is the term for the amplitude of the maximum positive excursion of a signal as viewed on an oscilloscope?
A. Peak-to-peak voltage
B. Inverse peak negative voltage
C. RMS voltage
D. Peak positive voltage

E8E02
What is the easiest voltage amplitude dimension to measure by viewing a pure sine wave signal on an oscilloscope?
A. Peak-to-peak voltage
B. RMS voltage
C. Average voltage
D. DC voltage

E8E03
What is the relationship between the peak-to-peak voltage and the peak voltage amplitude in a symmetrical waveform?
A. 1:1
B. 2:1
C. 3:1
D. 4:1

E8E04
What input-amplitude parameter is valuable in evaluating the signal-handling capability of a Class A amplifier?
A. Peak voltage
B. RMS voltage
C. An average reading power output meter
D. Resting voltage

E8E05
What is the PEP output of a transmitter that has a maximum peak of 30 volts to a 50-ohm load as observed on an oscilloscope?
A. 4.5 watts
B. 9 watts
C. 16 watts
D. 18 watts

E8E06
If an RMS reading AC voltmeter reads 65 volts on a sinusoidal waveform, what is the peak-to-peak voltage?
A. 46 volts
B. 92 volts
C. 130 volts
D. 184 volts

E8E07
What is the advantage of using a peak-reading voltmeter to monitor the output of a single-sideband transmitter?
A. It would be easy to calculate the PEP output of the transmitter
B. It would be easy to calculate the RMS output power of the transmitter
C. It would be easy to calculate the SWR on the transmission line
D. It would be easy to observe the output amplitude variations

E8E08
What are electromagnetic waves?
A. Alternating currents in the core of an electromagnet
B. A wave consisting of two electric fields at right angles to each other
C. A wave consisting of an electric field and a magnetic field at right angles to each other
D. A wave consisting of two magnetic fields at right angles to each other

E8E09
Why don't electromagnetic waves penetrate a good conductor for more than a fraction of a wavelength?
A. Electromagnetic waves are reflected by the surface of a good conductor
B. Oxide on the conductor surface acts as a magnetic shield
C. The electromagnetic waves are dissipated as eddy currents in the conductor surface
D. The resistance of the conductor surface dissipates the electromagnetic waves

E8E10
Which of the following best describes electromagnetic waves traveling in free space?
A. Electric and magnetic fields become aligned as they travel
B. The energy propagates through a medium with a high refractive index
C. The waves are reflected by the ionosphere and return to their source
D. Changing electric and magnetic fields propagate the energy across a vacuum

E8E05
(B)
Page 8-7

E8E06
(D)
Page 8-4

E8E07
(A)
Page 8-6

E8E08
(C)
Page 8-29

E8E09
(C)
Page 8-29

E8E10
(D)
Page 8-29

E8E11
(B)
Page 8-31

E8E11
What is meant by circularly polarized electromagnetic waves?
A. Waves with an electric field bent into a circular shape
B. Waves with a rotating electric field
C. Waves that circle the Earth
D. Waves produced by a loop antenna

E8E12
(D)
Page 8-29

E8E12
What is the polarization of an electromagnetic wave if its magnetic field is parallel to the surface of the Earth?
A. Circular
B. Horizontal
C. Elliptical
D. Vertical

E8E13
(A)
Page 8-29

E8E13
What is the polarization of an electromagnetic wave if its magnetic field is perpendicular to the surface of the Earth?
A. Horizontal
B. Circular
C. Elliptical
D. Vertical

E8E14
(D)
Page 8-31

E8E14
What is the primary source of noise that can be heard in an HF-band receiver with an antenna connected?
A. Detector noise
B. Man-made noise
C. Receiver front-end noise
D. Atmospheric noise

E8E15
(A)
Page 8-31

E8E15
What is the primary source of noise that can be heard in a VHF/UHF-band receiver with an antenna connected?
A. Receiver front-end noise
B. Man-made noise
C. Atmospheric noise
D. Detector noise

Subelement E9 — Antennas And Feed Lines

[5 Exam Questions — 5 Groups]

E9A Isotropic radiators: definition; used as a standard for comparison; radiation pattern; basic antenna parameters: radiation resistance and reactance (including wire dipole, folded dipole), gain, beamwidth, efficiency

E9A01
Which of the following describes an isotropic radiator?
A. A grounded radiator used to measure earth conductivity
B. A horizontal radiator used to compare Yagi antennas
C. A theoretical radiator used to compare other antennas
D. A spacecraft radiator used to direct signals toward the earth

E9A01
(C)
Page 9-2

E9A02
When is it useful to refer to an isotropic radiator?
A. When comparing the gains of directional antennas
B. When testing a transmission line for standing-wave ratio
C. When directing a transmission toward the tropical latitudes
D. When using a dummy load to tune a transmitter

E9A02
(A)
Page 9-2

E9A03
How much gain does a 1/2-wavelength dipole have over an isotropic radiator?
A. About 1.5 dB
B. About 2.1 dB
C. About 3.0 dB
D. About 6.0 dB

E9A03
(B)
Page 9-4

E9A04
Which of the following antennas has no gain in any direction?
A. Quarter-wave vertical
B. Yagi
C. Half-wave dipole
D. Isotropic radiator

E9A04
(D)
Page 9-2

E9A05
Which of the following describes the radiation pattern of an isotropic radiator?
A. A tear drop in the vertical plane
B. A circle in the horizontal plane
C. A sphere with the antenna in the center
D. Crossed polarized with a spiral shape

E9A05
(C)
Page 9-2

E9A06
Why would one need to know the radiation resistance of an antenna?
A. To match impedances for maximum power transfer
B. To measure the near-field radiation density from a transmitting antenna
C. To calculate the front-to-side ratio of the antenna
D. To calculate the front-to-back ratio of the antenna

E9A06
(A)
Page 9-5

E9A07
(B)
Page 9-5

E9A07
What factors determine the radiation resistance of an antenna?
A. Transmission-line length and antenna height
B. Antenna location with respect to nearby objects and the conductors' length/diameter ratio
C. It is a physical constant and is the same for all antennas
D. Sunspot activity and time of day

E9A08
(C)
Page 9-6

E9A08
What is the term for the ratio of the radiation resistance of an antenna to the total resistance of the system?
A. Effective radiated power
B. Radiation conversion loss
C. Antenna efficiency
D. Beamwidth

E9A09
(D)
Page 9-6

E9A09
What is included in the total resistance of an antenna system?
A. Radiation resistance plus space impedance
B. Radiation resistance plus transmission resistance
C. Transmission-line resistance plus radiation resistance
D. Radiation resistance plus ohmic resistance

E9A10
(C)
Page 9-6

E9A10
What is a folded dipole antenna?
A. A dipole one-quarter wavelength long
B. A type of ground-plane antenna
C. A dipole whose ends are connected by a one-half wavelength piece of wire
D. A hypothetical antenna used in theoretical discussions to replace the radiation resistance

E9A11
(A)
Page 9-14

E9A11
What is meant by antenna gain?
A. The numerical ratio relating the radiated signal strength of an antenna to that of another antenna
B. The numerical ratio of the signal in the forward direction to the signal in the back direction
C. The numerical ratio of the amount of power radiated by an antenna compared to the transmitter output power
D. The final amplifier gain minus the transmission-line losses (including any phasing lines present)

E9A12
(B)
Page 9-8

E9A12
What is meant by antenna bandwidth?
A. Antenna length divided by the number of elements
B. The frequency range over which an antenna can be expected to perform well
C. The angle between the half-power radiation points
D. The angle formed between two imaginary lines drawn through the ends of the elements

E9A13
How can the approximate beamwidth of a beam antenna be determined?
A. Note the two points where the signal strength of the antenna is down 3 dB from the maximum signal point and compute the angular difference
B. Measure the ratio of the signal strengths of the radiated power lobes from the front and rear of the antenna
C. Draw two imaginary lines through the ends of the elements and measure the angle between the lines
D. Measure the ratio of the signal strengths of the radiated power lobes from the front and side of the antenna

E9A13
(A)
Page 9-16

E9A14
How is antenna efficiency calculated?
A. (radiation resistance / transmission resistance) x 100%
B. (radiation resistance / total resistance) x 100%
C. (total resistance / radiation resistance) x 100%
D. (effective radiated power / transmitter output) x 100%

E9A14
(B)
Page 9-6

E9A15
How can the efficiency of an HF grounded vertical antenna be made comparable to that of a half-wave dipole antenna?
A. By installing a good ground radial system
B. By isolating the coax shield from ground
C. By shortening the vertical
D. By lengthening the vertical

E9A15
(A)
Page 9-6

E9B Free-space antenna patterns: E and H plane patterns (ie, azimuth and elevation in free-space); gain as a function of pattern; antenna design (computer modeling of antennas)

E9B01
What determines the free-space polarization of an antenna?
A. The orientation of its magnetic field (H Field)
B. The orientation of its free-space characteristic impedance
C. The orientation of its electric field (E Field)
D. Its elevation pattern

E9B01
(C)
Page 9-15

E9B02
Which of the following statements is true about the radiation pattern shown in Figure A9-1?
A. The pattern shows the beamwidth of the antenna
B. The pattern shows the azimuth or directional pattern of the antenna
C. The pattern shows the front to back pattern of the antenna
D. All of these choices are correct

E9B02
(D)
Page 9-16

E9B03
In the free-space H-Field radiation pattern shown in Figure A9-1, what is the 3-dB beamwidth?
A. 75 degrees
B. 50 degrees
C. 25 degrees
D. 30 degrees

E9B03
(B)
Page 9-16

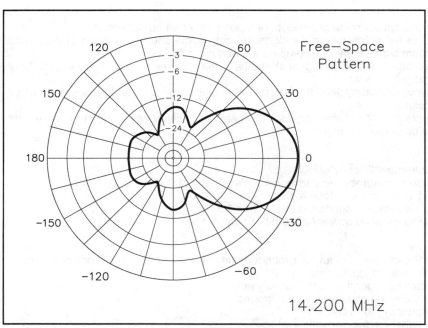

Figure A9-1 — Refer to questions E9B02 through E9B04 and E9B11.

E9B04
In the free-space H-Field pattern shown in Figure A9-1, what is the front-to-back ratio?
A. 36 dB
B. 18 dB
C. 24 dB
D. 14 dB

E9B05
What information is needed to accurately evaluate the gain of an antenna?
A. Radiation resistance
B. E-Field and H-Field patterns
C. Loss resistance
D. All of these choices

E9B06
Which is NOT an important reason to evaluate a gain antenna across the whole frequency band for which it was designed?
A. The gain may fall off rapidly over the whole frequency band
B. The feedpoint impedance may change radically with frequency
C. The rearward pattern lobes may vary excessively with frequency
D. The dielectric constant may vary significantly

E9B07
What usually occurs if a Yagi antenna is designed solely for maximum forward gain?
A. The front-to-back ratio increases
B. The feedpoint impedance becomes very low
C. The frequency response is widened over the whole frequency band
D. The SWR is reduced

E9B08
If the boom of a Yagi antenna is lengthened and the elements are properly retuned, what usually occurs?
A. The gain increases
B. The SWR decreases
C. The front-to-back ratio increases
D. The gain bandwidth decreases rapidly

E9B09
What type of computer program is commonly used for modeling antennas?
A. Graphical analysis
B. Method of Moments
C. Mutual impedance analysis
D. Calculus differentiation with respect to physical properties

E9B10
What is the principle of a "Method of Moments" analysis?
A. A wire is modeled as a series of segments, each having a distinct value of current
B. A wire is modeled as a single sine-wave current generator
C. A wire is modeled as a series of points, each having a distinct location in space
D. A wire is modeled as a series of segments, each having a distinct value of voltage across it

E9B11
In the free-space H-field pattern shown in Figure A9-1, what is the front-to-side ratio?
A. 12 dB
B. 14 dB
C. 18 dB
D. 24 dB

E9C Phased vertical antennas; radiation patterns; beverage antennas; rhombic antennas: resonant; nonresonant; radiation pattern; antenna patterns: elevation above real ground, ground effects as related to polarization, take-off angles as a function of height above ground

E9C01
What is the radiation pattern of two 1/4-wavelength vertical antennas spaced 1/2-wavelength apart and fed 180 degrees out of phase?
A. Unidirectional cardioid
B. Omnidirectional
C. Figure-8 broadside to the antennas
D. Figure-8 end-fire in line with the antennas

E9B07	(B) Page 9-21
E9B08	(A) Page 9-21
E9B09	(B) Page 9-15
E9B10	(A) Page 9-15
E9B11	(B) Page 9-17
E9C01	(D) Page 9-22

E9C02
(A)
Page 9-22

E9C02
What is the radiation pattern of two 1/4-wavelength vertical antennas spaced 1/4-wavelength apart and fed 90 degrees out of phase?
A. Unidirectional cardioid
B. Figure-8 end-fire
C. Figure-8 broadside
D. Omnidirectional

E9C03
(C)
Page 9-22

E9C03
What is the radiation pattern of two 1/4-wavelength vertical antennas spaced 1/2-wavelength apart and fed in phase?
A. Omnidirectional
B. Cardioid unidirectional
C. Figure-8 broadside to the antennas
D. Figure-8 end-fire in line with the antennas

E9C04
(D)
Page 9-22

E9C04
What is the radiation pattern of two 1/4-wavelength vertical antennas spaced 1/4-wavelength apart and fed 180 degrees out of phase?
A. Omnidirectional
B. Cardioid unidirectional
C. Figure-8 broadside to the antennas
D. Figure-8 end-fire in line with the antennas

E9C05
(D)
Page 9-22

E9C05
What is the radiation pattern for two 1/4-wavelength vertical antennas spaced 1/8-wavelength apart and fed 180 degrees out of phase?
A. Omnidirectional
B. Cardioid unidirectional
C. Figure-8 broadside to the antennas
D. Figure-8 end-fire in line with the antennas

E9C06
(B)
Page 9-22

E9C06
What is the radiation pattern for two 1/4-wavelength vertical antennas spaced 1/4-wavelength apart and fed in phase?
A. Substantially unidirectional
B. Elliptical
C. Cardioid unidirectional
D. Figure-8 end-fire in line with the antennas

E9C07
(B)
Page 9-24

E9C07
Which of the following is the best description of a resonant rhombic antenna?
A. Unidirectional; four-sided, each side a half-wavelength long; terminated in a resistance equal to its characteristic impedance
B. Bidirectional; four-sided, each side approximately one wavelength long; open at the end opposite the transmission line connection
C. Four-sided; an LC network at each vertex except for the transmission connection; tuned to resonate at the operating frequency
D. Four-sided, each side of a different physical length; traps at each vertex for changing resonance according to band usage

E9C08
What are the advantages of a nonresonant rhombic antenna?
A. Wide frequency range, high gain and high front-to-back ratio
B. High front-to-back ratio, compact size and high gain
C. Unidirectional radiation pattern, high gain and compact size
D. Bidirectional radiation pattern, high gain and wide frequency range

E9C09
What are the disadvantages of a nonresonant rhombic antenna?
A. A large area for proper installation and a narrow bandwidth
B. A large area for proper installation and a low front-to-back ratio
C. A large area and four sturdy supports for proper installation
D. A large amount of aluminum tubing and a low front-to-back ratio

E9C10
What is the effect of a terminating resistor on a rhombic antenna?
A. It reflects the standing waves on the antenna elements back to the transmitter
B. It changes the radiation pattern from essentially bidirectional to essentially unidirectional
C. It changes the radiation pattern from horizontal to vertical polarization
D. It decreases the ground loss

E9C11
What type of antenna pattern over real ground is shown in Figure A9-2?
A. Elevation pattern
B. Azimuth pattern
C. E-Plane pattern
D. Polarization pattern

E9C12
How is the far-field elevation pattern of a vertically polarized antenna affected by being mounted over seawater versus rocky ground?
A. The low-angle radiation decreases
B. The high-angle radiation increases
C. Both the high- and low-angle radiation decrease
D. The low-angle radiation increases

E9C13
If only a modest on-ground radial system can be used with an eighth-wavelength-high, inductively loaded vertical antenna, what would be the best compromise to minimize near-field losses?
A. 4 radial wires, 1 wavelength long
B. 8 radial wires, a half-wavelength long
C. A wire-mesh screen at the antenna base, an eighth-wavelength square
D. 4 radial wires, 2 wavelengths long

E9C08
(A)
Page 9-24

E9C09
(C)
Page 9-25

E9C10
(B)
Page 9-25

E9C11
(A)
Page 9-18

E9C12
(D)
Page 9-21

E9C13
(C)
Page 9-21

E9C14
(C)
Page 9-18

E9C14
In the antenna radiation pattern shown in Figure A9-2, what is the elevation angle of the peak response?
A. 45 degrees
B. 75 degrees
C. 7.5 degrees
D. 25 degrees

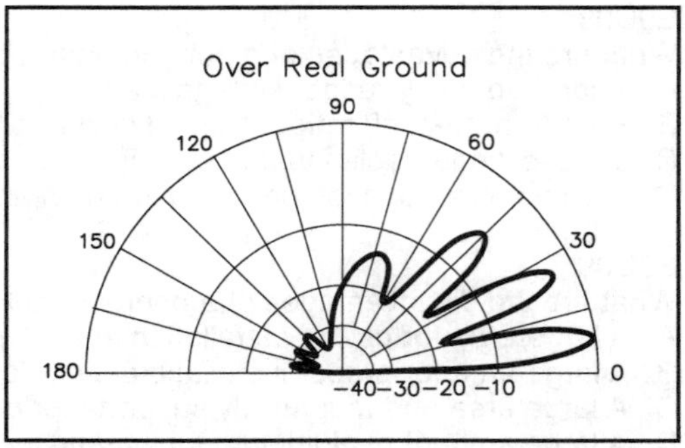

Over Real Ground

Figure A9-2 — Refer to questions E9C11, E9C14 and E9C15.

E9C15
(B)
Page 9-18

E9C15
In the antenna radiation pattern shown in Figure A9-2, what is the front-to-back ratio?
A. 15 dB
B. 28 dB
C. 3 dB
D. 24 dB

E9C16
(D)
Page 9-26

E9C16
What is one characteristic of a Beverage antenna?
A. For best performance it must not exceed 1/4 wavelength in length at the desired frequency
B. For best performance it must be mounted more than 1 wavelength above ground at the desired frequency
C. For best performance it should be configured as four-sided loop
D. For best performance it should be as long as possible

E9D Space and satellite communications antennas: gain; beamwidth; tracking; losses in real antennas and matching: resistivity losses, losses in resonating elements (loading coils, matching networks, etc. {ie, mobile, trap}); SWR bandwidth; efficiency

E9D01
(A)
Page 9-27

E9D01
What factors determine the receiving antenna gain required at an amateur satellite station in earth operation?
A. Height, transmitter power and antennas of satellite
B. Length of transmission line and impedance match between receiver and transmission line
C. Preamplifier location on transmission line and presence or absence of RF amplifier stages
D. Height of earth antenna and satellite orbit

E9D02
(A)
Page 9-27

E9D02
What factors determine the EIRP required by an amateur satellite station in earth operation?
A. Satellite antennas and height, satellite receiver sensitivity
B. Path loss, earth antenna gain, signal-to-noise ratio
C. Satellite transmitter power and orientation of ground receiving antenna
D. Elevation of satellite above horizon, signal-to-noise ratio, satellite transmitter power

E9D03
What is the beamwidth of a symmetrical pattern antenna with a gain of 20 dB as compared to an isotropic radiator?
A. 10.1 degrees
B. 20.3 degrees
C. 45.0 degrees
D. 60.9 degrees

E9D03
(B)
Page 9-20

E9D04
How does the gain of a parabolic dish antenna change when the operating frequency is doubled?
A. Gain does not change
B. Gain is multiplied by 0.707
C. Gain increases 6 dB
D. Gain increases 3 dB

E9D04
(C)
Page 9-26

E9D05
How is circular polarization produced using linearly polarized antennas?
A. Stack two Yagis, fed 90 degrees out of phase, to form an array with the respective elements in parallel planes
B. Stack two Yagis, fed in phase, to form an array with the respective elements in parallel planes
C. Arrange two Yagis perpendicular to each other, with the driven elements in the same plane, fed 90 degrees out of phase
D. Arrange two Yagis perpendicular to each other, with the driven elements in the same plane, fed in phase

E9D05
(C)
Page 9-29

E9D06
How does the beamwidth of an antenna vary as the gain is increased?
A. It increases geometrically
B. It increases arithmetically
C. It is essentially unaffected
D. It decreases

E9D06
(D)
Page 9-19

E9D07
Why does a satellite communications antenna system for earth operation need to have rotators for both azimuth and elevation control?
A. In order to track the satellite as it orbits the earth
B. Because the antennas are large and heavy
C. In order to point the antenna above the horizon to avoid terrestrial interference
D. To rotate antenna polarization along the azimuth and elevate the system towards the satellite

E9D07
(A)
Page 9-29

E9D08
For a shortened vertical antenna, where should a loading coil be placed to minimize losses and produce the most effective performance?
A. Near the center of the vertical radiator
B. As low as possible on the vertical radiator
C. As close to the transmitter as possible
D. At a voltage node

E9D08
(A)
Page 9-10

E9D09
(C)
Page 9-12

E9D09
Why should an HF mobile antenna loading coil have a high ratio of reactance to resistance?
A. To swamp out harmonics
B. To maximize losses
C. To minimize losses
D. To minimize the Q

E9D10
(A)
Page 9-10

E9D10
What is a disadvantage of using a trap antenna?
A. It will radiate harmonics
B. It can only be used for single-band operation
C. It is too sharply directional at lower frequencies
D. It must be neutralized

E9D11
(A)
Page 9-31

E9D11
How must the driven element in a 3-element Yagi be tuned to use a "hairpin" matching system?
A. The driven element reactance is capacitive
B. The driven element reactance is inductive
C. The driven element resonance is higher than the operating frequency
D. The driven element radiation resistance is higher than the characteristic impedance of the transmission line

E9D12
(C)
Page 9-31

E9D12
What is the equivalent lumped-constant network for a "hairpin" matching system on a 3-element Yagi?
A. Pi network
B. Pi-L network
C. L network
D. Parallel-resonant tank

E9D13
(B)
Page 9-10

E9D13
What happens to the bandwidth of an antenna as it is shortened through the use of loading coils?
A. It is increased
B. It is decreased
C. No change occurs
D. It becomes flat

E9D14
(D)
Page 9-14

E9D14
What is an advantage of using top loading in a shortened HF vertical antenna?
A. Lower Q
B. Greater structural strength
C. Higher losses
D. Improved radiation efficiency

E9E Matching antennas to feed lines; characteristics of open and shorted feed lines: 1/8 wavelength; 1/4 wavelength; 3/8 wavelength; 1/2 wavelength; 1/4 wavelength matching transformers; feed lines: coax versus open-wire; velocity factor; electrical length; transformation characteristics of line terminated in impedance not equal to characteristic impedance

E9E01
What system matches a high-impedance transmission line to a lower impedance antenna by connecting the line to the driven element in two places, spaced a fraction of a wavelength each side of element center?
A. The gamma matching system
B. The delta matching system
C. The omega matching system
D. The stub matching system

E9E01
(B)
Page 9-29

E9E02
What system matches an unbalanced feed line to an antenna by feeding the driven element both at the center of the element and at a fraction of a wavelength to one side of center?
A. The gamma matching system
B. The delta matching system
C. The omega matching system
D. The stub matching system

E9E02
(A)
Page 9-30

E9E03
What impedance matching system uses a short perpendicular section of transmission line connected to the feed line near the antenna?
A. The gamma matching system
B. The delta matching system
C. The omega matching system
D. The stub matching system

E9E03
(D)
Page 9-32

E9E04
What should be the approximate capacitance of the resonating capacitor in a gamma matching circuit on a 1/2-wavelength dipole antenna for the 20-meter wavelength band?
A. 70 pF
B. 140 pF
C. 200 pF
D. 0.2 pF

E9E04
(B)
Page 9-31

E9E05
What kind of impedance does a 1/4-wavelength transmission line present to a generator when the line is shorted at the far end?
A. A very high impedance
B. A very low impedance
C. The same as the characteristic impedance of the transmission line
D. The same as the generator output impedance

E9E05
(A)
Page 9-38

E9E06
(A)
Page 9-38

E9E06
What kind of impedance does a 1/2-wavelength transmission line present to a generator when the line is open at the far end?
A. A very high impedance
B. A very low impedance
C. The same as the characteristic impedance of the line
D. The same as the output impedance of the generator

E9E07
(D)
Page 9-35

E9E07
What is the velocity factor of a transmission line?
A. The ratio of the characteristic impedance of the line to the terminating impedance
B. The index of shielding for coaxial cable
C. The velocity of the wave on the transmission line multiplied by the velocity of light in a vacuum
D. The velocity of the wave on the transmission line divided by the velocity of light in a vacuum

E9E08
(C)
Page 9-35

E9E08
What determines the velocity factor in a transmission line?
A. The termination impedance
B. The line length
C. Dielectrics in the line
D. The center conductor resistivity

E9E09
(D)
Page 9-36

E9E09
Why is the physical length of a coaxial cable transmission line shorter than its electrical length?
A. Skin effect is less pronounced in the coaxial cable
B. The characteristic impedance is higher in the parallel feed line
C. The surge impedance is higher in the parallel feed line
D. RF energy moves slower along the coaxial cable

E9E10
(B)
Page 9-35

E9E10
What is the typical velocity factor for a coaxial cable with polyethylene dielectric?
A. 2.70
B. 0.66
C. 0.30
D. 0.10

E9E11
(C)
Page 9-36

E9E11
What would be the physical length of a typical coaxial transmission line that is electrically one-quarter wavelength long at 14.1 MHz? (Assume a velocity factor of 0.66.)
A. 20 meters
B. 2.33 meters
C. 3.51 meters
D. 0.25 meters

E9E12
What is the physical length of a parallel conductor feed line that is electrically one-half wavelength long at 14.10 MHz? (Assume a velocity factor of 0.95.)
A. 15 meters
B. 20.2 meters
C. 10.1 meters
D. 70.8 meters

E9E12
(C)
Page 9-36

E9E13
What parameter best describes the interactions at the load end of a mismatched transmission line?
A. Characteristic impedance
B. Reflection coefficient
C. Velocity factor
D. Dielectric Constant

E9E13
(B)
Page 9-38

E9E14
Which of the following measurements describes a mismatched transmission line?
A. An SWR less than 1:1
B. A reflection coefficient greater than 1
C. A dielectric constant greater than 1
D. An SWR greater than 1:1

E9E14
(D)
Page 9-37

E9E15
What characteristic will 450-ohm ladder line have at 50 MHz, as compared to 0.195-inch-diameter coaxial cable (such as RG-58)?
A. Lower loss in dB/100 feet
B. Higher SWR
C. Smaller reflection coefficient
D. Lower velocity factor

E9E15
(A)
Page 9-37

USEFUL TABLES

US Customary—Metric Conversion Factors

International System of Units (SI)—Metric Units

Prefix	Symbol		Multiplication Factor
exa	E	10^{18} =	1 000 000 000 000 000 000
peta	P	10^{15} =	1 000 000 000 000 000
tera	T	10^{12} =	1 000 000 000 000
giga	G	10^{9} =	1 000 000 000
mega	M	10^{6} =	1 000 000
kilo	k	10^{3} =	1 000
hecto	h	10^{2} =	100
deca	da	10^{1} =	10
(unit)		10^{0} =	1
deci	d	10^{-1} =	0.1
centi	c	10^{-2} =	0.01
milli	m	10^{-3} =	0.001
micro	μ	10^{-6} =	0.000001
nano	n	10^{-9} =	0.000000001
pico	p	10^{-12} =	0.000000000001
femto	f	10^{-15} =	0.000000000000001
atto	a	10^{-18} =	0.000000000000000001

Linear
1 metre (m) = 100 centimetres (cm) = 1000 millimetres (mm)

Area
$1\ m^2 = 1 \times 10^4\ cm^2 = 1 \times 10^6\ mm^2$

Volume
$1\ m^3 = 1 \times 10^6\ cm^3 = 1 \times 10^9\ mm^3$
$1\ litre\ (l) = 1000\ cm^3 = 1 \times 10^6\ mm^3$

Mass
1 kilogram (kg) = 1 000 grams (g)
 (Approximately the mass of 1 litre of water)
1 metric ton (or tonne) = 1 000 kg

US Customary Units

Linear Units
12 inches (in) = 1 foot (ft)
36 inches = 3 feet = 1 yard (yd)
1 rod = 5½ yards = 16½ feet
1 statute mile = 1 760 yards = 5 280 feet
1 nautical mile = 6 076.11549 feet

Area
$1\ ft^2 = 144\ in^2$
$1\ yd^2 = 9\ ft^2 = 1\ 296\ in^2$
$1\ rod^2 = 30¼\ yd^2$
$1\ acre = 4840\ yd^2 = 43\ 560\ ft^2$
$1\ acre = 160\ rod^2$
$1\ mile^2 = 640\ acres$

Volume
$1\ ft^3 = 1\ 728\ in^3$
$1\ yd^3 = 27\ ft^3$

Liquid Volume Measure
$1\ fluid\ ounce\ (fl\ oz) = 8\ fluidrams = 1.804\ in^3$
1 pint (pt) = 16 fl oz
$1\ quart\ (qt) = 2\ pt = 32\ fl\ oz = 57¾\ in^3$
$1\ gallon\ (gal) = 4\ qt = 231\ in^3$
1 barrel = 31½ gal

Dry Volume Measure
$1\ quart\ (qt) = 2\ pints\ (pt) = 67.2\ in^3$
1 peck = 8 qt
$1\ bushel = 4\ pecks = 2\ 150.42\ in^3$

Avoirdupois Weight
1 dram (dr) = 27.343 grains (gr) or (gr a)
1 ounce (oz) = 437.5 gr
1 pound (lb) = 16 oz = 7 000 gr
1 short ton = 2 000 lb, 1 long ton = 2 240 lb

Troy Weight
1 grain troy (gr t) = 1 grain avoirdupois
1 pennyweight (dwt) or (pwt) = 24 gr t
1 ounce troy (oz t) = 480 grains
1 lb t = 12 oz t = 5 760 grains

Apothecaries' Weight
1 grain apothecaries' (gr ap) = 1 gr t = 1 gr a
1 dram ap (dr ap) = 60 gr
1 oz ap = 1 oz t = 8 dr ap = 480 fr
1 lb ap = 1 lb t = 12 oz ap = 5 760 gr

Multiply →

Metric Unit = Conversion Factor × US Customary Unit

← **Divide**

Metric Unit ÷ Conversion Factor = US Customary Unit

Metric Unit =	Conversion Factor ×	US Unit		Metric Unit =	Conversion Factor ×	US Unit
(Length)				**(Volume)**		
mm	25.4	inch		mm³	16387.064	in³
cm	2.54	inch		cm³	16.387	in³
cm	30.48	foot		m³	0.028316	ft³
m	0.3048	foot		m³	0.764555	yd³
m	0.9144	yard		ml	16.387	in³
km	1.609	mile		ml	29.57	fl oz
km	1.852	nautical mile		ml	473	pint
				ml	946.333	quart
(Area)				l	28.32	ft³
mm²	645.16	inch²		l	0.9463	quart
cm²	6.4516	in²		l	3.785	gallon
cm²	929.03	ft²		l	1.101	dry quart
m²	0.0929	ft²		l	8.809	peck
cm²	8361.3	yd²		l	35.238	bushel
m²	0.83613	yd²				
m²	4047	acre		**(Mass)**	**(Troy Weight)**	
km²	2.59	mi²		g	31.103	oz t
				g	373.248	lb t
(Mass)	**(Avoirdupois Weight)**					
grams	0.0648	grains		**(Mass)**	**(Apothecaries' Weight)**	
g	28.349	oz		g	3.387	dr ap
g	453.59	lb		g	31.103	oz ap
kg	0.45359	lb		g	373.248	lb ap
tonne	0.907	short ton				
tonne	1.016	long ton				

Standard Resistance Values

Numbers in **bold** type are ± 10% values. Others are 5% values.

Ohms

										Megohms				
1.0	3.6	**12**	43	**150**	510	**1800**	6200	**22000**	75000	0.24	0.62	1.6	4.3	11.0
1.1	**3.9**	13	**47**	160	**560**	2000	**6800**	24000	**82000**	0.27	0.68	1.8	**4.7**	**12.0**
1.2	4.3	**15**	51	**180**	620	**2200**	7500	**27000**	91000	0.30	0.75	2.0	5.1	13.0
1.3	**4.7**	16	**56**	200	**680**	2400	**8200**	30000	**100000**	**0.33**	**0.82**	2.2	**5.6**	**15.0**
1.5	5.1	**18**	62	**220**	750	**2700**	9100	**33000**	110000	0.36	0.91	2.4	6.2	16.0
1.6	**5.6**	20	**68**	240	**820**	3000	**10000**	36000	**120000**	**0.39**	**1.0**	**2.7**	**6.8**	**18.0**
1.8	6.2	**22**	75	**270**	910	**3300**	11000	**39000**	130000	0.43	1.1	3.0	7.5	20.0
2.0	**6.8**	24	**82**	300	**1000**	3600	**12000**	43000	**150000**	**0.47**	**1.2**	**3.3**	**8.2**	**22.0**
2.2	7.5	**27**	91	**330**	1100	**3900**	13000	**47000**	160000	0.51	1.3	3.6	9.1	
2.4	**8.2**	30	**100**	360	**1200**	4300	**15000**	51000	**180000**	**0.56**	**1.5**	**3.9**	10.0	
2.7	9.1	**33**	110	**390**	1300	**4700**	16000	**56000**	200000					
3.0	**10.0**	36	**120**	430	**1500**	5100	**18000**	62000	**220000**					
3.3	11.0	**39**	130	**470**	1600	**5600**	20000	**68000**						

Resistor Color Code

Color	Sig. Figure	Decimal Multiplier	Tolerance (%)	Color	Sig. Figure	Decimal Multiplier	Tolerance (%)
Black	0	1		Violet	7	10,000,000	
Brown	1	10		Gray	8	100,000,000	
Red	2	100		White	9	1,000,000,000	
Orange	3	1,000		Gold	—	0.1	5
Yellow	4	10,000		Silver	—	0.01	10
Green	5	100,000		No color	—		20
Blue	6	1,000,000					

Standard Values for 1000-V Disc-Ceramic Capacitors

pF	pF	pF	pF
3.3	39	250	1000
5	47	270	1200
6	50	300	1500
6.8	51	330	1800
8	56	360	2000
10	68	390	2500
12	75	400	2700
15	82	470	3000
18	100	500	3300
20	120	510	3900
22	130	560	4700
24	150	600	5000
25	180	680	5600
27	200	750	6800
30	220	820	8200
33	240	910	10000

Common Values for Small Electrolytic Capacitors

μF	V*	μF	V*
33	6.3	10	35
33	10	22	35
100	10	33	35
220	10	47	35
330	10	100	35
470	10	220	35
10	16	330	35
22	16	470	35
33	16	1000	35
47	16	1	50
100	16	2.2	50
220	16	3.3	50
470	16	4.7	50
1000	16	10	50
2200	16	33	50
4.7	25	47	50
22	25	100	50
33	25	220	50
47	25	330	50
100	25	470	50
220	25	10	63
330	25	22	63
470	25	47	63
1000	25	1	100
2200	25	10	100
4.7	35	33	100

*Working voltage

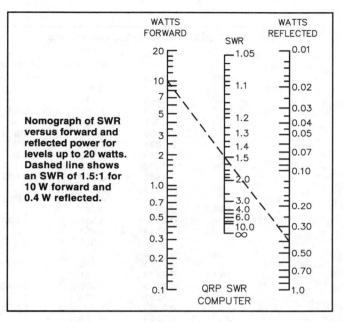

Nomograph of SWR versus forward and reflected power for levels up to 20 watts. Dashed line shows an SWR of 1.5:1 for 10 W forward and 0.4 W reflected.

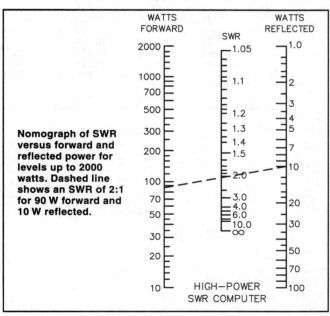

Nomograph of SWR versus forward and reflected power for levels up to 2000 watts. Dashed line shows an SWR of 2:1 for 90 W forward and 10 W reflected.

Fractions of an Inch with Metric Equivalents

Fractions Of An Inch		Decimals Of An Inch	Millimeters	Fractions Of An Inch		Decimals Of An Inch	Millimeters
	1/64	0.0156	0.397		33/64	0.5156	13.097
1/32		0.0313	0.794	17/32		0.5313	13.494
	3/64	0.0469	1.191		35/64	0.5469	13.891
1/16		0.0625	1.588	9/16		0.5625	14.288
	5/64	0.0781	1.984		37/64	0.5781	14.684
3/32		0.0938	2.381	19/32		0.5938	15.081
	7/64	0.1094	2.778		39/64	0.6094	15.478
1/8		0.1250	3.175	5/8		0.6250	15.875
	9/64	0.1406	3.572		41/64	0.6406	16.272
5/32		0.1563	3.969	21/32		0.6563	16.669
	11/64	0.1719	4.366		43/64	0.6719	17.066
3/16		0.1875	4.763	11/16		0.6875	17.463
	13/64	0.2031	5.159		45/64	0.7031	17.859
7/32		0.2188	5.556	23/32		0.7188	18.256
	15/64	0.2344	5.953		47/64	0.7344	18.653
1/4		0.2500	6.350	3/4		0.7500	19.050
	17/64	0.2656	6.747		49/64	0.7656	19.447
9/32		0.2813	7.144	25/32		0.7813	19.844
	19/64	0.2969	7.541		51/64	0.7969	20.241
5/16		0.3125	7.938	13/16		0.8125	20.638
	21/64	0.3281	8.334		53/64	0.8281	21.034
11/32		0.3438	8.731	27/32		0.8438	21.431
	23/64	0.3594	9.128		55/64	0.8594	21.828
3/8		0.3750	9.525	7/8		0.8750	22.225
	25/64	0.3906	9.922		57/64	0.8906	22.622
13/32		0.4063	10.319	29/32		0.9063	23.019
	27/64	0.4219	10.716		59/64	0.9219	23.416
7/16		0.4375	11.113	15/16		0.9375	23.813
	29/64	0.4531	11.509		61/64	0.9531	24.209
15/32		0.4688	11.906	31/32		0.9688	24.606
	31/64	0.4844	12.303		63/64	0.9844	25.003
1/2		0.5000	12.700	1		1.0000	25.400

Schematic Symbols Used in Circuit Diagrams

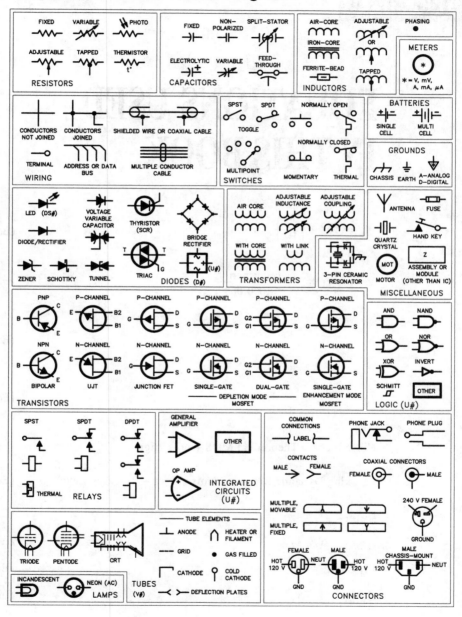

EQUATIONS USED IN THIS BOOK

$$dB = 10 \log\left(\frac{P_1}{P_2}\right)$$ (Equation 1-1)

$$Error = f(Hz) \times \frac{counter\ error}{1,000,000}$$ (Equation 4-1)

$$\frac{f_H}{f_V} = \frac{n_V}{n_H}$$ (Equation 4-2)

$$f_V = \frac{n_H}{n_V} f_H$$ (Equation 4-3)

Noise Floor = Theoretical MDS + noise figure (Equation 4-4)

Blocking Dynamic Range = | Noise Floor − Blocking Level | (Equation 4-5)

$$f_{IMD\ 1} = 2f_1 + f_2$$ (Equation 4-6)

$$f_{IMD\ 2} = 2f_1 - f_2$$ (Equation 4-7)

$$f_{IMD\ 3} = 2f_2 + f_1$$ (Equation 4-8)

$$f_{IMD\ 4} = 2f_2 - f_1$$ (Equation 4-9)

$$R = \frac{\rho \ell}{A}$$

(Equation 5-1)

$$\sigma = \frac{1}{\rho}$$

(Equation 5-2)

$$G = \frac{1}{R}$$

(Equation 5-3)

$$\tau = RC$$

(Equation 5-4)

$$R_T \text{ (series)} = R_1 + R_2 + R_3 + \dots + R_n$$

(Equation 5-5)

$$R_T \text{ (parallel)} = \frac{1}{\dfrac{1}{R_1} + \dfrac{1}{R_2} + \dfrac{1}{R_3} + \dots + \dfrac{1}{R_n}}$$

(Equation 5-6)

$$R_T \text{ (parallel)} = \frac{R_1 \times R_2}{R_1 + R_2}$$

(Equation 5-6A)

$$C_T \text{ (series)} = \frac{1}{\dfrac{1}{C_1} + \dfrac{1}{C_2} + \dfrac{1}{C_3} + \dots + \dfrac{1}{C_n}}$$

(Equation 5-7)

$$C_T \text{ (series)} = \frac{C_1 \times C_2}{C_1 + C_2}$$

(Equation 5-7A)

$$C_T \text{ (parallel)} = C_1 + C_2 + C_3 + \dots + C_n$$

(Equation 5-8)

$$V(t) = E\left(1 - e^{\frac{-t}{\tau}}\right)$$

(Equation 5-9)

$$V(t) = E\left(e^{\frac{-t}{\tau}}\right)$$

(Equation 5-10)

$$I(t) = \frac{E}{R}\left(1 - e^{\frac{-t}{\tau}}\right)$$

(Equation 5-11)

$$\tau = \frac{L}{R}$$

(Equation 5-12)

$$\tan\theta = \frac{\text{side opposite}}{\text{side adjacent}}$$

(Equation 5-13)

$$\sin\theta = \frac{\text{side opposite}}{\text{hypotenuse}}$$

(Equation 5-14)

$$C^2 = A^2 + B^2$$

(Equation 5-15)

$$C = \sqrt{A^2 + B^2}$$

(Equation 5-16)

$$Z = \frac{E}{I} = \frac{E_R + E_X}{I}$$

(Equation 5-17)

$$|Z| = \sqrt{R^2 + X^2}$$

(Equation 5-18)

$$\tan\theta = \frac{\text{side opposite to the angle}}{\text{side adjacent to the angle}}$$

(Equation 5-19)

$$Z = \frac{E}{I} = \frac{E}{I_R + I_X}$$

(Equation 5-20)

$$I_R = \frac{E_R}{R} = E_R \times G \qquad \text{(Equation 5-21)}$$

$$I_X = \frac{E_X}{X} = E_X \times B \qquad \text{(Equation 5-22)}$$

$$f_r = \frac{1}{2\pi \sqrt{LC}} \qquad \text{(Equation 5-23)}$$

$$L = \frac{1}{\left(2\pi f_r\right)^2 C} \qquad \text{(Equation 5-23A)}$$

$$C = \frac{1}{\left(2\pi f_r\right)^2 L} \qquad \text{(Equation 5-23B)}$$

$$Q = \frac{X}{R} \qquad \text{(Equation 5-24)}$$

$$Q = \frac{R}{X} \qquad \text{(Equation 5-25)}$$

$$\Delta f = \frac{f_r}{Q} \qquad \text{(Equation 5-26)}$$

$$P = IE \qquad \text{(Equation 5-27)}$$

$$P = I^2 R \qquad \text{(Equation 5-28)}$$

$$P = \frac{E^2}{R} \qquad \text{(Equation 5-29)}$$

$$\text{Power factor} = \frac{P_{REAL}}{P_{APPARENT}} \qquad \text{(Equation 5-30)}$$

$$P_{REAL} = P_{APPARENT} \times \text{Power factor} \qquad \text{(Equation 5-31)}$$

$$\text{Power factor} = \cos \theta \qquad \text{(Equation 5-32)}$$

$$dB = 10 \log \left(\frac{P_2}{P_1} \right) \qquad \text{(Equation 5-33)}$$

$$R_T = \frac{R_1 \times R_2}{R_1 + R_2} \qquad \text{(Equation 5-34)}$$

$$\beta = \frac{I_c}{I_b} \qquad \text{(Equation 6-1)}$$

$$\alpha = \frac{I_c}{I_e} \qquad \text{(Equation 6-2)}$$

$$V_{out} = - \frac{R_f}{R_1} V_{in} \qquad \text{(Equation 6-3)}$$

$$V_{gain} = \frac{V_{out}}{V_{in}} \qquad \text{(Equation 6-4)}$$

$$V_{gain} = \frac{\dfrac{R_f}{R_1} V_{in}}{V_{in}} = \frac{R_f}{R_1} \qquad \text{(Equation 6-5)}$$

$$\text{Gain (dB)} = 20 \log (V_{gain}) \qquad \text{(Equation 6-6)}$$

$$L = \frac{A_L \times N^2}{10,000} \quad \text{(For powdered-iron cores)}$$

(Equation 6-7)

$$N = 100 \sqrt{\frac{L}{A_L}} \quad \text{(For powdered-iron cores)}$$

(Equation 6-8)

$$L = \frac{A_L \times N^2}{1,000,000} \quad \text{(For ferrite cores)}$$

(Equation 6-9)

$$N = 1000 \sqrt{\frac{L}{A_L}} \quad \text{(For ferrite cores)}$$

(Equation 6-10)

$$T = 1.1 \, RC$$

(Equation 7-1)

$$f = \frac{1.46}{\left(R_1 + \left(2 \times R_2\right)\right)C_1}$$

(Equation 7-2)

$$R_1 = \frac{Q}{2\pi f_0 \, C_1}$$

(Equation 7-3)

$$R_2 = \frac{Q}{\left(2Q^2 - A_V\right)\left(2\pi f_0 \, C_1\right)}$$

(Equation 7-4)

$$R_3 = \frac{2Q}{2\pi f_0 \, C_1}$$

(Equation 7-5)

$$R_4 = R_5 \approx 0.02 \times R_3$$

(Equation 7-6)

$$R_{e-b} = \frac{26}{I_e}$$

(Equation 7-7)

$$A_V = \frac{R_L}{R_{e-b}}$$

(Equation 7-8)

$$R_b = \beta R_{e\text{-}b} \qquad \text{(Equation 7-9)}$$

$$A_V = \frac{R_L}{R_E} \qquad \text{(Equation 7-10)}$$

$$P = I \times E \qquad \text{(Equation 7-11)}$$

$$I = \frac{P}{E} \qquad \text{(Equation 7-12)}$$

$$Z = \frac{E}{I} \qquad \text{(Equation 7-13)}$$

$$F = N f_r \qquad \text{(Equation 7-14)}$$

$$f = \frac{1}{T} \qquad \text{(Equation 8-1)}$$

$$V_{peak} = V_{RMS} \times \sqrt{2} = V_{RMS} \times 1.414 \qquad \text{(Equation 8-2)}$$

$$V_{RMS} = \frac{V_{peak}}{\sqrt{2}} = V_{peak} \times 0.707 \qquad \text{(Equation 8-3)}$$

$$V_{avg} = V_{peak} \times 0.637 \qquad \text{(Equation 8-4)}$$

$$V_{avg} = V_{RMS} \times 0.900 \qquad \text{(Equation 8-5)}$$

peak-to-peak = peak positive − peak negative $\qquad$ (Equation 8-6)

peak-to-peak = 2 × peak positive = −2 × peak negative $\qquad$ (Equation 8-7)

$$V_{RMS} \times I_{RMS} = P_{avg}$$ (Equation 8-8)

$$V_{peak} \times I_{peak} = P_{peak} = 2 \times P_{avg}$$ (Equation 8-9)

$$PEP = \frac{\left(V_{RMS}\right)^2}{Z}$$ (Equation 8-10)

$$Efficiency = \frac{P_{OUT}}{P_{IN}} \times 100\%$$ (Equation 8-11)

$$P_{IN} = \frac{P_{OUT}}{Efficiency} \times 100\%$$ (Equation 8-12)

$$deviation\ ratio = \frac{D_{max}}{M}$$ (Equation 8-13)

$$modulation\ index = \frac{D_{max}}{m}$$ (Equation 8-14)

$$Bw = B \times K$$ (Equation 8-15)

$$\frac{1\ word}{minute} = \frac{50\ elements}{minute} = \frac{50\ elements}{60\ seconds} = 0.83\ bauds$$ (Equation 8-16)

$$\text{wpm} \times 0.83 = \text{bauds} \qquad \text{(Equation 8-17)}$$

$$\text{bauds} = \frac{\text{wpm}}{1.2} \qquad \text{(Equation 8-18)}$$

$$\text{Bw} = \left(\frac{\text{wpm}}{1.2}\right) \times \text{K} \qquad \text{(Equation 8-19)}$$

$$\text{Bw} = \left(\frac{\text{wpm}}{1.2}\right) \times 4.8 = \text{wpm} \times 4 \qquad \text{(Equation 8-20)}$$

$$\text{Bw} = (\text{K} \times \text{shift}) + \text{B} \qquad \text{(Equation 8-21)}$$

$$\text{dBd} = \text{dBi} - 2.14 \text{ dB} \qquad \text{(Equation 9-1)}$$

$$\text{dBi} = \text{dBd} + 2.14 \qquad \text{(Equation 9-2)}$$

$$\text{Efficiency} = \frac{R_R}{R_T} \times 100\% \qquad \text{(Equation 9-3)}$$

$$\text{Gain Ratio} = \frac{41,253}{\theta_E \times \theta_H} \quad \text{(For isotropic reference antenna)} \qquad \text{(Equation 9-4)}$$

$$\text{Gain Ratio} = \frac{25,000}{\theta_E \times \theta_H} \quad \text{(For dipole reference antenna)} \qquad \text{(Equation 9-5)}$$

$$\text{Gain Ratio} = \frac{41,253}{\theta^2} \quad \text{(For isotropic reference antenna)} \qquad \text{(Equation 9-6)}$$

$$\theta = \text{Beamwidth} = \sqrt{\frac{41,253}{\text{Gain Ratio}}} = \frac{203}{\sqrt{\text{Gain Ratio}}} \qquad \text{(Equation 9-7)}$$

$$dB = 10 \log\left(\frac{P_2}{P_1}\right) \qquad \text{(Equation 9-8)}$$

$$\frac{dB}{10} = \log(\text{Gain Ratio}) \qquad \text{(Equation 9-9)}$$

$$10^{\frac{dB}{10}} = \text{Gain Ratio} \qquad \text{(Equation 9-10)}$$

$$\lambda = \frac{v}{f} \qquad \text{(Equation 9-11)}$$

$$\lambda(\text{meters}) = \frac{300}{f(\text{MHz})} \qquad \text{(Equation 9-12)}$$

$$\lambda(\text{feet}) = \frac{984}{f(\text{MHz})} \qquad \text{(Equation 9-13)}$$

$$V = \frac{\text{speed of wave (in line)}}{\text{speed of light (in vacuum)}} \qquad \text{(Equation 9-14)}$$

$$V = \frac{1}{\sqrt{\varepsilon}}$$

(Equation 9-15)

$$\text{Length (meters)} = \frac{300}{f(\text{MHz})} \times V$$

(Equation 9-16)

$$\text{Length (feet)} = \frac{984}{f(\text{MHz})} \times V$$

(Equation 9-17)

GLOSSARY OF KEY WORDS

Absorption — The loss of energy from an electromagnetic wave as it travels through any material. The energy may be converted to heat or other forms. Absorption usually refers to energy lost as the wave travels through the ionosphere.

Accreditation — The process by which a **Volunteer Examiner Coordinator (VEC)** certifies that their **Volunteer Examiners (VEs)** are qualified to administer Amateur Radio license exams.

Adcock array — A radio direction finding antenna array consisting of two vertical elements fed 180° apart and capable of being rotated.

Admittance — The reciprocal of impedance, often used to aid the solution of a parallel-circuit impedance calculation.

Alpha (α) — The ratio of transistor collector current to emitter current. It is between 0.92 and 0.98 for a junction transistor.

Alpha cutoff frequency — A term used to express the useful upper frequency limit of a transistor. The point at which the gain of a common-base amplifier is 0.707 times the gain at 1 kHz.

Alternator whine — A common form of conducted interference typified by an audio tone being induced onto the received or transmitted signal. The pitch of the noise varies with alternator speed.

Amateur Satellite Service — A radiocommunication service using stations on Earth satellites for the same purpose as those of the amateur service.

Amplifier transfer function — A graph or equation that relates the input and output of an amplifier under various conditions.

Amplitude modulation — A method of superimposing an information signal on an RF carrier wave in which the amplitude of the RF envelope (carrier and sidebands) is varied in relation to the information signal strength.

Amplitude-compandored single sideband (ACSSB) — An SSB system that uses a logarithmic amplifier to compress voice signals at the transmitter and an inverse logarithmic amplifier to expand the voice signals in the receiver.

AND gate — A logic circuit whose output is 1 only when both of its inputs are 1.

Anode — The terminal that connects to the positive supply lead for current to flow through a device.

Antenna — An electric circuit designed specifically to radiate the energy applied to it in the form of electromagnetic waves. An antenna is reciprocal; a wave moving past it will induce a current in the circuit also. Antennas are used to transmit and receive radio waves.

Antenna bandwidth — A range of frequencies over which the antenna SWR will be below some specified value.

Antenna effect — One of two operational modes of a simple loop antenna wherein the antenna exhibits the characteristics of a small, nondirectional vertical antenna.

Antenna efficiency — The ratio of the radiation resistance to the total resistance of an antenna system, including losses.

Apogee — That point in a satellite's orbit (such as the Moon) when it is farthest from the Earth.

Apparent power — The product of the RMS current and voltage values in a circuit without consideration of the phase angle between them.

Ascending Pass — With respect to a particular ground station, a satellite pass during which the spacecraft is headed in a northerly direction while it is in range.

Ascending Pass — With respect to a particular ground station, a satellite pass during which the spacecraft is headed in a northerly direction while it is in range.

Astable (free-running) multivibrator — A circuit that alternates between two unstable states. This circuit could be considered as an oscillator that produces square waves.

ATV (amateur television) — A wideband TV system that can use commercial transmission standards. ATV is only permitted on the 70-cm band and higher frequencies.

ATV (amateur television) — A wideband TV system that can use commercial transmission standards. ATV is only permitted on the 70-cm band and higher frequencies.

Aurora — A disturbance of the atmosphere at high latitudes, which results from an interaction between electrically charged particles from the sun and the magnetic field of the Earth. Often a display of colored lights is produced, which is visible to those who are close enough to the magnetic-polar regions. Auroras can disrupt HF radio communication and enhance VHF communication. They are classified as visible auroras and radio auroras.

Authorization for alien reciprocal operation — FCC authorization to someone holding an amateur license issued by certain foreign governments to operate an amateur station in the US.

Automatic control — The operation of an amateur station without a control operator present at the control point. In §97.3 (a) (6), the FCC defines automatic control as "The use of devices and procedures for control of a station when it is transmitting so that compliance with the FCC Rules is achieved without the control operator being present at a control point."

Avalanche point — That point on a diode characteristic curve where the amount of reverse current increases greatly for small increases in reverse bias voltage.

Average power — The product of the RMS current and voltage values associated with a purely resistive circuit, equal to one half the peak power when the applied voltage is a sine wave.

Back EMF — An opposing electromotive force (voltage) produced by a changing current in a coil. It can be equal to (or greater than) the applied EMF under some conditions.

Balanced modulator — A circuit used in a single-sideband suppressed-carrier transmitter to combine a voice signal and an RF signal. The balanced modulator isolates the input signals from each other and the output, so that only the difference of the two input signals reaches the output.

Band-pass filter — A circuit that allows signals to go through it only if they are within a certain range of frequencies. It attenuates signals above and below this range.

Bandwidth — The frequency range (measured in hertz — Hz) over which a signal is stronger than some specified amount below the peak signal level. For example, a certain signal is at least half as strong as the peak power level over a range of ±3 kHz, so it has a 3-dB bandwidth of 6 kHz. As related to a transmitted signal, that frequency range that the signal occupies around a center frequency. Bandwidth increases with increasing information rate.

Base loading — The technique of inserting a coil at the bottom of an electrically short vertical antenna in order to cancel the capacitive reactance of the antenna, producing a resonant antenna system.

Baud — A unit of signaling speed equal to the number of discrete conditions or events per second. (For example, if the duration of a pulse is 3.33 milliseconds, the signaling rate is 300 bauds or the reciprocal of 0.00333 seconds.) One bit per second for single-channel binary-coded signals.

Baudot code — A digital code used for radioteletype operation, and also known as the International Telegraph Alphabet Number 2 (ITA2). Each character is represented by five data bits, plus additional start and stop bits.

Beamwidth — As related to directive antennas, the width (measured in degrees) of the major lobe between the two directions at which the relative power is one half (–3 dB) of the value at the peak of the lobe.

Bipolar junction transistor — A transistor made of two PN semiconductor junctions using two layers of similar-type material (N or P) with a third layer of the opposite type between them.

Bistable multivibrator — Another name for a flip-flop circuit that has two stable output states.

Blanking — Portion of a video signal that is "blacker than black," used to be certain that the return trace is invisible.

Blocking — A receiver condition in which reception of a desired weak signal is prevented because of a nearby, unwanted strong signal.

Butterworth filter — A filter whose passband frequency response is as flat as possible. The design is based on a Butterworth polynomial to calculate the input/output characteristics.

Capacitive coupling (of a dip meter) — A method of transferring energy from a dip-meter oscillator to a tuned circuit by means of an electric field.

Capture effect — An effect especially noticed with FM and PM systems whereby the strongest signal to reach the demodulator is the one to be received. You cannot tell whether weaker signals are present.

Cardioid radiation pattern — A heart-shaped antenna pattern characterized by a single, large lobe in one direction, and a deep, narrow null in the opposite direction.

Cathode — The terminal that connects to the negative supply lead for current to flow through a device.

Cathode-ray tube (CRT) — An electron beam tube in which the beam can be focused on a luminescent screen. The spot position can be varied to produce a pattern or picture on the screen.

Center loading — A technique for adding a series inductor at or near the center of an antenna element in order to cancel the capacitive reactance of the antenna. This technique is usually used with elements that are less than $1/4$ wavelength.

CEPT (European Conference of Postal and Telecommunications Administrations) agreement — A multilateral operating arrangement that allows US amateurs to operate in many European countries, and amateurs from many European countries to operate in the US.

Certificate of Successful Completion of Examination (CSCE) — A document issued by a Volunteer Examiner Team to certify that a candidate has passed specific exam elements at their test session. If the candidate qualified for a license upgrade at the exam session, this CSCE provides the authority to operate using the newly earned license privileges, with special identification procedures.

Certification — Equipment authorization granted by the FCC used to ensure that the equipment will function properly in the service for which it has been accepted.

Charge-coupled device (CCD) — An integrated circuit that uses a combination of analog and digital circuitry to sample and store analog signal voltage levels, passing the voltages through a capacitor string to the circuit output.

Chebyshev filter — A filter whose passband and stopband frequency response has an equal-amplitude ripple, and a sharper transition to the stop band than does a Butterworth filter. The design is based on a Chebyshev polynomial to calculate the input/output characteristics.

Circular polarization — Describes an electromagnetic wave in which the electric and magnetic fields are rotating. If the electric field vector is rotating in a clockwise sense, then it is called right-hand circular polarization and if the electric field is rotating in a counterclockwise sense, it is called left-hand circular polarization. Note that the polarization sense is determined by standing behind the antenna for a signal being transmitted, or in front of it for a signal being received.

Circulator — A passive device with three or more ports or input/output terminals. It can be used to combine the output from several transmitters to one antenna. A circulator acts as a one-way valve to allow radio waves to travel in one direction (to the antenna) but not in another (to the receiver).

Coaxial capacitor — A cylindrical capacitor used for noise-suppression purposes. The line to be filtered connects to the two ends of the capacitor, and a third connection is made to electrical ground. One side of the capacitor provides a dc path between the ends, while the other side of the capacitor connects to ground.

Compandoring — In an ACSSB system, the process of *com*pressing voice signals in a transmitter and ex*pand*ing them in a receiver.

Complementary metal-oxide semiconductor (CMOS) — A type of construction used to make digital integrated circuits. CMOS is composed of both N-channel and P-channel MOS devices on the same chip.

Complex number — A number that includes both a real and an imaginary part. Complex numbers provide a convenient way to represent a quantity (like impedance) that is made up of two different quantities (like resistance and reactance).

Conductance — The reciprocal of resistance. This is the real part of a complex admittance.

Conducted noise — Electrical noise that is imparted to a radio receiver or transmitter through the power connections to the radio.

Contest — An Amateur Radio operating activity in which operators try to contact as many other stations as possible. While each contest has its own particular rules and information to be exchanged, all contests have the common purpose of enhancing the communication and operating skills of amateurs, and helping ensure their readiness for emergency communications.

Control link — A device used by a control operator to manipulate the station adjustment controls from a location other than the station location. A control link provides the means of control between a control point and a remotely controlled station.

Counter (divider, divide-by-n counter) — A circuit that is able to change from one state to the next each time it receives an input signal. A counter produces an output signal every time a predetermined number of input signals have been received.

Crystal-controlled marker generator — An oscillator circuit that uses a quartz crystal to set the frequency, and which has an output rich in harmonics that can be used to determine band edges on a receiver. An output every 100 kHz or less is normally produced.

Crystal-lattice filter — A filter that employs piezoelectric crystals (usually quartz) as the reactive elements. They are most often used in the IF stages of a receiver or transmitter.

Decibel (dB) — One tenth of a bel, denoting a logarithm of the ratio of two power levels— dB = 10 log (P2/P1). Power gains and losses are expressed in decibels.

Delta match — A method for impedance matching between an open-wire transmission line and a half-wave radiator that is not split at the center. The feed-line wires are fanned out to attach to the antenna wire symmetrically around the center point. The resulting connection looks somewhat like a capital Greek delta.

Depletion mode — Type of operation in a JFET or MOSFET where current is reduced by reverse bias on the gate.

Descending pass — With respect to a particular ground station, a satellite pass during which the spacecraft is headed in a southerly direction while it is in range.

Desensitization — A reduction in receiver sensitivity caused by the receiver front end being overloaded by noise or RF from a local transmitter.

Detector — A circuit used in a receiver to recover the modulation (voice or other information) signal from the RF signal.

Deviation — The peak difference between an instantaneous frequency of the modulated wave and the unmodulated-carrier frequency in an FM system.

Deviation ratio — The ratio of the maximum frequency deviation to the maximum modulating frequency in an FM system.

Dielectric — An insulating material. A dielectric is a medium in which it is possible to maintain an electric field with little or no additional direct-current energy supplied after the field has been established.

Dielectric constant (ε) — Relative figure of merit for an insulating material. This is the property that determines how much electric energy can be stored in a unit volume of the material per volt of applied potential.

Digital IC — An integrated circuit whose output is either on (1) or off (0).

Dip meter — A tunable RF oscillator that supplies energy to another circuit resonant at the frequency that the oscillator is tuned to. A meter indicates when the most energy is being coupled out of the circuit by showing a dip in indicated current.

Dipole antenna — An antenna with two elements in a straight line that are fed in the center; literally, two poles. For amateur work, dipoles are usually operated at half-wave resonance.

Direct sequence — A spread-spectrum communications system where a very fast binary bit stream is used to shift the phase of an RF carrier.

Directive antenna — An antenna that concentrates the radiated energy to form one or more major lobes in specific directions. The receiving pattern is the same as the transmitting pattern.

Doping — The addition of impurities to a semiconductor material, with the intent to provide either excess electrons or positive charge carriers (holes) in the material.

Doppler shift — A change in the observed frequency of a signal, as compared with the transmitted frequency, caused by satellite movement toward or away from you.

Double-balanced mixer (DBM) — A mixer circuit that is balanced for both inputs, so that only the sum and the difference frequencies, but neither of the input frequencies, appear at the output. There will be no output unless both input signals are present.

Drain — The point at which the charge carriers exit an FET. Corresponds to the plate of a vacuum tube.

DX — Distance. On HF, often used to describe stations in countries outside your own.

Dynamic range — The ability of a receiver to tolerate strong signals outside the band-pass range. Blocking dynamic range and intermodulation distortion (IMD) dynamic range are the two most common dynamic range measurements used to predict receiver performance.

Earth station — An amateur station located on, or within 50 km of the Earth's surface intended for communications with space stations or with other Earth stations by means of one or more other objects in space.

Earth-Moon-Earth (EME) — A method of communicating with other stations by reflecting radio signals off the Moon's surface.

Effective isotropic radiated power (EIRP) — A measure of the power radiated from an antenna system. EIRP takes into account transmitter output power, feed-line losses and other system losses, and antenna gain as compared to an isotropic radiator.

Effective radiated power (ERP) — The relative amount of power radiated in a specific direction from an antenna, taking system gains and losses into account.

Electric field — A region through which an electric force will act on an electrically charged object.

Electric force — A push or pull exerted through space by on electrically charged object on another.

Electromagnetic waves — A disturbance moving through space or materials in the form of changing electric and magnetic fields.

Elliptical filter — A filter with equal-amplitude passband ripple and points of infinite attenuation in the stop band. The design is based on an elliptical function to calculate the input/output characteristics.

Emission designators — A method of identifying the characteristics of a signal from a radio transmitter using a series of three characters following the ITU system.

Emission types — A method of identifying the signals from a radio transmitter using a "plain English" format that simplifies the ITU **emission designators**.

Emissions — Any signals produced by a transmitter that reach the antenna connector to be radiated.

Enhancement mode — Type of operation in a MOSFET where current is increased by forward bias on the gate.

Equinoxes — One of two spots on the orbital path of the Earth around the sun, at which the Earth crosses a horizontal plane extending through the equator of the sun. The *vernal equinox* marks the beginning of spring and the *autumnal equinox* marks the beginning of autumn.

Examination Elements — Any of the telegraphy or written exam sections required for an Amateur Radio operator's license.

Exclusive OR gate — A logic circuit whose output is 1 when any single input is 1 and whose output is 0 when no input is 1 or when more than one input is 1.

Facsimile (fax) — The process of scanning pictures or images and converting the information into signals that can be used to form a likeness of the copy in another location. The pictures are often printed on paper for permanent display.

Faraday rotation — A rotation of the polarization plane of radio waves when the waves travel through the ionized magnetic field of the ionosphere.

Fast-scan TV (FSTV) — Another name for ATV, used because a new frame is transmitted every $1/30$ of a second, as compared to every 8 seconds for slow-scan TV.

Field — The region of space through which any of the invisible forces in nature, such as gravity, electric force or magnetic forces, act.

Field-effect transistor (FET) — A voltage-controlled semiconductor device. Output current can be varied by varying the input voltage. The input impedance of an FET is very high.

Flip-flop (bistable multivibrator) — A circuit that has two stable output states, and which can change from one state to the other when the proper input signals are detected.

Folded dipole — An antenna consisting of two (or more) parallel, closely spaced halfwave wires connected at their ends. One of the wires is fed at its center.

Forward bias — A voltage applied across a semiconductor junction so that it will tend to produce current.

Forward error correction (FEC) — One form of AMTOR operation, in which each character is sent twice. The receiving station compares the mark/space ratio of the characters to determine if any errors occurred in the reception.

Frequency, f — The number of complete cycles of a wave occurring in a unit of time.

Frequency counter — A digital-electronic device that counts the cycles of an electromagnetic wave for a certain amount of time and gives a digital readout of the frequency.

Frequency domain — A time-independent way to view a complex signal. The various component sine waves that make up a complex waveform are shown by frequency and amplitude on a graph or the CRT display of a spectrum analyzer.

Frequency hopping — A spread-spectrum communications system where the center frequency of a conventional carrier is altered many times a second in accordance with a pseudorandom list of channels.

Frequency modulation — A method of superimposing an information signal on an RF carrier wave in which the instantaneous frequency of an RF carrier wave is varied in relation to the information signal strength.

Frequency standard — A circuit or device used to produce a highly accurate reference frequency. The frequency standard may be a crystal oscillator in a marker generator or a radio broadcast, such as from WWV, with a carefully controlled transmit frequency.

Gain — An increase in the effective power radiated by an antenna in a certain desired direction. This is at the expense of power radiated in other directions.

Gamma match — A method for matching the impedance of a feed line to a half-wave radiator that is split in the center (such as a dipole). It consists of an adjustable arm that is mounted close to the driven element and in parallel with it near the feed point. The connection looks somewhat like a capital Greek gamma.

Gate — Control terminal of an FET. Corresponds to the grid of a vacuum tube. **Gate** also refers to a combinational logic element with two or more inputs and one output. The output state depends upon the state of the inputs.

Gray line — A band around the Earth that is the transition region between daylight and darkness.

Gray-line propagation — A generally north-south enhancement of propagation that occurs along the gray line, when D layer absorption is rapidly decreasing at sunset, or has not yet built up around sunrise.

Gray scale — A photographic term that defines a series of neutral densities (based on the percentage of incident light that is reflected from a surface), ranging from white to black.

Great circle — An imaginary circle around the surface of the Earth formed by the intersection of the surface with a plane passing through the center of the Earth.

Great-circle path — The shortest distance between two points on the surface of the Earth, which follows the arc of a great circle passing through both points.

Grid square locator — A 2° longitude by 1° latitude square, as part of a world wide numbering system. Grid square locators are exchanged in some contests, and are used as the basis for some VHF/UHF awards.

Ground-wave signals — Radio signals that are propagated along the ground rather than through the ionosphere or by some other means.

Half-power points — Those points on the response curve of a resonant circuit where the power is one half its value at resonance.

High-pass filter — A filter that allows signals above the cutoff frequency to pass through. It attenuates signals below the cutoff frequency.

Horizontal polarization — Describes an electromagnetic wave in which the electric field is horizontal, or parallel to the Earth's surface.

Hot-carrier diode — A type of diode in which a small metal dot is placed on a single semiconductor layer. It is superior to a point-contact diode in most respects.

IARP (International Amateur Radio Permit) — A multilateral operating arrangement that allows US amateurs to operate in many Central and South American countries, and amateurs from many Central and South American countries to operate in the US.

Image signal — An unwanted signal that mixes with a receiver local oscillator to produce a signal at the desired intermediate frequency.

Imaginary number — A value that sometimes comes up in solving a mathematical problem, equal to the square root of a negative number. Since there is no real number that can be multiplied by itself to give a negative result, this quantity is imaginary. In electronics work, the symbol j is used to represent . Other imaginary numbers are represented by j, where x is the positive part of the number. The reactance and susceptance of complex impedances and admittances are normally given in terms of j.

Inductive coupling (of a dip meter) — A method of transferring energy from a dip-meter oscillator to a tuned circuit by means of a magnetic field between two coils.

Integrated circuit — A device composed of many bipolar or field-effect transistors manufactured on the same chip, or wafer, of silicon.

Intermodulation distortion (IMD) — A type of interference that results from the unwanted mixing of two strong signals, producing a signal on an unintended frequency. The resulting mixing products can interfere with desired signals on those frequencies. "Intermod" usually occurs in a nonlinear stage or device.

Inverter — A logic circuit with one input and one output. The output is 1 when the input is 0, and the output is 0 when the input is 1.

Isolator — A passive attenuator in which the loss in one direction is much greater than the loss in the other.

Isotropic radiator — An imaginary antenna in free space that radiates equally in all directions (a spherical radiation pattern). It is used as a reference to compare the gain of various real antennas.

Joule — The unit of energy in the metric system of measure.

Junction field-effect transistor (JFET) — A field-effect transistor created by diffusing a gate of one type of semiconductor material into a channel of the opposite type of semiconductor material.

K index — A geomagnetic-field measurement that is updated every three hours at Boulder, Colorado. Changes in the K index can be used to indicate HF propagation conditions. Rising values generally indicate disturbed conditions while falling values indicate improving conditions.

L network — A combination of a capacitor and an inductor, one of which is connected in series with the signal lead while the other is shunted to ground.

Latch — Another name for a bistable multivibrator flip-flop circuit. The term **latch** reminds us that this circuit serves as a memory unit, storing a bit of information.

Libration fading — A fluttery, rapid fading of EME signals, caused by short-term motion of the Moon's surface relative to an observer on Earth.

Light-emitting diode — A device that uses a semiconductor junction to produce light when current flows through it.

Linear electronic voltage regulator — A type of voltage-regulator circuit that varies either the current through a fixed dropping resistor or the resistance of the dropping element itself. The conduction of the control element varies in direct proportion to the line voltage or load current.

Linear IC — An integrated circuit whose output voltage is a linear (straight line) representation of its input voltage.

Linear polarization — Describes the orientation of the electric-field component of an electromagnetic wave. The electric field can be vertical or horizontal with respect to the Earth's surface, resulting in either a vertically or a horizontally polarized wave. (Also called **plane polarization**.)

Lissajous figure — An oscilloscope pattern obtained by connecting one sine wave to the vertical amplifier and another sine wave to the horizontal amplifier. The two signals must be harmonically related to produce a stable pattern.

Loading coil — An inductor that is inserted in an antenna element or transmission line for the purpose of producing a resonant system at a specific frequency.

Logic probe — A simple piece of test equipment used to indicate high or low logic states (voltage levels) in digital-electronic circuits.

Long-path propagation — Propagation between two points on the Earth's surface that follows a path along the great circle between them, but is in a direction opposite from the shortest distance between them.

Loop antenna — An antenna configured in the shape of a loop. If the current in the loop, or in multiple parallel turns, is essentially uniform, and if the loop circumference is small compared with a wavelength, the radiation pattern is symmetrical, with maximum response in either direction of the loop plane.

Low-pass filter — A filter that allows signals below the cutoff frequency to pass through. It attenuates signals above the cutoff frequency.

Magnetic field — A region through which a magnetic force will act on a magnetic object.

Magnetic force — A push or pull exerted through space by one magnetically charged object on another.

Major lobe of radiation — A three-dimensional area that contains the maximum radiation peak in the space around an antenna. The field strength decreases from the peak level, until a point is reached where it starts to increase again. The area described by the radiation maximum is known as the major lobe.

Matching stub — A section of transmission line used to tune an antenna element to resonance or to aid in obtaining an impedance match between the feed point and the feed line.

Maximum average forward current — The highest average current that can flow through the diode in the forward direction for a specified junction temperature.

Metal-oxide semiconductor FET (MOSFET) — A field-effect transistor that has its gate insulated from the channel material. Also called an IGFET, or insulated gate FET.

Meteor — A particle of mineral or metallic material that is in a highly elliptical orbit around the Sun. As the Earth's orbit crosses the orbital path of a meteor, it is attracted by the Earth's gravitational field, and enters the atmosphere. A typical meteor is about the size of a grain of sand.

Meteor-scatter communication — A method of radio communication that uses the ionized trail of a meteor, which has burned up in the Earth's atmosphere, to reflect radio signals back to Earth.

Minimum discernible signal (MDS) — The smallest input signal level that can just be detected above the receiver internal noise. Also called **noise floor**.

Minor lobe of radiation — Those areas of an antenna pattern where there is some increase in radiation, but not as much as in the major lobe. Minor lobes normally appear at the back and sides of the antenna.

Mixer — A circuit that takes two or more input signals, and produces an output that includes the sum and difference of those signal frequencies.

Modulation index — The ratio of the maximum frequency deviation of the modulated wave to the instantaneous frequency of the modulating signal.

Modulator — A circuit designed to superimpose an information signal on an RF carrier wave.

Monolithic microwave integrated circuit (MMIC) — A small pill-sized amplifying device that simplifies amplifier designs for microwave-frequency circuits. An MMIC has an input lead, an output lead and two ground leads.

Monostable multivibrator (one shot) — A circuit that has one stable state. It can be forced into an unstable state for a time determined by external components, but it will revert to the stable state after that time.

Moonbounce — A common name for EME communication.

Multipath — A fading effect caused by the transmitted signal traveling to the receiving station over more than one path.

NAND (NOT AND) gate — A logic circuit whose output is 0 only when both inputs are 1.

Neutralization — Feeding part of the output signal from an amplifier back to the input so it arrives out of phase with the input signal. This negative feedback neutralizes the effect of positive feedback caused by coupling between the input and output circuits in the amplifier. The negative-feedback signal is usually supplied by connecting a capacitor from the output to the input circuit.

Node — A point where a satellite crosses the plane passing through the Earth's equator. It is an ascending node if the satellite is moving from south to north, and a descending node if the satellite is moving from north to south.

Noise figure — A ratio of the noise output power to the noise input power when the input termination is at a standard temperature of 290 K. It is a measure of the noise generated in the receiver circuitry.

Noise floor — The smallest input signal level that can just be detected above the receiver internal noise. Also called **minimum discernible signal (MDS)**.

Noninverting buffer — A logic circuit with one input and one output, and whose output level is the same as the input level.

Nonresonant rhombic antenna — A diamond-shaped antenna consisting of sides that are each at least one wavelength long. The feed line is connected to one end of the diamond, and there is a terminating resistance of approximately 800 Ω at the opposite end. The antenna has a unidirectional radiation pattern.

NOR (NOT OR) gate — A logic circuit whose output is 0 if either input is 1.

N-type material — Semiconductor material that has been treated with impurities to give it an excess of electrons. We call this a "donor material."

Offset voltage — As related to op amps, the differential amplifier output voltage when the inputs are shorted. It can also be measured as the voltage between the amplifier input terminals in a closed-loop configuration.

Operational amplifier (op amp) — A linear IC that can amplify dc as well as ac. Op amps have very high input impedance, very low output impedance and very high gain.

Optical shaft encoder — A device consisting of two pairs of photoemitters and photodetectors, used to sense the rotation speed and direction of a knob or dial. Optical shaft encoders are often used with the tuning knob on a modern radio to provide a tuning signal for the microprocessor controlling the frequency synthesizer.

Optocoupler (optoisolator) — A device consisting of a photoemitter and a photodetector used to transfer a signal between circuits using widely varying operating voltages.

OR gate — A logic circuit whose output is 1 when either input is 1.

Oscillator — A circuit built by adding positive feedback to an amplifier. It produces an alternating current signal with no input signal except the dc operating voltages.

Oscilloscope — A device using a cathode-ray tube to display the waveform of an electric signal with respect to time or as compared with another signal.

Packet cluster — A packet radio system that is devoted to serving a special interest group, such as DXers or contest operators.

Packet radio — A form of digital communication that includes error checking and correction, to ensure virtually error-free information exchange.

Parabolic (dish) antenna — An antenna reflector that is a portion of a parabolic curve. Used mainly at UHF and higher frequencies to obtain high gain and narrow beamwidth when excited by one of a variety of driven elements placed at the dish focus to illuminate the reflector.

Parallel-resonant circuit — A circuit including a capacitor, an inductor and sometimes a resistor, connected in parallel, and in which the inductive and capacitive reactances are equal at the applied-signal frequency. The circuit impedance is a maximum, and the current is a minimum at the resonant frequency.

Parasitics — Undesired oscillations or other responses in an amplifier.

Path loss — The total signal loss between transmitting and receiving stations relative to the total radiated signal energy.

Peak envelope power (PEP) — An expression used to indicate the maximum power level in a signal. It is found by squaring the RMS voltage at the envelope peak, and dividing by the load resistance. The average power of the RF envelope during a modulation peak. (Used for modulated RF signals.)

Peak envelope voltage (PEV) — The maximum peak voltage occurring in a complex waveform.

Peak inverse voltage (PIV) — The maximum instantaneous anode-to-cathode reverse voltage that is to be applied to a diode.

Peak negative value — On a signal waveform, the maximum displacement from the zero line in the negative direction.

Peak positive value — On a signal waveform, the maximum displacement from the zero line in the positive direction.

Peak power — The product of peak voltage and peak current in a resistive circuit. (Used with sine-wave signals.)

Peak voltage — A measure of voltage on an ac waveform taken from the centerline (0 V) and the maximum positive or negative level.

Peak-to-peak (P-P) value — On a signal waveform, the maximum displacement between the peak positive value and the peak negative value.

Peak-to-peak (P-P) voltage — A measure of the voltage taken between the negative and positive peaks on a cycle.

Pedersen ray — A high-angle radio wave that penetrates deeper into the F region of the ionosphere, so the wave is bent less than a lower-angle wave, and thus travels for some distance through the F region, returning to Earth at a distance farther than normally expected for single-hop propagation.

Perigee — That point in the orbit of a satellite (such as the Moon) when it is closest to the Earth.

Period — The time it takes for a complete orbit, usually measured from one equator crossing to the next. The higher the altitude, the longer the period.

Period, T — The time it takes to complete one cycle of an ac waveform.

Persistence — A property of a cathode-ray tube (CRT) that describes how long an image will remain visible on the face of the tube after the electron beam has been turned off.

Phase — A representation of the relative time or space between two points on a waveform, or between related points on different waveforms. Also the time interval between two events in a regularly recurring cycle.

Phase angle — If one complete cycle of a waveform is divided into 360 equal parts, then the phase relationship between two points or two waves can be expressed as an angle.

Phase-locked loop (PLL) — A servo loop consisting of a phase detector, low-pass filter, dc amplifier and voltage-controlled oscillator.

Phase modulation — A method of superimposing an information signal on an RF carrier wave in which the phase of an RF carrier wave is varied in relation to the information signal strength.

Phase modulator — A device capable of modulating an ac signal by varying the reactance of an amplifier circuit in response to the modulating signal. (The modulating signal may be voice, data, video or some other kind.) The circuit capacitance or inductance changes in response to an audio input signal. Used in PM (or FM) systems, this circuit acts as a variable reactance in an amplifier tank circuit.

Phase noise — Undesired variations in the phase of an oscillator signal. Phase noise is usually associated with phase-locked loop (PLL) oscillators.

Photocell — A solid-state device in which the voltage and current-conducting characteristics change as the amount of light striking the cell changes.

Photoconductive effect — A result of the photoelectric effect that shows up as an increase in the electric conductivity of a material. Many semiconductor materials exhibit a significant increase in conductance when electromagnetic radiation strikes them.

Photodetector — A device that produces an amplified signal that changes with the amount of light striking a light-sensitive surface.

Photoelectric effect — An interaction between electromagnetic radiation and matter resulting in photons of radiation being absorbed and electrons being knocked loose from the atom by this energy.

Phototransistor — A bipolar transistor constructed so the base-emitter junction is exposed to incident light. When light strikes this surface, current is generated at the junction, and this current is then amplified by transistor action

Pi network output-coupling circuits — A combination of two like reactances (coil or capacitor) and one of the opposite type. The single component is connected in series with the signal lead and the two others are shunted to ground, one on either side of the series element.

Piezoelectric effect — The physical deformation of a crystal when a voltage is applied across the crystal surfaces.

Pilot tone — In an ACSSB system, a 3.1-kHz tone transmitted with the voice signal to allow a mobile receiver to lock onto the signal. The pilot tone is also used to control the inverse logarithmic amplifier gain.

PIN diode — A diode consisting of a relatively thick layer of nearly pure semiconductor material (intrinsic semiconductor) with a layer of P-type material on one side and a layer of N-type material on the other.

Plane polarization — Describes the orientation of the electric-field component of an electromagnetic wave. The electric field can be vertical or horizontal with respect to the Earth's surface, resulting in either a vertically or a horizontally polarized wave. (Also called **linear polarization**.)

PN junction — The contact area between two layers of opposite-type semiconductor material.

Point-contact diode — A diode that is made by a pressure contact between a semiconductor material and a metal point.

Polar-coordinate system — A method of representing the position of a point on a plane by specifying the radial distance from an origin, and an angle measured counterclockwise from the 0° line.

Polarization — A property of an electromagnetic wave that tells whether the electric field of the wave is oriented vertically or horizontally with respect to earth. The polarization sense can change from vertical to horizontal under some conditions, and can even be gradually rotating either in a clockwise (right-hand-circular polarization) or a counterclockwise (left-hand-circular polarization) direction.

Potential energy — Stored energy. This stored energy can do some work when it is "released." For example, electrical energy can be stored as an electric field in a capacitor or as a magnetic field in an inductor. This stored energy can produce a current in a circuit when it is released.

Power — The time rate of transferring or transforming energy, or the rate at which work is done. In an electric circuit, power is calculated by multiplying the voltage applied to the circuit by the current through the circuit.

Power factor — The ratio of real power to apparent power in a circuit. Also calculated as the cosine of the phase angle between current and voltage in a circuit.

Prescaler — A divider circuit used to increase the useful range of a frequency counter.

Pseudonoise (PN) — A binary sequence designed to appear to be random (contain an approximately equal number of ones and zeros). Pseudonoise is generated by a digital circuit and mixed with digital information to produce a direct-sequence spread-spectrum signal.

P-type material — A semiconductor material that has been treated with impurities to give it an electron shortage. This creates excess positive charge carriers, or "holes," so it becomes an "acceptor material."

Pulse modulation — Modulation of an RF carrier by a series of pulses. These pulses convey the information that has been sampled from an analog signal.

Pulse-amplitude modulation (PAM) — A pulse-modulation system where the amplitude of a standard pulse is varied in relation to the information-signal amplitude at any instant.

Pulse-position modulation (PPM) — A pulse-modulation system where the position (timing) of the pulses is varied from a standard value in relation to the information-signal amplitude at any instant.

Pulse-width modulation (PWM) — A pulse-modulation system where the width of a pulse is varied from a standard value in relation to the information-signal amplitude at any instant.

Q — A quality factor describing how closely a practical coil or capacitor approaches the characteristics of an ideal component.

Radiated noise — Usually referring to a mobile installation, noise that is being radiated from the ignition system or electrical system of a vehicle and causing interference to the reception of radio signals.

Radiation resistance — The equivalent resistance that would dissipate the same amount of power as is radiated from an antenna. It is calculated by dividing the radiated power by the square of the RMS antenna current.

Radio Amateur Civil Emergency Service (RACES) — A radio service using amateur stations for civil defense communications during periods of local, regional or national civil emergencies.

Radio horizon — The position at which a direct wave radiated from an antenna becomes tangent to the surface of the Earth. Note that as the wave continues past the horizon, the wave gets higher and higher above the surface.

Reactance modulator — A device capable of modulating an ac signal by varying the reactance of an oscillator circuit in response to the modulating signal. (The modulating signal may be voice, data, video or some other kind.) The circuit capacitance or inductance changes in response to an audio input signal. Used in FM systems, this circuit acts as a variable reactance in an oscillator tank circuit.

Reactive power — The apparent power in an inductor or capacitor. The product of RMS current through a reactive component and the RMS voltage across it. Also called wattless power.

Real power — The actual power dissipated in a circuit, calculated to be the product of the apparent power times the phase angle between the voltage and current.

Rectangular-coordinate system — A method of representing the position of a point on a plane by specifying the distance from an origin in two perpendicular directions.

Reflection coefficient (ρ) — The ratio of the reflected voltage at a given point on a transmission line to the incident voltage at the same point. The reflection coefficient is also equal to the ratio of reflected and incident currents.

Remote control — The operation of an Amateur Radio station using a **control link** to manipulate the station operating adjustments from somewhere other than the station location.

Resonant frequency — That frequency at which a circuit including capacitors and inductors presents a purely resistive impedance. The inductive reactance in the circuit is equal to the capacitive reactance.

Resonant rhombic antenna — A diamond-shaped antenna consisting of sides that are each at least one wavelength long. The feed line is connected to one end of the diamond, and the opposite end is left open. The antenna has a bidirectional radiation pattern.

Reverse bias — A voltage applied across a semiconductor junction so that it will tend to prevent current.

Root-mean-square (RMS) voltage — A measure of the effective value of an ac voltage.

Sawtooth wave — A waveform consisting of a linear ramp and then a return to the original value. It is made up of sine waves at a fundamental frequency and all harmonics.

Scanning — The process of analyzing or synthesizing, in a predetermined manner, the light values or equivalent characteristics of elements constituting a picture area. Also the process of recreating those values to produce a picture on a CRT screen.

Selective fading — A variation of radio-wave field intensity that is different over small frequency changes. It may be caused by changes in the material that the wave is traveling through or changes in transmission path, among other things.

Selectivity — A measure of the ability of a receiver to distinguish between a desired signal and an undesired one at some different frequency. Selectivity can be applied to the RF, IF and AF stages.

Semiconductor material — A material with resistivity between that of metals and insulators. Pure semiconductor materials are usually doped with impurities to control the electrical properties.

Sensing antenna — An omnidirectional antenna used in conjunction with an antenna that exhibits a bidirectional pattern to produce a radio direction-finding system with a cardioid pattern.

Sensitivity — A measure of the minimum input-signal level that will produce a certain audio output from a receiver.

Sequential logic — A type of circuit element that has at least one output and one or more input channels, and in which the output state depends on the previous input states. A flip-flop is one sequential-logic element.

Series-resonant circuit — A circuit including a capacitor, an inductor and sometimes a resistor, connected in series, and in which the inductive and capacitive reactances are equal at the applied-signal frequency. The circuit impedance is at a minimum, and the current is a maximum at the resonant frequency.

Sine wave — A single-frequency waveform that can be expressed in terms of the mathematical sine function.

Single-sideband, suppressed-carrier signal — A radio signal in which only one of the two sidebands generated by amplitude modulation is transmitted. The other sideband and the RF carrier wave are removed before the signal is transmitted.

Skin effect — A condition in which ac flows in the outer portions of a conductor. The higher the signal frequency, the less the electric and magnetic fields penetrate the conductor and the smaller the effective area of a given wire for carrying the electrons.

Sky-wave signals — Radio signals that travel through the ionosphere to reach the receiving station. Sky-wave signals will cause a variation in the measured received-signal direction, resulting in an error with a radio direction-finding system.

Slow-scan television (SSTV) — A television system used by amateurs to transmit pictures within a signal bandwidth allowed on the HF bands by the FCC. It takes approximately 8 seconds to send a single black and white SSTV frame, and between 12 seconds and 4 1/2 minutes for the various color systems currently in use.

Smith Chart — A coordinate system developed by Phillip Smith to represent complex impedances on a graph. This chart makes it easy to perform calculations involving antenna and transmission-line impedances and SWR.

Solar wind — Electrically charged particles emitted by the sun, and traveling through space. The wind strength depends on how severe the disturbance on the sun was. These charged particles may have a sudden impact on radio communications when they arrive at the atmosphere of the Earth.

Source — The point at which the charge carriers enter an FET. Corresponds to the cathode of a vacuum tube.

Space station — An amateur station located more than 50 km above the Earth's surface.

Spectrum analyzer — A test instrument generally used to display the power (or amplitude) distribution of a signal with respect to frequency.

Spin modulation — Periodic amplitude fade-and-peak variations resulting from a Phase 3 satellite's 60 r/min spin.

Spread-spectrum (SS) communication — A communications method in which the RF bandwidth of the transmitted signal is much larger than that needed for traditional modulation schemes, and in which the RF bandwidth is independent of the modulation content. The frequency or phase of the RF carrier changes very rapidly according to a particular pseudo-random sequence. SS systems are resistant to interference because signals not using the same spreading sequence code are suppressed in the receiver.

Spread-spectrum modulation — A signal-transmission technique where the transmitted carrier is spread out over a wide bandwidth.

Spurious emissions — Any emission that is not part of the desired signal. The FCC defines this term as "an emission, on frequencies outside the necessary bandwidth of a transmission, the level of which may be reduced without affecting the information being transmitted."

Square wave — A periodic waveform that alternates between two values, and spends an equal time at each level. It is made up of sine waves at a fundamental frequency and all odd harmonics.

Surface-mount package — An electronic component without wire leads, designed to be soldered directly to copper-foil pads on a circuit board.

Susceptance — The reciprocal of reactance. This is the imaginary part of a complex admittance.

Switching regulator — A voltage-regulator circuit in which the output voltage is controlled by turning the pass element on and off at a high rate, often several kilohertz. The control-element duty cycle is proportional to the line or load conditions.

Telecommand operation — A one-way transmission to initiate, modify, or terminate functions of a device at a distance.

Telecommand station — An amateur station that transmits communications to initiate, modify, or terminate functions of a space station.

Telemetry — A one-way transmission of measurements at a distance from the measuring instrument.

Terminator — A band around the Earth that separates night from day.

Thevenin's Theorem — Any combination of voltage sources and impedances, no matter how complex, can be replaced by a single voltage source and a single impedance that will present the same voltage and current to a load circuit.

Time constant — The product of resistance and capacitance in a simple series or parallel RC circuit, or the inductance divided by the resistance in a simple series or parallel RL circuit. One time constant is the time required for a voltage across a capacitor or a current through an inductor to build up to 63.2% of its steady-state value, or to decay to 36.8% of the initial value. After a total of 5 time constants have elapsed, the voltage or current is considered to have reached its final value.

Time domain — A method of viewing a complex signal. The amplitude of the complex wave is displayed over changing time. The display shows only the complex waveform, and does not necessarily indicate the sine-wave signals that make up the wave.

Top loading — The addition of inductive reactance (a coil) or capacitive reactance (a capacitance hat) at the end of a driven element opposite the feed point. It is intended to increase the electrical length of the radiator.

Toroid — A coil wound on a donut-shaped ferrite or powdered-iron form.

Transequatorial propagation — A form of F-layer ionospheric propagation, in which signals of higher frequency than the expected MUF are propagated across the Earth's magnetic equator.

Transistor-transistor logic (TTL) — Digital integrated circuits composed of bipolar transistors, possibly as discrete components, but usually part of a single IC. Power supply voltage should be 5 V.

Transponder — A repeater aboard a satellite that retransmits, on another frequency band, any type of signals it receives. Signals within a certain receiver bandwidth are translated to a new frequency band, so many signals can share a transponder simultaneously.

Traps — Parallel LC networks inserted in an antenna element to provide multiband operation.

Triangulation — A radio direction-finding technique in which compass bearings from two or more locations are taken, and lines are drawn on a map to predict the location of a radio signal source.

Tropospheric ducting — A type of radio-wave propagation whereby the VHF communications range is greatly extended. Certain weather conditions cause portions of the troposphere to act like a duct or waveguide for the radio signals.

Truth table — A chart showing the outputs for all possible input combinations to a digital circuit.

Tunnel diode — A diode with an especially thin depletion region, so that it exhibits a negative resistance characteristic.

Unijunction transistor (UJT) — A three-terminal, single-junction device that exhibits negative resistance and switching characteristics unlike bipolar transistors.

Varactor diode — A component whose capacitance varies as the reverse-bias voltage is changed. This diode has a voltage-variable capacitance.

Velocity factor — An expression of how fast a radio wave will travel through a material. It is usually stated as a fraction of the speed the wave would have in free space (where the wave would have its maximum velocity). Velocity factor is also sometimes specified as a percentage of the speed of a radio wave in free space.

Vertical polarization — Describes an electromagnetic wave in which the electric field is vertical, or perpendicular to the Earth's surface.

Vestigial sideband (VSB) — A signal-transmission method in which one sideband, the carrier and part of the second sideband are transmitted. The bandwidth is not as wide as for a double-sideband AM signal, but not as narrow as a single-sideband signal.

Vidicon tube — A type of photosensitive vacuum tube widely used in TV cameras.

Volunteer Examiner (VE) — A licensed amateur who is accredited by a **Volunteer Examiner Coordinator (VEC)** to administer amateur license exams.

Volunteer Examiner Coordinator (VEC) — An organization that has entered into an agreement with the FCC to coordinate amateur license examinations.

Zener diode — A diode that is designed to be operated in the reverse-breakdown region of its characteristic curve.

Zener voltage — A reverse-bias voltage that produces a sudden change in apparent resistance across the diode junction, from a large value to a small value.

ABOUT THE ARRL

The seed for Amateur Radio was planted in the 1890s, when Guglielmo Marconi began his experiments in wireless telegraphy. Soon he was joined by dozens, then hundreds, of others who were enthusiastic about sending and receiving messages through the air—some with a commercial interest, but others solely out of a love for this new communications medium. The United States government began licensing Amateur Radio operators in 1912.

By 1914, there were thousands of Amateur Radio operators—hams—in the United States. Hiram Percy Maxim, a leading Hartford, Connecticut, inventor and industrialist, saw the need for an organization to band together this fledgling group of radio experimenters. In May 1914 he founded the American Radio Relay League (ARRL) to meet that need.

Today ARRL, with approximately 170,000 members, is the largest organization of radio amateurs in the United States. The ARRL is a not-for-profit organization that:

• promotes interest in Amateur Radio communications and experimentation
• represents US radio amateurs in legislative matters, and
• maintains fraternalism and a high standard of conduct among Amateur Radio operators.

At ARRL headquarters in the Hartford suburb of Newington, the staff helps serve the needs of members. ARRL is also International Secretariat for the International Amateur Radio Union, which is made up of similar societies in 150 countries around the world.

ARRL publishes the monthly journal *QST*, as well as newsletters and many publications covering all aspects of Amateur Radio. Its headquarters station, W1AW, transmits bulletins of interest to radio amateurs and Morse code practice sessions. The ARRL also coordinates an extensive field organization, which includes volunteers who provide technical information and other support for radio amateurs as well as communications for public-service activities. ARRL also represents US amateurs with the Federal Communications Commission and other government agencies in the US and abroad.

Membership in ARRL means much more than receiving *QST* each month. In addition to the services already described, ARRL offers membership services on a personal level, such as the ARRL Volunteer Examiner Coordinator Program and a QSL bureau.

Full ARRL membership (available only to licensed radio amateurs) gives you a voice in how the affairs of the organization are governed. ARRL policy is set by a Board of Directors (one from each of 15 Divisions). Each year, one-third of the ARRL Board of Directors stands for election by the full members they represent. The day-to-day operation of ARRL HQ is managed by an Executive Vice President and a Chief Financial Officer.

No matter what aspect of Amateur Radio attracts you, ARRL membership is relevant and important. There would be no Amateur Radio as we know it today were it not for the ARRL. We would be happy to welcome you as a member! (An Amateur Radio license is not required for Associate Membership.) For more information about ARRL and answers to any questions you may have about Amateur Radio, write or call:

ARRL—The national association for Amateur Radio
New Ham Desk
225 Main Street
Newington CT 06111-1494
860-594-0200

Prospective new amateurs call:
800-32-NEW HAM (800-326-3942)

You can also contact us via e-mail at **newham@arrl.org**
or check out *ARRLWeb* at **http://www.arrl.org/**

NOTES

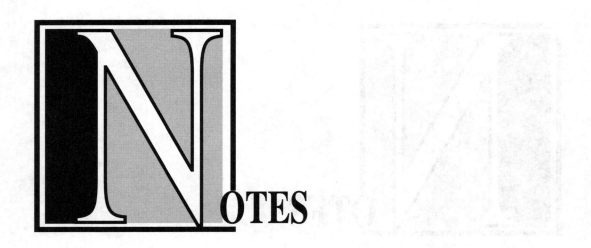

NOTES

NOTES

NOTES

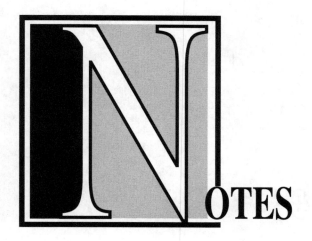

NOTES

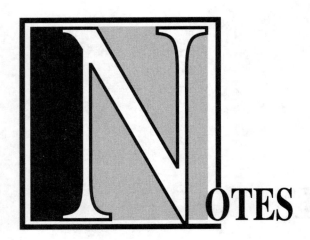

NOTES

INDEX

EXTRA CLASS
LICENSE MANUAL

PROOF OF
PURCHASE

FEEDBACK

Please use this form to give us your comments on this book and what you'd like to see in future editions, or e-mail us at **pubsfdbk@arrl.org** (publications feedback). If you use e-mail, please include your name, call, e-mail address and the book title, edition and printing in the body of your message. Also indicate whether or not you are an ARRL member.

Where did you purchase this book?
☐ From ARRL directly   ☐ From an ARRL dealer

Is there a dealer who carries ARRL publications within:
☐ 5 miles   ☐ 15 miles   ☐ 30 miles   of your location?   ☐ Not sure.

License class:
☐ Novice   ☐ Technician   ☐ Technician Plus   ☐ General   ☐ Advanced   ☐ Extra

---

Name _____

ARRL member?   ☐ Yes ☐ No

Call Sign _____

Daytime Phone   (   ) _____   Age _____

Address _____

City, State/Province, ZIP/Postal Code _____

If licensed, how long? _____   E-mail_____

Other hobbies _____

_____

Occupation _____

For ARRL use only	ECLM
Edition	7 8 9 10 11 12
Printing	2 3 4 5 6 7 8 9 10 11 12

From _____

_____

_____

EDITOR, EXTRA CLASS LICENSE MANUAL
ARRL—THE NATIONAL ASSOCIATION FOR AMATEUR RADIO
225 MAIN STREET
NEWINGTON CT 06111-1494

— — — — — — — — — — — — please fold and tape — — — — — — — — — — — — — —